THE FUNCTIONS, DISEASE-RELATED DYSFUNCTIONS, AND THERAPEUTIC TARGETING OF NEURONAL MITOCHONDRIA

Wiley Series on Neuropharmacology
Valentin K. Gribkoff and Leonard K. Kaczmarek, Series Editors

The Functions, Disease-Related Dysfunctions, and Therapeutic Targeting of Neuronal Mitochondria
Edited by Valentin K. Gribkoff, Elizabeth A. Jonas, and J. Marie Hardwick

THE FUNCTIONS, DISEASE-RELATED DYSFUNCTIONS, AND THERAPEUTIC TARGETING OF NEURONAL MITOCHONDRIA

Edited by

VALENTIN K. GRIBKOFF
Department of Internal Medicine, Section of Endocrinology
Yale University School of Medicine
New Haven, Connecticut, USA

ELIZABETH A. JONAS
Department of Internal Medicine, Section of Endocrinology
Yale University School of Medicine
New Haven, Connecticut, USA

J. MARIE HARDWICK
Department of Molecular Microbiology and Immunology, and Dept. of Neurology
Johns Hopkins University Bloomberg Schools of Public Health and Medicine
Baltimore, Maryland, USA

WILEY

Published by John Wiley & Sons, Inc., Hoboken, New Jersey
Published simultaneously in Canada

Library of Congress Cataloging-in-Publication Data:

The functions, disease-related dysfunctions, and therapeutic targeting of neuronal mitochondria / edited by Valentin K. Gribkoff, Elizabeth A. Jonas, J. Marie Hardwick.
pages cm – (Wiley series on neuropharmacology)
Includes bibliographical references and index.
ISBN 978-1-118-70923-8 (cloth)
1. Mitochondrial pathology. 2. Mitochondria–Formation. 3. Nervous system–Diseases–Treatment. I. Gribkoff, Valentin K., editor. II. Jonas, Elizabeth A., editor. III. Hardwick, J. Marie, editor.
RB147.5.F86 2015
572.8′69–dc23

2015020365

Cover Image: Courtesy of J. Marie Hardwick and Elizabeth A. Jonas

Set in 10/12pt Times by SPi Global, Pondicherry, India

10 9 8 7 6 5 4 3 2 1

1 2016

CONTENTS

CONTRIBUTORS

Kambiz N. Alavian, Division of Brain Sciences, Department of Medicine, Imperial College London, South Kensington Campus, London, UK

Giuseppe Arena, Institut de Recherche en Cancérologie de Montpellier, CSS-Université Montpellier, Montpellier, France

Sarah B. Berman, Department of Neurology, Pittsburgh Institute for Neurodegenerative Diseases, University of Pittsburgh, Pittsburgh, Pennsylvania, USA

Gisela Beutner, Department of Pediatrics (Cardiology), University of Rochester Medical Center, School of Medicine and Dentistry, Rochester, New York, USA

Laura Bonanni, Department of Neuroscience, Imaging and Clinical Sciences, University G.d'Annunzio of Chieti-Pescara, Chieti, Italy

Nickolay Brustovetsky, Department of Pharmacology and Toxicology, Stark Neuroscience Research Institute, Indiana University School of Medicine, Indianapolis, Indiana

Raymond Chuen-Chung Chang, Laboratory of Neurodegenerative Diseases, Department of Anatomy, Li Ka Shing Faculty of Medicine and Research Centre of Heart, Brain, Hormone and Healthy Aging, Medicine, *Li Ka Shing* Faculty of Medicine and State Key Laboratory of Brain and Cognitive Sciences, University of Hong Kong, Pokfulam, Hong Kong SAR, China

Yanmin Chen, Synaptic Functions Section, Porter Neuroscience Research Center, National Institute of Neurological Disorders and Stroke, National Institutes of Health, Bethesda, Maryland, USA

Sally Shuk-Yee Cheng, Laboratory of Neurodegenerative Diseases, Department of Anatomy, Li Ka Shing Faculty of Medicine, Hong Kong, China

Charleen T. Chu, Department of Pathology, University of Pittsburgh School of Medicine, Pittsburgh, Pennsylvania, USA

Michelle Dail, Department of Pathology, University of Pittsburgh School of Medicine, Pittsburgh, Pennsylvania, USA

Edward A. Fon, Montreal Neurological Institute, McGill University, Montreal, Quebec, Canada

Valerio Frazzini, Department of Neuroscience, Imaging and Clinical Sciences, University G.d'Annunzio of Chieti-Pescara, Chieti, Italy

Valentin K. Gribkoff, Department of Medicine, Section of Endocrinology, Yale University School of Medicine, New Haven, Connecticut, USA

Edward D. Hall, Department of Anatomy and Neurobiology, Spinal Cord and Brain Injury Research Center, University of Kentucky Chandler College of Medicine, Lexington, Kentucky

J. Marie Hardwick, W. Harry Feinstone Department of Molecular Microbiology and Immunology, and Department of Neurology, Johns Hopkins University Bloomberg Schools of Public Health and Medicine, Baltimore, Maryland, USA

Tamas L. Horvath, Interdepartmental Neuroscience Program, Program in Integrative Cell Signaling and Neurobiology of Metabolism, Section of Comparative Medicine, Yale University School of Medicine, New Haven, Connecticut, USA

Clara Hiu-Ling Hung, Laboratory of Neurodegenerative Diseases, Department of Anatomy, Li Ka Shing Faculty of Medicine, Hong Kong, China

Dakota Jackson, Center for Neuroscience, Departments of Physiology and Pharmacology, Robert C. Byrd Health Sciences Center, West Virginia University School of Medicine, Morgantown, West Virginia, USA

Renjie Jiao, State Key Laboratory of Brain and Cognitive Sciences, Institute of Biophysics, Chinese Academy of Sciences, Beijing, China

Elizabeth A. Jonas, Department Internal Medicine, Section of Endocrinology, Yale University School of Medicine, New Haven, Connecticut, USA

Kathleen W. Kinnally, Department of Basic Sciences, New York University College of Dentistry, New York, USA

Adam L. Knight, Synaptic Functions Section, Porter Neuroscience Research Center, National Institute of Neurological Disorders and Stroke, National Institutes of Health, Bethesda, Maryland, USA

Simon Ming-Yuen Lee, Institute of Chinese Medical Sciences, University of Macau, Macau SAR, China

Gian-Luca McLelland, Montreal Neurological Institute, McGill University, Montreal, Quebec, Canada

Nelli Mnatsakanyan, Department Internal Medicine, Section of Endocrinology, Yale University School of Medicine, New Haven, Connecticut, USA

Carole M. Nasrallah, Interdepartmental Neuroscience Program, Program in Integrative Cell Signaling and Neurobiology of Metabolism, Section of Comparative Medicine, Yale University School of Medicine, New Haven, Connecticut, USA

Marco Onofrj, Department of Neuroscience, Imaging and Clinical Sciences, University G.d'Annunzio of Chieti-Pescara, Chieti, Italy

Evgeny Pavlov, Department of Physiology and Biophysics, Dalhousie University Halifax, Nova Scotia

Pablo M. Peixoto, Department of Natural Sciences, Baruch College and Graduate Center at City University of New York, New York, USA

Guy A. Perkins, Center for Neuroscience, Departments of Physiology and Pharmacology, Robert C. Byrd Health Sciences Center, West Virginia University School of Medicine, Morgantown, West Virginia, USA

Shona Pfeiffer, Department of Physiology and Medical Physics, Centre for the Study of Neurological Disorders, Royal College of Surgeons in Ireland, Dublin, Ireland

George A. Porter, Jr., Department of Pediatrics (Cardiology), University of Rochester Medical Center, School of Medicine and Dentistry, Rochester, New York, USA

Jochen H. M. Prehn, Department of Physiology and Medical Physics, Centre for the Study of Neurological Disorders, Royal College of Surgeons in Ireland, Dublin, Ireland

Ann Saada, Monique and Jacques Roboh Department of Genetic Research and the Department of Genetic and Metabolic Diseases, Hadassah-Hebrew University Medical Center Jerusalem, Israel

Zu-Hang Sheng, Synaptic Functions Section, Porter Neuroscience Research Center, National Institute of Neurological Disorders and Stroke, National Institutes of Health, Bethesda, Maryland, USA

Devorah Soiferman, Monique and Jacques Roboh Department of Genetic Research and the Department of Genetic and Metabolic Diseases, Hadassah-Hebrew University Medical Center Jerusalem, Israel

George A. Spirou, Center for Neuroscience, Departments of Physiology and Pharmacology, Robert C. Byrd Health Sciences Center, West Virginia University School of Medicine, Morgantown, West Virginia, USA

Erin Steer, Department of Pathology, University of Pittsburgh School of Medicine, Pittsburgh, Pennsylvania, USA

Patrick G. Sullivan, Department of Anatomy and Neurobiology, Spinal Cord and Brain Injury Research Center, University of Kentucky Chandler College of Medicine, Lexington, Kentucky

Tao Sun, Synaptic Functions Section, Porter Neuroscience Research Center, National Institute of Neurological Disorders and Stroke, National Institutes of Health, Bethesda, Maryland, USA

Astrid Thomas, Department of Neuroscience, Imaging and Clinical Sciences, University G.d'Annunzio of Chieti-Pescara, Chieti, Italy

Yuriy M. Usachev, Department of Pharmacology, University of Iowa College of Medicine, Iowa City, Iowa

Enza Maria Valente, Section of Neuroscience, Department of Medicine and Surgery, University of Salerno, Salerno, Italy and Neurogenetics Unit, CSS-Mendel Institute, Rome, Italy

Petronela Weisová, Max F. Perutz Laboratories, University of Vienna, Austria

Yan Yan, State Key Laboratory of Brain and Cognitive Sciences, Institute of Biophysics, Chinese Academy of Sciences, Beijing, China

Heather M. Yonutas, Department of Anatomy and Neurobiology, Spinal Cord and Brain Injury Research Center, University of Kentucky Chandler College of Medicine, Lexington, Kentucky

PREFACE

Several years ago, when the idea first took hold to produce a collection of monographs on neuronal mitochondria, our hope was to survey this exciting field from both basic research and translational perspectives. Recent evidence, reviewed in this book, increasingly delineates the complexity and importance of mitochondrial function in neurons, including critical roles in synaptic function, and further implicates mitochondrial dysfunctions as either causal or contributory in a large number of central nervous system conditions and diseases. Although the successful targeting of dysfunctional mitochondria for neuronal disease has yet to be proven definitively in the clinic, the contributions in this volume indicate that significant progress is currently being made in this area.

The 18 chapters in this volume are presented by leading international research groups. In four sections, these contributions range from reviews of various aspects of mitochondrial structure and function focused on their specialized roles in neurons, the alterations in mitochondria dynamics that lead to mitochondrial dysfunction, the involvement of mitochondria in a range of acute and chronic neuronal disorders, and the development of model systems and proposed therapeutic approaches targeting these critical organelles. The chapters cover their subjects at depths appropriate for advanced students as well as researchers in the field; some are broader surveys of their subjects, and others are in-depth reviews of recent research from the authors' laboratories. All of the contributions will provide a solid foundation for further study. This has been a significant undertaking, and we hope the reader finds the contributions both enjoyable and informative.

This book is the first volume in a new series published by John Wiley and Sons, *The Wiley Series on Neuropharmacology*; the series co-editors are Valentin K. Gribkoff, Ph.D., and Leonard K. Kaczmarek, Ph.D., both of Yale University School of Medicine. Subsequent volumes will be focused, like the current one, on reviews of areas of significant current interest in the neurosciences that have resulted in considerable translational interest, leading to current or future promising therapeutic developments.

Valentin K. Gribkoff, Ph.D., Elizabeth A. Jonas, M.D., and
J. Marie Hardwick, Ph.D.

SECTION I

MITOCHONDRIAL STRUCTURE AND ION CHANNELS

1

MITOCHONDRIAL PERMEABILITY TRANSITION: A LOOK FROM A DIFFERENT ANGLE

NICKOLAY BRUSTOVETSKY

Department of Pharmacology and Toxicology, Stark Neuroscience Research Institute, Indiana University School of Medicine, Indianapolis, Indiana

1.1 REGULATION OF INTRACELLULAR CALCIUM IN NEURONS

Calcium ions occupy an exceptional niche in cell physiology and biochemistry because of their chemical properties and unique distribution across the plasma membrane. Ca^{2+} participates in diverse signaling pathways and regulates activity of various intracellular enzymes. Consequently, Ca^{2+} plays an important signaling and regulatory role in many different cell types. In neurons, fluctuations in cytosolic Ca^{2+} are involved in a variety of processes, including regulation of mitochondrial bioenergetics (Cardenas et al., 2010; Hansford, 1991). Under pathological conditions (e.g., prolonged activation of glutamate receptors), neurons can experience sustained elevation in cytosolic Ca^{2+}, or Ca^{2+} dysregulation, which is dangerous for cells and can lead to neuronal injury (Choi, 1988; Manev et al., 1989). The neuronal injury can occur as a result of activation of degradation enzymes such as calpains, which are Ca^{2+}-dependent proteases (Brustovetsky, Bolshakov, and Brustovetsky, 2010; Gerencser et al., 2009; Minger et al., 1998; Siman, Noszek, and Kegerise, 1989), and phospholipases A_2 (Bonventre, 1997; Kim et al., 1995). Ultimately, sustained elevation in cytosolic Ca^{2+} may lead to neuronal death (Tymianski et al., 1993a, 1993b). Therefore, precise regulation of cytosolic Ca^{2+} is critical for neuronal survival.

The Functions, Disease-Related Dysfunctions, and Therapeutic Targeting of Neuronal Mitochondria, First Edition. Edited by Valentin K. Gribkoff, Elizabeth A. Jonas and J. Marie Hardwick.

Ca^{2+} can enter neurons via different mechanisms such as transient receptor potential channels, Ca^{2+}-permeable AMPA/kainate receptors, and acid-sensing ion channels that may, to some extent, contribute to Ca^{2+} dysregulation and glutamate excitotoxicity (Aarts et al., 2003; Gao et al., 2005; Turetsky et al., 1994). However, there is substantial evidence that the N-methyl-D-aspartate (NMDA)-subtype of ionotropic glutamate receptors (NMDAR) plays a major role in massive Ca^{2+} influx into the neuron exposed to glutamate, and thus, is critical for glutamate-induced Ca^{2+} dysregulation (Tymianski et al., 1993b). In addition, the Na^+/Ca^{2+} exchanger (NCX), operating in the reverse mode and bringing Ca^{2+} into the cell in exchange for cytosolic Na^+, also significantly contributes to glutamate-induced Ca^{2+} dysregulation and excitotoxic neuronal death (Brittain et al., 2012; Czyz, Baranauskas, and Kiedrowski, 2002; Dietz, Kiedrowski, and Shuttleworth, 2007; Kiedrowski et al., 1994).

Fortunately, eukaryotic cells, including neurons, have elaborate defense mechanisms against unwanted elevations in cytosolic Ca^{2+} (Guerini, Coletto, and Carafoli, 2005). In neurons, NCX operating in the forward mode (extruding Ca^{2+} in an exchange with external Na^+) provides the highest transport capacity mechanism for extrusion of excessive Ca^{2+} at the expense of the Na^+ gradient across the plasma membrane (Annunziato, Pignataro, and Di Renzo, 2004). Consistent with the beneficial role of NCX in maintaining low cytosolic Ca^{2+}, down-regulation of different NCX isoforms (particularly, NCX2, NCX3, and NCKX2) exacerbates ischemic brain damage (Cuomo et al., 2008; Jeffs et al., 2008; Jeon et al., 2007), whereas BHK cells transfected with NCX3 become more resistant to hypoxia-reoxygenation injury (Secondo et al., 2007).

Mitochondrial Ca^{2+} uptake is another major mechanism aimed at lowering cytosolic Ca^{2+} under extreme conditions of Ca^{2+} overflow. Mitochondria possess Ca^{2+} channels in the inner membrane (historically called the "Ca^{2+} uniporter") that mediate Ca^{2+} influx into the organelles driven by high membrane potential across the inner mitochondrial membrane (Baughman et al., 2011; De et al., 2011; Kirichok, Krapivinsky, and Clapham, 2004; Perocchi et al., 2010). This allows mitochondria to accumulate significant amounts of Ca^{2+} and hence defer Ca^{2+} dysregulation in neurons induced by prolonged exposure to glutamate. This large Ca^{2+} uptake capacity of mitochondria stems from formation of biologically inactive Ca-phosphate precipitates in the mitochondrial matrix that was convincingly demonstrated by Brian Andrews' group (Pivovarova et al., 2004).

1.2 CALCIUM OVERLOAD AND MITOCHONDRIAL PERMEABILITY TRANSITION

Although the ability of mitochondria to accumulate Ca^{2+} is truly remarkable, mitochondrial Ca^{2+} uptake capacity is finite and mitochondrial Ca^{2+} overload comes with a hefty price. Under certain circumstances (ischemia-reperfusion, prolonged exposure to glutamate, and age-related neurodegenerative disorders such as Parkinson's, Alzheimer's, and Huntington's diseases), excessive Ca^{2+} accumulation

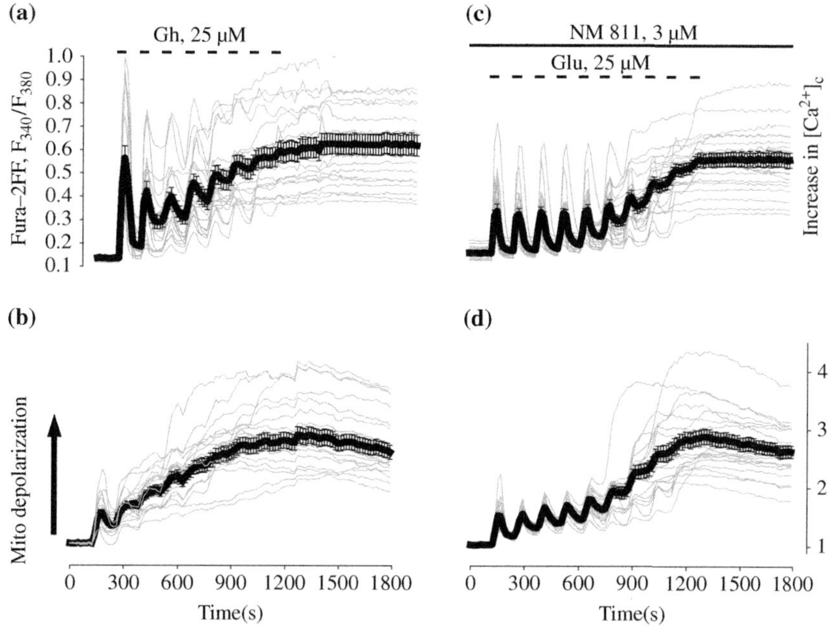

FIGURE 1.1 Glutamate-induced changes in cytosolic Ca^{2+} ($[Ca^{2+}]_c$) and mitochondrial membrane potential in rat cultured hippocampal neurons. Protection with NIM811. Neurons (12–14 days in vitro [12–14 DIV]) were coloaded with 2.6 μM Fura-2FF-AM and 1.7 μM Rhodamine 123 (Molecular Probes, Eugene, OR) to monitor changes in cytosolic Ca^{2+} and mitochondrial membrane potential, respectively. In **a** and **c**, the individual (*thin grey traces obtained from individual neurons*) and average Fura-2FF (F_{340}/F_{380}, *thick black traces*) fluorescence signals are shown. In **b** and **d**, the individual (*thin grey traces obtained from individual neurons*) and average Rhodamine 123 (Rh123, F/F_0, *thick black traces*) fluorescence signals are shown. Where indicated, neurons were treated with 25 μM glutamate (Glu, plus 10 μM glycine) for 30 seconds. In **c** and **d**, neurons were preincubated for 45 minutes with 3 μM NIM811. NIM811 (3 μM) was also present in the bath solution during the experiment.

in mitochondria can cause mitochondrial damage manifested in the activation of the mitochondrial permeability transition pore (PTP). The role of mitochondrial damage and permeability transition (PT) induction in disrupting Ca^{2+} homeostasis has been illustrated in our experiments with cultured neurons exposed to excitotoxic glutamate (Fig. 1.1). Multiple short exposures to glutamate induce transient increases in cytosolic Ca^{2+} accompanied by transient mitochondrial depolarizations. With increasing numbers of glutamate exposures, neuronal ability to lower cytosolic Ca^{2+} and restore mitochondrial membrane potential gradually vanished. Inhibition of the PTP with NIM811, a nonimmunosuppressive derivative of cyclosporin A (CsA) (Waldmeier et al., 2002), significantly improved recovery of both cytosolic Ca^{2+} and mitochondrial membrane potential, suggesting an important role for PTP induction in glutamate-induced Ca^{2+} dysregulation and mitochondrial depolarization. Inhibition of PTP

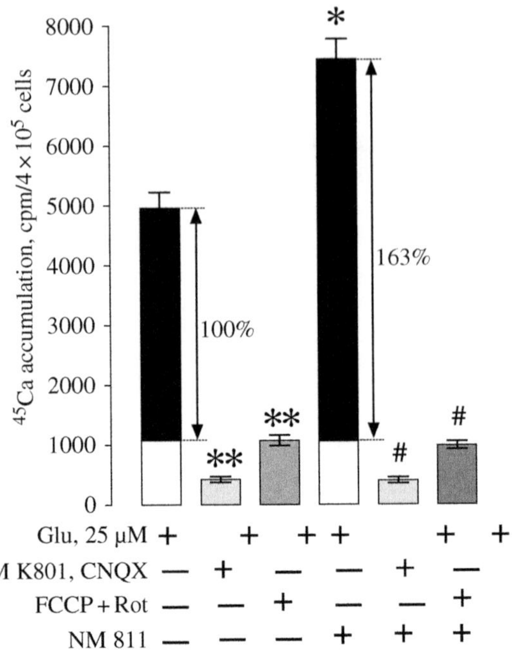

FIGURE 1.2 ^{45}Ca accumulation in rat cultured hippocampal neurons exposed to glutamate. NIM811 increases ^{45}Ca accumulation in mitochondria in situ. Neurons (12-14 days in vitro [DIV]) were incubated in the standard bath solution supplemented with 5 μCi of ^{45}Ca. ^{45}Ca accumulation was measured at 5 minutes after application of 25 μM glutamate (Glu) plus 10 μM glycine, as indicated. Nonspecific ^{45}Ca binding was measured in the presence of 10 μM MK801 (an inhibitor of NMDA receptor) and 10 μM CNQX (an inhibitor of AMPA/kainate receptors). Mitochondrial ^{45}Ca accumulation was evaluated by subtracting the amount of ^{45}Ca accumulated in neurons in the presence of 1 μM FCCP and 1 μM rotenone (Rot) from the amount of ^{45}Ca accumulated in the absence of inhibitors. This value was taken as 100%. Where indicated, neurons were preincubated for 45 minutes with 3 μM NIM811 before glutamate exposure. NIM811 (3 μM) was also present in the bath solution during the experiment. $^{*}p<0.05$ comparing neurons with and without NIM811 treatment; $^{**}p<0.01$ comparing neurons with and without mitochondrial inhibitors and antagonists of glutamate receptors; $^{\#}p<0.01$ comparing NIM811-treated neurons incubated with and without mitochondrial inhibitors and antagonists of glutamate receptors. Data are mean±SEM, N=3.

and protection of mitochondria in neurons resulted in an increase in mitochondrial Ca^{2+} uptake illustrated in the experiments with the ^{45}Ca isotope (Fig. 1.2). Neurons exposed for 5 minutes to 25 μM glutamate (plus 10 μM glycine) accumulated significant amounts of ^{45}Ca. Mitochondrial depolarization and inhibition of mitochondrial Ca^{2+} uptake with a combination of p-trifluoromethoxyphenylhydrazone (FCCP) and rotenone significantly decreased the amount of ^{45}Ca accumulated by neurons. This indicates that mitochondria store about 75% of all Ca^{2+} entering the neuron exposed to glutamate and thus play a pivotal role in regulating cytosolic Ca^{2+}

under stressful conditions. Inhibition of the PTP with NIM811 significantly increased the amount of $^{45}Ca^{2+}$ accumulated by mitochondria in neurons exposed to glutamate. Thus, inhibition of the PTP augments Ca^{2+} accumulation in mitochondria, stabilizes mitochondrial membrane potential, and defers Ca^{2+} dysregulation. Induction of the PTP, on the other hand, leads to mitochondrial depolarization, inhibition of ATP synthesis and Ca^{2+} uptake, mitochondrial swelling, and the rupture of the outer mitochondrial membrane. The latter may result in the release of mitochondrial apoptogenic proteins such as cytochrome *c*, Smac/DIABLO, endonuclease G, and AIF. It is not surprising, therefore, that studies of the PTP attract significant attention. Recently, several excellent review papers have been published summarizing current views on PTP structure and function (Bernardi, 2013; Elrod and Molkentin, 2013; Javadov and Kuznetsov, 2013; Siemen and Ziemer, 2013; Vianello et al., 2012). In this chapter, I will discuss only a few selected questions concerning PTP function and structure.

Early observations of PT manifestations were made in the 1950s and 1960s (for review see Bernardi, 2013). However, the concept of mitochondrial permeability transition (mPT) was formulated and introduced by Hunter and Haworth only in the late 1970s (Haworth and Hunter, 1979; Hunter and Haworth, 1979a, 1979b; Hunter, Haworth, and Southard, 1976.). In these papers, Hunter and Haworth used the term *Ca^{2+}-induced membrane transition* in mitochondria to signify the large increase in inner membrane permeability and the key role of Ca^{2+} in triggering this phenomenon (Hunter and Haworth, 1979a). In their seminal work, Hunter and Haworth demonstrated that the Ca^{2+}-induced membrane transition in mitochondria leads to mitochondrial uncoupling, release of previously accumulated Ca^{2+}, and mitochondrial swelling. The authors described major fundamental properties of mPT such as the requirement for Ca^{2+} entry into mitochondria, protection by reduced nicotinamide adenine dinucleotide (NADH) and mitochondrial energization (membrane polarization), facilitation by membrane depolarization, and inhibition by H^+, Mg^{2+}, and other cations (Haworth and Hunter, 1979; Hunter and Haworth, 1979a).

An early hypothesis concerning the mechanism of the Ca^{2+}-induced mPT postulated that the increase in inner membrane permeability occurs as a result of activation of mitochondrial Ca^{2+}-dependent phospholipase A_2 and accumulation of free fatty acids and lysophospholipids that disturb the phospholipid bilayer (Beatrice, Palmer, and Pfeiffer, 1980; Pfeiffer et al., 1979). However, the discovery that CsA inhibits mPT (Crompton, Ellinger, and Costi, 1988; Fournier, Ducet, and Crevat, 1987) but does not inhibit mitochondrial phospholipase A_2 (Broekemeier, Dempsey, and Pfeiffer, 1989) suggested that induction/activation of a proteinaceous pore in the inner mitochondrial membrane is most likely responsible for mPT. Since then, this hypothesis has dominated the field. Nevertheless, under conditions of massive Ca^{2+} accumulation in mitochondria, CsA inhibits mPT only transiently followed by CsA-insensitive PT, in which activated phospholipase A_2 may play an important role (Broekemeier and Pfeiffer, 1989). Moreover, palmitate, a free fatty acid that is predominately generated because of phospholipase A_2 activity, can induce a CsA-insensitive pore in the inner mitochondrial membrane (Sultan and Sokolove, 2001) and thus contribute to Ca^{2+}-induced mitochondrial damage beyond the classic CsA-sensitive mPTP.

1.3 THE MITOCHONDRIAL TRANSITION PORE

1.3.1 Evidence for ANT and VDAC as Components of the PTP

In early studies, Hunter and Haworth (1979–1980) described the inhibitory effect of adenosine diphosphate (ADP) acting on PT at the internal matrix side and opposite effects of atractyloside and bongkrekic acid (Haworth and Hunter, 1980; Hunter and Haworth, 1979a), adenine nucleotide translocator (ANT) inhibitors that stabilize ANT in *c*- (nucleotide binding site facing cytosol) and *m*- (nucleotide binding site facing mitochondrial matrix) conformational states, respectively (Buchanan et al., 1976). In ADP-depleted mitochondria, atractyloside stimulated, whereas bongkrekic acid antagonized, PT (Hunter and Haworth, 1979a). This suggested involvement of ANT in PT. Consistent with this, an earlier study by Asimakis and Sordahl (1977) showed that atractyloside and an endogenous ANT inhibitor, palmitoyl-coenzyme A, induced the release of previously accumulated Ca^{2+} from isolated cardiac mitochondria (Asimakis and Sordahl, 1977), apparently as a result of induction of mPT.

Shortly thereafter, Panov, Filippova, and Lyakhovich (1980) demonstrated that ADP, in a carboxyatractyloside-sensitive manner, inhibited H^+ and K^+ permeability in isolated liver mitochondria (Panov et al., 1980). This was interpreted as evidence for ANT being a pathway for H^+ and K^+ passive transport across the inner mitochondrial membrane. It has also been proposed that ANT operates as a gated pore (Panov et al., 1980). This is in line with the structural model of the ANT that was proposed soon after where ANT was suggested to operate as a gated pore as well (Klingenberg, 1981). Toninello, Siliprandi, and Siliprandi (1983) showed that ADP and ATP protect mitochondria from Ca^{2+}-induced membrane depolarization (Toninello et al., 1983). Bongkrekic acid also protected against depolarization, whereas atractyloside accelerated it. Interestingly, bongkrekic acid facilitated the protective effect of adenine nucleotides while atractyloside abrogated this protection. The authors agreed with the contention of Panov et al. (1980) that ANT, operating as a gating pore, might be responsible for the increase in membrane permeability (Toninello et al., 1983). Le Quoc and Le Quoc (1988) showed that carboxyatractyloside, palmitoyl coenzyme A, and pyridoxal phosphate, membrane-impermeant inhibitors of ANT, stimulated mPT, whereas bongkrekic acid and ADP impeded mPT (Le Quoc and Le Quoc, 1988). These authors also provided evidence suggesting that oxidation of sulfhydryl groups linked to mitochondrial pyridine nucleotide oxidation influenced the conformation of ANT, stabilizing it in its *c*-conformation (Le Quoc and Le Quoc, 1989). The latter was correlated with increased probability of mPT.

In the late 1980s, it was found that CsA binds to cyclophilin, a peptidyl-prolyl cis-trans isomerase involved in folding of different proteins (Fischer et al., 1989; Takahashi, Hayano, and Suzuki, 1989). Later, it was shown that mammalian mitochondria possess a distinct cyclophilin isoform, cyclophilin D (Connern and Halestrap, 1992). Importantly, oxidative stress facilitates recruitment of mitochondrial cyclophilin D to the inner membrane and promotes mPT (Connern and Halestrap, 1994). Consequently, the effect of CsA on mPT was shown to be mediated by CsA interaction with cyclophilin D (Halestrap and Davidson, 1990; Nicolli et al., 1996).

At the same time, studies of the ANT involvement in induction of mPT continued. In 1990, Halestrap and Davidson showed that Ca^{2+}-induced swelling of heart mitochondria was inhibited by ADP and bongkrekic acid, and the effect of ADP was reversed by carboxyatractyloside, a more potent derivative of atractyloside (Halestrap and Davidson, 1990). The authors suggested that Ca^{2+} interacts with ANT when it is in the *c*-conformation and proposed the model in which mitochondrial cyclophilin interacts with ANT in the presence of Ca^{2+}, leading to mPT. Later, Andrew Halestrap's group (Woodfield et al., 1998) experimentally proved this hypothesis by demonstrating mitochondrial cyclophilin D binding to ANT in the inner mitochondrial membrane as well as to purified ANT. However, these authors failed to find evidence for cyclophilin D binding to mitochondrial porin. Later, James Lechleiter's group (Lin and Lechleiter, 2002) also demonstrated binding of cyclophilin D to ANT.

In contrast to Halestrap's findings, Martin Crompton's group (Crompton, Virji, and Ward, 1998) reported that in their experiments, cyclophilin D binds to a complex of mitochondrial porin (voltage-dependent anion channel [VDAC]) and ANT to form mPTP (Crompton et al., 1998). The authors purified and reconstituted porin, ANT, and the glutathione S-transferase/cyclophilin D fusion protein into phophatidylcholine liposomes loaded with fluorescein. These proteoliposomes were permeabilized in the presence of Ca^{2+} plus P_i, and this permeabilization was inhibited by CsA. Thus, the authors concluded that the mPTP includes porin, ANT, and cyclophilin D, and that it is possibly formed at the contact sites between the outer and the inner mitochondrial membranes.

Currently, cyclophilin D is the only regulatory component of the permeability transition pore that is considered to be proven. In our early study, we found a correlation between augmented expression of cyclophilin D and increased sensitivity to the deleterious action of Ca^{2+} on mitochondria from the striatum compared to mitochondria isolated from the cortex (Brustovetsky et al., 2003). This suggested an increased propensity to mPTP activation that might underlie increased vulnerability of striatal cells in different neurodegenerative disorders. James Geddes' group (Brown, Sullivan, and Geddes, 2006) reported that synaptic mitochondria have higher cyclophilin D levels, resulting in increased susceptibility to induction of mPT (Naga, Sullivan, and Geddes, 2007). Consistent with this, the decreased sensitivity of brain mitochondria to Ca^{2+}-induced mPT compared to liver and heart mitochondria was explained by decreased expression of cyclophilin D (Eliseev et al., 2007). Experiments with cyclophilin D-knockout mice showed that genetic ablation of cyclophilin D suppresses mPT induction and increases Ca^{2+} uptake capacity in mitochondria two- to threefold (Baines et al., 2005; Basso et al., 2005; Nakagawa et al., 2005; Schinzel et al., 2005). Based on these studies, it was postulated that cyclophilin D sensitizes the pore to Ca^{2+} (Basso et al., 2005) and mediates neuronal death after focal cerebral ischemia (Schinzel et al., 2005). Shortly after that, we showed that genetic ablation of cyclophilin D improved Ca^{2+} handling in cultured hippocampal neurons exposed to glutamate and increased their survival rate (Li, Brustovetsky, and Brustovetsky, 2009). However, cyclophilin D does not always lead to increased cell vulnerability. In experiments with HEK293 and rat glioma C6 cells, overexpression of cyclophilin D appeared to be protective against oxidative stress and staurosporine-induced cell death (Lin and Lechleiter, 2002). This protection depended on the peptidyl-prolyl *cis-trans* isomerase (PPIase) activity of cyclophilin D and was independent from binding to ANT. Thus, depending on experimental conditions and applied stimuli, cyclophilin D can be protective because of its PPIase activity or detrimental as a result of sensitization of the mPTP to Ca^{2+}.

The findings that inhibitors and ligands of ANT can influence the mPT suggested ANT involvement in this phenomenon. However, for a long time it was not clear whether ANT could play the role of the pore or if its role was limited to modulatory action. The answer to this question came from experiments with reconstituted systems. Reinhard Krämer's group (Dierks et al., 1990a; Dierks, Salentin, and Kramer, 1990b) demonstrated that the aspartate/glutamate carrier and ANT modified by SH-reagents can be converted from an obligate exchange mechanism to a unidirectional transport mode, suggesting channel-like activity. Using electrophysiological patch-clamp techniques, this group also demonstrated that a P_i carrier from *Saccharomyces cerevisiae* mitochondria can behave as an ion channel, although with relatively low conductance ranging from 25 to 40 pS in the presence of Ca^{2+} and Mg^{2+} (Herick, Kramer, and Luhring, 1997).

In experiments with ANT reconstituted into the planar bilayer phospholipid membrane, Vladimir Skulachev's group (Tikhonova et al., 1994) detected large ion currents following application of the SH-reagent mersalyl. The channel appeared to be nonselective with multiple sublevels of conductance ranging from 200 pS up to 1.5 nS. This study demonstrated the long-suspected ability of modified ANT to operate as a pore. In studies with purified bovine heart ANT reconstituted into giant proteoliposomes, we found a large nonselective channel with multiple sublevels of conductance ranging from 300 to 600 pS in symmetrical 100 mM KCl (Brustovetsky and Klingenberg, 1996). Based on its behavior, the ANT-associated channel could be easily distinguished from the porin channel (VDAC). The ANT channel opening required high Ca^{2+} that we proposed to bind to cardiolipin molecules, which are tightly associated with the ANT (Beyer and Klingenberg, 1985). The ANT channel depended on pH and closed at pH 5.2. In addition, the ANT channel could be partly inhibited by bongkrekic acid and completely inhibited by a combination of bongkrekic acid and ADP. Carboxyatractyloside as well as CsA were without effect. The latter is not surprising because cyclophilin D was not present in the proteoliposomes. These experiments were based on an early assumption that ANT operates as a gated channel (Klingenberg, 1981) and showed that ANT indeed could be converted into a large nonselective channel by exposure to high Ca^{2+}. In the next study, we reconstituted recombinant ANT from *Neurospora crassa* into giant proteoliposomes and found a large nonselective channel with multiple conductance sublevels and an open state conductance of 500 to 700 pS in symmetrical 100 mM KCl (Brustovetsky et al., 2002b). The ANT channel required Ca^{2+} for activation and could be inhibited by ADP and bongkrekic acid. When purified cyclophilin from *Neurospora crassa* (a single form of cyclophilin is present in both cytosol and mitochondria [Tropschug et al., 1988]) was present, it stabilized the ANT channel in the open state. CsA antagonized the effect of cyclophilin. Thus, these experiments filled an important gap in our previous study by demonstrating the stabilizing effect of cyclophilin on the ANT channel open state and CsA-induced inhibition. Overall, the properties of the ANT channel strongly resembled the mPTP. This and other observations concerning ANT involvement in the mPT suggested that the ANT could be a key conducting component of the mPTP. This notion is supported by the crystal structure of ANT (Pebay-Peyroula et al., 2003) that revealed a 30Å deep and 20Å wide depression in the carrier molecule. When the ANT is stabilized in the *c*-conformation by carboxy-atractyloside (CAT), this depression appeared to be open to the cytosolic side, providing the structural basis for pore formation.

Dieter Brdiczka's group (Beutner et al., 1996) reconstituted an ANT-porin-hexokinase (or creatine kinase) complex in the planar bilayer phospholipid membrane and detected a large channel with a conductance of 6 nS. The ANT-porin-kinase complex was also reconstituted into phospholipid vesicles loaded with malate or ATP. Ca^{2+} induced the release of malate and ATP, whereas N-methylVal-4-cyclosporin, a derivative of CsA (Nicolli et al., 1996), inhibited the release of solutes. Based on these observations, the authors concluded that the ANT-porin-kinase complex constitutes the mPTP (Beutner et al., 1996). Although the results of this study to some extent resembled our findings (Brustovetsky and Klingenberg, 1996), there was one significant distinction. The ANT-porin-kinase pore detected in electrophysiological experiments did not require Ca^{2+} to operate as a pore, suggesting that porin interaction with ANT alters ANT channel behavior (although in metabolite efflux experiments, Ca^{2+} was necessary to stimulate malate and ATP leakage) (Beutner et al., 1996). In the subsequent study, Brdiczka's group (Beutner et al., 1998) showed that only ANT-porin-hexokinase but not ANT-porin-creatine kinase complex could be stimulated by Ca^{2+} to release malate and ATP from proteoliposomes in an N-methylVal-4-cyclosporin-sensitive manner. The authors also found mitochondrial cyclophilin D in the ANT-porin-hexokinase complex that could explain sensitivity to N-methylVal-4-cyclosporin. In the following study, Brdiczka's group (Beutner et al., 1998) showed that ANT purified from rat heart mitochondria and reconstituted into asolectin/cardiolipin vesicles loaded with ATP and malate allowed a release of the trapped compounds in response to increasing concentrations of Ca^{2+} (Ruck et al., 1998). This Ca^{2+}-induced solute leakage was not sensitive to N-methylVal-4-cyclosporin, but ADP inhibited the leakage. On the other hand, atractyloside and $HgCl_2$ stimulated release of solutes from ANT-reconstituted liposomes. These results strongly suggested that ANT itself is capable of adopting a porelike conformation under conditions known to induce mPT in mitochondria (Ruck et al., 1998).

Soon after these studies, Guido Kroemer's group (Brenner et al., 2000) reported that, consistent with our observations (Brustovetsky and Klingenberg, 1996), ANT reconstituted in the planar bilayer lipid membrane did not mediate ion currents unless Ca^{2+} or atractyloside are applied. Both Ca^{2+} and atractyloside stimulated ANT-associated channel activity, reaching in symmetrical 100 mM KCl a conductance of 250 and 30 pS, respectively. Thus, numerous experimental findings obtained in different laboratories with isolated mitochondria and reconstituted systems converged at the conclusion that ANT can be converted into a pore by Ca^{2+} or other factors promoting mPT, and therefore, modified ANT could be a conducting component of mPTP.

Although the ANT can be converted into a porelike structure without involvement of other proteins (Brenner et al., 2000; Brustovetsky and Klingenberg, 1996; Ruck et al., 1998), the observations that ANT can bind porin (Beutner et al., 1996) and that the ANT-porin complex binds mitochondrial cyclophilin D (Crompton et al., 1998), suggested a role for porin (VDAC) in the mPTP. In fact, for the first time the idea of VDAC involvement in mPT was introduced in the late 1980s and early 1990s by investigators performing electrophysiological experiments with mitoplasts, which are mitochondria with the outer membrane removed. In 1989, a group led by Mario Zoratti (Petronilli, Szabo, and Zoratti, 1989), using excised patches of the inner membrane, detected a large channel with multiple sublevels of conductance ranging from 30 pS up to 1.3 nS in symmetrical 150 mM KCl. These experiments were performed in the presence of 0.1 mM $CaCl_2$. At the same time, a similar channel in the inner membrane was found by Kinnally, Campo, and Tedeschi (1989). Interestingly, Ca^{2+} was not present in the solution in this study. Initially, the authors of both papers failed to make a connection between the large channels they observed and the mPTP. However, Szabo and Zoratti (1991) reported that CsA applied from the matrix side could inhibit the large channel detected in the inner mitochondrial membrane. Because it was reminiscent of the Ca^{2+}-dependent CsA-sensitive mPTP, the authors concluded that the channel detected in their study was the mPTP. The molecular identity of this channel remained unknown. At the same time, Kinnally's group (1991) found that Ca^{2+} significantly increases the probability of detection of the large channel in patches of the inner membrane excised from mitoplasts (the authors called it the "multiconductance channel" [MCC]), and although they did not demonstrate CsA sensitivity, the authors discussed this channel in the context of the mPTP. This study showed that MCC activity could be induced by Ca^{2+} interacting with the outer side of the inner membrane, and that application of voltage of both signs (>±60 mV) stimulated channel activity. Both observations appeared to be incompatible with the already known regulation of the mPTP: Ca^{2+} activates the pore from inside of mitochondria and membrane polarization inhibits the pore (Hunter and Haworth, 1979a). In the next study, Szabo, Bernardi, and Zoratti (1992) showed that the large channel (the authors now called it "megachannel") can be activated by Ca^{2+} and inhibited by other divalent cations. In addition, the authors found that low pH favors the closed state of the megachannel. These observations echoed findings made by Haworth and Hunter (1979) and Bernardi's group (1992) that revealed regulation of the mPTP by H^+ and divalent cations. In the following paper, Szabo and Zoratti (1992) continued to characterize the megachannel. Here they showed that the nonselective megachannel is activated by Ca^{2+} and inhibited by Mg^{2+}, CsA, and ADP. Based on these observations, the authors proposed that the megachannel is indeed the mPTP. Simultaneously, Kinnally's group (Zorov, Kinnally, and Tedeschi, 1992) made significant advances in their studies and reported that the MCC in excised patches from heart inner mitochondrial membrane is also inhibited by CsA. In this study, the authors again reported activation of the MCC by voltage that contrasted with mPTP regulation in mitochondria, where the pore is inhibited by voltage and activated by membrane depolarization (Bernardi, 1992; Hunter and Haworth, 1979a). Zoratti and co-workers (Szabo and Zoratti, 1993) reinvestigated voltage dependence of the megachannel and again concluded that the megachannel corresponds to the mPTP, but molecular identity of the pore remained unknown. Based on a comparison of voltage dependence of the megachannel and VDAC, the authors concluded that "a VDAC dimer may be the channel-forming component of the mitochondrial permeability transition pore." In the companion paper, the authors further analyzed electrophysiological properties of VDAC and came to the conclusion that "the PTP may consist of two cooperating VDAC channels, plus presumably an adenine nucleotide carrier dimer and a third component known to be a part of the mitochondrial benzodiazepine receptor" (Szabo, De, and Zoratti, 1993). The latter conclusion was based on the effect of Alpidem, a ligand for the mitochondrial benzodiazepine receptor that elicited ion currents from silent mitoplast patches. Because VDAC is the channel of the outer membrane and the outer membrane with VDAC was found to be present on mitoplasts (Schwaiger Herzog, and Neupert, 1987), the authors hypothesized that the mPTP might be preferentially localized at the contact sites between the outer and the inner membranes (Szabo et al., 1993). However, at that time it was already evident that the mPT is an inner membrane event. Correspondingly, the authors hypothesized that, at the contact sites, some VDAC molecules could be translocated from the outer to the inner membrane (Szabo et al., 1993). In the subsequent study the authors tested this hypothesis by trying to insert VDAC into mitoplast membrane, but they failed to obtain conclusive evidence supporting VDAC insertion (Zoratti, Szabo, and De Pinto,, 1994). Later, Bernardi's group (Krauskopf et al., 2006) showed that in VDAC1 knockout mice (*VDAC1*$^{-/-}$ mice) the permeability transition in liver mitochondria has the same properties as in mitochondria from wild-type animals. However, mitochondria from *VDAC1*$^{-/-}$ mice contained VDAC2 and VDAC3 isoforms, and therefore, the authors concluded that this study cannot rule out involvement of VDAC in the permeability transition. Jeffrey Molkentin's group (Baines et al., 2007) showed that mitochondria from VDAC1-, VDAC3-, and VDAC1- to VDAC3-null mice have a Ca^{2+} and oxidative stress-induced mPT that is indistinguishable from wild-type mitochondria. Moreover, the authors found that the mPT in fibroblast mitochondria lacking all three VDAC isoforms was identical to mPT in wild-type mitochondria. The authors concluded that VDAC is dispensable for mPT (Baines et al., 2007). Consistent with this, in our study with isolated rat brain mitochondria, inhibition of VDAC by König's polyanion (Konig et al., 1982) exacerbated mPT rather than inhibited it, arguing against VDAC involvement (Brustovetsky et al., 2002a).

The stimulatory effect of König's polyanion on mPT (Brustovetsky et al., 2002a) may also illustrate a possible role of membrane surface potential in regulating the inner membrane permeability. It was postulated that making membrane surface potential more negative is associated with facilitated pore opening, whereas making surface potential more positive hinders pore opening (Bernardi, Broekemeier, and Pfeiffer, 1994). Free fatty acid- or König's polyanion-induced stimulation of mPT (Bernardi, et al., 1994; Brustovetsky et al., 2002a; Schonfeld and Bohnensack, 1997; Wieckowski and Wojtczak, 1998; Wieckowski, Brdiczka, and Wojtczak, 2000) illustrates the former case, whereas inhibition with spermine and other polycations (Bernardi et al., 1994; Lapidus and Sokolove, 1992, 1993; Rigobello et al., 1995) exemplifies the latter case. However, there is an exclusion from this rule that cast some doubt on the surface potential hypothesis: a positively charged ceramide appeared to be a potent inducer of mPT in liver mitochondria and mitochondria from HepG2 cells (Novgorodov et al., 2005). Nevertheless, an interesting assumption was drawn from the surface potential hypothesis: the effects of ANT inhibitors and ligands on mPT might result from changes in surface membrane potential associated with stabilizing ANT in different conformational states rather than from changes in conductance of ANT itself (Bernardi et al., 1994).

Indeed, stabilization of the ANT in *c*- or *m*-conformation with atractyloside or bongkrekic acid and ADP, respectively, significantly influenced morphology of isolated bovine heart mitochondria, and it was speculated that the changes in membrane surface potential could be one of the causes of these morphological transformations (Klingenberg, 1985; Lee, Xu, and Colombini, 1996). However, measurements of membrane surface potential were not performed in these studies. This was done by Hagai Rotenberg and Miriam Marbach (1989, 1990). In these studies, the authors found a significant effect of ANT conformation on the apparent surface potential of the *matrix* side of the inner membrane: locking the ANT in *c*-conformation with atractyloside increased the positive surface charge density, whereas locking the ANT in the *m*-conformation with ADP and bongkrekic acid greatly increased the negative surface potential (Rottenberg and Marbach, 1990). From this study, however, it remained unclear how changes in ANT conformation could affect surface potential on the *outer* side of the inner membrane, which is obviously the most susceptible for membrane surface potential modifications by exogenous modulators. Schonfeld and Bohnensack (1997) investigated the mechanism of free fatty acid-mediated stimulation of mPT and concluded that an increase in the negative surface charge by free fatty acids seems to be of minor importance for their ability to promote mPT (Schonfeld and Bohnensack, 1997). Nevertheless, a possible regulation of mPTP by membrane surface potential cannot be ruled out. On the other hand, the observations that membrane surface potential changes occur following locking the ANT in *m*- or *c*-conformation cannot rule out a scenario in which the ANT can be converted into a large channel and operate as a conducting pathway in the mPTP.

Although over the years the ANT hypothesis received significant experimental support from many independent investigators, some studies produced results that seem difficult to reconcile with the key role of ANT as the mPTP. In early studies, Hunter and Haworth (1979a) found the ADP inhibitory effect on the mPTP is mediated by ADP acting at two different binding sites. One of these sites appeared to be high affinity and atractyloside-sensitive, belonging to the ANT, whereas the other is a low affinity and carboxyatractyloside-insensitive site of unknown molecular identity. Novgorodov et al. (1991, 1992) showed that in liver and heart mitochondria the inhibitory effect of CsA was augmented by low concentrations of ADP in a carboxyatractyloside-sensitive manner. In these studies, however, higher concentrations of ADP in combination with CsA inhibited the pore even in the presence of CAT. Based on these findings, the authors proposed that the pore is probably formed not directly by the ANT, and that another ADP-binding protein with low affinity to ADP could be involved in the pore induction/activation. Gizatullina et al. (2005) also reported that high ADP (2 mM) could inhibit the pore in the presence of CAT (Gizatullina et al., 2005). In this study, the authors provided some evidence that the low affinity regulatory ADP binding site is located on the outer side of the inner membrane. Halestrap's group (Halestrap, Woodfield, and Connern, 1997) reported that in liver mitochondria, K_i for the inhibitory effect of ADP in the absence and presence of CAT is 1.9±0.4 µM and 27.2±5.2 µM, respectively. The authors speculated that this may reflect the existence of two ADP binding sites of high and low affinity on the ANT, of which the former is sensitive and the latter insensitive to CAT. However, it is possible that CAT-insensitive ADP inhibition of mPTP occurs because of ADP binding to a protein different from the ANT, for example, to mitochondrial Ca^{2+} uniporter that also interacts with ATP and ADP (Litsky and Pfeiffer, 1997) and is required for Ca^{2+} uptake by mitochondria to activate mPTP (Hunter and Haworth, 1979a).

The role of the ANT as the mPTP was also questioned in experiments with mitochondria isolated from embryos of the crustacean *Artemia franciscana* (Menze et al., 2005). These mitochondria possessed ANT but apparently did not undergo classic large amplitude swelling in response to Ca^{2+}, and therefore, appeared to be resistant to classic mPT induction. Later, it was determined that the ANT in *Artemia franciscana* has a stretch of amino acids in the 198 to 225 region with a significant difference compared to ANT from other species (only 48–56% similarity and deletion of three amino acids at position 211, 212, and 219) (Konrad et al., 2011). This correlated with the unusual effects of ANT ligands: adenine nucleotides decreased, whereas CAT increased Ca^{2+} uptake capacity, and bongkrekic acid failed to influence Ca^{2+} uptake capacity and ADP-ATP exchange. In the authors' view, these findings illuminate a correlation between the lack of sensitivity of the ANT to bongkrekic acid and absence of the classic Ca^{2+}- induced mPT (Konrad et al., 2011). In the following study with mitochondria isolated from brown shrimp (*Crangpon crangon*) and common prawn (*Palaemon serratus*), the authors found bongkrekic acid-sensitive adenine nucleotide transport without apparent manifestations of the Ca^{2+}-inducible mPT (Konrad et al., 2012). Mitochondria from these species have a large (but finite) Ca^{2+} uptake capacity, which was not sensitive to CsA and bongkrekic acid and did not undergo swelling measured as a change of light scattering at 660 nm. Thus, despite the sensitivity of the ANT to bongkrekic acid, these mitochondria have increased resistance to Ca^{2+} and do not undergo the classic mPT accompanied by large-amplitude mitochondrial swelling. On the other hand, following a larger Ca^{2+} load these mitochondria become damaged and, similarly to mammalian mitochondria, could not further take up Ca^{2+}. This suggests that these mitochondria can undergo some form of mPT that precludes infinite Ca^{2+} uptake, but this mPT is not accompanied by mitochondrial morphological changes and is not sensitive to classic mPTP inhibitors. Interestingly, mitochondria from *Artemia franciscana* undergo large amplitude swelling in response to $HgCl_2$, which is also an activator of mPT in mammalian mitochondria (Zoratti and Szabo, 1995), but this swelling is insensitive to CsA (Menze et al., 2005). Overall, these findings cannot rule out a key role of the ANT in the mPT in other species, but they suggest existence of multiple mechanisms underlying this phenomenon.

The ultimate test of the ANT hypothesis became possible with the advent of ANT-knockout mice. Doug Wallace's group (Kokoszka et al., 2004) generated mice, lacking both ANT1 and ANT2 in liver mitochondria. In a series of elegant experiments, the authors obtained results that argued against an essential role of ANT in mPT. In these experiments the authors, using a light scattering assay, showed that Ca^{2+}-preloaded liver mitochondria from ANT-knockout mice responded to depolarization with FCCP with the same amplitude of light scattering changes as mitochondria from wild-type animals. In experiments with tetraphenyl phosphonium-sensitive electrodes (used to measure changes in mitochondrial membrane potential), the authors found that mitochondria from ANT-knockout mice required a three times higher Ca^{2+} load to completely lose membrane potential, suggesting a threefold larger Ca^{2+} uptake capacity of ANT-deficient mitochondria. In experiments with hepatocytes from ANT-knockout mice, the authors found mitochondrial swelling in response to Ca^{2+} influx into the cell mediated by A23187, a Ca^{2+}-ionophore. Overall, these results were interpreted as strong evidence supporting a nonessential role of ANT as the mPTP. However, the data shown by the authors raise some important questions. First, in experiments with light scattering measurements (see Figure 3**a,b** in Kokoszka et al., 2004) it is not clear why the initial slow decline in light scattering is much steeper with ANT-deficient mitochondria than with mitochondria from wild-type animals. If all conditions were the same, this slope should be the same for both types of mitochondria (for comparison, see Figure 2A in Brustovetsky Shalbuyeva, and Brustovetsky, 2005). Second, it is not clear why the fast drop in light scattering in response to FCCP is much larger with ANT-deficient mitochondria than with mitochondria from wild-type animals (Figure 3**a,b** in Kokoszka et al., 2004). In fact, this fast light scattering response to FCCP is most likely an artifact produced by addition of clear, ethanol-based FCCP solution to the turbid mitochondrial suspension. A similar initial response to ethanol was shown by other investigators (see Figure 2A in Yamada et al., 2009). If all conditions were the same, the fast response to FCCP should have the same amplitude. Third, the authors claim that A23187 induces mPT in ANT-deficient hepatocyte mitochondria, resulting in mitochondrial swelling. However, on the electron micrograph the ANT-deficient mitochondria look enlarged but not swollen. On electron micrographs, swollen mitochondria are characterized not only by an increase in size, but most importantly by decreased electron density (increased transparency) of the matrix (for comparison, see Guerrieri et al., 2002). This was not the case in the experiment shown by the authors in Figure 4**a** (Kokoszka et al., 2004). Moreover, on the micrograph that shows ionophore-treated hepatocytes with enlarged ANT-deficient mitochondria, the black inclusions of unidentified nature appeared to be proportionally enlarged compared to similar inclusions on the micrograph with wild-type hepatocytes. Fourth, it is not clear to what extent the conclusions from this study with liver mitochondria are applicable to mitochondria from other tissues. For example, there is a report that skeletal muscle mitochondria, lacking ANT, require eight times larger Ca^{2+} load to induce mPT (Panov, Andreeva, and Greenamyre, 2004). Such greatly increased resistance to Ca^{2+} hardly can be viewed as nonessential. How much Ca^{2+} would be necessary to induce permeability transition in brain mitochondria lacking ANT remains unknown. However, it is known that ANT1 deficiency in mitochondria of cortical and hippocampal neurons makes these cells much more resistant to glutamate excitotoxicity (Lee, Schriner, and Wallace, 2009), in which mPT plays an important role (Li et al., 2009; Schinder et al., 1996; White and Reynolds, 1996). Thus, the results reported by Wallace's group (Kokoszka et al., 2004) raise some concerns about the methodologies used and the interpretation of results. Until these concerns are resolved, these results hardly can be considered as unequivocal evidence against an essential role for ANT in forming the mPTP.

1.3.2 Alternative Hypotheses of mPTP Composition

In recent years, new hypotheses regarding composition of the PTP have emerged. Andrew Halestrap (Leung, Varanyuwatana, and Halestrap, 2008) proposed that a P_i transporter in the inner mitochondrial membrane could be a conducting pathway that forms mPTP. This is in line with findings made by Reinhard Krämer's group (Schroers, Kramer, and Wohlrab, 1997) that a P_i transporter can be converted into a channel-like structure following treatment with mercurials. In another recent study, investigators provided evidence suggesting that the *c*-subunit of F_0 complex in F_0F_1-ATP synthase could be involved in mPTP functioning (Bonora et al., 2013). Consistent with this, Tamara Azarashvili's group (2002) found decreased phosphorylation of the 3.5 kDa peptide related to subunit *c* of F_0F_1-ATP synthase following induction of the permeability transition by Ca^{2+}. Incorporation of this peptide into a planar bilayer membrane resulted in increased membrane conductivity as a result of channel activity that was inhibited by antibodies to subunit *c*. However, in contrast to ANT that under normal conditions translocates bulk ADP^{3-} and ATP^{4-} anions with molecular weights of around 500 Da (Klingenberg, 2008), a single subunit *c* provides only the H^+ translocating pathway. Therefore, it is not clear whether subunit *c* of the F_0 component of the F_0F_1-ATP synthase could be converted into mPTP, permeable for solutes with molecular weights up to 1 to 1.5 kDa (Brustovetsky and Dubinsky, 2000; Haworth and Hunter, 1979). Interestingly, ion currents mediated by subunit *c* incorporated into a lipid bilayer were recorded in previous studies (McGeoch et al., 2000). However, these currents were found to be activated by cyclic guanosine monophosphate (cGMP) that does not induce the mPTP, and surprisingly, suppressed by Ca^{2+} that is the major activator of the mPTP.

In another recent study, it was shown that mitochondrial cyclophilin D interacts with the lateral stalk subunits of F_0F_1-ATP synthase, including oligomycin-sensitivity conferring protein (OSCP), subunits *b* and *d* (Giorgio et al., 2009). This binding required P_i and partially inhibited F_0F_1-ATP synthase activity. CsA disrupted this interaction and increased F_0F_1-ATP synthase activity (Giorgio et al., 2009). The authors showed that mouse liver mitochondria energized by respiration in the presence of stable levels of ADP maintained by hexokinase plus glucose, an ATP-consuming system, had lower Ca^{2+} uptake capacity than mitochondria energized by ATP hydrolysis in the presence of stable levels of ATP maintained by creatine kinase plus phosphocreatine, an ATP-generating system (Giorgio et al., 2013). Based on these observations, the authors concluded that the catalytic activity of F_0F_1-ATP synthase (synthesis versus hydrolysis) affects the Ca^{2+} sensitivity of the mPTP.

Interestingly, in these experiments lower Ca^{2+} uptake capacity was observed with respiring mitochondria, whereas mitochondria lacking respiratory substrates, and therefore, having negligible respiratory activity, had higher Ca^{2+} uptake capacity. It is also notable that Ca^{2+} uptake capacity in mitochondria energized by ATP hydrolysis was significantly diminished when their respiration was activated by adding respiratory substrates. The reason for that was not provided but a higher level of reactive oxygen species (ROS) production in respiring mitochondria might contribute to this difference.

In the experiments with digitonin-permeabilized HQB17 cells incubated with ATP and an ATP-regenerating system (phosphocreatine and creatine kinase in the absence of respiratory substrates), the investigators found that OSCP knockdown slightly increased mitochondrial Ca^{2+} uptake capacity (by about 30%) (Giorgio et al., 2013). This slight increase in Ca^{2+} uptake capacity was interpreted as evidence for F_0F_1-ATP synthase involvement in mPTP formation. However, it is also conceivable that OSCP down-regulation could affect F_0F_1-ATP synthase activity and increase ADP/ATP ratio, resulting in a slight increase in mitochondrial Ca^{2+} uptake capacity. Indeed, in the presence of oligomycin mitochondria accumulated about 25% more Ca^{2+} when incubated with ADP compared to ATP (Giorgio et al., 2013).

Obviously, the most remarkable finding in this study is that dimers of F_0F_1-ATP synthase can form a large channel in the planar lipid bilayer (Giorgio et al., 2013). This channel was Ca^{2+}-dependent, could be inhibited by ADP and Mg^{2+}, and was insensitive to CsA and ANT inhibitors. To some extent this channel resembled the megachannel described in other studies and attributed to the mPTP (Petronilli et al., 1989; Szabo and Zoratti, 1991; Szabo et al., 1992). Further, it was proposed that this mPTP-like channel is formed at the membrane interface between two adjacent F_0 complexes (Bernardi, 2013). Importantly, the channel formed by F_0F_1-ATP synthase dimers required activation by benzodiazepine-423 (Bz-423), whereas the mPTP induction in mitochondria or Ca^{2+}-activated conversion of the reconstituted ANT do not require this agent. This could be explained by the fact that both Bz-423 and CyD bind to OSCP (Giorgio et al., 2009; Johnson et al., 2005), and according to the hypothesis proposed by Giorgi et al. (2013), this binding favors opening of the large channel formed by F_0F_1-ATP synthase dimers. Although the large channel formed by dimers of reconstituted F_0F_1-ATP synthase resembles the megachannel described previously and linked to the PTP (Petronilli et al., 1989; Szabo and Zoratti, 1991; Szabo et al., 1992), the hypothesis that this F_0F_1-ATP synthase conversion into the pore takes place in mitochondria needs additional experimental proof.

In this regard, interesting findings pertinent to the possible role of F_0F_1-ATP synthase in the PTP were reported by Vladimir Skulachev's group (Shchepina et al., 2002). They found that CsA and oligomycin, but not aurovertin B, suppress TNFα-induced cell death in HeLa cells (Shchepina et al., 2002). The inhibitory effect of oligomycin on the mPTP is usually attributed to the change in ADP/ATP ratio. ADP is more effective in inhibition of the mPTP than ATP (Novgorodov et al., 1992). In rat liver mitochondria, oligomycin causes an almost 20-fold increase in ADP/ATP ratio: from 0.15 ± 0.05 to 2.82 ± 0.99, without a change in total concentration of adenine nucleotides (Novgorodov et al., 1994). Both oligomycin and aurovertin B inhibit ATP synthase, but oligomycin interacts with F_0, whereas aurovertin interacts

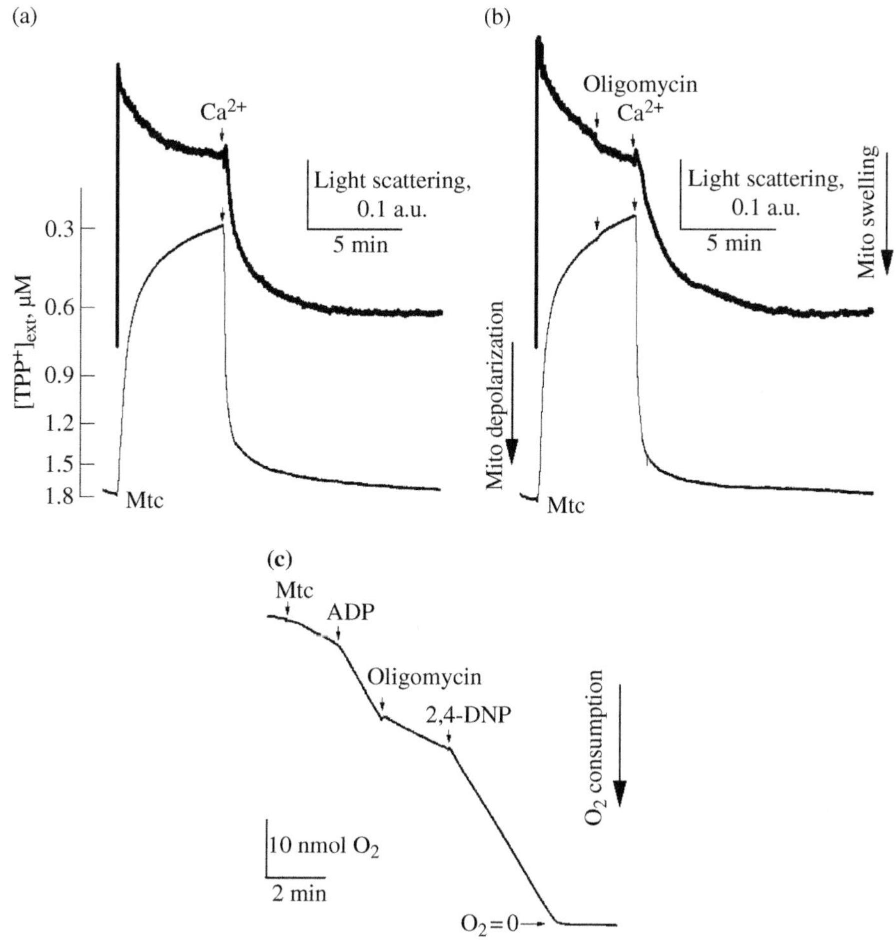

FIGURE 1.3 Oligomycin failed to inhibit Ca^{2+}-induced permeability transition in isolated nonsynaptic brain mitochondria. Isolated Percoll gradient-purified rat brain mitochondria were incubated at 37° C under continuous stirring in the incubation medium containing 210 mM mannitol, 50 mM sucrose, 3 mM succinate, 3 mM glutamate, 3 mM KH_2PO_4, 0.5 mM $MgCl_2$, 10 mM EGTA, 0.1% BSA (free from fatty acids), 10 mM Hepes, pH 7.4. In **a** and **b**, mitochondrial swelling (*thick traces*) was evaluated by following changes in light scattering of mitochondrial suspension at 525 nm with an incident light beam under 180° C in a 0.3-ml chamber at 37° C and continuous stirring. Simultaneously, $\Delta\psi$ was monitored by following the distribution of tetraphenylphosphonium cation (TPP^+) between the external medium (initially 1.8 μM TPP^+-Cl) and the mitochondrial matrix with a TPP^+-sensitive electrode (Kamo et al., 1979). Where indicated, 2.4 μmol Ca^{2+}/mg protein and 1 μM oligomycin were added. In **c**, recording of mitochondrial oxygen consumption, with oligomycin inhibiting ADP-stimulated respiration, is shown as a positive control. Additions: ADP, 300 μM; oligomycin, 1 μM; 2,4-dinitrophenol (DNP), 60 μM. Mitochondrial respiration was recorded in the 0.3-ml incubation chamber equipped with a Clark-type oxygen electrode and a tightly closed lid. The slope of the O_2 electrode trace corresponds to the respiration rate.

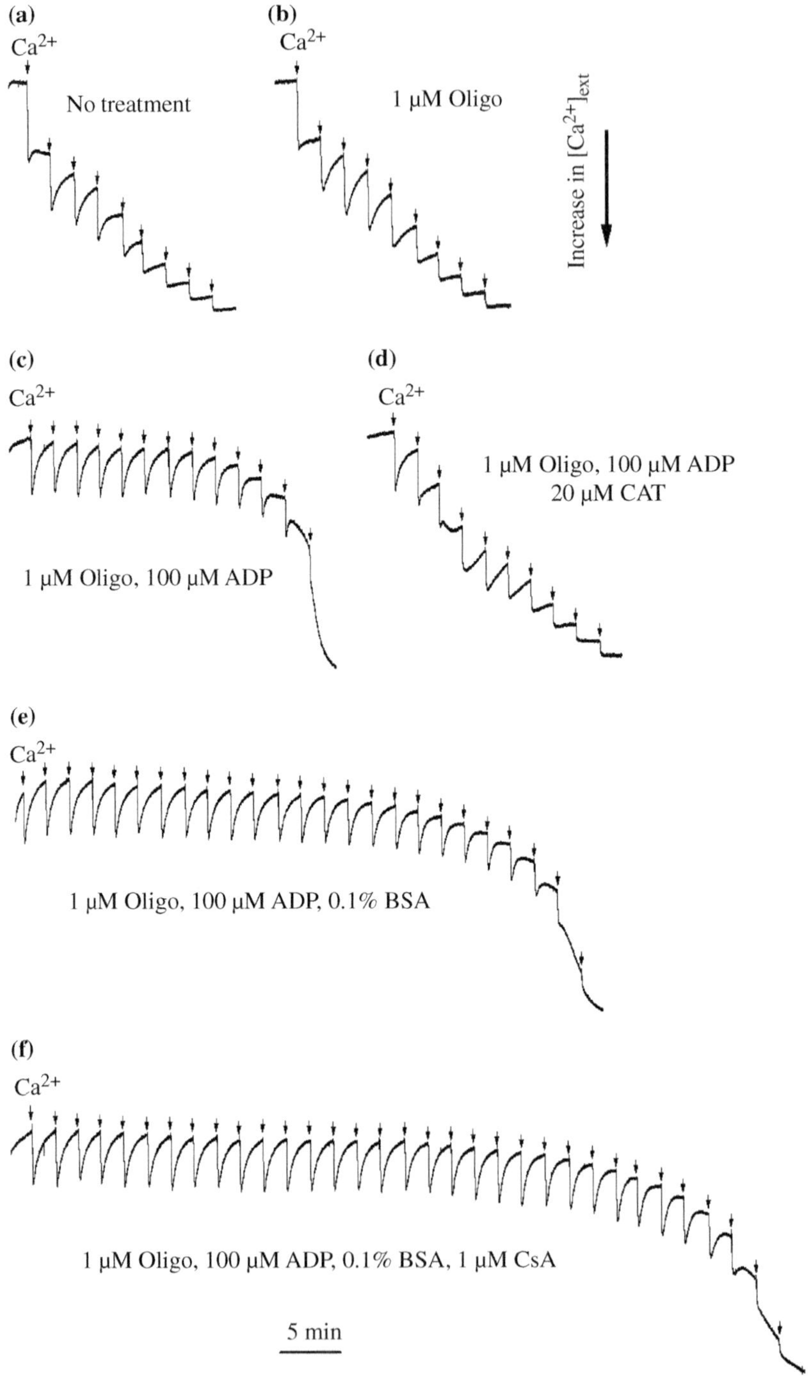

FIGURE 1.4 Ca^{2+} uptake capacity of isolated nonsynaptic brain mitochondria. The effects of oligomycin (Oligo), adenosine diphosphate (ADP), carboxyatractyloside (CAT), bovine

with the F_1 complex of ATP synthase. The mPTP was implicated in cell death induced by TNFα (Shchepina et al., 2002), and it was shown previously that application of oligomycin to isolated rat liver mitochondria inhibits the pore (Novgorodov et al., 1990). Correspondingly, the authors speculated that the protective effect of oligomycin is related to its interaction with the oligomycin-sensitive subunit of F_0 and inhibition of the pore (Shchepina et al., 2002). However, in our experiments with isolated, Percoll gradient-purified brain mitochondria oligomycin failed to inhibit Ca^{2+}-induced permeability transition (Fig. 1.3), suggesting that either F_0 does not form a pore in brain mitochondria or oligomycin does not inhibit the pore formed by F_0. In addition, oligomycin failed to increase Ca^{2+} uptake capacity of brain mitochondria (Fig. 1.4**a**,**b**), whereas 100 μM ADP augmented Ca^{2+} accumulation (Fig. 1.4**c**) and this ADP effect was abolished by 20 μM CAT (Fig. 1.4**d**). Bovine serum albumin (BSA, 0.1%, free from fatty acids) significantly increased Ca^{2+} uptake capacity of mitochondria incubated with oligomycin and ADP (Fig. 1.4**e**). CsA further increased Ca^{2+} accumulation in mitochondria (Fig. 1.4**f**). Thus, the known inhibitors of the mPTP, ADP, BSA, and CsA, increase Ca^{2+} uptake capacity of brain mitochondria, while oligomycin does not.

Taken together, all these observations do not disprove a key role of ANT in the mPT. Consequently, all these data cannot rule out a pore-forming function of ANT in the mPTP. However, the recent data strongly suggest a multiplicity of the mechanisms underlying mPTP. This idea has already been introduced by Zoratti Szabo, and De Marchi (2005) in their excellent review paper on the mPTP and by Martin Klingenberg (2008), a world-renowned expert on ANT, in his brilliant review paper on ANT and adenine nucleotide transport in mitochondria. It seems likely that the multiplicity of mechanisms contributing to mPT is a key to answering many smoldering questions about structure, function, and the pharmacological profile of the mPTP. Consequently, accepting the idea that multiple mechanisms underlie the mPT may help to better understand the complex nature of this phenomenon and resolve many controversial issues concerning molecular identity of the mPTP. Future studies will show whether diversity of the mechanisms underlying the mPTP is a viable idea.

FIGURE 1.4 (*Continued*) serum albumin (BSA), and cyclosporin A (CsA). Mitochondria were incubated at 37° C in 0.3-ml thermostatically regulated chamber under continuous stirring. Ca^{2+} uptake by mitochondria was followed by measuring Ca^{2+} concentration in the incubation medium ($[Ca^{2+}]_{ext}$) with a miniature, custom-made Ca^{2+}-selective electrode. The standard incubation medium contained 125 mM KCl, 0.5 mM $MgCl_2$, 3 mM KH_2PO_4, 10 mM Hepes, pH 7.4, 10 μM EGTA, 3 mM succinate, and 3 mM glutamate. Where indicated, 10 μM Ca^{2+} pulses were applied. In **a**, mitochondria were incubated in the standard incubation medium without other additions. In **b**, the incubation medium was supplemented with 1 μM oligomycin (Oligo); in **c**, with 1 μM oligomycin and 100 μM ADP; in **d**, with 1 μM oligomycin, 100 μM ADP, and 20 μM carboxyatractyloside (CAT), in **e**, with 1 μM oligomycin, 100 μM ADP, and 0.1% BSA (free from fatty acids), in **f**, with 1 μM oligomycin, 100 μM ADP, 0.1% BSA (free from fatty acids), and 1 μM cyclosporin A (CsA). Time bar in **f** is applicable to all panels.

ACKNOWLEDGMENTS

The author is grateful to James Hamilton for performing Ca^{2+} uptake experiments. This work was supported by NIH/NINDS grant R01 NS078008.

REFERENCES

Aarts, M., et al., 2003. A key role for TRPM7 channels in anoxic neuronal death. *Cell,* 115, pp.863–877.

Annunziato, L., Pignataro, G. and Di Renzo, G. F., 2004. Pharmacology of brain Na+/Ca2+ exchanger: from molecular biology to therapeutic perspectives. *Pharmacol Rev,* 56, pp.633–654.

Asimakis, G. K. and Sordahl, L. A., 1977. Effects of atractyloside and palmitoyl coenzyme A on calcium transport in cardiac mitochondria. *Arch Biochem Biophys,* 179, pp.200–210.

Azarashvili, T. S., et al., 2002. Phosphorylation of a peptide related to subunit c of the F0F1-ATPase/ATP synthase and relationship to permeability transition pore opening in mitochondria. *J Bioenerg Biomembr,* 34, pp.279–284.

Baines, C. P., et al., 2005. Loss of cyclophilin D reveals a critical role for mitochondrial permeability transition in cell death. *Nature,* 434, pp.658–662.

Baines, C. P., et al., 2007. Voltage-dependent anion channels are dispensable for mitochondrial-dependent cell death. *Nat Cell Biol,* 9, pp.550–555.

Basso, E., et al., 2005. Properties of the permeability transition pore in mitochondria devoid of Cyclophilin D. *J Biol Chem,* 280, pp.18558–18561.

Baughman, J. M., et al., 2011. Integrative genomics identifies MCU as an essential component of the mitochondrial calcium uniporter. *Nature,* 476, pp.341–345.

Beatrice, M. C., Palmer, J. W. and Pfeiffer, D. R., 1980. The relationship between mitochondrial membrane permeability, membrane potential, and the retention of Ca2+ by mitochondria. *J Biol Chem* 255,8663–8671.

Bernardi, P., 1992. Modulation of the mitochondrial cyclosporin A-sensitive permeability transition pore by the proton electrochemical gradient. Evidence that the pore can be opened by membrane depolarization. *J Biol Chem,* 267, pp.8834–8839.

Bernardi, P., 2013. The mitochondrial permeability transition pore: a mystery solved? *Front Physiol,* 4, p.95.

Bernardi, P., Broekemeier, K. M. and Pfeiffer, D.R., 1994. Recent progress on regulation of the mitochondrial permeability transition pore; a cyclosporin-sensitive pore in the inner mitochondrial membrane. *J Bioenerg Biomembr,* 26, pp.509–517.

Bernardi, P., et al., 1992. Modulation of the mitochondrial permeability transition pore. Effect of protons and divalent cations. *J Biol Chem,* 267, pp.2934–2939.

Beutner, G., et al., 1998. Complexes between porin, hexokinase, mitochondrial creatine kinase and adenylate translocator display properties of the permeability transition pore. Implication for regulation of permeability transition by the kinases. *Biochim Biophys Acta,* 1368, pp.7–18.

Beutner, G., et al., 1996. Complexes between kinases, mitochondrial porin and adenylate translocator in rat brain resemble the permeability transition pore. *FEBS Lett,* 396, pp.189–195.

Beyer, K. and Klingenberg, M., 1985. ADP/ATP carrier protein from beef heart mitochondria has high amounts of tightly bound cardiolipin, as revealed by 31P nuclear magnetic resonance. *Biochemistry,* 24, pp.3821–3826.

Bonora, M., et al., 2013. Role of the c subunit of the FO ATP synthase in mitochondrial permeability transition. *Cell Cycle,* 12, pp.674–683.

Bonventre, J. V., 1997. Roles of phospholipases A2 in brain cell and tissue injury associated with ischemia and excitotoxicity. *J Lipid Mediat Cell Signal,* 17, pp.71–79.

Brenner, C., et al., 2000. Bcl-2 and Bax regulate the channel activity of the mitochondrial adenine nucleotide translocator. *Oncogene,* 19, pp.329–336.

Brittain, M. K., et al., 2012. Delayed calcium dysregulation in neurons requires both the NMDA receptor and the reverse Na+/Ca2+ exchanger. *Neurobiol Dis,* 46, pp.109–117.

Broekemeier, K. M., Dempsey, M. E. and Pfeiffer, D. R., 1989. Cyclosporin A is a potent inhibitor of the inner membrane permeability transition in liver mitochondria. *J Biol Chem,* 264, pp.7826–7830.

Broekemeier, K. M. and Pfeiffer, D. R., 1989. Cyclosporin A-sensitive and insensitive mechanisms produce the permeability transition in mitochondria. *Biochem Biophys Res Commun,* 163, pp.561–566.

Brown, M. R., Sullivan, P. G. and Geddes, J. W., 2006. Synaptic mitochondria are more susceptible to Ca2+overload than nonsynaptic mitochondria. *J Biol Chem,* 281, pp.11658–11668.

Brustovetsky, N. and Dubinsky, J. M., 2000. Limitations of cyclosporin A inhibition of the permeability transition in CNS mitochondria. *J Neurosci,* 20, pp.8229–8237.

Brustovetsky, N. and Klingenberg, M., 1996. Mitochondrial ADP/ATP carrier can be reversibly converted into a large channel by Ca2+. *Biochemistry,* 35, pp.8483–8488.

Brustovetsky, N., et al., 2002a. Calcium-induced cytochrome c release from CNS mitochondria is associated with the permeability transition and rupture of the outer membrane. *J Neurochem,* 80, pp.207–218.

Brustovetsky, N., et al., 2002b. A large Ca2+-dependent channel formed by recombinant ADP/ATP carrier from Neurospora crassa resembles the mitochondrial permeability transition pore. *Biochemistry,* 41, pp.11804–11811.

Brustovetsky, N., et al., 2003. Increased susceptibility of striatal mitochondria to calcium-induced permeability transition. *J Neurosci,* 23, pp.4858–4867.

Brustovetsky, T., Bolshakov, A. and Brustovetsky, N., 2010. Calpain activation and Na(+)/Ca(2+) exchanger degradation occur downstream of calcium deregulation in hippocampal neurons exposed to excitotoxic glutamate. *J Neurosci Res,* 88, pp.1317–1328.

Brustovetsky, T., Shalbuyeva, N. and Brustovetsky. N., 2005. Lack of manifestations of diazoxide/5-hydroxydecanoate-sensitive KATP channel in rat brain nonsynaptosomal mitochondria. *J Physiol,* 568, pp.47–59.

Buchanan, B. B., et al., 1976. Antibody evidence for different conformational states of ADP, ATP translocator protein isolated from mitochondria. *Proc Natl Acad Sci U S A,* 73, pp.2280–2284.

Cardenas, C., et al., 2010. Essential regulation of cell bioenergetics by constitutive InsP3 receptor Ca2+ transfer to mitochondria. *Cell,* 142, pp.270–283.

Choi, D. W., 1988. Glutamate neurotoxicity and diseases of the nervous system. *Neuron,* 1, pp.623–634.

Connern, C. P. and Halestrap, A. P., 1992. Purification and N-terminal sequencing of peptidyl-prolyl cis-trans-isomerase from rat liver mitochondrial matrix reveals the existence of a distinct mitochondrial cyclophilin. *Biochem J,* 284 (Pt 2), pp.381–385.

Connern, C. P. and Halestrap, A, P., 1994. Recruitment of mitochondrial cyclophilin to the mitochondrial inner membrane under conditions of oxidative stress that enhance the opening of a calcium-sensitive non-specific channel. *Biochem J,* 302 (Pt 2), pp.321–324.

Crompton, M., Ellinger, H. and Costi, A., 1988. Inhibition by cyclosporin A of a Ca2+-dependent pore in heart mitochondria activated by inorganic phosphate and oxidative stress. *Biochem J,* 255, pp.357–360.

Crompton, M., Virji, S. and Ward, J. M., 1998. Cyclophilin-D binds strongly to complexes of the voltage-dependent anion channel and the adenine nucleotide translocase to form the permeability transition pore. *Eur J Biochem,* 258, pp.729–735.

Cuomo, O., et al., 2008. A critical role for the potassium-dependent sodium-calcium exchanger NCKX2 in protection against focal ischemic brain damage. *J Neurosci,* 28, pp.2053–2063.

Czyz, A., Baranauskas, G. and Kiedrowski, L., 2002. Instrumental role of Na+ in NMDA excitotoxicity in glucose-deprived and depolarized cerebellar granule cells. *J Neurochem,* 81, pp.379–389.

De, S. D., et al., 2011. A forty-kilodalton protein of the inner membrane is the mitochondrial calcium uniporter. *Nature,* 476, pp.336–340.

Dierks, T., et al., 1990a. The mitochondrial aspartate/glutamate and ADP/ATP carrier switch from obligate counterexchange to unidirectional transport after modification by SH-reagents. *Biochim Biophys Acta* 1028, 268–280.

Dierks, T., Salentin, A. and Kramer, R., 1990b. Pore-like and carrier-like properties of the mitochondrial aspartate/glutamate carrier after modification by SH-reagents: evidence for a performed channel as a structural requirement of carrier-mediated transport. *Biochim Biophys Acta,* 1028, pp.281–288.

Dietz, R. M., Kiedrowski, L. and Shuttleworth, C. W., 2007. Contribution of Na(+)/Ca(2+) exchange to excessive Ca(2+) loading in dendrites and somata of CA1 neurons in acute slice. *Hippocampus,* 17, pp.1049–1059.

Eliseev, R. A., et al., 2007. Role of cyclophilin D in the resistance of brain mitochondria to the permeability transition. *Neurobiol Aging* 28, 1532–1542.

Elrod, J. W. and Molkentin, J. D., 2013. Physiologic functions of cyclophilin D and the mitochondrial permeability transition pore. *Circ J,* 77, pp.1111–1122.

Fischer, G., et al., 1989. Cyclophilin and peptidyl-prolyl cis-trans isomerase are probably identical proteins. *Nature,* 337, pp.476–478.

Fournier, N., Ducet, G. and Crevat, A., 1987. Action of cyclosporine on mitochondrial calcium fluxes. *J Bioenerg Biomembr,* 19, pp.297–303.

Gao. J., et al., 2005. Coupling between NMDA receptor and acid-sensing ion channel contributes to ischemic neuronal death. *Neuron,* 48, pp.635–646.

Gerencser, A. A., et al., 2009. Real-time visualization of cytoplasmic calpain activation and calcium deregulation in acute glutamate excitotoxicity. *J Neurochem,* 110, pp.990–1004.

Giorgio, V., et al., 2009. Cyclophilin D modulates mitochondrial F0F1-ATP synthase by interacting with the lateral stalk of the complex. *J Biol Chem,* 284, pp.33982–33988.

Giorgio, V., et al., 2013. Dimers of mitochondrial ATP synthase form the permeability transition pore. *Proc Natl Acad Sci U S A,* 110, pp.5887–5892.

Gizatullina, Z. Z., et al., 2005. Effects of extramitochondrial ADP on permeability transition of mouse liver mitochondria. *Biochim Biophys Acta,* 1706, pp.98–104.

Guerini, D., Coletto, L. and Carafoli, E., 2005. Exporting calcium from cells. *Cell Calcium,* 38, pp.281–289.

Guerrieri, F., et al., 2002. Changes in ultrastructure and the occurrence of permeability transition in mitochondria during rat liver regeneration. *Eur J Biochem,* 269, pp.3304–3312.

Halestrap, A. P. and Davidson, A. M., 1990. Inhibition of Ca2(+)-induced large-amplitude swelling of liver and heart mitochondria by cyclosporin is probably caused by the inhibitor binding to mitochondrial-matrix peptidyl-prolyl cis-trans isomerase and preventing it interacting with the adenine nucleotide translocase. *Biochem J,* 268, pp.153–160.

Halestrap, A. P., Woodfield, K. Y. and Connern, C. P., 1997. Oxidative stress, thiol reagents, and membrane potential modulate the mitochondrial permeability transition by affecting nucleotide binding to the adenine nucleotide translocase. *J Biol Chem,* 272, pp.3346–3354.

Hansford, R. G., 1991. Dehydrogenase activation by Ca2+ in cells and tissues. *J Bioenerg Biomembr,* 23, pp.823–854.

Haworth, R. A. and Hunter, D. R., 1979. The Ca2+-induced membrane transition in mitochondria. II. Nature of the Ca2+ trigger site. *Arch Biochem Biophys,* 195, pp.460–467.

Haworth, R. A. and Hunter, D. R., 1980. Allosteric inhibition of the Ca2+-activated hydrophilic channel of the mitochondrial inner membrane by nucleotides. *J Membr Biol,* 54, pp.231–236.

Herick, K., Kramer, R. and Luhring, H., 1997. Patch clamp investigation into the phosphate carrier from Saccharomyces cerevisiae mitochondria. *Biochim Biophys Acta,* 1321, pp.207–220.

Hunter, D. R. and Haworth, R. A., 1979a. The Ca2+-induced membrane transition in mitochondria. I. The protective mechanisms. *Arch Biochem Biophys,* 195, pp.453–459.

Hunter, D. R. and Haworth, R. A., 1979b. The Ca2+-induced membrane transition in mitochondria. III. Transitional Ca2+ release. *Arch Biochem Biophys,* 195, pp.468–477.

Hunter, D. R., Haworth, R. A. and Southard, J. H., 1976. Relationship between configuration, function, and permeability in calcium-treated mitochondria. *J Biol Chem,* 251, pp.5069–5077.

Javadov, S. and Kuznetsov, A., 2013. Mitochondrial permeability transition and cell death: the role of cyclophilin d. *Front Physiol,* 4, p.76.

Jeffs, G. J., et al., 2008. NCX3 knockout mice exhibit increased hippocampal CA1 and CA2 neuronal damage compared to wild-type mice following global cerebral ischemia. *Exp Neurol,* 210, pp.268–273.

Jeon, D., et al., 2007. Na(+)/Ca(2+) exchanger 2 is neuroprotective by exporting Ca(2+) during a transient focal cerebral ischemia in the mouse. *Cell Calcium,* 43, pp.482–491.

Johnson, K. M., et al., 2005. Identification and validation of the mitochondrial F1F0-ATPase as the molecular target of the immunomodulatory benzodiazepine Bz-423. *Chem Biol,* 12, pp.485–496.

Kamo, N., et al., 1979. Membrane potential of mitochondria measured with an electrode sensitive to tetraphenyl phosphonium and relationship between proton electrochemical potential and phosphorylation potential in steady state. *J Membr Biol,* 49, pp.105–121.

Kiedrowski, L., et al., 1994. Glutamate impairs neuronal calcium extrusion while reducing sodium gradient. *Neuron,* 12, pp.295–300.

Kim, D. K., et al., 1995. Glutamate stably enhances the activity of two cytosolic forms of phospholipase A2 in brain cortical cultures. *Biochem J,* 310 (Pt 1), pp.83–90.

Kinnally, K. W., Campo, M. L. and Tedeschi, H., 1989. Mitochondrial channel activity studied by patch-clamping mitoplasts. *J Bioenerg Biomembr,* 21, pp.497–506.

Kinnally, K. W., et al., 1991. Calcium modulation of mitochondrial inner membrane channel activity. *Biochem Biophys Res Commun* 176, 1183–1188.

Kirichok, Y., Krapivinsky, G. and Clapham, D. E., 2004. The mitochondrial calcium uniporter is a highly selective ion channel. *Nature,* 427, pp.360–364.

Klingenberg, M. 1981. Membrane protein oligomeric structure and transport function. *Nature,* 290, pp.449–454.

Klingenberg, M., 1985. The ADP/ATP carrier in mitochondrial membranes. In: A. N. Martonosi, ed. 1985. *The enzymes of biological membranes.* New York: Plenum Press, pp.511–553.

Klingenberg, M., 2008. The ADP and ATP transport in mitochondria and its carrier. *Biochim Biophys Acta,* 1778, pp.1978–2021.

Kokoszka, J. E., et al., 2004. The ADP/ATP translocator is not essential for the mitochondrial permeability transition pore. *Nature,* 427, pp.461–465.

Konig, T., et al., 1982. Inhibition of mitochondrial substrate anion translocators by a synthetic amphipathic polyanion. *J Bioenerg Biomembr,* 14, pp.297–305.

Konrad, C, et al., 2012. Absence of Ca2+-induced mitochondrial permeability transition but presence of bongkrekate-sensitive nucleotide exchange in C. crangon and P. serratus. *PLoS ONE,* 7, e39839.

Konrad, C., et al., 2011. A distinct sequence in the adenine nucleotide translocase from Artemia franciscana embryos is associated with insensitivity to bongkrekate and atypical effects of adenine nucleotides on Ca2+ uptake and sequestration. *FEBS J,* 278, pp.822–836.

Krauskopf, A., et al., 2006. Properties of the permeability transition in VDAC1(-/-) mitochondria. *Biochim Biophys Acta,* 1757, pp.590–595.

Lapidus, R. G. and Sokolove, P. M., 1992. Inhibition by spermine of the inner membrane permeability transition of isolated rat heart mitochondria. *FEBS Lett,* 313, pp.314–318.

Lapidus, R. G. and Sokolove, P. M, 1993. Spermine inhibition of the permeability transition of isolated rat liver mitochondria: an investigation of mechanism. *Arch Biochem Biophys,* 306, pp.246–253.

Le Quoc, D. and Le Quoc, K., 1989. Relationships between the NAD(P) redox state, fatty acid oxidation, and inner membrane permeability in rat liver mitochondria. *Arch Biochem Biophys,* 273, pp.466–478.

Le Quoc, K. and Le Quoc, D., 1988. Involvement of the ADP/ATP carrier in calcium-induced perturbations of the mitochondrial inner membrane permeability: importance of the orientation of the nucleotide binding site. *Arch Biochem Biophys,* 265, pp.249–257.

Lee, A. C., Xu, X. and Colombini, M., 1996. The role of pyridine dinucleotides in regulating the permeability of the mitochondrial outer membrane. *J Biol Chem,* 271, pp.26724–26731.

Lee, J., Schriner, S. E. and Wallace, D. C., 2009. Adenine nucleotide translocator 1 deficiency increases resistance of mouse brain and neurons to excitotoxic insults. *Biochim Biophys Acta,* 1787, pp.364–370.

Leung, A. W., Varanyuwatana, P. and Halestrap, A. P., 2008. The mitochondrial phosphate carrier interacts with cyclophilin D and may play a key role in the permeability transition. *J Biol Chem,* 283, pp.26312–26323.

Li, V., Brustovetsky, T. and Brustovetsky, N., 2009. Role of cyclophilin D-dependent mitochondrial permeability transition in glutamate-induced calcium deregulation and excitotoxic neuronal death. *Exp Neurol,* 218, pp.171–182.

Lin, D. T. and Lechleiter, J. D., 2002. Mitochondrial targeted cyclophilin D protects cells from cell death by peptidyl prolyl isomerization. *J Biol Chem,* 277, pp.31134–31141.

Litsky, M. L. and Pfeiffer, D. R., 1997. Regulation of the mitochondrial Ca2+ uniporter by external adenine nucleotides: the uniporter behaves like a gated channel which is regulated by nucleotides and divalent cations. *Biochemistry,* 36, pp.7071–7080.

Manev, H., et al., 1989. Delayed increase of Ca2+ influx elicited by glutamate: role in neuronal death. *Mol Pharmacol,* 36, pp.106–112.

McGeoch, J. E., et al., 2000. Opposing actions of cGMP and calcium on the conductance of the F(0) subunit c pore. *Biochem Biophys Res Commun,* 274, pp.835–840.

Menze, M. A., et al., 2005. Mitochondrial permeability transition in the crustacean Artemia franciscana: absence of a calcium-regulated pore in the face of profound calcium storage. *Am J Physiol Regul Integr Comp Physiol,* 289, pp.R68–R76.

Minger, S. L., et al., 1998. Glutamate receptor antagonists inhibit calpain-mediated cytoskeletal proteolysis in focal cerebral ischemia. *Brain Res,* 810, pp.181–199.

Naga, K. K., Sullivan, P. G. and Geddes, J. W., 2007. High cyclophilin D content of synaptic mitochondria results in increased vulnerability to permeability transition. *J Neurosci,* 27, pp.7469–7475.

Nakagawa, T., et al., 2005. Cyclophilin D-dependent mitochondrial permeability transition regulates some necrotic but not apoptotic cell death. *Nature,* 434, pp.652–658.

Nicolli, A., et al., 1996. Interactions of cyclophilin with the mitochondrial inner membrane and regulation of the permeability transition pore, and cyclosporin A-sensitive channel. *J Biol Chem,* 271, pp.2185–2192.

Novgorodov, S. A., et al., 1990. Effect of cyclosporine A and oligomycin on non-specific permeability of the inner mitochondrial membrane. *FEBS Lett,* 270, pp.108–110.

Novgorodov, S. A., et al., 1991. The nonspecific inner membrane pore of liver mitochondria: modulation of cyclosporin sensitivity by ADP at carboxyatractyloside-sensitive and insensitive sites. *Biochem Biophys Res Commun,* 180, pp.33–38.

Novgorodov, S. A., et al., 1992. The permeability transition in heart mitochondria is regulated synergistically by ADP and cyclosporin A. *J Biol Chem,* 267, pp.16274–16282.

Novgorodov, S. A., et al., 1994. Magnesium ion modulates the sensitivity of the mitochondrial permeability transition pore to cyclosporin A and ADP. *Arch Biochem Biophys,* 311, pp.219–228.

Novgorodov, S. A., et al., 2005. Positively charged ceramide is a potent inducer of mitochondrial permeabilization. *J Biol Chem,* 280, 16096–16105.

Panov, A., Filippova, S. and Lyakhovich, V., 1980. Adenine nucleotide translocase as a site of regulation by ADP of the rat liver mitochondria permeability to H+ and K+ ions. *Arch Biochem Biophys,* 199, pp.420–426.

Panov, A. V., Andreeva, L. and Greenamyre, J. T., 2004. Quantitative evaluation of the effects of mitochondrial permeability transition pore modifiers on accumulation of calcium phosphate: comparison of rat liver and brain mitochondria. *Arch Biochem Biophys,* 424, pp.44–52.

Pebay-Peyroula, E., et al., 2003. Structure of mitochondrial ADP/ATP carrier in complex with carboxyatractyloside. *Nature,* 426, pp.39–44.

Perocchi, F., et al., 2010. MICU1 encodes a mitochondrial EF hand protein required for Ca(2+) uptake. *Nature,* 467, pp.291–296.

Petronilli, V., Szabo, I. and Zoratti, M., 1989. The inner mitochondrial membrane contains ion-conducting channels similar to those found in bacteria. *FEBS Lett,* 259, pp.137–143.

Pfeiffer, D. R., et al., 1979. Intramitochondrial phospholipase activity and the effects of Ca2+ plus N-ethylmaleimide on mitochondrial function. *J Biol Chem,* 254, pp.11485–11494.

Pivovarova, N. B., et al., 2004. Excitotoxic calcium overload in a subpopulation of mitochondria triggers delayed death in hippocampal neurons. *J Neurosci,* 24, pp.5611–5622.

Rigobello, M. P., et al., 1995. Effect of polycation peptides on mitochondrial permeability transition. *Biochem Biophys Res Commun,* 217, pp.144–149.

Rottenberg, H. and Marbach, M., 1989. Adenine nucleotides regulate Ca2+ transport in brain mitochondria. *FEBS Lett,* 247, pp.483–486.

Rottenberg, H. and Marbach, M., 1990. Regulation of Ca2+ transport in brain mitochondria. II. The mechanism of the adenine nucleotides enhancement of Ca2+ uptake and retention. *Biochim Biophys Acta,* 1016, pp.87–98.

Ruck, A., et al., 1998. Reconstituted adenine nucleotide translocase forms a channel for small molecules comparable to the mitochondrial permeability transition pore. *FEBS Lett,* 426, pp.97–101.

Schinder, A. F., et al., 1996. Mitochondrial dysfunction is a primary event in glutamate neurotoxicity. *J Neurosci,* 16, pp.6125–6133.

Schinzel, A. C., et al., 2005. Cyclophilin D is a component of mitochondrial permeability transition and mediates neuronal cell death after focal cerebral ischemia. *Proc Natl Acad Sci U S A,* 102, pp.12005–12010.

Schonfeld, P. and Bohnensack, R., 1997. Fatty acid-promoted mitochondrial permeability transition by membrane depolarization and binding to the ADP/ATP carrier. *FEBS Lett,* 420, pp.167–170.

Schroers, A., Kramer, R. and Wohlrab, H. (1997) The reversible antiport-uniport conversion of the phosphate carrier from yeast mitochondria depends on the presence of a single cysteine. *J Biol Chem,* 272, pp.10558–10564.

Schwaiger, M., Herzog, V. and Neupert, W., 1987. Characterization of translocation contact sites involved in the import of mitochondrial proteins. *J Cell Biol,* 105, pp.235–246.

Secondo, A., et al., 2007. BHK cells transfected with NCX3 are more resistant to hypoxia followed by reoxygenation than those transfected with NCX1 and NCX2: Possible relationship with mitochondrial membrane potential. *Cell Calcium,* 42, pp.521–535.

Shchepina, L. A., et al., 2002. Oligomycin, inhibitor of the F0 part of H+-ATP-synthase, suppresses the TNF-induced apoptosis. *Oncogene,* 21, pp.8149–8157.

Siemen, D. and Ziemer, M., 2013. What is the nature of the mitochondrial permeability transition pore and what is it not? *IUBMB Life,* 65, pp.255–262.

Siman, R., Noszek, J. C. and Kegerise, C., 1989. Calpain I activation is specifically related to excitatory amino acid induction of hippocampal damage. *J Neurosci,* 9, pp.1579–1590.

Sultan, A. and Sokolove, P. M., 2001. Palmitic acid opens a novel cyclosporin A-insensitive pore in the inner mitochondrial membrane. *Arch Biochem Biophys,* 386, pp.37–51.

Szabo, I., Bernardi, P. and Zoratti, M., 1992. Modulation of the mitochondrial megachannel by divalent cations and protons. *J Biol Chem,* 267, pp.2940–2946.

Szabo, I., De, P. and Zoratti, M., 1993. The mitochondrial permeability transition pore may comprise VDAC molecules. II. The electrophysiological properties of VDAC are compatible with those of the mitochondrial megachannel. *FEBS Lett,* 330, pp.206–210.

Szabo, I. and Zoratti, M., 1991. The giant channel of the inner mitochondrial membrane is inhibited by cyclosporin A. *J Biol Chem,* 266, pp.3376–3379.

Szabo, I. and Zoratti, M., 1992. The mitochondrial megachannel is the permeability transition pore. *J Bioenerg Biomembr,* 24, pp.111–117.

Szabo, I. and Zoratti, M., 1993. The mitochondrial permeability transition pore may comprise VDAC molecules. I. Binary structure and voltage dependence of the pore. *FEBS Lett,* 330, pp.201–205.

Takahashi, N., Hayano, T. and Suzuki, M., 1989. Peptidyl-prolyl cis-trans isomerase is the cyclosporin A-binding protein cyclophilin. *Nature,* 337, pp.473–475.

Tikhonova, I. M., et al., 1994. Ion permeability induced in artificial membranes by the ATP/ADP antiporter. *FEBS Lett* 337, 231–234.

Toninello, A., Siliprandi, D. and Siliprandi, N., 1983. On the mechanism by which Mg2+ and adenine nucleotides restore membrane potential in rat liver mitochondria deenergized by Ca2+ and phosphate. *Biochem Biophys Res Commun,* 111, pp.792–797.

Tropschug, M., et al., 1988. Cyclosporin A-binding protein (cyclophilin) of Neurospora crassa. One gene codes for both the cytosolic and mitochondrial forms. *J Biol Chem,* 263, pp.14433–14440.

Turetsky, D. M., et al., 1994. Cortical neurones exhibiting kainate-activated Co2+ uptake are selectively vulnerable to AMPA/kainate receptor-mediated toxicity. *Neurobiol Dis,* 1, pp.101–110.

Tymianski, M., et al., 1993a. Secondary Ca2+ overload indicates early neuronal injury which precedes staining with viability indicators. *Brain Res,* 607, pp.319–323.

Tymianski, M., et al., 1993b. Source specificity of early calcium neurotoxicity in cultured embryonic spinal neurons. *J Neurosci,* 13, pp.2085–2104.

Vianello, A., et al., 2012. The mitochondrial permeability transition pore (PTP) - an example of multiple molecular exaptation? *Biochim Biophys Acta,* 1817, pp.2072–2086.

Waldmeier, P. C., 2002. Inhibition of the mitochondrial permeability transition by the nonimmunosuppressive cyclosporin derivative NIM811. *Mol Pharmacol,* 62, pp.22–29.

White, R. J. and Reynolds, I. J., 1996. Mitochondrial depolarization in glutamate-stimulated neurons: an early signal specific to excitotoxin exposure. *J Neurosci,* 16, pp.5688–5697.

Wieckowski, M. R. and Wojtczak, L., 1998. Fatty acid-induced uncoupling of oxidative phosphorylation is partly due to opening of the mitochondrial permeability transition pore. *FEBS Lett,* 423, pp.339–342.

Wieckowski, M. R., Brdiczka, D. and Wojtczak, L., 2000. Long-chain fatty acids promote opening of the reconstituted mitochondrial permeability transition pore. *FEBS Lett,* 484, pp.61–64.

Woodfield, K., et al., 1998. Direct demonstration of a specific interaction between cyclophilin-D and the adenine nucleotide translocase confirms their role in the mitochondrial permeability transition. *Biochem J,* 336 (Pt 2), pp.287–290.

Yamada, A., et al., 2009. Ca2+-induced permeability transition can be observed even in yeast mitochondria under optimized experimental conditions. *Biochim Biophys Acta,* 1787, pp.1486–1491.

Zoratti, M. and Szabo, I., 1995. The mitochondrial permeability transition. *Biochim Biophys Acta,* 1241, pp.139–176.

Zoratti, M., Szabo, I. and De Marchi, U., 2005. Mitochondrial permeability transitions: how many doors to the house? *Biochim Biophys Acta,* 1706, pp.40–52.

Zoratti, M., Szabo, I. and De Pinto, V., 1994. The mitochondrial permeabilization pore. In: M. Forte and M. Colombini, eds. Molecular biology of mitochondrial transport systems. Berlin: Springer-Verlag, pp.153–168.

Zorov, D. B., Kinnally, K. W. and Tedeschi, H., 1992. Voltage activation of heart inner mitochondrial membrane channels. *J Bioenerg Biomembr,* 24, pp.119–124.

2

THE MITOCHONDRIAL PERMEABILITY TRANSITION PORE, THE c-SUBUNIT OF THE F_1F_O ATP SYNTHASE, CELLULAR DEVELOPMENT, AND SYNAPTIC EFFICIENCY

Elizabeth A. Jonas

Department Internal Medicine, Section of Endocrinology, Yale University School of Medicine, New Haven, Connecticut, USA

George A. Porter, Jr.

Department of Pediatrics (Cardiology), University of Rochester Medical Center, School of Medicine and Dentistry, Rochester, New York, USA

Gisela Beutner

Department of Pediatrics (Cardiology), University of Rochester Medical Center, School of Medicine and Dentistry, Rochester, New York, USA

Nelli Mnatsakanyan

Department Internal Medicine, Section of Endocrinology, Yale University School of Medicine, New Haven, Connecticut, USA

Kambiz N. Alavian

Division of Brain Sciences, Department of Medicine, Imperial College London, South Kensington Campus, London, UK

The Functions, Disease-Related Dysfunctions, and Therapeutic Targeting of Neuronal Mitochondria, First Edition. Edited by Valentin K. Gribkoff, Elizabeth A. Jonas and J. Marie Hardwick.

2.1 INTRODUCTION

Ion transport across the mitochondrial inner and outer membranes is central to mitochondrial function, including regulation of oxidative phosphorylation and cell death. Although required for adenosine triphosphate (ATP) production by mitochondria, recent findings have confirmed that the c-subunit of the ATP synthase also houses a large conductance uncoupling channel, the mitochondrial permeability transition pore (mPTP), the persistent opening of which produces osmotic dysregulation of the inner mitochondrial membrane and cell death. This chapter will discuss recent advances in understanding the molecular components of mPTP and its regulatory mechanisms during cell death. In contrast to increased mitochondrial inner membrane uncoupling, decreased uncoupling occurs in states of enhanced mitochondrial efficiency; relative closure of mPTP therefore contributes to cell functions as diverse as cardiac development and synaptic efficacy.

2.2 MITOCHONDRIA AT THE CENTER OF CELL METABOLISM AND CELL DEATH

Mitochondria are complex organelles responsible for producing energy in the form of ATP for most eukaryotic cells. They are also at the center of several other essential processes including Ca^{2+} homeostasis and heme and steroid biosynthesis. Additionally, the mitochondrion regulates cellular responses to stress and controls cell death.

To produce ATP, mitochondria use substrates produced in the cytosol by carbohydrate, lipid, and protein metabolic pathways. Products such as acetyl co-enzyme A enter the tricarboxylic acid (TCA) cycle. The TCA cycle synthesizes nicotinamide adenine dinucleotide (NADH) and flavin adenine dinucleotide ($FADH_2$), reducing agents that donate their electrons to the electron transport chain and H^+ ions to the electron transport complexes to be pumped out of the matrix. This creates a proton motive force that in turn drives the F_1F_O ATP synthase (Mitchell, 1961). Upon kinetic repositioning of the ATP synthase rotor, ATP is synthesized from adenosine diphosphate (ADP) and inorganic phosphate (P_i) (Watt et al., 2010). The machinery required for ADP/ATP exchange between the cytoplasm and the matrix includes the outer membrane voltage dependent anion channel (VDAC) and the adenine nucleotide transporter (ANT) at the inner membrane. These proteins are intimately linked to the ATP synthase (Chen et al., 2004).

2.3 MITOCHONDRIAL INNER MEMBRANE LEAK: REGULATOR OF METABOLIC RATE AND UNCOUPLING

Two currents complete the current loop of the proton pumping activity of the electron transport complexes. First, ATP is formed by hydrogen ion (H+) translocation through the ATP synthase in the opposite direction to that of the electron transport complexes. Second, an apparently wasteful leak in the inner mitochondrial membrane uncouples oxidation from phosphorylation as H+ ions enter the matrix through channels independent of ATP production. Classically, uncoupling proteins generate heat for organisms with large surface-to-volume ratio, depolarize mitochondria to temper oxidative damage, and regulate metabolic rate during hibernation and at other times (Andrews, Diano, and Horvath, 2005; Divakaruni and Brand, 2011; Fedorenko, Lishko, and Kirichok, 2012; Nicholls and Rial, 1999). In addition to uncoupling proteins, however, intrinsic uncoupling exists within several inner mitochondrial membrane channels and transporters including the F_1F_O ATP synthase (Caviston et al., 1998; D'Alessandro, Turina, and Melandri, 2008).

2.4 MITOCHONDRIAL INNER MEMBRANE CHANNELS AND EXCHANGERS ARE NECESSARY FOR Ca^{2+} CYCLING AND CELLULAR Ca^{2+} DYNAMICS

Mitochondrial inner membrane depolarization occurs not only through proton movement but also via the flux of other ions including Ca^{2+} across mitochondrial membranes. Calcium movement into the mitochondrial matrix is a physiological event that takes place in response to increased cytosolic Ca^{2+} levels. Ca^{2+} buffering is frequently employed by mitochondria in cells that experience rapidly changing cytosolic Ca^{2+} levels such as those of excitable tissues.

Mitochondria regulate cytosolic levels of Ca^{2+} and the release of Ca^{2+} and metabolites using several ion channels and exchangers. The Ca^{2+} uniporter ion channel, MCU, located at the mitochondrial inner membrane participates in mitochondrial Ca^{2+} uptake within the cell body of many types of cells and also in the presynaptic terminals of neurons (Billups and Forsythe, 2002; Chouhan et al., 2012; Kang et al., 2008; Rizzuto et al., 2012). Isoforms of mitochondrial calcium uniporter (MCU) and its helper MICU confer tissue specificity and add complexity to the mechanisms of activity dependent energy production by mitochondria (De Stefani and Rizzuto, 2014; Raffaello, De Stefani, and Rizzuto, 2012). Mitochondrial Ca^{2+} release also appears to be highly regulated, but unlike the MCU, the molecular components of a Ca^{2+} release channel were only recently discovered and form the main focus of this chapter.

Ca^{2+} re-release from mitochondria prolongs the elevation of cytosolic Ca^{2+} after influx across the plasma membrane (Zucker and Regehr, 2002). During neuronal activity, Ca^{2+} influx across the plasma membrane through voltage-gated Ca^{2+} channels and other cation channels is cleared by the actions of Ca^{2+} ATPases at the plasma membrane and by buffering through uptake by intracellular stores including endoplasmic reticulum (ER) and mitochondria (Lopreiato, Giacomello, and Carafoli, 2014; Rizzuto et al., 2012); these processes reset the normally low Ca^{2+} levels present in resting cells. The buffered Ca^{2+} is eventually re-released. In synaptic endings, residual Ca^{2+} increases the Ca^{2+} available for synaptic vesicle fusion, enhancing the amount of neurotransmitter released for a given stimulus (Jonas, 2006; Neher and Sakaba, 2008).

Ca^{2+} influx into cells and into mitochondria also regulates TCA cycle enzymes (Denton, 2009) and enzymes in the electron transport chain (Gellerich et al., 2010) to speed the process of ATP synthesis. As cytosolic Ca^{2+} increases during cellular activity, mitochondrial TCA cycle enzymatic activity is enhanced, increasing the levels of mitochondrial NADH and ATP. Ca^{2+} influx into mitochondria enhances the proportion of pyruvate dehydrogenase complex in its active, dephosphorylated form. Ca^{2+} also increases the ability of isocitrate dehydrogenase and oxoglutarate dehydrogenase to bind their substrates. The result is regulation of mitochondrial TCA enzymes by cytosolic Ca^{2+} or "external regulation" as opposed to "internal regulation" caused by changes in NAD/NADH ratio and the ratio of ADP/ATP in the mitochondrial matrix. In neurons, increases in cytosolic Ca^{2+} arise from depolarization of the plasma membrane with resultant increase in permeability to Ca^{2+} through glutamate receptors and voltage-gated Ca^{2+} channels. In addition, mitochondria receive Ca^{2+} from the ER, which partners with mitochondria to respond to cytosolic Ca^{2+} loads. Elevated cytosolic Ca^{2+} enhances the release of Ca^{2+} from IP3 receptors directly into mitochondria within a tightly coupled space between the two organelles (Rizzuto et al., 2012). Oscillatory Ca^{2+} release from ER provided by IP3 receptors causes influx of Ca^{2+} into the mitochondria and drives TCA cycle enzymes even more effectively than sustained elevations of mitochondrial Ca^{2+} (Hajnoczky et al., 2000), most likely because sustained Ca^{2+} allows re-equilibration of the signal in the matrix.

The interaction between cellular activity, Ca^{2+} influx into mitochondria and energy generation was clarified recently in a study in *Drosophila* neuromuscular junction (Chouhan et al., 2012). Using a complex set of imaging techniques, including genetically encoded Ca^{2+}/pH indicators, it was shown that neuronal activity enhanced Ca^{2+} uptake by mitochondria, increased mitochondrial NAD(P)H levels and hyperpolarized the mitochondrial inner membrane potential. These events were inhibited by pharmacological agents that blocked mitochondrial Ca^{2+} uptake. Levels of cytosolic Ca^{2+} remained similar in different neurons despite their different firing rates, suggesting that a certain level of cytosolic Ca^{2+} is optimum for energy production during activity. An exciting implication of these novel findings is that changes in cytosolic Ca^{2+} levels could theoretically produce plasticity in mitochondrial responses to adjust to changes in energy demands in neurons undergoing long-term changes in excitability.

Mitochondrial management of Ca^{2+} to decrease cytosolic Ca^{2+} burden was observed in the large mammalian brain stem presynaptic terminals of the Calyx of Held. In this synapse, mitochondria remove Ca^{2+} during rises in cytosolic Ca^{2+} produced by Ca^{2+} influx though voltage-gated channels. Mitochondrial Ca^{2+} uptake rapidly dampens the overall rise in Ca^{2+} in the presynaptic terminal. The effect of this is to prevent synaptic depression by attenuating vesicle depletion and allowing for continued effective synaptic transmission during a train of action potentials (Billups and Forsythe, 2002).

2.5 MITOCHONDRIAL INNER AND OUTER MEMBRANE CHANNEL ACTIVITY REGULATES Ca^{2+} RE-RELEASE FROM MITOCHONDRIA AFTER BUFFERING

Ca^{2+} buffering and Ca^{2+} re-release from mitochondria regulate cytosolic Ca^{2+} levels to affect neuronal activity and plasticity, Ca^{2+} re-release from mitochondria occurs through exchangers and channels located on the mitochondrial inner membrane and via channels in the outer membrane. Ca^{2+}-sensitive ligand-gated mitochondrial channels, which are widely conserved and found in species from invertebrates to mammals, open in response to elevated Ca^{2+} within the mitochondrial matrix. In the squid presynaptic terminal, opening of a Ca^{2+}-activated mitochondrial channel is correlated with enhanced neurotransmitter release (Jonas, Buchanan, and Kaczmarek, 1999). Electrophysiological recordings (Jonas, Knox, and Kaczmarek, 1997) demonstrated that within the resting presynaptic terminal, the conductance of mitochondrial membranes is low (Jonas et al., 1999). In contrast, during high-frequency electrical stimulation of the presynaptic nerve, a large increase in mitochondrial membrane ion channel activity takes place (Jonas et al., 1999). The delay in onset of the mitochondrial activity, the dependence of the mitochondrial channel activity on Ca^{2+} uptake across the plasma membrane and the persistence of the mitochondrial activity after stimulation are in keeping with the role of a channel or exchanger in re-releasing Ca^{2+} from mitochondria during certain forms of short-term plasticity (Friel and Tsien, 1994; Jonas et al., 1999; Tang and Zucker, 1997). Furthermore, mitochondrial channel activity and short-term increases in Ca^{2+}-dependent poststimulation synaptic transmitter release are both abrogated by applying the uncoupler carbonyl cyanide *p*-trifluoromethoxyphenylhydrazone (FCCP), which depolarizes mitochondria, preventing Ca^{2+} handling (Jonas et al., 1999).

Although exchangers as well as channels regulate the release of ions and metabolites across the inner membrane, the role of ion and metabolite release by the mitochondrial outer membrane is played in large part by the outer mitochondrial membrane channel known as the VDAC, which transfers Ca^{2+}, ATP, and other metabolites across the outer membrane into the cytosol (Gottlieb, Armour, and Thompson, 2002; Vander Heiden et al., 2001). VDAC also regulates the uptake of ADP and other metabolites during normal and pathological cell activities (Mannella and Kinnally, 2008; Pang et al., 2010; Rostovtseva and Colombini, 1997) and is regulated by cytoskeletal elements, in particular tubulin in its dimeric form (Rostovtseva and Bezrukov, 2012).

2.6 BCL-2 FAMILY PROTEINS REGULATE PATHOLOGICAL OUTER MITOCHONDRIAL MEMBRANE PERMEABILIZATION (MOMP)

As discussed previously, extrinsic control of energy production occurs via calcium influx into the matrix, which activates oxidative phosphorylation, but Ca^{2+} influx can also activate Ca^{2+} release from the matrix. Although this release via channels and exchangers in the mitochondrial membranes occurs during normal physiological processes; the channel opening that accompanies it can be either physiologic or pathologic. One important regulator of pathological channel activity in response to cell stress is the Bcl-2 family of proteins. Programmed cell death (apoptosis) in vertebrate cells may be initiated by signaling at the plasma membrane or by intracellular pathways that lead to changes in mitochondria (Kinnally et al., 2006). The final common pathway for programmed cell death in many cells is mitochondrial outer membrane permeabilization (MOMP) (Adams and Cory, 2007; Antonsson et al., 2000; Dejean et al., 2005, 2006). Pro-apoptotic

Bcl-2 family members such as Bax regulate MOMP by inducing the formation of large outer membrane pores comprised of activated oligomerized proteins, aided by other pro-apoptotic moieties (Antonsson et al., 2000; Dejean et al., 2005; Kim et al., 2006). In their canonical role, the anti-apoptotic Bcl-2 family proteins such as Bcl-xL protect cells against MOMP by interacting with, and inhibiting the pore forming properties of, the pro-apoptotic family members (Adams and Cory, 2007; Galonek and Hardwick, 2006).

MOMP leads to the release of several intermembrane space proteins such as cytochrome *c* (Green and Kroemer, 2004; Martinez-Caballero et al., 2005). The resultant decrease in cytochrome *c* levels compromises the ability of mitochondria to maintain the mitochondrial inner membrane potential and to produce ATP (Gottlieb et al., 2002). In addition, cytochrome *c* released into the cytoplasm activates downstream cytosolic enzyme pathways including effector caspases that execute cell death (Youle and Strasser, 2008).

2.7 PATHOLOGICAL INNER MEMBRANE DEPOLARIZATION: MITOCHONDRIAL PERMEABILITY TRANSITION

An increase in mitochondrial outer membrane permeability may also be triggered by an acute inner membrane depolarization (Galluzzi, Blomgren, and Kroemer, 2009), particularly after cytosolic and mitochondrial Ca^{2+} overload. Although, as described, Ca^{2+} uptake and re-release from mitochondria is a normal physiological event in cells, accumulation of Ca^{2+} in the matrix can diminish energy production by the ATP synthase (Budd and Nicholls, 1996) once the availability of other ions for exchange with Ca^{2+} has become exhausted. Ca^{2+} overload then produces an uncoupling process described historically as a rapid increase in permeability of the mitochondrial inner membrane to solutes and the halting of ATP production (Haworth and Hunter, 1979; Hunter and Haworth, 1979a, 1979b). This phenomenon is termed *permeability transition* (PT).

PT can be reversible or irreversible (Crompton, 1999; Hausenloy et al., 2004; Haworth and Hunter, 1979; Huser and Blatter, 1999; Korge et al., 2011; Petronilli et al., 1999; Wang et al., 2008). If not reversed, PT leads to an even more extreme form of catastrophic PT characterized by structural breakdown of the mitochondrial matrix accompanied by outer mitochondrial membrane rupture and cell death. The difference between this kind of mitochondrial cell death and apoptotic death produced by MOMP seems to be that pathological PT is associated with necrotic cell death such as is found in ischemia or injury, whereas MOMP occurring in the presence of sufficient amounts of ATP may have a more important role in developmental and genetically predetermined death (Baines, 2011; Bonora and Pinton, 2015). Intermembrane space pro-apoptotic factors such as cytochrome *c* and Smac/DIABLO are released during both forms of cell death. In MOMP, outer membrane permeabilization alone leads to release of these factors, whereas in prolonged PT, rupture of the outer membrane after inner membrane swelling releases pro-apoptotic factors into the cytosol (Galluzzi et al., 2009).

PT has been extensively studied for its role in ischemic injury in brain, heart, and other organs as well as in neurodegenerative conditions (Bonora et al., 2015). In addition to Ca^{2+} influx into the matrix, PT is also induced by reactive oxygen species (ROS), elevated P_i, and intracellular acidification (Giorgio et al., 2013; Szabo, Bernardi, and Zoratti, 1992). In contrast, it is inhibited by ATP/ADP and Mg^{2+} (Crompton, 1999; Kowaltowski et al., 1998). The pharmacological agent most efficient in inhibiting PT is cyclosporine A (CsA), an immunosuppressant drug, which binds to cyclophilin D (CypD) and inhibits the channel activity associated with PT (see below). Inhibitors of ANT can either attenuate (bongkrekic acid) or enhance (atractyloside) PT opening (Giorgio et al., 2009; Hunter and Haworth, 1979b; Szabo and Zoratti, 1991). Recent reports have also confirmed increased activity of PT by polyphosphates, chains of 10s to 100s of repeating phosphates linked by ATP-like high energy bonds (Abramov et al., 2007; Holmstrom et al., 2013; Seidlmayer et al., 2012; Stotz et al., 2014). The actions of Ca^{2+} may also require polyhydroxybutyrate (PHB), which enters mitochondria and enhances the ability of Ca^{2+} to induce PT (Elustondo et al., 2013).

2.8 THE QUEST FOR AN INNER MEMBRANE Ca^{2+}-SENSITIVE UNCOUPLING CHANNEL: THE PT PORE

2.8.1 Electrophysiologic Properties of the mPTP

Thus, PT is an important event that performs both physiologic and pathophysiologic functions. PT most likely begins as the opening of a Ca^{2+}-sensitive ion channel in the inner mitochondrial membrane similar to the ion channel activity initiated by physiological mitochondrial Ca^{2+} influx. The Ca^{2+} release channel is heavily regulated; therefore it is assumed that only after prolonged opening does pathological PT (with MOMP) occur (Bernardi, 1999). The conversion of a physiological Ca^{2+} extrusion mechanism into a pathological channel opening is perhaps correlated with energy failure as a result of arrest of ATP synthesizing activity and slowing of energy dependent Ca^{2+} extrusion mechanisms. All the factors that regulate the transition from physiological to pathophysiological events are not completely understood. Therefore, identification of the molecular structure of the pore will provide the target for regulatory activities, allowing for a greater understanding of PT modulation during health and disease.

Description of the biophysical properties of the pore that opens in the inner membrane during PT (the mPTP) provided the earliest indication that PT was initiated by the opening of an ion channel. The first electrophysiological studies of mitochondrial inner membrane were published in 1987. This early report described a ~100 pS channel recorded by patch-clamping giant mouse liver mitochondria produced by cuprizone application (Sorgato, Keller, and Stuhmer, 1987). In the late 1980s, a putative mPTP was recorded by patch-clamping mitochondrial inner membrane or mitoplast preparations (Petronilli, Szabo, and Zoratti, 1989). The activity occurred at positive potentials of the patch pipette and was found either in whole-organelle mode or in single-channel recordings in the organelle-attached configuration. The activity was slightly anion over cation selective with multiple subconductance states ranging from 30 pS to a peak single channel conductance of 1.3 nS. Lower conductances were attributed to substates of the larger channel openings because of long periods lacking activity followed by periods of multiconductance behavior (Petronilli et al., 1989). Conductances of 550 pS were frequently observed at positive potentials. Gating was less common at negative potentials but this observation was consistent with the presence of prolonged openings and fewer subconductance steps at negative patch potentials contrasted with increased flickering at positive potentials. The authors concluded that conductance levels were not sharply defined, consistent with the existence of many varied conductance levels of the channel.

Kinnally, Campo, and Tedeschi (1989) recorded a similar mitochondrial multiconductance channel (MMC) in mouse liver mitoplasts. This channel changed over time, with low activity at the onset of the recording followed by progressively higher activity at later times during the recording. The channels were sometimes open more frequently at negative potentials, but at times channel activity was more frequent at positive potentials (Fig. 2.1). Channel activity displayed multiple conductances

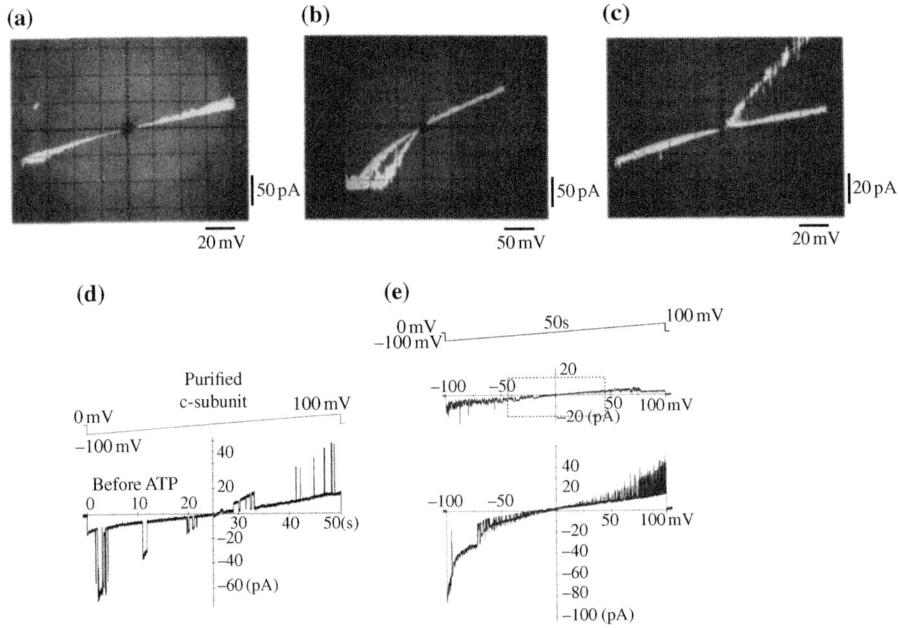

FIGURE 2.1 The current-voltage relationship (I-V) of proteoliposome recordings of the purified c-subunit mimics that of mitoplast recordings of mPTP. **a–c**, Graphs showing examples of the different I-V relationships for the mPTP found in different mitoplast (isolated inner mitochondrial membrane) recordings (reproduced from Kinnally et al., 1989). **d–e**, Graphs showing examples of the different I-V relationships found in different proteoliposome recordings of the purified c-subunit (reproduced from Alavian et al., 2014a, 2014b). Shown in **e** are a low and high conductance mode of the channel (Alavian et al., 2014a). Portions of this figure reprinted with the kind permission of Springer Science and Business Media.

ranging from 10 to 1000 pS and was weakly cation-selective. These early studies began to establish expected criteria for activity of mPTP.

Shortly after the first recordings of the putative mPTP were performed, similar inner membrane activity was found to be inhibited by CsA. In patch-clamp experiments performed in liver mitochondria, channel activity was rapidly inhibited by submicromolar concentrations of CsA in a manner consistent with the expression of the binding site on the matrix side of the inner membrane. Ca^{2+}-activated large conductance channel activity up to 1.3 nS was inhibitable, but a 107-pS inner membrane conductance similar to the first recorded inner mitochondrial membrane channel was also observed in the recordings. This smaller conductance was resistant to CsA, suggesting that this activity might be the result of a separate ion channel (Szabo and Zoratti, 1991). The large conductance channel was sensitive to Mg^{2+}, Mn^{2+}, Ba^{2+}, and Sr^{2+} in that order, which inhibited the activity in a competitive manner with Ca^{2+}, the main activator of the channel (Szabo et al., 1992).

2.8.2 Characterization of a Molecular Complex Regulating the Pore

The recent identification of the molecular structure matching the biophysical properties of mPTP was aided by seemingly unrelated sets of findings. One was that Bcl-xL enhances metabolic efficiency (decreases uncoupling) by binding to the β-subunit of the ATP synthase. The second finding was that CypD, which had been known for many years to regulate PT, binds to the stator arm of ATP synthase, specifically on the oligomycin-sensitivity conferring protein (OSCP) subunit. The third finding suggested that closure of the mPTP is related to the level of CypD expression in a developmentally regulated manner because it falls at the onset of respiration in mammalian heart. The final project found that ATP synthase assembles into a large complex with other proteins that may regulate the mPTP. These findings and their convergence that led to the definition of the molecular nature of the mPTP will now be discussed in greater detail.

2.8.3 Bcl-xL Regulates Metabolic Efficiency by Binding to the β-Subunit of the ATP Synthase

Although Bcl-2 family proteins regulate mitochondrial outer membrane permeability to produce or inhibit cell death (Galonek and Hardwick, 2006; Jonas, 2009; Jonas, Porter, and Alavian, 2014a, 2014b; Park et al., 2015), Bcl-xL also alters the metabolic properties of cells, even in the absence of cell death signals. Bcl-xL increases the release of ATP through enhanced VDAC opening. Enhanced ATP release by mitochondria decreases the probability of MOMP in cancer cell lines by providing extra ATP to overcome cell death stimuli (Gottlieb et al., 2002; Vander Heiden et al., 2001). In the neuronal synapse, injection of either Bcl-xL or ATP enhances synaptic transmitter release (Jonas et al., 2003), suggesting that Bcl-xL helps increase ATP levels in the synapse during synaptic activity (Hickman et al., 2008).

Inefficiency of metabolism is correlated with cell death under conditions of neurodegeneration or acute cellular injury such as occurs during PT (Beal, 2007;

Brand, 2005; Dodson and Guo, 2007). In contrast, a highly efficient state of metabolism requires maximally decreased uncoupling of the inner membrane (Alavian et al., 2011; Chen et al., 2011; Hockenbery et al., 1990). Therefore, Bcl-2 family proteins could form part of a large protein complex that regulates inner membrane coupling; Bcl-xL might regulate not only the release but also the production of ATP. In support of a role for Bcl-xL in the manufacture of ATP, hippocampal neurons overexpressing Bcl-xL show a large increase in cytoplasmic ATP levels. Surprisingly, this increase in ATP accompanies a decrease in neuronal oxygen uptake and on aerobic glycolysis, consistent with an effect of Bcl-xL overexpression on mitochondrial bioenergetic efficiency (Alavian et al., 2011; Chen et al., 2011). Bcl-xL depletion reverses these effects on metabolism (Alavian et al., 2011; Li et al., 2008). Furthermore, direct interaction of Bcl-xL with the β-subunit of the ATP synthase maximizes the efficiency of ATP production (Alavian et al., 2011; Chen et al., 2011).

Closure of the uncoupling leak within the inner membrane in the presence of Bcl-xL aids actively firing neurons to increase neurotransmitter release (Li et al., 2008, 2013), consistent with a correlation between the increase in metabolic efficiency and the long-term higher efficacy of synaptic transmission found in Bcl-xL expressing neurons. In contrast, opening of the Bcl-xL-regulated inner membrane leak decreases metabolic efficiency and predisposes neurons to death. Neurons lacking Bcl-xL display a fluctuating mitochondrial inner membrane potential and a marked depolarization in the presence of the ATP synthase inhibitor oligomycin (Chen et al., 2011). These data support the idea that Bcl-xL regulates inner membrane coupling, preventing energy failure and cell death via direct effects on F_1F_O ATP synthase.

2.8.4 CypD Binds to ATP Synthase and Regulates Permeability Transition

Another piece of the puzzle that helped determine the molecular components of mPTP was the discovery of the interaction between CypD and ATP synthase (Bernardi, 2013). It has long been known that CypD, a chaperone protein and peptidyl-prolyl cis-trans isomerase that resides in the mitochondrial matrix, regulates the mPTP by enhancing is sensitivity to Ca^{2+}. Once it was found that CsA inhibits the PT, CsA rapidly became one of the major tools to study the mPTP. Later, it was found that CsA does this by inhibiting CypD (Crompton, Ellinger, and Costi, 1988; Halestrap and Davidson, 1990; McGuinness et al., 1990).

In 2005, experiments using CypD-null mice demonstrated that CypD was not itself the pore of the mPTP but that it played an important regulatory role in the modulation of the mPTP by Ca^{2+} (Baines et al., 2005; Basso et al., 2005; Nakagawa et al., 2005; Schinzel et al., 2005). Four groups showed that deletion of CypD decreased sensitivity to ischemia-reperfusion injury in the heart and brain (Baines et al., 2005; Basso et al., 2005; Nakagawa et al., 2005) and the Molkentin group (Elrod et al., 2010) suggested that the physiologic function of mPTP is to maintain "homeostatic mitochondrial Ca^{2+} levels to match metabolism with alterations in myocardial workload."

CypD expression varies widely among cell types. CypD is more highly expressed in aged hearts, and these changes in expression may regulate its association with a complex of proteins that increases mPTP opening during reperfusion (Zhu et al., 2013).

Furthermore, CypD activity appears to be regulated by cell signaling and metabolic pathways (Di Lisa et al., 2011; Hafner et al., 2010; Nguyen et al., 2011; Shulga and Pastorino, 2010; Shulga, Wilson-Smith, and Pastorino, 2010) and by developmental cues in differentiating myocytes (see below).

Although the exact mechanism of CypD's action in the mitochondrial matrix has not been clarified, recent data support the idea that this requires its binding to F_1F_O ATP synthase. First, the propensity toward PT is regulated by ATP hydrolysis and synthesis in as much as this regulates the membrane potential. Therefore, PT requires twice the Ca^{2+} load in mitochondria that are in the process of hydrolyzing ATP (making a membrane potential) versus synthesizing ATP (dissipating the membrane potential) (Giorgio et al., 2013; Hunter and Haworth, 1979a, 1979b). Second, CypD binds to proteins on the stator of ATP synthase (OSCP, b and d-subunits [Giorgio et al., 2009]) as well as to the F_1F_O ATP synthase binding partners ANT (Crompton, Virji, and Ward, 1998; Woodfield et al., 1998) and the phosphate carrier ([PiC] (Elrod et al., 2010; Gutierrez-Aguilar et al., 2014; Leung and Halestrap, 2008). Recent work on the binding of CypD to ATP synthase indicates that this occurs exclusively with OSCP, and depletion of OSCP using siRNA also depletes CypD (Giorgio et al., 2013) suggesting that the association regulates the expression of each protein. Interestingly, CsA decreases the propensity toward PT to the same degree (half) that decreases in OSCP expression increase the propensity toward PT (double), suggesting an inhibitory effect of OSCP on PT (Giorgio et al., 2013). Third, purified ATP synthase dimers produce current consistent with mPTP in artificial lipid membranes (Giorgio et al., 2013). The single-channel activity has a maximal conductance of 1 to 1.3 nS with subconductance states. The current is not stimulated by Ca^{2+} or inhibited by CsA presumably because CypD is not present in the purified dimer preparation although the enzymatic activity of the F_1F_O ATP synthase is preserved. The ATPase dimer current is also not sensitive to bongkrekic acid or atractyloside, agents that primarily affect ANT activity, making it unlikely that ANT directly forms part of the pore of the channel (Giorgio et al., 2013). These data emphasize the idea that the regulation of the mPTP may occur via the interaction of CypD and other molecules with F_1F_O ATP synthase and its binding partners.

2.8.5 PT Activity Regulates Cardiac Development

A third line of evidence that helped unravel the identity of the mPTP was a series of studies of mitochondrial function during cardiac development. Until recently, the general consensus was that opening of the mPTP was a devastating event that triggers cell death. However, over the last two decades, data have emerged suggesting that transient opening of the mPTP could serve a physiologic purpose. Some of these data in neurons are discussed in this chapter, but additional data demonstrate that transient mPTP opening occurs in many other cell types (Crompton, 1999; Hausenloy et al., 2004; Huser and Blatter, 1999; Korge et al., 2011; Petronilli et al., 1999; Wang et al., 2008).

In the heart, physiologic variations in mPTP activity play a critical role in cardiac myocyte differentiation and cardiac development. Mitochondria in cardiac myocytes from adult hearts display transient depolarizations that may be associated with "superoxide flashes," and these depolarizations occur more often in neonatal myocytes (Wang et al., 2008). Interestingly, developmental studies demonstrate that the mPTP is open in myocytes in the early embryonic mouse heart, and this opening is not associated with any form of cell death. However, by the mid-embryonic stage, the mPTP is closed (Hom et al., 2011). This closure coincides with activation of complex I of the electron transport chain, assembly of electron transport chain supercomplexes called *respirasomes*, and activation of oxidative phosphorylation (Beutner, Eliseev, and Porter, 2014). These changes cause a fall in mitochondrial-derived ROS that signals the myocyte to undergo further differentiation (Hom et al., 2011). Furthermore, pharmacologically inhibiting mPTP or genetically inhibiting CypD enhances myocyte differentiation, whereas opening mPTP inhibits differentiation (Hom et al., 2011). These findings have been confirmed in cardiac stem cells (Cho et al., 2014; Fujiwara et al., 2011), and various reports have stressed the importance of the mPTP during cardiac development and myocyte differentiation (Drenckhahn, 2011; Folmes et al., 2012).

2.8.6 Regulatory Molecules Do Not Form the Pore of mPTP

The F_1F_O ATP synthase interacts with a large number of proteins many of which have been candidates for mPTP. ANT was an early candidate to form the mPTP because atractyloside and bongkrekic acid, which inhibit ANT, affect the mPTP (Hunter and Haworth, 1979b) and ANT was found to interact with CypD (Halestrap and Davidson, 1990). VDAC was also an early candidate to form the mPTP because of its high conductance and its association with ANT in immunoprecipitation experiments (Crompton et al., 1998). In addition, it was shown that a complex of ANT, VDAC, hexokinase, and mitochondrial creatine kinase (mtCK) could form high conductance pores when reconstituted into membranes (Beutner et al., 1996, 1998). Finally, the PiC is a more recent candidate to form the mPTP (Leung and Halestrap, 2008).

However, genetic deletion of ANT1 and ANT2 and of the PiC demonstrated that these proteins were not essential to mPTP formation, but that they served regulatory roles (Gutierrez-Aguilar et al., 2014; Kokoszka et al., 2004; Kwong et al., 2014). Furthermore, deletion of VDAC did not affect pore formation (Kinnally et al., 2011). Additional data suggest that the conformation of ANT may be important for regulation of the mPTP (Gunter and Sheu, 2009). Atractyloside induces mPTP opening and is known to stabilize the *c* conformation of ANT, such that the adenine nucleotide transport site faces the *c*ytoplasmic, or intermembrane space, side of the inner mitochondrial membrane (Gunter and Sheu, 2009). In contrast, bongkrekic acid prevents mPTP opening and stabilizes ANT in its *m*, or *m*atrix facing conformation (Gunter and Sheu, 2009). However, because both ATR and BKA inhibit ANT, it is unlikely that specific effects on ADP/ATP translocation regulate the mPTP and a more likely scenario is that the conformation of ANT itself can regulate the mPTP.

The details of how these candidate molecules regulate the mPTP are not yet fully known, but evidence suggests that they participate as part of a large macromolecular structure with F_1F_O ATP synthase in the inner mitochondrial membrane. ANT and PiC form a complex with F_1F_O ATP synthase called the *synthasome* (Chen et al., 2004).

In addition, the complex of ANT, VDAC, hexokinase, and mtCK is likely also involved in the regulation of ATP synthesis (Beutner et al., 1996, 1998). Therefore, each regulatory molecule may alter the structure or activity of F_1F_O ATP synthase, and in so doing, modulate the opening of the mPTP.

2.9 THE mPTP: A MOLECULAR DEFINITION

2.9.1 The c-Subunit of F_1F_O ATP Synthase Comprises the PT Pore

These various reports all agree that F_1F_O ATP synthase is a major factor in the formation of the mPTP, and recent evidence suggests that the F_O or membrane portion of F_1F_O ATP synthase in fact forms the pore (Alavian et al., 2014a; Azarashvili et al., 2014; Bonora et al., 2013, 2015; Chinopoulos and Szabadkai, 2014; Karch and Molkentin, 2014) (Figs. 2.2 and 2.3). Mammalian F_1F_O ATP synthase is a ~600 kDa complex of 15 subunits. The membrane portion, or F_O, contains a ring of eight hydrophobic c-subunits and subunits a, b, e, f, g, and A6L. A stalk composed of the γ, δ, and ε subunits connects the c-subunit ring to the catalytic F_1 component made of a hexamer of alternating α and β subunits, where ATP synthesis and hydrolysis occur. Finally, a stator containing the b, d, F6, and OSCP subunits connects the lateral portion of F_O to the top of the F_1. Movement of protons between the c-subunit and the a-subunit causes rotation of the c-subunit ring, the energy of which is transferred to F_1 to synthesize ATP (Carbajo et al., 2005; Jonckheere et al., 2012; Pedersen, 1994; Walker, 2013; Wittig and Schagger, 2009).

Rat brain and heart mitochondrial F_1F_O ATP synthase complexes undergo age-dependent structural and functional alterations (Guerrieri et al., 1992). The F_O portion

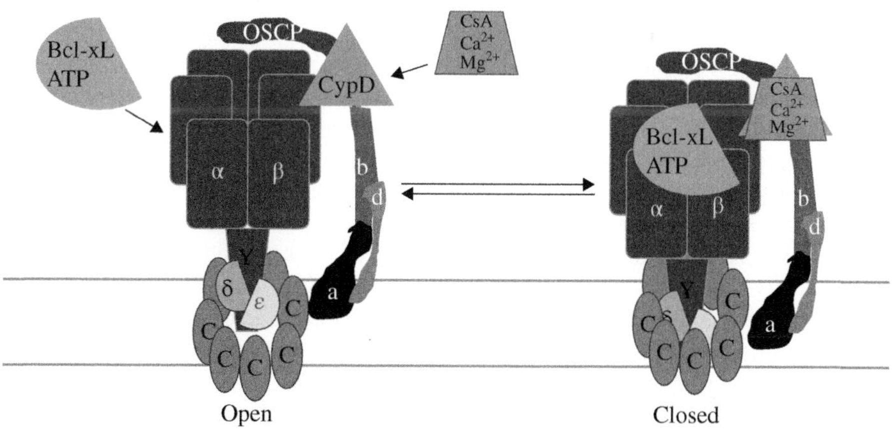

FIGURE 2.2 Model of physiological mPTP activity. β subunit, which binds Bcl-xL and ATP, blocks c-subunit conductance. CsA, inhibitor of CypD, blocks c-subunit conductance. Both Bcl-xL and CsA bind to F1 components during pore closure. The pore diameter also decreases during pore closure. The channel can switch reversibly between these open and closed states. (*See insert for color representation of the figure.*)

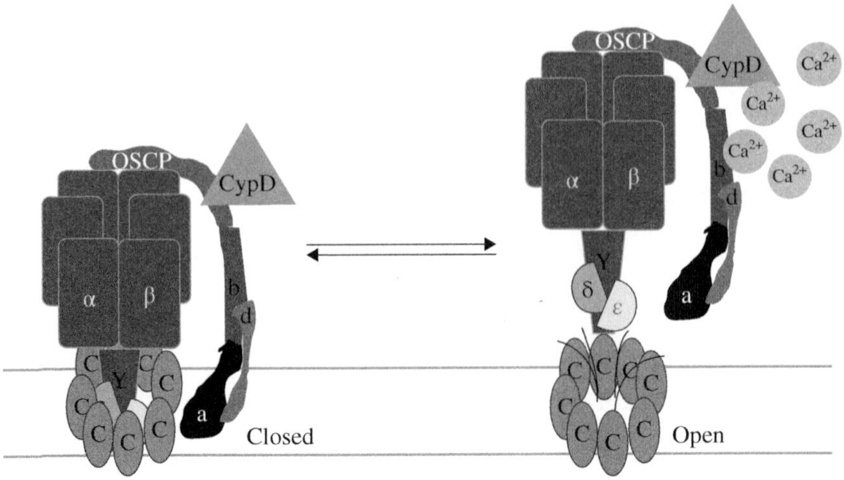

FIGURE 2.3 Model of pathophysiological mPTP activity. Ca^{2+} induces separation of F_1 from F_O during mPTP opening. The pore diameter also expands during Ca^{2+}-induced increases in mPTP conductance. (*See insert for color representation of the figure.*)

is present in the absence of an equivalent complement of F_1 at 3 months of age in heart mitochondria. Levels of F_1 then increase from 3 to 12 months as ATPase activity increases and this accompanies a decrease in proton leak secondary to binding of F_1 to F_O. Decreases of F_1 content with respect to that observed for F_O are detected in aging heart of 12- to 24-months-old animals, suggesting the presence of F_1-lacking F_O leak channels in the membranes of these aging mitochondria (Guerrieri et al., 1992).

To determine if the membrane portion of the F_1F_O ATP synthase comprises a cell death channel that could produce PT under stress, cell death was measured after depletion of all three isoforms of the c-subunit or after overexpression of a tagged version of the wild type c-subunit in HeLa cells (Bonora et al., 2013). Cell lines are able to use glycolysis for ATP production in normal glucose-containing medium; therefore, depleting the c-subunit did not diminish ATP levels in the cells. In contrast, c-subunit depletion prevented CsA-sensitive PT measured by calcein-cobalt quench and prevented mitochondrial morphological changes typical for PT. Furthermore, c-subunit depletion markedly attenuated H_2O_2-induced cell death and glutamate-induced excitotoxicity in neurons, suggesting that the c-subunit was required for PT.

Although the study indicated that the F_1F_O ATP synthase c-subunit was an important component of the mPTP (Bonora et al., 2013), this work did not directly determine what portion of ATP synthase could form the pore of the mPTP. CypD and Bcl-xL interact with the stator and β subunit of ATPase, respectively, and these proteins are membrane-attached but not embedded. The channel or pore forming subunits, however, must span the mitochondrial inner membrane. Nine polypeptides form the F_0 and the stator, but only three, a, b and c, are required for proton translocation and are evolutionarily highly conserved, like PT. In addition, ρ0 cells that lack mitochondrial DNA do not contain an a-subunit but do undergo PT. Attention focused on the c-subunit, and not the b-subunit because the c-subunit had been shown previously to express ion channel activity (McGeoch and Guidotti, 1997). The mammalian c-subunit undergoes conformational changes from an α-helix to a β-sheet when in contact with water, forming the walls of ion channels (McGeoch and McGeoch, 2008) with a diameter of 2.3 nm allowing molecules up to 1.5 kDa to pass, similarly to PT.

Homologues of ATP synthase c-subunit are also present in the V_O subcomplex of various vacuolar H^+-ATPases, and V_OV_1-ATPase c-subunits line a water-accessible pore structure (Harrison et al., 2003; Jones et al., 1995) found both on the plasma membrane and on intracellular membranes (Finbow and Harrison, 1997; Mandel et al., 1988). V_O c-subunits have also been shown to be involved in formation of mega-channels in gap junctions between cells (Peters et al., 2001). Therefore, pore-forming ability is an important feature of all homologous c-subunits that share similar amino acid sequence (Azarashvili et al., 2014).

2.9.2 The c-Subunit of ATP Synthase Creates the High Conductance mPTP Pore

More recent experiments have directly tested the hypothesis that the main membrane embedded portion of mammalian F_1F_O ATP synthase, that is, the c-subunit ring, forms the pore of the mPTP (Alavian et al., 2014b; Azarashvili et al., 2014). Indeed, electrophysiologic recordings of the purified mitochondrial c-subunit yield a multiconductance, voltage-dependent channel with prominent subconductance states (Alavian et al., 2014a). Patches contain a ~100-pS conductance, which appears to be a subconductance state of the larger activity rather than a separate conductance. Recordings reveal peak single channel conductances of ~1.5 to 2 nS, similarly to activity described previously for the mitochondrial multiple conductance channel (MCC) (Kinnally et al., 1989). Also consistent with MCC, channel activity often but not always demonstrates negative rectification. At extreme positive patch pipette potentials of higher than 100 mV, single-channel conductances of ~1.5 nS and ~2 nS are also consistently observed. Single-channel events and gating are more likely to be discerned at positive potentials most likely because of the negative rectification, in keeping with previous reports for the mPTP (Petronilli et al., 1989).

Voltage dependence is an inherent property of the channel (Borjesson and Elinder, 2008) and is not dependent on the mitochondrial inner membrane potential or on the solutions used to record the currents. When measured by varying the voltage across the membrane (the command or holding voltage), recent I-V plots on the purified c-subunit prove consistent with previously reported I-V plots of the mPTP currents (Alavian et al., 2014a; Kinnally et al., 1989; Petronilli et al., 1989; see Fig. 2.1).

2.9.3 F_1 Regulates Biophysical Characteristics of the Purified c-Subunit

Purified c-subunit protein reconstituted into liposomes clearly lacks extrinsic regulatory moieties that are important for mPTP opening. The effects of Ca^{2+}, a critical regulatory molecule, demonstrate this property. Ca^{2+} binding sites have not been detected in mitochondrial F_O, perhaps because mammalian c-subunits lack the formyl

Met at the N-terminus to which Ca^{2+} binds to *Escherichia coli* or chloroplast F_1F_O ATP synthase c-subunits (Zakharov et al., 1996). In contrast, Ca^{2+} binds to specific, low affinity, and moderate capacity sites on the β-subunit of F1 ATP synthase (Hubbard and McHugh, 1996). Other sites in other ATP synthase-interacting molecules may also be important for regulation by Ca^{2+} and other agents (Beutner et al., 1996, 1998). Therefore, although the new models of the mPTP pore must account for all inducers and inhibitors, these molecules may not interact directly with the c-subunit pore itself, but may instead bind to sites in the F_1 or to other molecules such as ANT and CypD that undergo structural rearrangements on opening and closing of the pore.

To determine the location of the regulators, mitochondrial recordings were carried out using purified mitochondrial and F_1F_O ATP synthase preparations. In these studies, the absence of an effect of a modulator was taken as an indication that the ligand or binding site for that modulator had been removed by the purification process. For example, because in mitochondria or isolated inner membrane preparations, Ca^{2+} activates the c-subunit leak channel, whereas CsA and ATP/ADP inhibit it, the Ca^{2+}- and CsA-sensitive sites must be present in these preparations. In contrast, removal of the F_1 and other peripheral membrane proteins by urea treatment of the inner membrane or removal of CypD by purification of ATP synthase monomers abrogates regulation of the c-subunit channel by CsA and Ca^{2+} and greatly diminishes sensitivity to ATP/ADP. These studies suggest that the CypD/calcium binding site is contained within or associated with the F_1 portion of the ATP synthase and that a second, low-affinity ATP binding site exists in the F_O (Alavian et al., 2014a). These results are consistent with reports identifying the binding site of CypD and benzodiazepine 423, an ATP synthase-inhibitory and mPTP-sensitizing agent, on OSCP (Giorgio et al., 2009, 2013), and suggest that the assembly of F_1F_O ATP synthase into monomers, dimers, and higher-order oligomers can regulate the formation of the mPTP.

Channel activity of the purified c-subunit is inhibited by the purified β-subunit of F_1, suggesting a structural rearrangement whereby F_1 of the ATP synthase inhibits opening of the c-subunit channel, aided by ADP/ATP/Bcl-xL binding to the β-subunit and opposed by CypD/Ca^{2+} interaction with OSCP (Fig. 2.2). To demonstrate this further, mitochondria were treated with Ca^{2+} following which the F_1F_O ATP synthase was immunoprecipitated using an antibody directed at the F_1. These studies determined that Ca^{2+} exposure destabilizes the connection between the stalk and the c-subunit, disrupting protein/protein interaction between the c-subunit and F_1 (Fig. 2.3). The model suggests that the channel of the mPTP forms within the c-subunit ring itself upon reversible CypD and Ca^{2+}-dependent movement of the stalk away from the c-subunit (Alavian et al., 2014a).

Loss of protein/protein interaction between F_1 and F_O requires fairly mild conditions such as 60 μM Ca^{2+} in the bath to initiate what may indeed be reversible PT (Alavian et al., 2014a). This concentration is well within the range of physiological Ca^{2+} concentrations found within the mitochondrial matrix (Csordas, Thomas, and Hajnoczky, 2001; Rizzuto and Pozzan, 2006; Rizzuto, Bernardi, and Pozzan, 2000) or concentrations measured adjacent to the mitochondria in Ca^{2+} microdomains at the plasma membrane or ER membrane (Rizzuto et al., 2009; Schneggenburger and Neher, 2005). That loss of protein/protein interaction between F_1 and F_O is likely to be reversible has been shown upon chelation of Ca^{2+} in mitoplasts (Beutner et al., 1998), intact mitochondria (Roestenberg et al., 2012), intact neurons (Jonas et al., 1999) as well as in reconstituted dimers of F-ATP synthase (Gomez et al., 2007). Therefore the F_1 and the c-subunit may recombine to close the mPTP, reforming intact F_1F_O ATP synthase and reinitiating enzymatic function (Pedersen and Hullihen, 1978). However, under certain conditions, this separation may become irreversible, forming pathophysiologic PT (with MOMP).

Another possible regulatory molecule is the ATPase Inhibitory Factor 1 (IF1), a small, nuclear-encoded endogenous polypeptide, which is involved in the regulation of the oligomeric state of the F_1F_O ATP synthase by facilitating enzyme dimerization of two F1 domains (Garcia et al., 2006). IF1 is also known as an intrinsic inhibitor of the ATP synthase. It interacts with the catalytic β subunit of ATP synthase and is responsible for inhibiting ATP hydrolysis but not synthesis. The binding of IF1 is pH dependent and below neutrality its inhibitory activity increases (Cabezon et al., 2001). IF1 plays a protective role during hypoxic/ischemic conditions, when the electrochemical gradient across the membrane collapses and the enzyme switches from ATP synthesis to hydrolysis (Jennings, Reimer, and Steenbergen, 1991; Pullman and Monroy, 1963; Rouslin, 1991). The absence of IF1 has been linked with the human pathological condition known as Luft's disease, characterized by densely packed mitochondrial cristae and significantly high ATPase rate (DiMauro et al., 1976; Yamada and Huzel, 1992). IF1 expression is increased in many human carcinomas, supporting its anti-apoptotic and potentially tumorigenic role (Bravo et al., 2004; Sanchez-Cenizo et al., 2010). IF1 protects cells from necrotic and apoptotic cell death by regulating mitochondrial cristae morphology, promoting the dimerization of ATP synthase, limiting release of Cyt c, and preventing activation of Drp1 and oligomerization of pro-apoptotic Bax (Campanella et al., 2008; Faccenda et al., 2013a, 2013b). It perhaps also controls ATP synthase function during physiological as well as pathological conditions, but its role in daily metabolism is unknown.

Additional regulation of the mPTP may be the result of the association of other molecules with F_1F_O ATP synthase. As discussed previously, F_1F_O ATP synthase may complex with ANT and PiC. In the study of F_1F_O ATP synthase dimers, bongkrekic acid, which inhibits ANT, fails to attenuate the mPTP channel activity (Giorgio et al., 2013); this is recapitulated in studies of submitochondrial vesicles (SMVs) (Alavian et al., 2011). Therefore, the regulation of the mPTP by ANT and PiC may occur through their association with the peripheral membrane components of F_1F_O ATP synthase. Furthermore, the association of F_1F_O ATP synthase and ANT with PiC, mtCK, VDAC, and hexokinase may explain why mPTP regulation can occur via these molecules. The layers of regulation that control formation of the c-subunit ring/mPTP are clearly complex and involve many known and also undiscovered proteins that associate directly or indirectly with F_1F_O ATP synthase (Bernardi and Di Lisa, 2015; Chen et al., 2004; Halestrap and Richardson, 2015).

Some other known regulators of the mPTP also work directly on the F_1F_O ATP synthase itself. F1 has binding sites that accommodate the effects of Ca^{2+}, Mg^{2+}, adenine nucleotides Pi, Bcl-xL and through CypD (un)binding those of H+, CsA, and possibly of oxidants (Kruse et al., 2008). Therefore, in summary, the new model of mPTP explains either direct or indirect interaction with all known inducers, inhibitors, and modulators of pore function (see Figs. 2.2 and 2.3).

2.9.4 Structural Location of the Pore within the c-Subunit Ring

The exact location of the ion-conducting pore of the c-subunit is becoming increasingly understood. Although it has been proposed that the pore sits between the two lateral stalks of a dimer of F_1F_O ATP synthases and not within the c-subunit ring (Giorgio et al., 2013), there is currently no electrophysiological evidence for the formation of such a channel, and regulation of the mPTP by components of F_1 (Alavian et al., 2014a, 2015; Giorgio et al., 2013) argues against this. For F_1 to regulate channel activity if the pore were located between dimeric membrane-associated subunits, it would need to be positioned over the dimeric link, outside of the F_O and directed to the side of the complex. This seems unlikely to happen, and thus, a model of the pore forming in the region between monomers of F_1F_O ATP synthase requires further hypothesis testing.

Rather, it is likely that the leak is located either within the central portion of the c-subunit ring, between the individual c-subunit monomers, or between the c-subunit and the other F_O subunits, although the latter is less likely given the presence of PT in ρ0 cells that lack both mitochondrial DNA and the a-subunit (Bonora et al., 2013). In two separate experiments, it has been demonstrated that the c-subunit ring expands when it conducts ions, making it likely that the pore is formed by the c-subunit ring. The first experiment used fluorescent tetracysteine display with the placement of cysteine pairs on all c-subunit monomers within the ring. These studies showed that Ca^{2+} influx into cells causes a decrease in fluorescence consistent with expansion of the diameter of the c-subunit ring, whereas CsA increases fluorescence consistent with a decrease in ring diameter (Alavian et al., 2014a). Mutagenesis to increase the diameter of the c-subunit ring also demonstrated that ring expansion is a means to increase conductance. Mutations targeted to four highly conserved glycines within the first (N terminus) alpha-helical region of the c-subunit (Norris, Karp, and Fimmel, 1992) decrease the tight packing (increase the radius) of the ring and increase average single-channel conductance compared with WT c-subunit rings (Alavian et al., 2014a).

These findings support the hypothesis that the c-subunit is necessary and sufficient to produce the pore of mPTP. When viewed from the intermembrane space, the denuded c-subunit oligomer appears as a ring with a central porelike structure that is normally obscured by the F_1 stalk components γ, δ, and ε, suggesting that the pore may form within the center of the ring given the proper conformation. Although it has been suggested that phospholipids occupy the central cavity of the c-subunit ring in F_1F_O ATP synthases from different species (Matthies et al., 2009; Meier et al., 2001; Oberfeld, Brunner, and Dimroth, 2006) other evidence provides for formation of a proteolipid or proteophospholipid channel structure within the central lipid region (Abramov et al., 2007; Azarashvili et al., 2014; Elustondo et al., 2013; McGeoch and McGeoch, 2008; Pavlov et al., 2005). Data suggest a working model whereby the c-subunit pore forms within the proteolipid milieu on activation of mPTP (e.g., by elevated matrix Ca^{2+}) whereupon the ring expands and F_1 shifts; the pore is closed by a decrease in diameter of the ring and by binding of the F_1 components to the ring (see Figs. 2.2 and 2.3). The details of these changes and their regulation remain a work in progress.

2.10 CLOSING OF THE mPTP MAY ENHANCE MITOCHONDRIAL METABOLIC PLASTICITY AND REGULATE SYNAPTIC PROPERTIES IN HIPPOCAMPAL NEURONS

Unlike in the high fidelity synapse of the squid in which long-term synaptic responses are relatively unvaried, in the mammalian hippocampus, long-term changes in synaptic transmission in both directions occur, and these changes may underlie learning and memory formation. In cultured hippocampal neurons, overexpression of the anti-apoptotic protein Bcl-xL produces an increase in mitochondrial targeting to synaptic sites, and an enhancement in spontaneous and evoked synaptic responses (Li et al., 2008, 2013). Furthermore, Bcl-xL overexpression is correlated with structural alterations in the synapse including enhanced expression of synaptic vesicle numbers, increased synaptic vesicle markers, and an increase in postsynaptic markers, consistent with an increase in size and number of synapses (Li et al., 2008). The Bcl-xL-dependent increase in localization of mitochondria to the synapse suggests a link between mitochondrial activity and the marked alterations of synaptic structure found in developing and plastic synapses. If the probability of closure of mPTP is increased by binding of Bcl-xL to the β-subunit of F_1F_O ATP synthase, then the mitochondria will follow cytosolic Ca^{2+} transients more faithfully (Chen et al., 2011), the efficiency of energy production will be enhanced and energy will be provided more efficiently where it is needed for active neurons (Alavian et al., 2011; Li et al., 2013). This hypothesis also suggests that Bcl-xL translocates to mitochondria during neuronal activity (Li et al., 2013), and that Bcl-xL repositions mitochondria for actively enlarging or developing synapses (Berman et al., 2009; Li et al., 2008) (Fig. 2.4).

2.11 mPTP OPENING CORRELATES WITH CELL DEATH IN ACUTE ISCHEMIA, ROS DAMAGE, OR GLUTAMATE EXCITOTOXICITY

In contrast to enhanced neuronal excitability and synaptic efficacy in healthy neurons, opening of the PT pore under stress may be sensitive to a decrease in Bcl-xL expression or interaction with the F_1 β-subunit thereby preventing an increase in probability of closing of the c-subunit pore (Alavian et al., 2014a). PT occurs acutely during such events as OS brain or cardiac ischemia (Baines, 2009), predicting that genetic depletion of the c-subunit may prevent PT. Indeed, excitotoxic and ROS-induced cell death are greatly attenuated on depletion of the c-subunit by shRNA in neurons and other cells (Alavian et al., 2014a; Bonora et al., 2013);

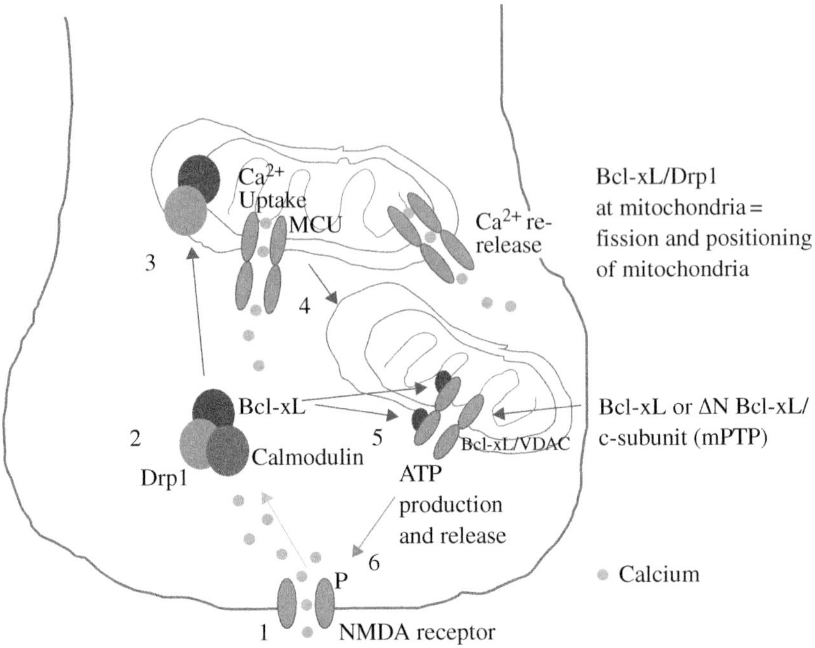

FIGURE 2.4 Proposed role of Bcl-xL in mitochondrial metabolism during synaptic plasticity. Events at the synapse after stimuli that produce long-term changes in synaptic responsiveness. (1) A change in level of activity produces a change in Ca^{2+} influx through NMDA receptors. (2) Ca^{2+} binds to calmodulin, which forms a complex with Drp1 and Bcl-xL. (3) The Drp1-Bcl-xL-complex translocates to mitochondria in a calmodulin-dependent manner. At the same time, Ca^{2+} is being taken up into mitochondria, TCA cycle is stimulated, mitochondria buffer and re-release Ca^{2+}. (4) Bcl-xL-Drp1 complex enhances mitochondrial positioning to postsynaptic sites (close to the postsynaptic density). (5) Bcl-xL enhances the efficiency of production and release of ATP over the long term. ΔN Bcl-xL decreases the efficiency of ATP production. (6) Changes in phosphorylation of downstream targets occurs, producing long-lasting changes in synaptic structure and function, including, for example, the expression level of postsynaptic glutamate receptors.

in addition, cell death protection by c-subunit depletion is not further attenuated by CsA, suggesting that the c-subunit forms the inner mitochondrial membrane target of the CsA-sensitive complex (Alavian et al., 2014b). In contrast to c-subunit depletion, overexpression of the wild type c-subunit (Bonora et al., 2013) or mutation of the c-subunit to form a high-conductance leaky c-subunit (Alavian et al., 2014b) predisposes to enhanced cell death on excitotoxic or ROS stimulation. Death under these conditions is not sensitive to CsA, presumably because the leaky pore prevents normal regulation by components of F_1. Upon unrelieved Ca^{2+} or ROS exposure, opening of the pore produces mitochondrial membrane depolarization, abrogation of the H+ gradient across the inner membrane, and cell death.

2.12 PRO-APOPTOTIC PROTEOLYTIC CLEAVAGE FRAGMENT OF BCL-XL CAUSES LARGE CONDUCTANCE MITOCHONDRIAL ION CHANNEL ACTIVITY CORRELATED WITH HYPOXIC SYNAPTIC FAILURE: OUTER MITOCHONDRIAL CHANNEL MEMBRANE ACTIVITY ALONE OR MPTP?

In the synapse, ROS or excitotoxicity may not produce death of the soma, but rather severe neuronal stress may result in long-lasting synaptic depression or a decline in neuronal excitability. A decline in neurotransmitter release or recycling can subsequently mark a synapse for elimination, followed by somatic death if many synapses are sequentially eliminated. Before the onset of decline, a set of changes occurs in mitochondrial membrane activity that negatively affects synaptic function.

Under pro-apoptotic conditions in growth-factor deprived cancer cell lines, Bcl-2 family proteins activate large mitochondrial outer membrane channel activity that participates in release of pro-apoptotic factors from mitochondria (Antonsson et al., 2000; Dejean et al., 2005) either in the absence of any change to the properties of the inner membrane or, as may occur during ischemia, accompanying induction of calcium or ROS-induced depolarization and loss of osmotic regulation (PT) of the inner mitochondrial membrane (D'Orsi et al., 2012; Ozden and Isenmann, 2004). In the synapse, the effects of hypoxia provide an opportunity to study the role of Bcl-xL in mitochondrial changes taking place during severe neuronal injury (Jonas et al., 2004, 2005b) and to ask questions about which mitochondrial ion channels are activated. The giant presynaptic terminal in squid is sensitive to hypoxia, which attenuates synaptic transmission over 10 to 30 minutes. Patch-clamp recordings of mitochondrial membranes at rest during hypoxia reveal large conductance activity not found frequently in controls (Jonas et al., 2004; Ofengeim et al., 2012). The channel activity is larger than that induced by pipette-mediated application of recombinant full length Bcl-xL protein and is mimicked by activity of recombinant proteolytically altered Bcl-xL (ΔN Bcl-xL) that is known to form a large conductance (pro-apoptotic or MOMP-like) channel activity in the outer mitochondrial membrane (Jonas et al., 2004); furthermore the large conductance activity is eliminated by a pan-caspase/calpain inhibitor that prevents the cleavage of Bcl-xL. Appearance of the channel associated with ΔN Bcl-xL during hypoxia most likely arises from specific proteolysis of Bcl-xL and not from general injury, because levels of VDAC are preserved in both caspase/calpain inhibitor-treated and untreated hypoxic synapses (Jonas et al., 2004, 2005b).

In contrast to the response to injection into the synapse of full length Bcl-xL, which causes synaptic potentiation, injection of ΔN Bcl-xL protein produces a marked synaptic depression (Hickman et al., 2008; Jonas et al., 2003). The time course of rundown of synaptic responses matches that of hypoxia, suggesting a correlation between the two types of synaptic decline.

More evidence that Bcl-xL protein can produce two different conductance level channel activities in mitochondria comes from *in vivo* studies with the Bcl-xL inhibitor ABT-737. When applied to mitochondria within the squid presynaptic terminal just before healthy synaptic transmission, ABT-737 inhibits the channel activity of mitochondrial membranes induced by synaptic stimulation, suggesting that full length Bcl-xL is necessary for this activity (Hickman et al., 2008). ABT-737, however, also attenuates the channel activity of ΔN Bcl-xL (Hickman et al., 2008). ABT-737 thereby reverses the decline in synaptic responses produced by hypoxia or by direct injection of ΔN Bcl-xL into the terminal, resulting in enhanced synaptic function, suggesting that the amplitude of activity at mitochondrial outer membranes may determine the direction of changes in synaptic strength. In other words, large mitochondrial channel activity produces a decline in synaptic strength, whereas small channel activity produces an enhancement of synaptic strength. In mammalian systems as well, ABT-737 attenuates the channel activity of ΔN Bcl-xL in an *in vivo* model of transient global ischemia and effectively prevents delayed cell death of hippocampal CA1 neurons (Ofengeim et al., 2012). Ischemic death of CA1 neurons is also prevented in a KI mouse containing a form of Bcl-xL resistant to caspase/calpain cleavage (Ofengeim et al., 2012) confirming the specific role of ΔN Bcl-xL in the onset of cell death in hippocampal CA1 neurons during global ischemic brain injury. Although these findings implicate ΔN Bcl-xL in mitochondrial outer membrane activity, they also raise the possibility that mPTP is activated by cleavage of Bcl-xL to form ΔN Bcl-xL (see Fig. 2.4). Ongoing experiments will illuminate this possibility.

2.13 SYNAPTIC RESPONSES DECLINE DURING LONG-TERM DEPRESSION IN ASSOCIATION WITH BCL-2 FAMILY-REGULATED MITOCHONDRIAL CHANNEL ACTIVITY

The enhancement of synaptic responses by full length Bcl-xL could be related to a difference not only in size of the conductance of its mitochondrial ion channel compared to that of pro-apoptotic molecules such as Bax or ΔN Bcl-xL, but also to a difference in function of the activity. A key characteristic of the ion channel activity regulated by full length Bcl-xL is that it can induce ATP production and exchange across mitochondrial membranes (Alavian et al., 2014a; Vander Heiden et al., 2000, 2001), whereas activity of ΔN Bcl-xL or Bax causes release of pro-apoptotic factors such as cytochrome c, caspase activation, and now may be implicated in mPTP activation (Karch et al., 2013). The delicate balance between pro- and anti-apoptotic Bcl-2-related activities, in addition to regulation of MOMP, may also regulate mitochondrial metabolism. These changes at times of stress may control the timing of a decline in synaptic capabilities or neuronal excitability, presaging death of the soma if neuronal processes become overwhelmed (Jonas, Hardwick, and Kaczmarek, 2005a; Plas and Thompson, 2002). Release of factors such as cytochrome c not only activates downstream caspases that inactivate cellular processes, but also directly compromises function by depriving mitochondria of electron transport carriers; furthermore, if mitochondria are Ca^{2+}-overloaded and hampered in ATP production, mPTP may be activated, and somatic death may be initiated. These features may all contribute to the decline in synaptic responses found after full length Bcl-xL cleavage or Bax activation and activation of MOMP or mPTP in synaptic mitochondria.

2.14 SYNAPSE LOSS DURING NEURODEGENERATIVE DISEASE MAY REQUIRE MITOCHONDRIAL CHANNEL ACTIVITY

Long-term synaptic depression (LTD) caused by low-frequency stimulation or by cell signaling is a normal mechanism of synaptic plasticity opposite in some ways to long-term potentiation (LTP) brought on by high-frequency stimulation (Malenka and Bear, 2004). Despite its role in normal synaptic plasticity, however, LTD can also serve as a marker for a predegenerative synaptic state. In hippocampal CA3 to CA1 synapses, low synaptic activity leads to a long-lasting decline in synaptic efficacy, brought about in part by removal of postsynaptic receptors (Malenka and Bear, 2004; Malinow and Malenka, 2002); this state can be quite stable and may never lead to somatic demise. It has been described recently that mitochondria are important for a form of LTD associated with normal synaptic plasticity in hippocampal CA1 neurons. In the CA1 dendrite, low frequency activity causes Bcl-xL-sensitive mitochondrially mediated release of cytochrome c followed by low level activation of caspase 3, which leads to the removal of postsynaptic glutamate receptors from the plasma membrane, resulting in a form of LTD (Li et al., 2010). In addition to these findings, however, age- or developmentally related degenerative changes may be set in motion by prolonged synaptic depression. In synapses from Bax –/– mice treated with the toxic Aβ protein, the prevention of LTP by Aβ was attenuated, implying that Bax actions at mitochondria may be necessary for the relative loss of synapses during Aβ-induced toxicity (Erturk, Wang, and Sheng, 2014; Olsen and Sheng, 2012). In a model of developmental axonal targeting in spinal neurons, both mitochondrial Bax and caspase 6 activation were found to control axonal loss in response to nerve growth factor withdrawal (Nikolaev et al., 2009). In this scenario, the N-terminus of amyloid precursor protein (APP) bound to death receptor 6 (DR6) to initiate an intracellular cascade resulting in mitochondrial-dependent axonal demise.

The difference between decreased activity and damage is made clearer in a recent study of neurite growth arrest in cultured neurons depleted of Bcl-xL by siRNA. The work shows that declining Bcl-xL levels prevent normal outgrowth and branching of neuronal processes over 4 weeks in culture before any somatic death occurs (Park et al., 2015). The neurite growth arrest is associated with up-regulation of both DR6 and Bax. In contrast to the slowly occurring growth arrest found on Bcl-xL depletion, loss of neurites takes place with a much more rapid timescale after a hypoxic stimulus. This hypoxia-induced loss is greatly attenuated in neurons with normal endogenous Bcl-xL levels and neurons in which DR6 has been previously depleted. These findings in hypoxic cells raise the interesting question of whether the up-regulation of DR6, Bax, and neurite growth arrest in Bcl-xL–depleted neurons is triggered by metabolic changes and relative opening of mPTP in dendritic and axonal mitochondria. These events could occur in hypoxic or in depressed synapses.

To support the role of metabolism in developing synapses, it was shown that in healthy or developing neurons, Bcl-xL collaborates with Drp1 by using mitochondrial fission to distribute mitochondria to active synapses (Berman et al., 2009; Li et al., 2008). Recent studies also suggest that axonal branching may be regulated by mitochondrial positioning: The MARK and SAD family member kinase NUAK1 helps capture mitochondria at axonal branch points (Courchet et al., 2013; Lewis, Courchet, and Polleux, 2013), possibly related to increased activity at these sites. Mitochondria are known to pause at sites where activity is high and they continue to move past sites where activity is low, regulated by specific docking proteins, kinases, and external growth factor cues (Chada and Hollenbeck, 2003; Colbourne, Sutherland, and Auer, 1999; Kang et al., 2008; Vaccaro and Kittler, 2013); these processes may also contribute to synaptic plasticity (Obashi and Okabe, 2013; Sun et al., 2013). The sites where mitochondria dock in preparation for axon or dendritic branching are sites that may form hot spots for mRNA accumulation, protein translation, and vesicular trafficking. These events result in insertion of receptors for neurotransmitters and growth factors to support the new functions of growing neuritic processes. These studies emphasize that cues from the extracellular environment during neuronal development or neuronal plasticity may regulate mitochondrial positioning and metabolism as well as protein translation and vesicular trafficking to determine the eventual outcome of neuronal network formation.

2.15 CONCLUSIONS

For many years investigators have sought the identity of the molecular structure underlying acute alterations in mitochondrial morphology and inner membrane conductance associated with acute cell death and known collectively as mitochondrial PT. Early evidence asserted that PT was caused by opening of an inner membrane ion channel. More recent data have shown that the c-subunit of the F_1F_O ATP synthase forms a channel with similar biophysical characteristics to mPTP but whose Ca^{2+}, CsA, CypD, and Bcl-xL regulatory sites are contained in the F_1 including the stator and the catalytic portions. Depletion of c-subunit isoforms in cells blocks CsA-dependent PT and subsequent cell death. Inhibitors and activators may also work through peripheral regulatory moieties such as ANT, PiC, and VDAC that exist in a large complex of proteins with the F_1F_O ATP synthase. Lipids and polyphosphates also may play an important role in pore-gating or formation.

In this model, activators of the mPTP open the pore in a gating mechanism in which F_1 moves away from the mouth of the c-subunit ring while the ring expands (see Fig. 2.3). The process is reversible, perhaps due to binding of F_1 components like the β-subunit to the ring or by the re-association of the entire F_1 onto the ring (Fig. 2.2). Although a wealth of information regarding the molecular structure and regulation of mPTP has been unearthed recently, there is still much to be learned about this fascinating complex.

The role of mPTP in enhanced inner membrane coupling during development of oxidative phosphorylation in aging and in supercomplex formation is rapidly changing fields. In neurons, it is now becoming clear that the metabolic efficiency of mitochondria regulates neuronal survival, neurotransmitter vesicle recycling, and synaptic development. During neurodegeneration, metabolic inefficiency or disuse of neurons or neuronal processes may lead to depression of synaptic responses, pro-apoptotic mitochondrial conductance changes, metabolic compromise, and eventual loss of neurite outgrowth. During these synaptic changes, increases in the probability of mPTP opening may occur. Synaptic depression is also found acutely in hypoxic brain damage and may also be associated with opening of the c-subunit pore in its role as mPTP. Ongoing studies will illuminate the molecular structural changes associated with mitochondrial channel activity during cell development, plasticity, and during stressful or degenerative events.

ACKNOWLEDGMENTS

E. A. Jonas and G. A. Porter, Jr. are considered to have made equal contributions to this chapter.

REFERENCES

Abramov, A. Y., et al., 2007. Targeted polyphosphatase expression alters mitochondrial metabolism and inhibits calcium-dependent cell death. *Proc Natl Acad Sci U S A*, 104, pp.18091–18096.

Adams, J. M. and Cory, S., 2007. The Bcl-2 apoptotic switch in cancer development and therapy. *Oncogene*, 26, pp.1324–1337.

Alavian, K. N., et al., 2011. Bcl-xL regulates metabolic efficiency of neurons through interaction with the mitochondrial F1FO ATP synthase. *Nat Cell Biol*, 13, pp.1224–1233.

Alavian, K. N., et al., 2014a. An uncoupling channel within the c-subunit ring of the F1FO ATP synthase is the mitochondrial permeability transition pore. *Proc Natl Acad Sci U S A*, 111, pp.10580–10585.

Alavian, K. N., et al., 2014b. An uncoupling channel within the c-subunit ring of the F1FO ATP synthase is the mitochondrial permeability transition pore. *Proc Natl Acad Sci U S A*, 111, pp.10580–10585.

Alavian, K. N., et al., 2015. The mitochondrial complex v-associated large-conductance inner membrane current is regulated by cyclosporine and dexpramipexole. *Mol Pharmacol*, 87, pp.1–8.

Andrews, Z. B., Diano, S. and Horvath, T. L., 2005. Mitochondrial uncoupling proteins in the CNS: in support of function and survival. *Nature Reviews Neuroscience*, 6, pp.829–840.

Antonsson, B., et al., 2000. Bax oligomerization is required for channel-forming activity in liposomes and to trigger cytochrome c release from mitochondria. *Biochemical Journal*, 345 Pt 2, pp.271–278.

Azarashvili, T., et al., 2014. Potential role of subunit c of F0F1-ATPase and subunit c of storage body in the mitochondrial permeability transition. Effect of the phosphorylation status of subunit c on pore opening. *Cell Calcium*, 55, pp.69–77.

Baines, C. P., 2009. The mitochondrial permeability transition pore and ischemia-reperfusion injury. *Basic Res Cardiol*, 104, pp.181–188.

Baines, C. P., 2011. The mitochondrial permeability transition pore and the cardiac necrotic program. *Pediatr Cardiol*, 32, pp.258–262.

Baines, C. P., et al., 2005. Loss of cyclophilin D reveals a critical role for mitochondrial permeability transition in cell death.[see comment]. *Nature*, 434, pp.658–662.

Basso, E., et al., 2005. Properties of the permeability transition pore in mitochondria devoid of Cyclophilin D. *J Biol Chem*, 280, pp.18558–18561.

Beal, M. F., 2007. Mitochondria and neurodegeneration. *Novartis Foundation Symposium*, 287, pp.183–192; discussion 192–196.

Berman, S. B., et al., 2009. Bcl-x L increases mitochondrial fission, fusion, and biomass in neurons. *J Cell Biol*, 184, pp.707–719.

Bernardi, P., 1999. Mitochondrial transport of cations: channels, exchangers, and permeability transition. *Physiological Reviews*, 79, pp.1127–1155.

Bernardi, P., 2013. The mitochondrial permeability transition pore: a mystery solved? *Front Physiol*, 4, p.95.

Bernardi, P. and Di Lisa, F., 2015. The mitochondrial permeability transition pore: Molecular nature and role as a target in cardioprotection. *J Mol Cell Cardiol*, 78, pp.100–106.

Beutner, G., et al., 1996. Complexes between kinases, mitochondrial porin and adenylate translocator in rat brain resemble the permeability transition pore. *FEBS Lett*, 396, 189–195.

Beutner, G., et al., 1998. Complexes between porin, hexokinase, mitochondrial creatine kinase and adenylate translocator display properties of the permeability transition pore. Implication for regulation of permeability transition by the kinases. *Biochim Biophys Acta*, 1368, pp.7–18.

Beutner, G., Eliseev, R. A. and Porter, G. A., Jr., 2014. Initiation of electron transport chain activity in the embryonic heart coincides with the activation of mitochondrial complex 1 and the formation of supercomplexes. *PLoS One*, 9, p.e113330.

Billups, B. and Forsythe, I. D., 2002. Presynaptic mitochondrial calcium sequestration influences transmission at mammalian central synapses. *J Neuroscience*, 22, 5840–5847.

Bonora, M., et al., 2013. Role of the c subunit of the FO ATP synthase in mitochondrial permeability transition. *Cell Cycle*, 12, pp.674–683.

Bonora, M. and Pinton, P., 2015. Shedding light on molecular mechanisms and identity of mPTP. *Mitochondrion*, 21, p.11.

Bonora, M., et al., 2015. Molecular mechanisms of cell death: central implication of ATP synthase in mitochondrial permeability transition. *Oncogene*, 34, pp.1475–1486.

Borjesson, S. I. and Elinder, F., 2008. Structure, function, and modification of the voltage sensor in voltage-gated ion channels. *Cell Biochem Biophys*, 52, pp.149–174.

Brand, M. D., 2005. The efficiency and plasticity of mitochondrial energy transduction. *Biochemical Society Transactions*, 33, pp. 897–904.

Bravo, C., et al., 2004. Overexpression of the inhibitor protein IF(1) in AS-30D hepatoma produces a higher association with mitochondrial F(1)F(0) ATP synthase compared to normal rat liver: functional and cross-linking studies. *J Bioenerg Biomembr*, 36, 257–264.

Budd, S. L. and Nicholls, D. G., 1996. Mitochondria, calcium regulation, and acute glutamate excitotoxicity in cultured cerebellar granule cells. *Journal of Neurochemistry*, 67, 2282–2291.

Cabezon, E., et al., 2001. The structure of bovine IF(1), the regulatory subunit of mitochondrial F-ATPase. *EMBO J*, 20, 6990–6996.

Campanella, M., et al., 2008. Regulation of mitochondrial structure and function by the F1FO-ATPase inhibitor protein, IF1. *Cell Metab*, 8, pp.13–25.

Carbajo, R. J., et al., 2005. Structure of the F1-binding domain of the stator of bovine F1FO-ATPase and how it binds an alpha-subunit. *J Mol Biol*, 351, pp.824–838.

Caviston, T. L., et al., 1998. Identification of an uncoupling mutation affecting the b subunit of F1F0 ATP synthase in Escherichia coli. *FEBS Letters*, 429, pp.201–206.

Chada, S. R. and Hollenbeck, P. J., 2003. Mitochondrial movement and positioning in axons: the role of growth factor signaling. *Journal of Experimental Biology*, 206, pp.1985–1992.

Chen, C., et al., 2004. Mitochondrial ATP synthasome: three-dimensional structure by electron microscopy of the ATP synthase in complex formation with carriers for Pi and ADP/ATP. *Journal of Biological Chemistry*, 279, pp.31761–31768.

Chen, Y. B., et al., 2011. Bcl-xL regulates mitochondrial energetics by stabilizing the inner membrane potential. *Journal of Cell Biology*, 195, pp.263–276.

Chinopoulos, C. and Szabadkai, G., 2014. What Makes You Can also Break You, Part III: Mitochondrial Permeability Transition Pore Formation by an Uncoupling Channel within the C-Subunit Ring of the F1FO ATP Synthase? *Front Oncol*, 4, p. 235.

Cho, S. W., et al., 2014. Dual modulation of the mitochondrial permeability transition pore and redox signaling synergistically promotes cardiomyocyte differentiation from pluripotent stem cells. *J Am Heart Assoc*, 3, p. e000693.

Chouhan, A. K., et al., 2012. Cytosolic calcium coordinates mitochondrial energy metabolism with presynaptic activity. *J Neurosci*, 32, pp.1233–1243.

Colbourne, F., Sutherland, G. R. and Auer, R. N., 1999. Electron microscopic evidence against apoptosis as the mechanism of neuronal death in global ischemia. *Journal of Neuroscience*, 19, pp.4200–4210.

Courchet, J., et al., 2013. Terminal axon branching is regulated by the LKB1-NUAK1 kinase pathway via presynaptic mitochondrial capture. *Cell*, 153, pp.1510–1525.

Crompton, M., Ellinger, H. and Costi, A., 1988. Inhibition by cyclosporin A of a Ca2+-dependent pore in heart mitochondria activated by inorganic phosphate and oxidative stress. *Biochem J*, 255, pp.357–360.

Crompton, M., Virji, S. and Ward, J. M., 1998. Cyclophilin-D binds strongly to complexes of the voltage-dependent anion channel and the adenine nucleotide translocase to form the permeability transition pore. *European Journal of Biochemistry*, 258, 729–735.

Crompton, M., 1999. The mitochondrial permeability transition pore and its role in cell death. *Biochemical Journal*, 341, pp.233–249.

Csordas, G., Thomas, A. P. and Hajnoczky, G., 2001. Calcium signal transmission between ryanodine receptors and mitochondria in cardiac muscle. *Trends Cardiovasc Med*, 11, pp.269–275.

D'Alessandro, M., Turina, P. and Melandri, B. A., 2008. Intrinsic uncoupling in the ATP synthase of Escherichia coli. *Biochim Biophys Acta*, 1777, pp.1518–1527.

D'Orsi, B., et al., 2012. Calpains are downstream effectors of bax-dependent excitotoxic apoptosis. *J Neurosci*, 32, pp.1847–1858.

De Stefani, D. and Rizzuto, R., 2014. Molecular control of mitochondrial calcium uptake. *Biochem Biophys Res Commun*, 449, pp.373–376.

Dejean, L. M., et al., 2005. Oligomeric Bax is a component of the putative cytochrome c release channel MAC, mitochondrial apoptosis-induced channel. *Molecular Biology of the Cell*, 16, pp.2424–2432.

Dejean, L. M., et al., 2006. Regulation of the mitochondrial apoptosis-induced channel, MAC, by BCL-2 family proteins. *Biochimica et Biophysica Acta*, 1762, pp.191–201.

Denton, R. M., 2009. Regulation of mitochondrial dehydrogenases by calcium ions. *Biochim Biophys Acta*, 1787, 1309–1316.

Di Lisa, F., et al., 2011. The mitochondrial permeability transition pore and cyclophilin D in cardioprotection. *Biochim Biophys Acta*, 1813, pp.1316–1322.

DiMauro, S., et al., 1976. Luft's disease. Further biochemical and ultrastructural studies of skeletal muscle in the second case. *J Neurol Sci*, 27, pp.217–232.

Divakaruni, A. S. and Brand, M. D., 2011. The regulation and physiology of mitochondrial proton leak. *Physiology (Bethesda)*, 26, pp.192–205.

Dodson, M. W. and Guo, M., 2007. Pink1, Parkin, DJ-1 and mitochondrial dysfunction in Parkinson's disease. *Current Opinion in Neurobiology*, 17, pp.331–337.

Drenckhahn, J. D., 2011. Heart development: mitochondria in command of cardiomyocyte differentiation. *Dev Cell*, 21, pp.392–393.

Elrod, J. W., et al., 2010. Cyclophilin D controls mitochondrial pore-dependent Ca(2+) exchange, metabolic flexibility, and propensity for heart failure in mice. *J Clin Invest*, 120, pp.3680–3687.

Elustondo, P. A., et al., 2013. Polyhydroxybutyrate targets mammalian mitochondria and increases permeability of plasmalemmal and mitochondrial membranes. *PLoS One*, 8, p.e75812.

Erturk, A., Wang, Y. and Sheng, M., 2014. Local pruning of dendrites and spines by caspase-3-dependent and proteasome-limited mechanisms. *J Neurosci*, 34, pp.1672–1688.

Faccenda, D., et al., 2013a. Mitochondrial IF(1) preserves cristae structure to limit apoptotic cell death signaling. *Cell Cycle*, 12, pp.2530–2352.

Faccenda, D., et al., 2013b. IF1 limits the apoptotic-signalling cascade by preventing mitochondrial remodelling. *Cell Death Differ*, 20, pp.686–697.

Fedorenko, A., Lishko, P. V. and Kirichok, Y., 2012. Mechanism of fatty-acid-dependent UCP1 uncoupling in brown fat mitochondria. *Cell*, 151, pp.400–413.

Finbow, M. E. and Harrison, M. A., 1997. The vacuolar H+-ATPase: a universal proton pump of eukaryotes. *Biochem J*, 324 (Pt 3), pp.697–712.

Folmes, C. D., et al., 2012. Mitochondria in control of cell fate. *Circ Res*, 110, pp.526–529.

Friel, D. D. and Tsien, R. W., 1994. An FCCP-sensitive Ca2+ store in bullfrog sympathetic neurons and its participation in stimulus-evoked changes in [Ca2+]i. *Journal of Neuroscience*, 14, pp.4007–4024.

Fujiwara, M., et al., 2011. Induction and enhancement of cardiac cell differentiation from mouse and human induced pluripotent stem cells with cyclosporin-A. *PLoS One*, 6, p.e16734.

Galluzzi, L., Blomgren, K. and Kroemer, G., 2009. Mitochondrial membrane permeabilization in neuronal injury. *Nature Reviews Neuroscience*, 10, pp.481–494.

Galonek, H. L. and Hardwick, J. M., 2006. Upgrading the BCL-2 network.[comment]. *Nature Cell Biology*, 8, pp.1317–1319.

Garcia, J. J., et al., 2006. The inhibitor protein (IF1) promotes dimerization of the mitochondrial F1F0-ATP synthase. *Biochemistry*, 45, pp.12695–12703.

Gellerich, F. N., et al., 2010. The regulation of OXPHOS by extramitochondrial calcium. *Biochim Biophys Acta*, 1797, 1018–1027.

Giorgio, V., et al., 2009. Cyclophilin D modulates mitochondrial F0F1-ATP synthase by interacting with the lateral stalk of the complex. *J Biologic Chem*, 284, pp.33982–33988.

Giorgio, V., et al., 2013. Dimers of mitochondrial ATP synthase form the permeability transition pore. *Proc Natl Acad Sci U S A*, 110, pp.5887–5592.

Gomez, L., et al., 2007. Inhibition of mitochondrial permeability transition improves functional recovery and reduces mortality following acute myocardial infarction in mice. *Am J Physiol Heart Circ Physiol*, 293, H1654–H1661.

Gottlieb, E., Armour, S. M. and Thompson, C. B., 2002. Mitochondrial respiratory control is lost during growth factor deprivation. *Proc Natl Acad Sci U S A*, 99, pp.12801–12806.

Green, D. R. and Kroemer, G., 2004. The pathophysiology of mitochondrial cell death. *Science*, 305, pp.626–629.

Guerrieri, F., et al., 1992. Age-related changes of mitochondrial F0F1 ATP synthase. *Ann N Y Acad Sci* 671, pp.395–402.

Gunter, T. E. and Sheu, S. S., 2009. Characteristics and possible functions of mitochondrial Ca(2+) transport mechanisms. *Biochim Biophys Acta*, 1787, pp.1291–1308.

Gutierrez-Aguilar, M., et al., 2014. Genetic manipulation of the cardiac mitochondrial phosphate carrier does not affect permeability transition. *J Mol Cell Cardiol*, 72, pp.316–325.

Hafner, A. V., et al., 2010. Regulation of the mPTP by SIRT3-mediated deacetylation of CypD at lysine 166 suppresses age-related cardiac hypertrophy. *Aging (Albany NY)*, 2, pp.914–923.

Hajnoczky, G., et al., 2000. Mitochondrial calcium signaling driven by the IP3 receptor. *J Bioenerg Biomembr*, 32, pp.15–25.

Halestrap, A. P. and Davidson, A. M., 1990. Inhibition of Ca2(+)-induced large-amplitude swelling of liver and heart mitochondria by cyclosporin is probably caused by the inhibitor binding to mitochondrial-matrix peptidyl-prolyl cis-trans isomerase and preventing it interacting with the adenine nucleotide translocase. *Biochem J*, 268, pp.153–160.

Halestrap, A. P. and Richardson, A. P., 2015. The mitochondrial permeability transition: A current perspective on its identity and role in ischaemia/reperfusion injury. *J Mol Cell Cardiol*, 78, pp.129–141.

Harrison, M., et al., 2003. Structure and function of the vacuolar H+-ATPase: moving from low-resolution models to high-resolution structures. *J Bioenerg Biomembr*, 35, pp.337–345.

Hausenloy, D., et al., 2004. Transient mitochondrial permeability transition pore opening mediates preconditioning-induced protection. *Circulation*, 109, pp.1714–1717.

Haworth, R. A. and Hunter, D. R., 1979. The Ca2+-induced membrane transition in mitochondria. II. Nature of the Ca2+ trigger site. *Arch Biochem Biophys*, 195, pp.460–467.

Hickman, J. A., et al., 2008. Bcl-xL inhibitor ABT-737 reveals a dual role for Bcl-xL in synaptic transmission. *Journal of Neurophysiology*, 99, 1515–1522.

Hockenbery, D., et al., 1990. Bcl-2 is an inner mitochondrial membrane protein that blocks programmed cell death. *Nature*, 348, pp.334–336.

Holmstrom, K. M., et al., 2013. Signalling properties of inorganic polyphosphate in the mammalian brain. *Nat Commun*, 4, p.1362.

Hom, J. R., et al., 2011. The permeability transition pore controls cardiac mitochondrial maturation and myocyte differentiation. *Dev Cell*, 21, pp.469–478.

Hubbard, M. J. and McHugh, N. J., 1996. Mitochondrial ATP synthase F1-beta-subunit is a calcium-binding protein. *FEBS Lett*, 391, pp.323–329.

Hunter, D. R. and Haworth, R. A., 1979a. The Ca2+-induced membrane transition in mitochondria. III. Transitional Ca2+ release. *Arch Biochem Biophys*, 195, pp.468–477.

Hunter, D. R. and Haworth, R.A., 1979b. The Ca2+-induced membrane transition in mitochondria. I. The protective mechanisms. *Arch Biochem Biophys*, 195, pp.453–459.

Huser, J. and Blatter, L. A., 1999. Fluctuations in mitochondrial membrane potential caused by repetitive gating of the permeability transition pore. *Biochem J*, 343 (Pt 2), pp.311–317.

Jennings, R. B., Reimer, K. A. and Steenbergen, C., 1991. Effect of inhibition of the mitochondrial ATPase on net myocardial ATP in total ischemia. *J Mol Cell Cardiol*, 23, pp.1383–1395.

Jonas, E., 2006. BCL-xL regulates synaptic plasticity. *Molecular Interventions*, 6, pp.208–222.

Jonas, E. A., 2009. Molecular participants in mitochondrial cell death channel formation during neuronal ischemia. *Experimental Neurology*, 218, pp.203–212.

Jonas, E. A., Buchanan, J., and Kaczmarek, L. K., 1999. Prolonged activation of mitochondrial conductances during synaptic transmission. *Science*, 286, pp.1347–1350.

Jonas, E. A., Hardwick, J. M. and Kaczmarek, L. K., 2005a. Actions of BAX on mitochondrial channel activity and on synaptic transmission. *Antioxidants & Redox Signaling*, 7, pp.1092–1100.

Jonas, E. A., Knox, R. J. and Kaczmarek, L. K., 1997. Giga-ohm seals on intracellular membranes: a technique for studying intracellular ion channels in intact cells. *Neuron*, 19, 7–13.

Jonas, E. A., Porter, G. A. and Alavian, K. N., 2014. Bcl-xL in neuroprotection and plasticity. *Front Physiol*, 5, p.355.

Jonas, E. A., et al., 2003. Modulation of synaptic transmission by the BCL-2 family protein BCL-xL. *Journal of Neuroscience*, 23, pp.8423–8431.

Jonas, E. A., et al., 2004. Proapoptotic N-truncated BCL-xL protein activates endogenous mitochondrial channels in living synaptic terminals. *Proc Natl Acad Sci U S A*, 101, pp.13590–13595.

Jonas, E. A., et al., 2005b. Exposure to hypoxia rapidly induces mitochondrial channel activity within a living synapse. *Journal of Biological Chemistry*, 280, pp.4491–4497.

Jonckheere, A. I., Smeitink, J. A. and Rodenburg, R. J., 2012. Mitochondrial ATP synthase: architecture, function and pathology. *J Inherit Metab Dis*, 35, pp.211–225.

Jones, P. C., et al., 1995. The first putative transmembrane helix of the 16 kDa proteolipid lines a pore in the Vo sector of the vacuolar H(+)-ATPase. *Biochem J*, 312 (Pt 3), pp.739–747.

Kang, J. S., et al., 2008. Docking of axonal mitochondria by syntaphilin controls their mobility and affects short-term facilitation. *Cell*, 132, pp.137–148.

Karch, J., et al., 2013. Bax and Bak function as the outer membrane component of the mitochondrial permeability pore in regulating necrotic cell death in mice. *Elife*, 2, p.e00772.

Karch, J. and Molkentin, J. D., 2014. Identifying the components of the elusive mitochondrial permeability transition pore. *Proc Natl Acad Sci U S A*, 111, pp.10396–10397.

Kim, H., et al., 2006. Hierarchical regulation of mitochondrion-dependent apoptosis by BCL-2 subfamilies [see comment]. *Nature Cell Biology*, 8, pp.1348–1358.

Kinnally, K. W., Campo, M. L. and Tedeschi, H., 1989. Mitochondrial channel activity studied by patch-clamping mitoplasts. *Journal of Bioenergetics & Biomembranes*, 21, pp.497–506.

Kinnally, K. W., Martinez-Caballero, S. and Dejean, L. M., 2006. Detection of the mitochondrial apoptosis-induced channel (MAC) and its regulation by Bcl-2 family proteins. *Current Protocols in Toxicology,* Chapter 2, Unit2 12.

Kinnally, K. W., et al., 2011. Is mPTP the gatekeeper for necrosis, apoptosis, or both? *Biochim Biophys Acta*, 1813, pp.616–622.

Kokoszka, J. E., et al., 2004. The ADP/ATP translocator is not essential for the mitochondrial permeability transition pore [see comment]. *Nature*, 427, pp.461–465.

Korge, P., et al., 2011. Protective role of transient pore openings in calcium handling by cardiac mitochondria. *J Biol Chem*, 286, pp.34851–34857.

Kowaltowski, A. J., et al., 1998. Ca2+-stimulated mitochondrial reactive oxygen species generation and permeability transition are inhibited by dibucaine or Mg2+. *Arch Biochem Biophys*, 359, pp.77–81.

Kruse, S. E., et al., 2008. Mice with mitochondrial complex I deficiency develop a fatal encephalomyopathy. *Cell Metab*, 7, pp.312–320.

Kwong, J. Q., et al., 2014. Genetic deletion of the mitochondrial phosphate carrier desensitizes the mitochondrial permeability transition pore and causes cardiomyopathy. *Cell Death Differ*, 21, pp.1209–1217.

Leung, A. W. and Halestrap, A. P., 2008. Recent progress in elucidating the molecular mechanism of the mitochondrial permeability transition pore. *Biochim Biophys Acta*, 1777, pp.946–952.

Lewis, T. L., Jr., Courchet, J. and Polleux, F., 2013. Cell biology in neuroscience: Cellular and molecular mechanisms underlying axon formation, growth, and branching. *J Cell Biol*, 202, pp.837–848.

Li, H., et al., 2008. Bcl-xL induces Drp1-dependent synapse formation in cultured hippocampal neurons. *Proc Natl Acad Sci U S A*, 105, pp.2169–2174.

Li, H., et al., 2013. A Bcl-xL-Drp1 complex regulates synaptic vesicle membrane dynamics during endocytosis. *Nat Cell Biol*, 15, pp.773–785.

Li, Z., et al., 2010. Caspase-3 activation via mitochondria is required for long-term depression and AMPA receptor internalization. *Cell*, 141, pp.859–871.

Lopreiato, R., Giacomello, M. and Carafoli, E., 2014. The plasma membrane calcium pump: new ways to look at an old enzyme. *J Biol Chem*, 289, pp.10261–10268.

Malenka, R. C. and Bear, M. F., 2004. LTP and LTD: an embarrassment of riches. *Neuron*, 44, pp.5–21.

Malinow, R. and Malenka, R. C., 2002. AMPA receptor trafficking and synaptic plasticity. *Annual Review of Neuroscience*, 25, pp.103–126.

Mandel, M., et al., 1988. cDNA sequence encoding the 16-kDa proteolipid of chromaffin granules implies gene duplication in the evolution of H+-ATPases. *Proc Natl Acad Sci U S A*, 85, pp.5521–5524.

Mannella, C. A. and Kinnally, K. W., 2008. Reflections on VDAC as a voltage-gated channel and a mitochondrial regulator. *Journal of Bioenergetics & Biomembranes*, 40, pp.149–155.

Martinez-Caballero, S., et al., 2005. The role of the mitochondrial apoptosis induced channel MAC in cytochrome c release. *Journal of Bioenergetics & Biomembranes*, 37, pp.155–164.

Matthies, D., et al., 2009. The c13 ring from a thermoalkaliphilic ATP synthase reveals an extended diameter due to a special structural region. *J Mol Biol*, 388, pp.611–618.

McGeoch, J. E. and Guidotti, G., 1997. A 0.1-700 Hz current through a voltage-clamped pore: candidate protein for initiator of neural oscillations. *Brain Res*, 766, pp.188–194.

McGeoch, J. E. and McGeoch, M. W., 2008. Entrapment of water by subunit c of ATP synthase. *J R Soc Interface*, 5, pp.311–318.

McGuinness, O., et al., 1990. The presence of two classes of high-affinity cyclosporin A binding sites in mitochondria. Evidence that the minor component is involved in the opening of an inner-membrane Ca(2+)-dependent pore. *Eur J Biochem*, 194, pp.671–679.

Meier, T., et al., 2001. The central plug in the reconstituted undecameric c cylinder of a bacterial ATP synthase consists of phospholipids. *FEBS Lett*, 505, pp.353–356.

Mitchell, P., 1961. Coupling of phosphorylation to electron and hydrogen transfer by a chemi-osmotic type of mechanism. *Nature*, 191, pp.144–148.

Nakagawa, T., et al., 2005. Cyclophilin D-dependent mitochondrial permeability transition regulates some necrotic but not apoptotic cell death [see comment]. *Nature*, 434, pp.652–658.

Neher, E. and Sakaba, T., 2008. Multiple roles of calcium ions in the regulation of neurotransmitter release. *Neuron*, 59, pp.861–872.

Nguyen, T. T., et al., 2011. Cysteine 203 of cyclophilin D is critical for cyclophilin D activation of the mitochondrial permeability transition pore. *J Biol Chem*, 286, pp.40184–40192.

Nicholls, D. G. and Rial, E., 1999. A history of the first uncoupling protein, UCP1. *J Bioenerg Biomembr*, 31, pp.399–406.

Nikolaev, A., et al., 2009. APP binds DR6 to trigger axon pruning and neuron death via distinct caspases. *Nature*, 457, pp.981–989.

Norris, U., Karp, P. E. and Fimmel, A. L., 1992. Mutational analysis of the glycine-rich region of the c subunit of the Escherichia coli F0F1 ATPase. *J Bacteriol*, 174, 4496–4499.

Obashi, K. and Okabe, S., 2013. Regulation of mitochondrial dynamics and distribution by synapse position and neuronal activity in the axon. *Eur J Neurosci*, 38, 2350–2363.

Oberfeld, B., Brunner, J. and Dimroth, P., 2006. Phospholipids occupy the internal lumen of the c ring of the ATP synthase of Escherichia coli. *Biochemistry*, 45, pp.1841–1851.

Ofengeim, D., et al., 2012. N-terminally cleaved Bcl-xL mediates ischemia-induced neuronal death. *Nat Neurosci*, 15, pp.574–580.

Olsen, K. M. and Sheng, M., 2012. NMDA receptors and BAX are essential for Abeta impairment of LTP. *Sci Rep*, 2, p.225.

Ozden, S. and Isenmann, S., 2004. Neuroprotective properties of different anesthetics on axotomized rat retinal ganglion cells in vivo. *J Neurotrauma*, 21, pp.73–82.

Pang, Z. P., et al., 2010. Calmodulin controls synaptic strength via presynaptic activation of calmodulin kinase II. *Journal of Neuroscience*, 30, pp.4132–4142.

Park, H. A., et al., 2015. Bcl-xL is necessary for neurite outgrowth in hippocampal neurons. *Antioxid Redox Signal*, 22, pp.93–108.

Pavlov, E., et al., 2005. A large, voltage-dependent channel, isolated from mitochondria by water-free chloroform extraction. *Biophys J*, 88, pp.2614–2625.

Pedersen, P. L. and Hullihen, J., 1978. Adenosine triphosphatase of rat liver mitochondria. Capacity of the homogeneous F1 component of the enzyme to restore ATP synthesis in urea-treated membranes. *J Biol Chem*, 253, pp.2176–2183.

Pedersen, P. L., 1994. ATP synthase. The machine that makes ATP. *Curr Biol*, 4, pp.1138–1141.

Peters, C., et al., 2001. Trans-complex formation by proteolipid channels in the terminal phase of membrane fusion. *Nature*, 409, pp.581–588.

Petronilli, V., Szabo, I. and Zoratti, M., 1989. The inner mitochondrial membrane contains ion-conducting channels similar to those found in bacteria. *FEBS Letters*, 259, pp.137–143.

Petronilli, V., et al., 1999. Transient and long-lasting openings of the mitochondrial permeability transition pore can be monitored directly in intact cells by changes in mitochondrial calcein fluorescence. *Biophys J*, 76, pp.725–734.

Plas, D. R. and Thompson, C.B., 2002. Cell metabolism in the regulation of programmed cell death. *Trends in Endocrinology & Metabolism*, 13, pp.75–78.

Pullman, M. E. and Monroy, G. C., 1963. A naturally occurring inhibitor of mitochondrial adenosine triphosphatase. *J Biol Chem*, 238, pp.3762–3769.

Raffaello, A., De Stefani, D. and Rizzuto, R., 2012. The mitochondrial Ca(2+) uniporter. *Cell Calcium*, 52, pp.16–21.

Rizzuto, R., Bernardi, P. and Pozzan, T., 2000. Mitochondria as all-round players of the calcium game. *J Physiol*, 529 (Pt 1), pp.37–47.

Rizzuto, R. and Pozzan, T., 2006. Microdomains of intracellular Ca2+: molecular determinants and functional consequences. *Physiol Rev*, 86, pp.369–408.

Rizzuto, R., et al., 2009. Ca(2+) transfer from the ER to mitochondria: when, how and why. *Biochimica et Biophysica Acta*, 1787, pp.1342–1351.

Rizzuto, R., et al., 2012. Mitochondria as sensors and regulators of calcium signalling. *Nat Rev Mol Cell Biol*, 13, pp.566–578.

Roestenberg, P., et al., 2012. Pharmacological targeting of mitochondrial complex I deficiency: the cellular level and beyond. *Mitochondrion*, 12, pp.57–65.

Rostovtseva, T. and Colombini, M., 1997. VDAC channels mediate and gate the flow of ATP: implications for the regulation of mitochondrial function. *Biophysical Journal*, 72, pp.1954–1962.

Rostovtseva, T. K. and Bezrukov, S. M., 2012. VDAC inhibition by tubulin and its physiological implications. *Biochim Biophys Acta*, 1818, pp.1526–1535.

Rouslin, W., 1991. Regulation of the mitochondrial ATPase in situ in cardiac muscle: role of the inhibitor subunit. *J Bioenerg Biomembr*, 23, pp.873–888.

Sanchez-Cenizo, L., et al., 2010. Up-regulation of the ATPase inhibitory factor 1 (IF1) of the mitochondrial H+-ATP synthase in human tumors mediates the metabolic shift of cancer cells to a Warburg phenotype. *J Biol Chem*, 285, pp.25308–25313.

Schinzel, A. C., et al., 2005. Cyclophilin D is a component of mitochondrial permeability transition and mediates neuronal cell death after focal cerebral ischemia. *Proc Natl Acad Sci U S A*, 102, pp.12005–12010.

Schneggenburger, R. and Neher, E., 2005. Presynaptic calcium and control of vesicle fusion. *Curr Opin Neurobiol*, 15, pp.266–274.

Seidlmayer, L. K., et al., 2012. Inorganic polyphosphate—an unusual suspect of the mitochondrial permeability transition mystery. *Channels (Austin)*, 6, 463–467.

Shulga, N. and Pastorino, J. G., 2010. Ethanol sensitizes mitochondria to the permeability transition by inhibiting deacetylation of cyclophilin-D mediated by sirtuin-3. *J Cell Sci*, 123, 4117–4127.

Shulga, N., Wilson-Smith, R. and Pastorino, J. G., 2010. Sirtuin-3 deacetylation of cyclophilin D induces dissociation of hexokinase II from the mitochondria. *J Cell Sci*, 123, pp.894–902.

Sorgato, M. C., Keller, B. U. and Stuhmer, W., 1987. Patch-clamping of the inner mitochondrial membrane reveals a voltage-dependent ion channel. *Nature*, 330, pp.498–500.

Stotz, S. C., et al., 2014. Inorganic polyphosphate regulates neuronal excitability through modulation of voltage-gated channels. *Mol Brain*, 7, p.42–.

Sun, T., et al., 2013. Motile axonal mitochondria contribute to the variability of presynaptic strength. *Cell Rep*, 4, pp.413–419.

Szabo, I. and Zoratti, M., 1991. The giant channel of the inner mitochondrial membrane is inhibited by cyclosporin A. *Journal of Biological Chemistry*, 266, pp.3376–3379.

Szabo, I., Bernardi, P. and Zoratti, M., 1992. Modulation of the mitochondrial megachannel by divalent cations and protons. *Journal of Biological Chemistry*, 267, pp.2940–2946.

Tang, Y. and Zucker, R. S., 1997. Mitochondrial involvement in post-tetanic potentiation of synaptic transmission. *Neuron*, 18, pp.483–491.

Vaccaro, V. and Kittler, J., 2013. Shedding light on mitochondrial movements in axons (commentary on Obashi & Okabe). *Eur J Neurosci*, 38, p.2349.

Vander Heiden, M. G., et al., 2000. Outer mitochondrial membrane permeability can regulate coupled respiration and cell survival. *Proc Natl Acad Sci U S A*, 97, pp.4666–4671.

Vander Heiden, M. G., et al., 2001. Bcl-xL promotes the open configuration of the voltage-dependent anion channel and metabolite passage through the outer mitochondrial membrane. *Journal of Biological Chemistry*, 276, pp.19414–19419.

Walker, J. E., 2013. The ATP synthase: the understood, the uncertain and the unknown. *Biochem Soc Trans*, 41, pp.1–16.

Wang, W., et al., 2008. Superoxide flashes in single mitochondria. *Cell*, 134, pp.279–290.

Watt, I. N., et al., 2010. Bioenergetic cost of making an adenosine triphosphate molecule in animal mitochondria. *Proc Natl Acad Sci U S A*, 107, pp.16823–16827.

Wittig, I. and Schagger, H., 2009. Supramolecular organization of ATP synthase and respiratory chain in mitochondrial membranes. *Biochim Biophys Acta*, 1787, pp.672–680.

Woodfield, K., et al., 1998. Direct demonstration of a specific interaction between cyclophilin-D and the adenine nucleotide translocase confirms their role in the mitochondrial permeability transition. *Biochem J*, 336 (Pt 2), 287–290.

Yamada, E. W. and Huzel, N. J., 1992. Distribution of the ATPase inhibitor proteins of mitochondria in mammalian tissues including fibroblasts from a patient with Luft's disease. *Biochim Biophys Acta*, 1139, 143–147.

Youle, R. J. and Strasser, A., 2008. The BCL-2 protein family: opposing activities that mediate cell death. *Nature Reviews Molecular Cell Biology*, 9, pp.47–59.

Zakharov, S. D., et al., 1996. Calcium binding to the subunit c of E. coli ATP-synthase and possible functional implications in energy coupling. *J Bioenerg Biomembr*, 28, pp.483–494.

Zhu, J., et al., 2013. Interactions of GSK-3beta with mitochondrial permeability transition pore modulators during preconditioning: age-associated differences. *J Gerontol A Biol Sci Med Sci*, 68, pp.395–403.

Zucker, R. S. and Regehr, W. G., 2002. Short-term synaptic plasticity. *Annual Review of Physiology*, 64, pp.355–405.

3

MITOCHONDRIAL CHANNELS IN NEURODEGENERATION

PABLO M. PEIXOTO

Department of Natural Sciences, Baruch College and Graduate Center at City University of New York, New York, USA

KATHLEEN W. KINNALLY

Department of Basic Sciences, New York University College of Dentistry, New York, USA

EVGENY PAVLOV

Department of Physiology and Biophysics, Dalhousie University, Halifax, Nova Scotia

3.1 INTRODUCTION

Open a textbook in neuroscience and you may not be surprised to find that neuronal activities such as action potentials and synaptic transmission require large amounts of energy. This requirement may be the reason why mitochondria are present in large amounts in the brain relative to other organs (e.g., the gut). What you may not know is that in, addition to delivering the energy necessary for generation of action potentials and synaptic transmission, mitochondria regulate calcium signaling and plasticity at the synaptic terminal and gene transcription at the nucleus. Channels across the inner and outer membranes of mitochondria mediate these emergent functions by providing the pathway for delivery of adenosine triphosphate (ATP), reactive oxygen species (ROS), calcium, and other factors to neuronal compartments. In this chapter, you will

The Functions, Disease-Related Dysfunctions, and Therapeutic Targeting of Neuronal Mitochondria, First Edition. Edited by Valentin K. Gribkoff, Elizabeth A. Jonas and J. Marie Hardwick.

become acquainted with the currently known mitochondrial channels, some of which have unknown functions and whose study in neurons may inspire potential approaches to treatment of neurodegenerative diseases. We should note that many of the channel properties discussed here were investigated using model mammalian or even yeast mitochondria. However, although fine regulation of the channels might vary from organ to organ, it is likely that general properties of these channels will be similar and thus can be applied to neurons.

3.2 MITOCHONDRIAL CHANNELS IN THE HEALTHY NEURON

3.2.1 Voltage Dependent Anion-Selective Channel is the Food Channel

Voltage dependent anion-selective channel (VDAC) was the first mitochondrial channel to be identified and is a highly abundant protein in the outer membrane (Schein, Colombini, and Finkelstein, 1976). We dubbed VDAC the "food channel" because it provides the aqueous pathway across the outer membrane for the exchange of metabolites between the cytosol and the mitochondria. These metabolites are used by mitochondria to generate ATP through oxidative phosphorylation and therefore must pass through VDAC on their way to the electron transport chain in the inner membrane. In keeping with this channel's bidirectional trafficking, the ATP and hemes generated by mitochondria exit through VDAC to the cytosol. Interestingly, VDAC interacts with many other proteins (e.g., hexokinase and creatine kinase) to trap the newly generated ATP in high-energy storage forms like glucose 6-phosphate and creatine phosphate in brain and muscle, respectively.

Although the structure of VDAC is still a matter of debate, there is a consensus that the channel has a single pore that is ~3 nm wide when fully open (Colombini, 2009; Hiller et al., 2008; Ujwal et al., 2008). Generally, VDAC's pore has slight anion-selectivity that shifts to cation-selectivity when it partially closes (Colombini, 1980; Pavlov et al., 2005a). These dynamic gating properties of fully open VDAC allow regulation of the flux across the outer membrane of anionic metabolites like ATP, adenosine diphosphate (ADP), and inorganic phosphate (Pi), along with cations like Ca^{2+}, K+, and Na+ (Rostovtseva and Colombini, 1996, 1997).

Recent advances in resolving VDAC structure arose from NMR and crystallography studies (Bayrhuber et al., 2008; Hiller et al., 2008; Ujwal et al., 2008). These agree with previous electron microscopy studies that report a slightly elliptical 2.7- to 3.5- × 2.4-to 3.1-nm pore. In the three-dimensional (3D) models, this pore straightens in the middle into 1.5- to 2.7-× 1- to 1.7-nm (Guo et al., 1995; Verschoor, Tivol, and Mannella, 2001). These dimensions are likely to represent those of a fully open VDAC pore. According to these models, the channel would insert up to 19 β-strands and one α-helix into the lipid phase. A consensus has yet to be achieved about the role and location of this positively charged α-helix. The 3D models locate this N-terminal helix either next to the barrel's wall or outside in the cytoplasm. Electrophysiological studies are used to examine pore structure, where conductance is a reflection of pore diameter. Current recordings of VDAC lacking the α-helix show lower conductances, consistent with a smaller pore and a placement of the helix within the wall (De Pinto et al., 2008; Runke et al., 2006). In a recent opinion article, Colombini suggested that these and other discrepancies regarding 3D and "functionally-derived" structures are inherent to the refolding and reconstitution procedures used to obtain the 3D structures. Interestingly, and in contrast to atomic force microscopy (AFM) images (Hoogenboom et al., 2007), the 3D structures lack superficial protrusions. These protrusions could be formed by the α-helix (De Pinto et al., 2008), by a β-strand (Colombini, 2009), or possibly by a putative accessory protein(s).

Despite the many structure-function studies of VDAC, its gating in the outer membrane of isolated mitochondria has never been reconstituted. We hypothesize that a putative modulator that controls VDAC gating in isolated mitochondria could possibly be lost in reconstituted systems. Indeed, the list of potential VDAC modulators amounts to 55 cytoplasmic factors, and counting (Roman et al., 2006; Table 3.1). For example, nicotinamide adenine dinucleotide (NADH) and Mg^{2+}-NAD(P)H decrease the ADP flux through VDAC (Zizi et al., 1994). Moreover, an intermembrane space (IMS) protein (Holden and Colombini, 1993) and tBid (a cleaved form of Bid; Rostovtseva et al., 2004) have been reported as putative endogenous inhibitors of VDAC. Finally, VDAC can be phosphorylated with antagonistic effects by hexokinase I and hexokinase II, whereas the first induces closing, the latter induces opening of VDAC (Azoulay-Zohar et al., 2004; Vander Heiden et al., 2000).

Is there a relationship between VDAC conductance and modulation of metabolite flux in intact cells? The Jonas lab at Yale directly surveyed currents on the outer membrane of mitochondria within neurons using an elegant variation of patch-clamp techniques. Once inserted into intact neurons, a patch pipette was protruded out of a microinjection needle to patch mitochondria. They showed that a large outer membrane channel activity with a voltage dependence typical of VDAC can be activated in intact neurons by a train of synaptic action potentials (Jonas, Buchanan, and Kaczmarek, 1999) or by a full length forms of Bcl-xL. In a later study, the group demonstrated that large outer membrane channel activity can be inhibited by NADH or activated by hypoxia (Jonas et al., 2004). Interestingly, an ethanol-induced accumulation of NADH (reduced NAD^+) inhibited ADP flux across the outer membrane of hepatocytes (Lemasters and Holmuhamedov, 2006). Also, one study reported that the molecular cut-off size and conductance of VDAC both vary with physiological Ca^{2+} oscillations in the cytosol (Bathori et al., 2006). Finally, Ca^{2+} pretreatment of permeabilized cells enhanced subsequent ATP uptake, implying a VDAC-mediated regulation of metabolism. All these studies show that the best studied mitochondrial channel VDAC has many faces that are just now beginning to be revealed.

3.2.2 Protein Import Channels

3.2.2a Translocase Outer Membrane

The theory of symbiotic evolution of mitochondria states that almost all of this organelle's genes were transferred to the nucleus. For example, the human mitochondria retained the nucleic acid recipe to only 13 of the ~1500 proteins found within the mitochondrial compartments (Andreoli et al., 2004). This symbiotic evolution must then

TABLE 3.1 Biophysical Properties of Mitochondrial Channels. Otherwise Stated, All Parameters Were Measured in Mitochondrial Membrane Preparations or Mitoplasts in 0.15 M KCl.

	Single channel peak conductance (nS)	Ion selectivity (P_K^+/ P_{Cl}^-) or (cation/anion)	Voltage dependence	Molecular identity of the pore	Modulators
Ceramide (Siskind et al., 2000; 2002; 2008)	1 – 200	unknown	unknown	C2 and C16 ceramide	Bcl-2, Bcl-$_{XL}$
MAC (Pavlov et al., 2001; Martinez-Caballero et al., 2004; 2009; Hetz et al., 2005; Peixoto et al., 2009;)	1.5 – 5	3.1	absent	Bax and/or BaK	positive: tBid, apoptosis inducers; negative: Bcl-2, dibucaine, trifluoperazine, propranolol, Bcis, iMACs
SAM (Kutik et al., 2008)	0.3 (0.25 M KCl in lipid bylayers)	4		Sam50	Beta-signal peptide
TOM (Pavlov et al., 2001)	1 (0.5 each pore)	3.6	Open between ± 60 mV	Tom40	presequences
VDAC (Pavlov et al., 2005)	0.7 – 0.9	0.2 fully open 2.4 – 10 half open	Open between ± 20 mV	VDAC1 VDAC2 VDAC3	negative: NADH, DIDS, tubulin, G3839, tBid; interactions: hexokinase, Bak, Bcl-xL, Bcl-2
AAA (Antonenko et al., 1991; 1994)	0.045	0.3 – 0.5	Straight I x V between ± 50 mV in alkaline pH	unknown	positive: matrix alkalinization; negative: Mg^{2+}
ACA (Antonenko et al., 1991; 1994)	0.015	10	Straight I x V between ± 50 mV in alkaline pH	unknown	positive: matrix alkalinization; negative: Mg^{2+}
mitoKv1.3 (Szabo et al., 2005; Bednarczyk et al., 2010)	0.025 – 0.1	cationic	Open between ± 60 mV	pmKv1.3?	Positive: negative: margatoxin, ShK

IMAC (Beavis, 1992; Beavis and Garlid, 1988; Beavis and Powers, 1989; Beavis and Vercesi, 1992)	No electrophysiology data available			unknown	negative: DIDS, quinine, matrix Mg^{2+} and H^+
mCS (Sorgato, Keller and Stuhmer, 1987; Klitsch and Siemen, 1991; Borecky, Jezek and Siemen, 1997; Antonenko et al., 1991)	0.1	0.22		unknown	negative: purine nucleotides, amiodarone, propanolol, nifedipine
MCU (Baughman et al., 2011; DeStefani et al., 2011; Kirichok, Krapivinsky and Clapham, 2004; Michels et al., 2009)	0.002 – 0.007	High Ca^{2+} selectivity		MCU	positive: spermine; negative: ruthenium red, Ru360
mitoBKCa^{2+} (Siemen et al., 1999; Xu et al., 2002; O'Rourke, 2007; Skalska et al., 2009; Cheng et al., 2008; Sakamoto et al., 2008; Heinen et al., 2007a; 2007b; Gu et al., 2007; Ohya et al., 2005)	~0.3		Straight I x V between ± 60 mV	pmBKCa^{2+}?	positive: Ca^{2+}, diCl-DHAA, NS1619, 17-beta-estradiol, hypoxia; negative: 5-hydroxydecanoate, charybdotoxin, iberiotoxin, paxillin
mitoKATP (Inoue et al., 1991)	0.015 – 0.08			pmKATP?	positive: Diazoxide; negative: glibenclamide, 5-hydroxydecanoate, ATP
mRyR (Ryu et al., 2010; 2011)	0.22 (150 mM CsCl)	$P_{Ca}^{2+}/P_K^+ = 6$	Straight I x V between ± 40 mV (in CsCl)	RyR?	ryanodine

(Continued)

TABLE 3.1 (*Continued*)

	Single channel peak conductance (nS)	Ion selectivity (P_K^+/ P_{Cl}^-) or (cation/anion)	Voltage dependence	Molecular identity of the pore	Modulators
Mrs2 (Schindl et al., 2007)	0.15 (105 mM $MgCl_2$/150 mM Na^+ gluconate	Impermeable to Ca^{2+}, Mn^{2+}, and Co^{2+}	Closed at positive voltages		
mtCLIC (Fernandez-Salas et al., 2002; Suh et al., 2004)	No direct electrophysiology data available			p64?	unknown
PTP (Bernardi et al., 2006; Szabo and Zoratti, 1991; 1992; Zoratti and Szabo, 1995; Bonora et al., 2013; Giorgio et al., 2013; Alavian et al., 2014; Pavlov et al., 2005; Seidlmayer et al., 2012a; 2012b)	1 – 1.5	7	Once active, open between ± 50 mV	ATP synthase subunit c PolyP/Ca^{2+}/ PHB?To be determined	positive: atractyloside, ROS, Ca^{2+}, Zn^{2+}; negative: Mg^{2+}, ADP, Pi, bongkrekate, NIM811, cyclosporine A
TIM23 (Lohret, Jensen and Kinnally, 1997; Martinez-Caballero et al., 2007; Peixoto et al., 2007)	1 (0.5 each pore)	5.7	Open between ± 60 mV	Tim23 Tim17	presequences
TIM22 (Peixoto et al., 2007)	1.1 (~0.5 each pore, 0.06 smaller states)	7.2	Open between ± 40 mV in presence of signal peptide	Tim22	positive: P2 (fragment of a signal peptide)

have relied on protein import. Properly tagged proteins en route to mitochondria must pass through the translocase outer member (TOM) complex, the main gate located in the outer membrane. The tag can be a cysteine motif (Herrmann and Hell, 2005), a lysine-rich presequence (Roise and Schatz, 1988) or an internal targeting element (Brix et al., 1999). All these targeting sequences are recognized by the receptor proteins Tom70 and Tom20, which then deliver the precursor protein to Tom22. Finally, the precursor is handed off to the translocation pore of the TOM complex (Abe et al., 2000; Meisinger et al., 2001; Saitoh et al., 2007; van Wilpe et al., 1999; Young, Hoogenraad, and Hartl, 2003). Tom40 forms the protein-conducting pore (Hill et al., 1998; Kunkele et al., 1998), whereas the small accessory proteins Tom5, Tom6, and Tom7 regulate stability of the complex (Schmitt et al., 2005; Sherman, Go, and Nargang, 2005). From this point, the general import pathway diverges into a few different routes to which access is controlled by TIM23 and TIM22, protein import channels of the inner membrane that are discussed in this chapter.

Electrophysiology and imaging studies determined that TOM has two or three pores that likely flicker as a protein is passing through. Conductance changes in the presence of dextran nanobeads indicated that this channel indeed has a double barrel pore that is 1.5 to 2.2 nanometers in diameter (Hill et al., 1998; Peixoto, Ryu, and Kinnally, 2010). It is also speculated that a slight cation selectivity (see Table 3.1) observed in patch-clamp experiments could be important for transport of positively charged presequences. However, a more recent analysis of TOM conductance levels reports up to six different states and suggests the double pore structure may be an oversimplified view of the channel (Poynor, Eckert, and Nussberger, 2008).

Although TOM is considered the general import pathway to mitochondria, its participation in transport of some outer membrane proteins is still a matter of debate. For example, the insertion of the peripheral benzodiazepine receptor seems to rely exclusively on Tom70 (Otera et al., 2007). A similar role of Tom70 was suggested for the docking of peripheral outer membrane proteins Mfb1 and Mcl-1 to the mitochondrial surface (Becker et al., 2008; Chou Lee, and Yang-Yen, 2006; Kondo-Okamoto, Shaw, and Okamoto, 2008). Controversy also lies in the possible involvement of the TOM complex in docking of pro-apoptotic Bax (Bellot et al., 2007; Cartron et al., 2008; Colin et al., 2009; Ott et al., 2007; Ross, Rudel, and Kozjak-Pavlovic, 2009; Sanjuan Szklarz et al., 2007). Finally, the assembly of the peripheral outer membrane protein Sam37 into the SAM complex was found to occur independently of any TOM component (Habib et al., 2005). Hence, the processes underlying the insertion of proteins containing membrane-spanning α-helices and the association of peripherally attached proteins with the outer membrane need further investigation.

3.2.2b Sorting and Assembly Machinery

Sorting and assembly machinery (SAM) incorporates beta-barrel proteins like VDAC into the outer membrane, a process that has been conserved from yeast to mammals (Kozjak-Pavlovic et al., 2007). An interesting feature of the complex is its varying molecular composition, which may include members of the TOM complex (Thornton et al., 2010). For the most part, the main composition of SAM includes an integral pore-forming protein, Sam50, and two smaller subunits, Sam35 and Sam37 (Becker et al., 2009; Endo and Yamano, 2010). The electrophysiological properties of SAM have yet to be described in native systems. However, both the recombinant protein and the purified complex can reconstitute a channel activity in planar lipid bilayers (Kutik et al., 2008; see Table 3.1). Intriguingly, the peak conductance (measured in 250 mM KCl) of yeast mitochondrial preparations was 160 pS and increased to 320 pS and, occasionally, 640 pS in the presence of a putative signal peptide. Assuming a depth of 5.5 nm across the 7-nm thick outer membrane (Perkins et al., 1997), the maximum (and infrequent) conductance of 640 pS would correspond to a pore around 1.4 nanometers-wide. It remains to be determined if this diameter (which is approximately half that of TOM's pore) can accommodate protein import.

3.2.2c Translocase Inner Membrane 23

The mitochondrial matrix is the final destination for most of the nuclear-encoded precursor proteins imported through TOM. These proteins carry a presequence tag that clears their way through translocase inner membrane 23 (TIM23), one of the protein translocases across the inner membrane. The TIM23 complex is an intricate machine that could contain up to 10 subunits, depending on the final destination of the precursor protein en route (Chacinska et al., 2009). The core components are the proteins Tim17, Tim23 (pore), and Tim50 (receptor), which coordinate a membrane-potential-dependent translocation *across* the inner membrane. An additional subunit, Tim21, regulates the lateral release of proteins *into* the inner membrane. In absence of Tim21, the core rather interacts with the presequence translocase-associated motor (PAM) complex to allow translocation of proteins into the matrix. PAM is an ATP-consuming motor formed by Pam16-18, Pam44, Ssc1 (yeast homologue is Hsp70), and Mge1. Although the mechanism of lateral release awaits further investigation, translocation into the matrix is better understood. Through ATP cleavage cycles, Mge1 assists Ssc1 to grab and pull the precursor protein into the matrix. Once translocation is complete, a mitochondrial peptidase cuts off the presequence tag and matrix chaperones assist in protein refolding (Chacinska et al., 2005; Mokranjac et al., 2005; Popov-Celeketic et al., 2008; van der Laan et al., 2005).

Patch-clamp studies in mitoplasts (mitochondria stripped of the outer membrane) isolated from yeast and rat liver identified a channel whose high conductance (see Table 3.1) was activated by signal peptides (Kinnally et al., 1991, 1996; Kinnally, Antonenko, and Zorov, 1992). Recombinant Tim23 protein showed a similar activity in bilayer systems, but signal peptides were not needed for activation. Interestingly, reconstitution of isolated inner membranes into liposomes had a similar effect, suggesting a regulating factor present in native preparations had been lost after reconstitution. Site-specific mutations and antibody inhibition later confirmed that the previously identified large conductances indeed belonged to Tim23 (Lohret, Jensen, and Kinnally, 1997). This translocase channel activity was essentially identical to that of TOM and reflected the cooperative and presequence-sensitive gating of twin pores. However, recombinant Tim23 had a smaller conductance than the fully assembled TIM23 complex, suggesting additional components molded the pore. In agreement, mutation or knockdown of Tim17 modified the conductance. In the absence of Tim17, polymer exclusion experiments showed that the twin pores became singlet with aberrant or absent gating properties, signal peptide sensitivity, and protein import activity. This seminal work demonstrated that, rather than an accessory protein, Tim17 was a core component of the translocation pore (Martinez-Caballero et al., 2007).

Although TIM23 is classically known to transport proteins into the matrix, studies revealed its ability to insert proteins into the inner membrane by two different mechanisms (Neupert and Herrmann, 2007). One group of inner membrane proteins are transported into mitochondria by a conservative sorting pathway that involves N-terminal presequence-driven translocation into the matrix via TIM23 and subsequent insertion into the inner membrane from the matrix side. Another group uses a stop-transfer mechanism for mitochondrial import and sorting. A presequence signals translocation into the matrix but a hydrophobic stop-transfer sequence stalls translocation, leading to lateral release of the precursor, much like in the endoplasmic reticulum. A third group of inner membrane proteins that use the TIM23 complex have an internal targeting signal that is composed of a transmembrane domain directly followed by a presequence-like element. Whether switching between these different transport modes involves conformational changes of the complex or if assembly and disassembly processes are needed is currently under investigation (Chacinska et al., 2005; Mokranjac et al., 2005; van der Laan et al., 2005).

Most of the structural and functional features currently attributed to TIM23 were obtained from yeast models, although electrophysiological recordings were also performed in mammalian mitoplasts. In a pioneering effort, Ahting et al. (2009) generated the first mouse knockout model of the Tim23 protein. Although homozygous Tim23$^{-/-}$ mice were not viable, heterozygous F1 mutants showed a 50% reduction of Tim23 protein in Western blots, a neurological phenotype and a markedly reduced life span. These studies illustrate the critical role of the protein import machinery for maintaining mitochondrial function in mammals. Electrophysiological, biochemical, and morphological characterization of mitochondria from this knockout model may reveal novel intricacies of protein import, mitochondrial dynamics, and other essential functions.

3.2.2d TIM22

Unlike matrix proteins, inner membrane proteins like the phosphate carrier are inserted by the TIM22 translocase in an energy-independent manner (for a review, see Neupert and Herrmann, 2007). The 300 kDa protein import complex consists of three integral membrane proteins Tim22, Tim54, and Tim18 and three soluble proteins, Tim9, Tim10, and Tim12 in the intermembrane space. Tim22 is the core of the complex and is related in amino acid sequence to Tim17 and Tim23 (Sirrenberg et al., 1996). Tim54 seems to function as a scaffold protein that would hold the complex together and is essential for protein import (Kerscher et al., 1997; Kovermann et al., 2002). Tim18 is a distant homologue of subunit 4 of the succinate dehydrogenase, Sdh4. Interestingly, another complex II subunit, Sdh3, has been found to be a part of yeast Tim22 complex via interaction with Tim18 (Gebert et al., 2011). Knock down of Sdh3 in yeast prevents Tim18 insertion into the complex. The precise functions of Tim54 and Tim18 are not known, and to date, their presence in mammalian mitochondria has yet to be confirmed. A complex consisting of the small Tim subunits Tim9, Tim10, and Tim12 is permanently tethered to the IMS side of the TIM22 complex. Tim54 might contribute to the binding of this Tim9-10-12 complex because its association with Tim22 is destabilized in Tim54 deletion mutants (Kovermann et al., 2002). Tim12 is an essential protein, and the Tim9-10-12 complex may play a vital role in substrate recognition by the TIM22 complex.

Although the channel activity of TIM23 in both artificial and native systems was identified almost concurrently, the channel activity of TIM22 in native mitochondrial membranes was only recently described (Peixoto et al., 2007). TIM22 is a quiescent channel that requires signal sequence interaction for its activation in both reconstituted and native mitochondrial inner membranes (see Table 3.1). Once open, TIM22 channel has activity remarkably similar to the TIM23 and TOM complexes, except for its lower opening probability and the presence of smaller subcondanctance levels whose functional relevance have yet to be determined.

3.2.3 Mitochondrial Ca^{2+} Channels

Calcium is the most important neuronal second messenger because this divalent cation translates action potentials into neurotransmitter release and synaptic plasticity. Not surprisingly, neurons have developed extensive and intricate Ca^{2+} signaling pathways. Mitochondria are situated in the cross roads of these pathways, contributing to the spatiotemporal tuning of the Ca^{2+} concentration in the cytosol. Through uptake and release mechanisms, mitochondria control how much of this ion enters the cell, its concentration in cytoplasmic microdomains, the frequency of cytosolic spikes, and the rate of propagation of a calcium signal. In turn, mitochondria use calcium transport to modulate the rate of ATP synthesis through activation of Krebs cycle (tricarboxylic acid) dehydrogenases, regulation of the supply of oxidizable substrates, and regulation of the enzymatic activity of the ATP synthase (Balaban, 2009; Beutner et al., 2005; Cali, Ottolini, and Brini, 2012; Denton, 2009; Mammucari et al., 2011). Therefore, Ca^{2+} channels allow mitochondria to sense and coordinate changes in cytosolic energy needs. In pathological conditions, however, mitochondrial Ca^{2+} overload contributes to the generation of reactive oxygen species (Brookes et al., 2004; Feissner et al., 2009), opening of permeability transition pore (PTP), which will be discussed in Part II (Bernardi et al., 2006; Halestrap, 2009), and cell death (Hajnoczky et al., 2006). Moreover, dysregulation of mitochondrial Ca^{2+} handling is associated with contractile dysfunction during heart failure (Maack and O'Rourke, 2007).

It has been known for more than five decades that mitochondria transport Ca^{2+} into the matrix (Deluca and Engstrom, 1961; Vasington and Murphy, 1962). That was long before the demonstration in 1983 that the endoplasmic reticulum (ER) is the main Ca^{2+} store mobilized during cell activation (Streb et al., 1983). In the case of the ER, discovery of the calcium channel, the IP3 receptor, was almost concomitant (Furuichi et al., 1989; Mikoshiba, Huchet, and Changeux, 1979). However, it was only recently that the identity of the mitochondrial calcium uniporter (MCU) was revealed (Baughman et al., 2011; De Stefani et al., 2011).

Equally elusive was the identification of the exchanger or antiport proteins that mediate efflux of Ca^{2+} from mitochondria (reviewed in Drago, Pizzo, and Pozzan, 2011; Gunter and Sheu, 2009), which are not canonical ion channels and will therefore not be discussed here.

Electrophysiological and biochemical approaches have identified many calcium channel activities associated with the functions described previously and they seem to belong to at least six seemingly different entities including MCU, MiCa, mCa1, mCa2, RaM, and the mitochondrial ryanodine receptor mRyR. However, the molecular basis of only MCU and mRyR has been identified and much debate exists around the idea that the remaining four activities represent different kinetic states of MCU. The biophysical properties summarized in Table 3.1 combine these possible kinetic states. These channels and the mentioned debate are thoroughly addressed in Ryu et al. (2010). The hallmark common feature among the four channels is their sensitivity to nM Ru360, although mCa2 only partially closes at ~10 μM (Kirichok, Krapivinsky, and Clapham, 2004; Michels et al., 2009). MCU and MiCa also share a similar permeability to some divalent cations ($Ca^{2+} \approx Sr^{2+} >> Mn^{2+} \approx Ba^{2+}$) and impermeability to Mg^{2+}. However, MiCa has a much higher Ca^{+2} flux rate than previously reported for MCU (Gunter et al., 2004; Kirichok et al., 2004). Like MiCa, mCa1 and mCa2 have inward rectification currents for Ca^{+2} at negative potentials. Furthermore, mCa1 has a smaller overall open probability and ~10-fold higher conductance than mCa2 (Michels et al., 2009). Finally, RaM provides a rapid kinetic mode of mitochondrial Ca^{2+} uptake that, like the mRyR, is both activated and inhibited by Ca^{+2} (Ryu et al., 2010, 2011). RaM is inactive at mM Ca^{+2} concentrations and less sensitive to Ru360 than MCU is (Gunter et al., 2004). At this moment, it is not known whether RaM is mediated by the same molecular complex as that of MCU. Although it is possible that all these channel activities represent different molecular entities, differences could also be the result of experimental conditions and cell types used.

3.2.4 Mrs2 – Mg^{2+} Channel

The membrane potential is a driving force that allows mitochondria to store not only Ca^{2+}, but also Mg^{2+} in amounts double that of the cytoplasm (Kubota et al., 2005). Although significant, this concentration gradient does not span orders of magnitude (unlike that of Ca^{2+}), which led many to initially dismiss the possible existence of Mg^{2+} channels. However, patch-clamp analysis on inner membrane vesicles suggested yeast Mrs2 protein forms a Mg^{2+} selective channel of high conductance (155 pS in 150 mM sodium gluconate) that is inhibited by increased Mg^{2+} on the matrix side of the channel (Schindl et al., 2007). Mutational changes in key domains confirmed the electrophysiological properties and supported the mitochondrial localization of yeast Mrs2. Although no patch-clamp data is available from mammalian mitochondria, the ortholog conserves a GMN domain that has been shown to be essential for Mg^{2+} selectivity of the pentameric channel (Palombo, Daley, and Rapp, 2013). Importantly, expression of the human ortholog rescued Mrs2 activity in yeast knockouts (Zsurka Gregan, and Schweyen, 2001). Furthermore, RNA silencing in PC12 cells suggested that the channel was indeed the main route for Mg^{2+} import into the mitochondrial matrix (Shindo et al., 2011). The study did not report on the physiological effects of knocking down Mrs2.

3.2.5 Mitochondrial K^+ Channels

Electrogenic mechanisms for K^+ entry into mitochondria have been reported for 40 years and are thought to play a key role in the regulation of matrix volume (Bernardi, 1999; Diwan and Tedeschi, 1975; Drahota et al., 1969). A number of channel activities associated with K^+ flux through the inner membrane were reported, but again the molecular identities remain elusive (Szewczyk, Jarmuszkiewicz, and Kunz, 2009). The first strong evidence of a K^+ channel activity was obtained from patch-clamp studies on fused giant mitoplasts from rat liver (Inoue et al., 1991). The properties resembled those of the plasma membrane cardiac KATP channel, except that the blocker HMR1098 had a lower effect on this putative mitochondrial counterpart (Sato et al., 2000). Another plasma membrane channel, BK^+_{Ca2}, was found in mitoplast preparations, but this time no significant pharmacological differences were found (Siemen et al., 1999; Xu et al., 2002), although these channels were found in highly pure mitochondrial fractions (Douglas et al., 2006). Finally, a margatoxin-sensitive voltage-gated Kv1.3 channel was identified in T lymphocyte mitoplasts and immunodetected in mitochondria. Like other mitochondrial potassium channels, Kv1.3 is similar in mitochondrial and plasma membranes. Importantly, the mitochondrial membranes used in these studies lacked plasma membrane and endoplasmic reticulum markers (Szabo et al., 2005). A recent study from the same group has suggested that Kv1.3 interacts with pro-apoptotic Bax in a mechanism that precludes K^+ flux and induces a ROS-mediated detachment of cytochrome c from the inner membrane (Szabo et al., 2008). These and other emerging potassium channels including TASK3, SKCa, and IKCa, with possible mitochondrial localization are extensively discussed elsewhere (Szabo and Zoratti, 2014).

3.2.6 Mitochondrial Centum Pico-Siemens

Another elusive channel activity happens to be the first ever identified by directly patch clamping mitoplasts, mitochondrial Centum pico-Siemens (mCS) (Sorgato, Keller, and Stuhmer, 1987). Subsequent recordings have been confirmed in several mammalian cells and tissues, such as liver, heart, brain, pancreas, adrenal glands, brown adipose tissue, as well as in mitoplasts from human tissue culture cells (Inoue et al., 1991; Kinnally, Campo, and Tedeschi, 1989; Klitsch and Siemen, 1991; Moran et al., 1990; Murphy et al., 1998; Petronilli, Szabo, and Zoratti, 1989). Interestingly, mCS activity has not yet been found in yeast. The channel is slightly anion-selective ($P^-_{Cl}/P^+_K = 4.5$) and has a strong voltage dependence. At least two closed and two open states of similar conductance are commonly seen: a 50% subconductance level (Campo, Kinnally, and Tedeschi, 1992), as well as infrequent channel openings one third and two thirds of the most frequent conductance level of 107 pS (Klitsch and Siemen, 1991). Additional peak conductances >140 pS in the presence of amiodarone have also been reported (Antonenko et al., 1991). All these conductances had the same voltage dependence, and at least the 50 pS substate showed the same anionic selectivity. mCS was largely unaffected by variations of pH from 6 to 9, on the matrix or cytoplasmic side of the membrane (Sorgato et al., 1987). Although Sorgato et al.recorded the channel activity without any specific activation procedure, we found the channel was normally quiescent, but the detection frequency was increased by chelating divalent cations (Kinnally et al., 1991).

Pharmacology studies have enabled the elimination of a variety of candidate proteins that might underlie mCS activity. Adenine nucleotide transporter (ANT) was discarded because mCS was not affected by carboxyatractylate and brongkrekate (see discussion in this chapter). Similarly, mCS was not associated with the F_O region of the ATP-synthase because its activity was insensitive to oligomycin and dicyclohexylcarbodiimide (DCCD) (Sorgato et al., 1989). Klitsch and Siemen reported that mCS was inhibited by submillimolar levels of di- and trinucleoside phosphates, as well as guanosine monophosphate (GMP), when added in the patch pipette (Klitsch and Siemen, 1991). These results excluded the candidate uncoupling protein, thermogenin, which is known to be insensitive to GMP. Furthermore, Inoue et al. (Inoue et al., 1991) reported that mCS was not affected when millimolar Mg^{2+} or ATP and micromolar ADP were applied on the matrix side of excised patches. Finally, the mitochondrial inner membrane anion channel (IMAC) is a channel inferred from light scattering (matrix swelling) studies, which had broad anion selectivity and conducted mono-, di-, and trivalent anions (Beavis, 1992). This channel is believed to be an important component of the volume homeostatic mechanism of mitochondria and is maintained closed or inactive by matrix Mg^{2+} and H^+. The lack of effect of Mg^{2+} (Szabo, Bernardi, and Zoratti, 1992), DIDS (4,4′-diisothiocyanostilbene-2,2′-disulfonic acid), quinine, or DCCD (Sorgato et al., 1989) on mCS activity, which are all well-known blockers of IMAC (Beavis and Garlid, 1988; Beavis and Powers, 1989; Beavis and Vercesi, 1992), eliminated this candidate from the list. Of note, mCS activity has also been recorded in mitoplasts lacking mitochondrial DNA (Murphy et al., 1998), which suggests the gene might be encoded in the nucleus. With the surge of more mitochondrial channels in the 80s and 90s, the arduous search for mCS' identity and function dissipated.

3.2.7 Alkaline-Induced Anion-Selective Activity and Alkaline-Induced Anion-Selective Activity

Two other pH-sensitive channel activities were reported in liver mitoplasts. Both of them displayed greater open probability upon alkalinization of the matrix side of the membrane and were activated by depletion of Mg^{2+} (Antonenko et al., 1991, 1994). They were referred to as alkaline-induced cation-selective activity (ACA, 15 pS) and alkaline-induced anion-selective activity (AAA, 45 pS), respectively. AAA was initially thought to correspond to IMAC because of their similar induction by alkaline pH or Mg^{2+} depletion. However, IMAC is ~100-fold more selective than AAA for Cl^- over glucouronate (Antonenko et al., 1994; Beavis and Vercesi, 1992). The conductance and voltage dependence of ACA were similar to those of the K^+ATP channel. However ACA was insensitive to 4- aminopyridine and glibenclamide plus

ATP. The selectivity and inhibition by Mg^{2+} suggested that ACA channel activity could correspond to one of the cation uniporters implicated in volume homeostasis, whose existence was inferred from passive swelling experiments (Jung and Brierley, 1999; Kim He, and Lemasters, 2003).

3.2.8 Chloride Intracellular Channels

Chloride intracellular channels (CLICs) represent a new class of intracellular anion channels that have been identified by their homology to the p64 protein. A mitochondrial homolog, mtCLIC, has been identified from differential display analysis of differentiating mouse keratinocytes from $p53^{+/+}$ and $p53^{-/-}$ mice. mtCLIC colocalized with cytochrome oxidase in keratinocyte mitochondria but was also detected in the cytoplasmic compartment. This p53-regulated putative channel has been associated with apoptosis (Fernandez-Salas et al., 2002) and may exist as either a soluble or a transmembrane form that may translocate to the nucleus in response to cell stress (Suh et al., 2004).

3.2.9 Alternative Ion Transport Pathways

It should be noted that although a number of ion channels have been detected in the mitochondrial inner membrane it is impossible to experimentally confirm if and how these channels behave in the intact cells. One of the defining features of the ion channel is its ability to transport ions at high rates, which are several orders of magnitude higher than rates of membrane transporters. In this respect, one of the interesting considerations is the fact that the overall kinetics of mitochondrial ion transport are expected to be rather slow. The major driving force for the ion flux across mitochondrial inner membranes is the electrochemical potential generated by the respiratory chain.

As postulated by Mitchell (1961) during the introduction of his chemiosmotic theory, the presence of the mitochondrial membrane potential suggests that under physiological conditions mitochondria are not expected to have open ion channels. Rates of ion flux measured in the experiments with intact mitochondria are up to 100 times slower than what can be expected from functional ion channels (Kane and Pavlov, 2013). Recent progress in identification of the proteins involved in formation of ion channels further highlights the importance of the understanding of relationships between mitochondrial channel activities and physiologically relevant ion fluxes in the intact mitochondria and cells. For example, genetic knock-out of the protein coding the mitochondrial calcium channel (MCU) leads to elimination of the channel activity (Chaudhuri et al., 2013) as well as reduction of calcium uptake in the intact mitochondria. However, knocking out MCU does not lead to significant changes in mitochondrial and cellular function under physiologically relevant conditions (Pan et al., 2013). This observation has led to the hypothesis that alternative ion transporting pathways of critical significance exist but currently remain undetected (Pan et al., 2013). One possibility is that there exist endogenous compounds that can provide pathways for ion transport by slower nonchannel mechanisms (Blondin, 1975). Some of the possible candidates might include endogenous ionophores as ferutinine (Abramov et al., 2001) and polyhydorxybutyrate (PHB) (Elustondo et al., 2013).

3.3 MITOCHONDRIAL CHANNELS IN THE DYING CELL

Some time ago, a novel notion was proposed that mitochondria are "deathsomes" that compartmentalize a myriad of death-inducing factors. At the time, this proposal was quite shocking to many in the field. Some of these factors include bifunctional proteins like cytochrome c, which is an electron transport component in mitochondria and triggers caspase activation in the cytosol, and apoptosis-inducing factor (AIF), which plays a role in oxidation of NADH in mitochondria and DNA degradation in the nucleus. Mitochondrial permeabilization and release of such death factors is generally considered the rate-limiting step in apoptosis. It has become clear that mitochondrial channels play significant roles in both apoptotic and necrotic mechanisms of neuronal death (Duchen, 2012; Kinnally and Antonsson, 2007). In fact, activation of mitochondrial channels is believed to play a role in the commitment steps that lead to both apoptosis and necrosis. The mitochondrial channels with established roles in neuronal death include mitochondrial apoptosis-induced channel (MAC) and VDAC of the outer membrane and the mitochondrial permeability transition pore (mPTP) of the inner membrane. Studies that show association of mitochondrial channels with neurodegenerative diseases are summarized in Table 3.2.

3.3.1 Apoptosis

MAC plays a central role during apoptosis by permeabilizing mitochondria, that is, by providing an aqueous pathway for release of folded proteins, for example, cytochrome c from the IMS to the cytosol (Dejean et al., 2010). Although the exact composition is not yet established, MAC is likely composed of an oligomeric form of Bax or Bak. These proteins are Bcl-2 family members and are pivotal regulators of the apoptotic cascade. During apoptotic signaling, Bax incorporates into the mitochondrial outer membrane where Bak already resides, bound to VDAC2 (Gross, McDonnell, and Korsmeyer, 1999). Oligomerization of Bax or Bak leads to the formation of MAC, which is a large pore in the mitochondrial outer membrane. MAC can be detected as a high conductance channel by patch-clamping mitochondria isolated from apoptotic cells (Pavlov et al., 2001). The MAC pore is sufficiently wide to provide a pathway for the escape of cytochrome c into the cytosol, which leads to the execution of the apoptotic pathway. Expression of either Bax or Bak is requisite for MAC formation and cytochrome c release. Interestingly, mitochondrial fragmentation and neuronal cell death have been observed in response to the MAC agonist ABT-737 (Dohm et al., 2006; Jeon et al., 2012; Young et al., 2010).

Oligomerized forms of VDAC have been proposed to form an alternative pathway for cytochrome c release during apoptosis (Keinan et al., 2013). However, MAC activity has been detected in VDAC knockout cells, suggesting that VDAC is not essential for MAC formation and cytochrome c release (Pavlov et al., 2001). Furthermore, others have shown apoptosis is not suppressed when all three isoforms of VDAC are lacking (Baines et al., 2007).

In a seminal novel study, lipids in lieu of proteins were found to form enormous channels, which might play an important role in regulation of the cellular apoptosis

TABLE 3.2 Possible involvement of Mitochondrial Channels in Neurodegenerative Diseases.

	ALS	Alzheimer's	Parkinson's	other CNS dysfunctions
MAC	(Kostic et al., 1997; Vukosavic et al., 1999; Gould et al., 2006; Reyes et al., 2010)	(Biswas et al., 2007; Giovanni et al., 2000; Kudo et al., 2012; Culmsee et al., 2005; Zhu et al., 2004)	(Vila et al., 2001; Hassouna et al., 1996; Akao et al., 2002)	
VDAC	(Israelson et al., 2010)	(Manczak and Reddy, 2012; Cuadrado-Tejedor et al., 2011; Hur et al., 2012)	(Lu et al., 2013; Sun et al., 2012; Okatsu et al., 2012)	
PTP				
TOM	(Duchen, 2012)	(Devi et al., 2006)	(Bender et al., 2013; Li et al., 2010)	
TIM23		(Devi et al., 2006)		
mitoKATP		(Liu et al., 2010; Ma et al., 2009; Wang et al., 2012)	(Tai, McCrossan and Abbott, 2003; Xie et al., 2010; Zeng, wang and Chen, 2007)	
Mrs2			(Shindo et al., 2011; Hashimoto et al., 2008; Oyanagi et al., 2006; Yasui, Kihira and Ota, 1992)	(Kuramoto et al., 2011)

(Colombini, 2013). The sphingolipid ceramide is present in the outer membranes and is capable of mitochondrial permeabilization (Richter and Ghafourifar, 1999; Siskind, Kolesnick, and Colombini, 2002). In experiments using model membranes, ceramide formed large conductance ion channels (Siskind and Colombini, 2000). According to this model, Bcl-2 family proteins regulate apoptosis through modulation of ceramide channel activity (Perera et al., 2012).

3.3.2 Necrosis

Although necrosis can be induced by a variety of conditions, several stress pathways are thought to converge on mitochondria. In fact, the commitment step of necrotic cell death seems to be primarily initiated by activation of the channel called *mPTP*. The most direct pathological consequence of mPTP opening is mitochondrial depolarization, uncoupling of the oxidative phosphorylation process and ensuing disruption of ATP generation by the mitochondria. Furthermore, the high protein concentration within the matrix facilitates swelling of that space on opening of the mPTP. Ultimately, the unfolding of the cristae leads to rupture of the outer membrane and spilling of death factors into the cytosol. These events set the cell up for necrotic death (Kinnally et al., 2011).

Electrophysiological investigations established that mPTP is a weakly selective cation channel with a peak conductance state of 1 to 1.5 nS in 150 mM KCl. This voltage-dependent channel allows the passage of molecules up to molecular weight of ~1000 Da. mPTP also possesses a number of subconductance states and hence was originally dubbed the multiple conductance channel (MCC) (Kinnally et al., 1989, 1996). Similarly, this giant channel activity was also referred to as the mitochondrial megachannel (MMC) (Petronilli et al., 1989). Further regulatory and pharmacological characterization established that MCC, MMC, and mPTP channels represent the activity of the same entity (Szabo and Zoratti, 1991, 1992). In the intact cell, mPTP activation can be modulated by a large variety of natural as well as pharmacological compounds. Calcium and ROS are two of the most potent activators of mPTP that are directly linked to the cellular pathological states (Bernardi et al., 2006; Zoratti and Szabo, 1995).

The molecular composition of the mPTP has been the subject of extensive investigations but remains elusive. Structurally, the channel is envisioned as a multiprotein complex that likely spans both mitochondrial membranes. The current count of proteins and protein complexes (e.g., ATP synthase complex) putatively involved in the structure of mPTP is at 15 (Bonora et al., 2015) and rising. The puzzle of mPTP structure is further complicated by the fact that some of these components are themselves formed by multiprotein complexes. The most striking example is the ATP synthase, which is an assembly of over two dozen single proteins. The cumulative evidence for involvement of so many proteins has made it difficult to conclusively answer the compelling question of the precise molecular identity of the pore. The notable candidate protein members of mPTP pore include VDAC (Szabo and Zoratti, 1993), ANT (Brustovetsky et al., 2002), and a phosphate carrier (Halestrap, 2009). Interestingly, although all these proteins are capable of forming ion channels, knock-out studies show that none of them seems to be indispensable for the function of mPTP (Baines et al., 2007; Kokoszka et al., 2004; Kwong et al., 2014). Cyclophilin D, a non–pore-forming peripheral protein located in the inner membrane is the only currently undisputed regulator component of mPTP (reviewed in Javadov and Kuznetsov, 2013). Inhibition of cyclophilin D either by gene knock out or use of the well-known antagonist cyclosporine A alters mPTP sensitivity to Ca^{2+} in a number of cell and animal models (Javadov and Kuznetsov, 2013).

As mentioned previously, the most recent addition to the list of putative mPTP components is the ATP synthase. A significant portion of this complex is located within the mitochondrial inner membrane, where it is hypothesized that the pore of mPTP might be located (Bonora et al., 2013; Giorgio et al., 2013; Alavian et al., 2014). There are currently three models describing the relationship between the ATP synthase and the pore. One model postulates that opening of the mPTP requires dimerization of the ATP synthase. The ion-conducting pore is presumed to be located in the space between the two ATP synthase complexes (Bernardi, 2013). This model does not specify the putative molecular nature of this conducting pathway but suggests close involvement of some components of the ATP synthase.

A second model proposes that the c-subunits of the ATP synthase fill the essential role of ion transport for the mPTP. The c-subunit is an integral part of the ATP synthase and forms an oligomeric, barrel-like structure (c-ring structure) in the inner membrane (Baker et al., 2012). Purified c-subunit is capable of forming ion channels in model membranes and changes in expression levels of the c-subunit correlate with changes in mPTP activity in the intact cells. Thus, it is conceivable that the c-subunit oligomer provides the passageway for ions within the mPTP. We should note that these two models are not mutually exclusive and that, hypothetically, conformational changes caused by the dimerization could lead to the opening of the putative c-subunit pore.

A third model offers an alternative to the "proteo-centric" views of the mPTP pore. The model postulates an essential role of nonprotein components (Pavlov et al., 2005b), namely two polymers of inorganic polyphosphate (polyP) and PHB. A channel activity is elicited in mitochondrial lipid preparations and synthetic membranes, likely because of complex formation between PolyP, Ca^{2+}, and PHB. When purified from the mitochondria and reconstituted into model lipid membranes this complex demonstrates properties that match those of native mPTP (Pavlov et al., 2005b). These include size, selectivity and voltage-dependence. Involvement of polyP in the process of mPTP formation is supported by observations that mitochondria depleted of polyP either do not develop mPTP or have mPTP activity reduced to the levels seen in the presence of the cyclophilin D inhibitor, cyclosporine A (Abramov et al., 2007; Seidlmayer et al., 2012a, 2012b). Furthermore, recent studies performed on intact cells and isolated mitochondria demonstrate that the other putative component of mPTP, PHB is likely involved in regulation of mitochondrial ion transport, although its relationship with mPTP is not yet clear (Elustondo et al., 2013; Smithen et al., 2013). Although the polyP/Ca^{2+}/PHB complex is sufficient to form channel activity even in the absence of proteins (Das et al., 1997), it is likely that its formation and membrane incorporation in the live cells will require the presence of certain membrane proteins. For example, polyP/Ca^{2+}/PHB channels isolated from rat liver mitochondria also contained traces of the ATP synthase c-subunit (Pavlov et al., 2005b). However, the exact relationship between polyP, PHB, and protein components of PTP remains to be established.

Evidently, mitochondrial channels are intimately involved in the commitment step of cell death because they can effectuate release of the IMS protein cytochrome c into the cytosol. The major channels involved are the MAC and mPTP. Although MAC is in the outer and mPTP is in the inner membrane, opening of either channel results in cytochrome c release, committing the cell to die. Cytochrome c can directly exit to the cytosol through the water-filled pore of the open state of MAC. Opening of PTP, whose pore is much smaller, causes a massive swelling of the matrix space and the ensuing unfolding of the cristae ruptures the outer membrane effectively dumping cytochrome c into the cytosol. Most of the present evidence is consistent with oligomers of Bax or Bak underlying the molecular identity of MAC, although ceramide and VDAC may play a role. mPTP is a multiprotein assembly whose debated pore-forming candidates include the ANT, VDAC, the ATP synthase complex, the c-subunit oligomers, and the nonprotein polyP/Ca^{2+}/PHB complex.

3.4 MITOCHONDRIAL CHANNELS IN NEURODEGENERATIVE DISEASES

Premature death of neurons occurs in a cohort of neurodegenerative disorders that includes Huntington's, Parkinson's, Alzheimer's, amyotrophic lateral sclerosis (ALS), Friedrich's ataxia, hereditary spastic paraplegia, dominant optic atrophy, Charcot-Marie-Tooth type 2A, neuropathy ataxia, retinitis pigmentosa, and Leber's hereditary optic atrophy. It is believed that common mechanisms may underlie the pathogenesis of many of these neurodegenerative diseases. In fact, many mitochondrial myopathies lead to neurodegeneration, for example, mutation of AIF (Ghezzi et al., 2010). However, these common mechanisms are not yet well understood.

Mitochondrial dysfunction that impacts calcium handling, the electron transport chain, ATP synthesis and delivery, fusion, and fission seems to be the trigger of neuronal death in a variety of disease models (Neumann et al., 2006). As discussed previously, these mitochondrial functions/dysfunctions rely on transport across the inner and outer membranes, which raises the intriguing possibility that neurodegeneration and death may be caused by mitochondrial "channelopathies" (Table 3.2). This possibility is supported by the increasing number of studies that show interactions between mitochondrial ion channels and well-known markers of neurodegeneration. For example, the food channel VDAC seems to interact with Parkinson's markers alpha-synuclein (Lu et al., 2013) and parkin (Okatsu et al., 2012; Sun et al., 2012), Alzheimer's markers amyloid beta, phosphorylated tau, and gamma-secretase (Cuadrado-Tejedor et al., 2011; Hur et al., 2012; Manczak and Reddy, 2012), and the ALS marker SOD1 (Israelson et al., 2010). In each of these studies, modulations in VDAC's interactions with these markers were associated with mitochondrial dysfunction and neuronal death. In certain studies, VDAC has been linked with neuronal death because of its putative role in regulation of mPTP (Ghosh et al., 2007; Jonas, 2009; Perez Velazquez et al., 2000, 2003; Rao, Carlson, and Yan, 2013). However, understanding the physiological meaning of VDAC interactions during neurodegeneration will only be possible through discovery of specific inhibitors, and of course, identification of the molecular nature of the mPTP.

Calcium homeostasis is a major function of mitochondria in neurons, and there is ample evidence that intracellular calcium is deregulated, for example, in ALS (Appel et al., 2000, 2001; Grosskreutz, Van Den Bosch, and Keller, 2010; Ionov, 2007; Kawamata and Manfredi, 2010; von Lewinski and Keller, 2005). Mitochondrial calcium uptake increases oxidative phosphorylation under physiological conditions (Jouaville et al., 1999) through activation of pyruvate dehydrogenase and dehydrogenases of the Krebs cycle (Hansford, 1991). For that reason, calcium transfer from the cytosol to the mitochondria may act as a signal coordinating production of ATP with metabolic demand. Neuronal mitochondria regulate their axonal movement in response to intracellular calcium (Macaskill et al., 2009; Saotome et al., 2008; Wang and Schwarz, 2009; Yi, Weaver, and Hajnoczky, 2004). This mechanism coordinates retention of mitochondria in presynaptic terminals and postsynaptic dendritic spines, where calcium influx is dynamic and the ATP availability is critical.

But what happens during the onset of pathology? Recent studies in ALS mouse models (overexpressing a G93A mutant of superoxide dismutase 1 [G93A-SOD1]) found defects in neuronal mitochondrial transport that may be associated with abnormalities in calcium signaling (Magrane et al., 2012; Vande Velde et al., 2011). These studies and others argue that recycling of mitochondria in presynaptic terminals might be compromised by these transport defects. Interestingly, enhancement of mitochondrial calcium buffering capacity in the same model protected mitochondrial function and reduced motor neuron death but did not prevent muscle denervation (Parone et al., 2013). In this later study, calcium-buffering capacity in the spinal cord was protected after crossing the ALS mouse model G93A-SOD1 with a cyclophilin D knockout, which may link the pathological results to mPTP activation. As a cautionary note, spinal cord preparations include mitochondria from several cell types and intracellular locations. Although it is challenging to obtain sufficient amounts of mitochondria from specific sites, for example, the synaptic terminals of motor neurons, results on heterogeneous preparations could mask functional diversity of mitochondrial subpopulations. This could account for the persistent muscle denervation despite the global restoration of calcium buffering observed in heterogeneous preparations. Future studies of mitochondrial regulation of calcium signaling in the dying neuron may benefit from the recent identification of the MCU and its two regulatory subunits (MICU1 and MICU2) in the inner membrane (De Stefani et al., 2011; Perocchi et al., 2010; Plovanich et al., 2013).

Although no direct role has been demonstrated for Mrs2 and neurodegeneration, studies link Mg^{2+} deregulation with migraines (Boska et al., 2002) and demyelination (Kuramoto et al., 2011). Mrs2 may be involved in the pathophysiology of Parkinson's as evidenced by a deficiency in Mg^{2+} uptake in PC12 cells treated with the 1-methyl-4-phenylpyridinium ion MPP+ (Shindo et al., 2011). Interestingly, Parkinson's-like dopaminergic neuronal loss was observed in the rat brain after exposure to low Mg^{2+} intake (Hashimoto et al., 2008; Oyanagi et al., 2006). In agreement with these conclusions, a lower concentration of Mg^{2+} has been found in regions of the brain in patients with Parkinson's and Alzheimer's (Andrasi et al., 2000; Yasui et al., 1992a, 1992b), supporting a pathological role for disruption of Mg^{2+} homeostasis in mitochondria.

There is increasing interest in the role of mitochondrial ATP-dependent potassium channels (KATP) in neuronal degeneration, especially because of their established role in hypoxic preconditioning in the heart, which also seems to be conserved in the brain (Busija et al., 2008; Dirnagl and Meisel, 2008). Accordingly, activation of brain mitochondrial KATP channels with chemical agonists like diazoxide elicits preconditioning-associated protective effects, while inhibitors like glibenclamide and 5-hydroxydecanoate (5-HD) abrogate preconditioning (Adamczyk et al., 2010; Busija et al., 2008; Jonas, 2009; Nakagawa et al., 2005; Robin et al., 2011; Teshima et al., 2003; Virgili et al., 2013; Wang et al., 2011). Interestingly, studies on the modulation of mitochondrial KATP suggest these channels are involved in the pathophysiological mechanisms of other neurodegenerative diseases like Alzheimer's and Parkinson's (Liu et al., 2010; Ma et al., 2009; Tai, McCrossan, and Abbott, 2003; Wang et al., 2012; Xie et al., 2010; Zeng, Wang, and Chen, 2007), as well as with microglial activation (Rodriguez et al., 2013). Finally, a recent study shows that the metabolic dysfunctions observed in an animal model of Charcot-Marie-Tooth type 2A can be mimicked or reversed by agonists and antagonists of mitochondrial KATP, respectively. The study also suggests a dysfunction in two proposed components of these channels, the respiratory complexes II and V (Guillet et al., 2011).

The protein import channels have also been implicated in disease models of neurodegeneration. A well-known example in the field, the Mohr-Tranebjaerg syndrome (or deafness dystonia syndrome), is characterized by progressive hearing loss, dystonia, mental deficiency, and visual disability (Jin et al., 1996; Kokotas, Petersen, and Willems, 2007). The disease is caused by a mutation of the deafness/dystonia protein1 gene DDP1, or Timm8A, which is a component of the TIM23 complex. Transport of proteins into the matrix is the principal role for TIM23 and knocking out Tim23 is lethal in mice as well as in yeast (Neupert and Herrmann, 2007). Furthermore, another study suggests that abnormal accumulation of amyloid precursor proteins across mitochondrial import channels is a hallmark of Alzheimer's. Mitochondria-associated amyloid precursor proteins formed stable complexes with Tom40 and Tim23 (Devi et al., 2006) and compromised mitochondrial function. Finally, protein import deficiencies have been observed in ALS, Huntington's Parkinson's disease models (Bender et al., 2013; Li et al., 2010; Yano et al., 2014). In the later example, Tom40 protein levels were specifically reduced in Parkinson's patients and in transgenic mice overexpressing alpha-synuclein. Lentiviral expression of Tom40 in the brain of these transgenic mice ameliorated energy deficits as well as oxidative burden, suggesting that alterations in the protein transport machinery might contribute to mitochondrial impairment in alpha-synucleinopathies.

Although the mitochondrial channels described are involved in dysfunction that may lead to neurodegeneration, opening MAC or mPTP invokes a sentence of neuronal death. These channels control mitochondrial permeability changes that lead to release of apoptotic transducers into the cytosol. An overview of MAC and PTP structures and functions has been detailed previously. Here is a description of their relevance to neurodegeneration and the currently available small molecule inhibitors with potential therapeutic applications.

MAC involvement in neurodegenerative diseases was initially inferred by a previous study that showed overexpression of its main antagonistic regulator, Bcl-2, prolonged life of a mouse model of ALS (Kostic et al., 1997). A subsequent study showed that the same mouse models had intrinsically lower levels of Bcl-2 and increased levels of one of MAC's main component, Bax, and an agonistic regulator that sequesters Bcl-2, Bad (Vukosavic et al., 1999). Interestingly, studies of Bax deletion in this ALS model prolonged survival but did not prevent neuromuscular denervation (Gould et al., 2006). Noting that Bax and Bak are often partly redundant in triggering apoptosis as they are in MAC formation (Lindsten et al., 2000; Wei et al., 2001), others found the double deletion of Bax and Bak halts neuronal loss, prevents axonal degeneration, delays symptom onset, and extends survival (Reyes et al., 2010).

MAC's involvement in neurodegenerative diseases is further supported by several studies. In particular, one investigation found that Bax levels are increased in the substantia nigra pars compacta of mice treated with 1-Methyl-4-phenyl-1,2,3,6-tetrahydropyridine

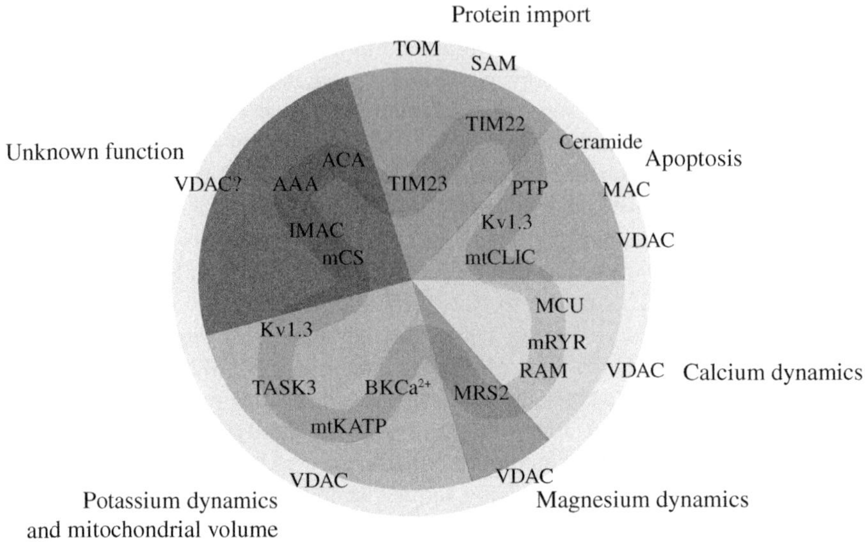

FIGURE 3.1 Mitochondrial ion channels by location and function. The pie chart represents the spatial and functional distribution of ion channels in the outer *(grey circle)* and inner *(shadowed cristae)* mitochondrial membranes. Detailed information and references are compiled in the text. (*See insert for color representation of the figure.*)

(MPTP) to induce Parkinson's disease symptoms (Hassouna et al., 1996; Vila et al., 2001). Bax-deficient mice are resistant to MPTP intoxication (Vila et al., 2001). These results elicited many in vitro studies, not discussed here, in which MPTP-protective treatments directly or indirectly involve MAC regulation. To cite one example, an anti-Parkinson's disease drug, rasagiline, enhanced expression of Bcl-2 in human dopaminergic SH-SY5Y cells (Akao et al., 2002). Finally, evidence of MAC's role in yet another neurodegenerative disease, Alzheimer's, arose from studies of overexpression of anti-apoptotic proteins, Bcl-w, or genetic ablation of a pro-apoptotic effector, Bim. Either approach significantly protects neurons against fibrillar Aβ-induced apoptosis in neuroblastoma cell lines and primary neuron culture (Biswas et al., 2007; Giovanni et al., 2000). In agreement, another study reports that fibrillar Aβ-induced neuronal death is dependent on Bax. Furthermore, treatment with oligomeric Aβ increases Bim levels but decreases Bcl-2 levels, leading to the activation of Bax and neuronal cell death both in vitro and in vivo. Inhibition of Bax activity either by Bax-inhibiting peptide or *bax* gene knockout significantly prevents oligomeric Aβ-induced neuronal cell death (Kudo et al., 2012).

MAC inhibition is particularly attractive and relatively unexplored. Two small Bax inhibitors prevented cytochrome c release, mitochondrial depolarization, and apoptosis in an animal model of ischemic brain injury (Hetz et al., 2005). Interestingly, Bax inhibitor-1 protects neurons from oxygen-glucose deprivation and enhances survival and neuronal differentiation of embryonic stem cells. Stronger inhibitors of MAC (iMACs) are now available (Peixoto et al., 2009, 2010), but their effects in neuronal cell death await further investigation. Another promising study approaches MAC inhibition by targeting its activator tBid (Becattini et al., 2004). Small 4-phenylsulfanyl-phenylamine derivatives are capable of occupying a deep hydrophobic crevice on the surface of Bid and inhibit tBid-induced Smac release, caspase 3 activation, and cell death. In neuronal cell cultures, these novel inhibitors attenuate cell death induced by glutamate, glucose-oxygen deprivation, or β-amyloid in a dose-dependent manner (Culmsee et al., 2005). An interesting approach to MAC inhibition could be the targeting of anti-apoptotic Bcl-2 in neurons. Interestingly, a novel mechanism of apoptosis in Alzheimer's disease has been proposed, where presenilins destroy Bcl-2 (Wang et al., 2005). This raises the possibility that presenilin dysfunction may cause neurons to be more susceptible to degeneration in Alzheimer's disease.

3.5 CONCLUSIONS

Mitochondrial (dys)function is a key determinant in neuronal cell death during acute stress as well as during neurodegeneration. The neuronal death sentence proclaimed by mitochondria is broadcasted through inner and outer membrane channels whose structure-function relationships were covered in this chapter. Recent advances have uncovered the identities of a few of these channels or at least offered promising candidates. However, many channels remain anonymous despite having well-known activities—the elusive mCS was the first channel ever recorded in mitoplasts. Other channels seem to have sprouted from their conventional plasma membrane and ER locations to broadcast from the outer and even the inner mitochondrial membranes! There is clearly a need to uncover the elusive mitochondrial channel identities and understand what plasma membrane and ER channels are doing in mitochondria. Equally pressing is the need look inside intact cells, especially neurons, and revisit some channels we think we know.

REFERENCES

Abe, Y., et al., 2000. Structural basis of presequence recognition by the mitochondrial protein import receptor Tom20. *Cell,* 100, pp.551–560.

Abramov, A. Y., et al., 2001. Influence of plant terpenoids on the permeability of mitochondria and lipid bilayers. *Biochim Biophys Acta,* 1512, pp.98–110.

Abramov, A. Y., et al., 2007. Targeted polyphosphatase expression alters mitochondrial metabolism and inhibits calcium-dependent cell death. *Proc Natl Acad Sci U S A,* 104, pp.18091–18096.

Adamczyk, S., et al., 2010. Sevoflurane pre- and post-conditioning protect the brain via the mitochondrial K ATP channel. *Br J Anaesth,* 104, pp.191–200.

Ahting, U., et al., 2009. Neurological phenotype and reduced lifespan in heterozygous Tim23 knockout mice, the first mouse model of defective mitochondrial import. *Biochim Biophys Acta,* 1787, pp.371–376.

Akao, Y., et al., 2002. An anti-Parkinson's disease drug, N-propargyl-1(R)-aminoindan (rasagiline), enhances expression of anti-apoptotic bcl-2 in human dopaminergic SH-SY5Y cells. *Neurosci Lett,* 326, pp.105–108.

Alavian K. N., 2014. An uncoupling channel within the c-subunit ring of the F1FO ATP synthase is the mitochondrial permeability transition pore. *Proc Natl Acad Sci U S A,* 111, pp.10580–10585.

Andrasi, E., et al., 2000. Disturbances of magnesium concentrations in various brain areas in Alzheimer's disease. *Magnes Res,* 13, pp.189–196.

Andreoli, C., et al., 2004. MitoP2, an integrated database on mitochondrial proteins in yeast and man. *Nucleic Acids Res,* 32, pp.D459–D462.

Antonenko, Y. N., et al., 1991. Selective effect of inhibitors on inner mitochondrial membrane channels. *FEBS Lett,* 285, pp.89–93.

Antonenko, Y. N., et al., 1994. Single-channel activity induced in mitoplasts by alkaline pH. *Biochim Biophys Acta,* 1194, pp.247–254.

Appel, S. H., et al., 2000. Altered calcium homeostasis in ALS as a target for therapy. *Amyotroph Lateral Scler Other Motor Neuron Disord,* 1 Suppl 4, pp.27–32.

Appel, S. H., et al., 2001. Calcium: the Darth Vader of ALS. *Amyotroph Lateral Scler Other Motor Neuron Disord,* 2 Suppl 1, pp.S47–S54.

Azoulay-Zohar, H., et al., 2004. In self-defence: hexokinase promotes voltage-dependent anion channel closure and prevents mitochondria-mediated apoptotic cell death. *Biochem J,* 377, pp.347–355.

Baines, C. P., et al., 2007. Voltage-dependent anion channels are dispensable for mitochondrial-dependent cell death. *Nat Cell Biol,* 9, pp.550–555.

Baker, L. A., et al., 2012. Arrangement of subunits in intact mammalian mitochondrial ATP synthase determined by cryo-EM. *Proc Natl Acad Sci U S A,* 109, pp.11675–11680.

Balaban, R. S., 2009. The role of Ca(2+) signaling in the coordination of mitochondrial ATP production with cardiac work. *Biochim Biophys Acta,* 1787, pp.1334–1341.

Bathori, G., et al., 2006. Ca2+-dependent control of the permeability properties of the mitochondrial outer membrane and voltage-dependent anion-selective channel (VDAC). *J Biol Chem,* 281, pp.17347–17358.

Baughman, J. M., et al., 2011. Integrative genomics identifies MCU as an essential component of the mitochondrial calcium uniporter. *Nature,* 476, pp.341–345.

Bayrhuber, M., et al., 2008. Structure of the human voltage-dependent anion channel. *Proc Natl Acad Sci U S A,* 105, pp.15370–15375.

Beavis, A. D. and Garlid, K. D., 1988. Inhibition of the mitochondrial inner membrane anion channel by dicyclohexylcarbodiimide. Evidence for a specific transport pathway. *J Biol Chem,* 263, pp.7574–7580.

Beavis, A. D. and Powers, M. F., 1989. On the regulation of the mitochondrial inner membrane anion channel by magnesium and protons. *J Biol Chem,* 264, pp.17148–17155.

Beavis, A. D., 1992. Properties of the inner membrane anion channel in intact mitochondria. *J Bioenerg Biomembr,* 24, pp.77–90.

Beavis, A. D. and Vercesi, A.E., 1992. Anion uniport in plant mitochondria is mediated by a Mg(2+)-insensitive inner membrane anion channel. *J Biol Chem,* 267, pp.3079–3087.

Becattini, B., et al., 2004. Targeting apoptosis via chemical design: inhibition of bid-induced cell death by small organic molecules. *Chem Biol,* 11, pp.1107–1117.

Becker, T., et al., 2008. Sorting and assembly of mitochondrial outer membrane proteins. *Biochim Biophys Acta,* 1777, pp.557–563.

Becker, T., et al., 2009. Biogenesis of mitochondrial membrane proteins. *Curr Opin Cell Biol,* 21, pp.484–493.

Bellot, G., et al., 2007. TOM22, a core component of the mitochondria outer membrane protein translocation pore, is a mitochondrial receptor for the proapoptotic protein Bax. *Cell Death Differ,* 14, pp.785–794.

Bender, A., et al., 2013. TOM40 mediates mitochondrial dysfunction induced by alpha-synuclein accumulation in Parkinson's disease. *PLoS One,* 8, p.e62277.

Bernardi, P., 1999. Mitochondrial transport of cations: channels, exchangers, and permeability transition. *Physiol. Rev,* 79, pp.1127–1155.

Bernardi, P., et al., 2006. The mitochondrial permeability transition from in vitro artifact to disease target. *FEBS J,* 273, pp.2077–2099.

Bernardi, P., 2013. The mitochondrial permeability transition pore: a mystery solved? *Front Physiol,* 4, p.95.

Beutner, G., et al., 2005. Type 1 ryanodine receptor in cardiac mitochondria: transducer of excitation-metabolism coupling. *Biochim Biophys Acta,* 1717, pp.1–10.

Biswas, S. C., et al., 2007. Bim is elevated in Alzheimer's disease neurons and is required for beta-amyloid-induced neuronal apoptosis. *J Neurosci,* 27, pp.893–900.

Blondin, G. A., 1975. Isolation, properties, and structural features of divalent cation ionophores derived from beef heart mitochondria. *Ann N Y Acad Sci,* 264, pp.98–111.

Bonora, M., et al., 2013. Role of the c subunit of the FO ATP synthase in mitochondrial permeability transition. *Cell Cycle,* 12, pp.674–683.

Bonora, M., et al., 2015. Molecular mechanisms of cell death: central implication of ATP synthase in mitochondrial permeability transition. *Oncogene,* 34, pp.1475–1486.

Boska, M. D., et al., 2002. Contrasts in cortical magnesium, phospholipid and energy metabolism between migraine syndromes. *Neurology,* 58, pp.1227–1233.

Brix, J., et al., 1999. Distribution of binding sequences for the mitochondrial import receptors Tom20, Tom22, and Tom70 in a presequence-carrying preprotein and a non-cleavable preprotein. *J Biol Chem,* 274, pp.16522–16530.

Brookes, P. S., et al., 2004. Calcium, ATP, and ROS: a mitochondrial love-hate triangle. *Am J Physiol Cell Physiol,* 287, pp.C817–C833.

Brustovetsky, N., et al., 2002. A large Ca2+-dependent channel formed by recombinant ADP/ATP carrier from Neurospora crassa resembles the mitochondrial permeability transition pore. *Biochemistry,* 41, pp.11804–11811.

Busija, D. W., et al., 2008. Mitochondrial-mediated suppression of ROS production upon exposure of neurons to lethal stress: mitochondrial targeted preconditioning. *Adv Drug Deliv Rev,* 60, pp.1471–1477.

Cali, T., Ottolini, D. and Brini, M., 2012. Mitochondrial Ca(2+) as a key regulator of mitochondrial activities. *Adv Exp Med Biol,* 942, pp.53–73.

Campo, M. L., Kinnally, K. W. and Tedeschi, H., 1992. The effect of antimycin A on mouse liver inner mitochondrial membrane channel activity. *J Biol Chem*, 267, pp.8123–8127.

Cartron, P. F., et al., 2008. Bax inserts into the mitochondrial outer membrane by different mechanisms. *FEBS Lett,* 582, 3045–3051.

Chacinska, A., et al., 2005. Mitochondrial presequence translocase: switching between TOM tethering and motor recruitment involves Tim21 and Tim17. *Cell,* 120, pp.817–829.

Chacinska, A., et al., 2009. Importing mitochondrial proteins: machineries and mechanisms. *Cell,* 138, 628–644.

Chaudhuri, D., et al., 2013. MCU encodes the pore conducting mitochondrial calcium currents. *Elife,* 2, p.e00704.

Chou, C. H., Lee, R. S. and Yang-Yen, H. F., 2006. An internal EELD domain facilitates mitochondrial targeting of Mcl-1 via a Tom70-dependent pathway. *Mol Biol Cell,* 17, pp.3952–3963.

Colin, J., et al., 2009. The mitochondrial TOM complex modulates bax-induced apoptosis in Drosophila. *Biochem Biophys Res Commun,* 379, pp.939–943.

Colombini, M., 1980. Structure and mode of action of a voltage dependent anion-selective channel (VDAC) located in the outer mitochondrial membrane. *Ann N Y Acad Sci,* 341, 552–563.

Colombini, M., 2009. The published 3D structure of the VDAC channel: native or not? *Trends Biochem Sci,* 34, pp.382–389.

Colombini, M., 2013. Membrane channels formed by ceramide. *Handb Exp Pharmacol,* pp.109–126.

Cuadrado-Tejedor, M., et al., 2011. Enhanced expression of the voltage-dependent anion channel 1 (VDAC1) in Alzheimer's disease transgenic mice: an insight into the pathogenic effects of amyloid-beta. *J Alzheimers Dis,* 23, pp.195–206.

Culmsee, C., et al., 2005. Apoptosis-inducing factor triggered by poly(ADP-ribose) polymerase and Bid mediates neuronal cell death after oxygen-glucose deprivation and focal cerebral ischemia. *J Neurosci,* 25, pp.10262–10272.

Das, S., et al., 1997. Proof for a nonproteinaceous calcium-selective channel in Escherichia coli by total synthesis from (R)-3-hydroxybutanoic acid and inorganic polyphosphate. *Proc Natl Acad Sci U S A,* 94, pp.9075–9079.

De Pinto, V., et al., 2008. Structure of the voltage dependent anion channel: state of the art. *J Bioenerg Biomembr,* 40, pp.139–147.

De Stefani, D., et al., 2011. A forty-kilodalton protein of the inner membrane is the mitochondrial calcium uniporter. *Nature,* 476, pp.336–340.

Dejean, L. M., et al., 2010. MAC and Bcl-2 family proteins conspire in a deadly plot. *Biochim Biophys Acta,* 1797, pp.1231–1238.

Deluca, H. F. and Engstrom, G. W., 1961. Calcium uptake by rat kidney mitochondria. *Proc Natl Acad Sci U S A,* 47, pp.1744–1750.

Denton, R. M., 2009. Regulation of mitochondrial dehydrogenases by calcium ions. *Biochim Biophys Acta* ,1787, pp.1309–1316.

Devi, L., et al., 2006. Accumulation of amyloid precursor protein in the mitochondrial import channels of human Alzheimer's disease brain is associated with mitochondrial dysfunction. *J Neurosci,* 26, 9057–9068.

Dirnagl, U. and Meisel, A., 2008. Endogenous neuroprotection: mitochondria as gateways to cerebral preconditioning? *Neuropharmacology,* 55, pp.334–44.

Diwan, J. J. and Tedeschi, H., 1975. K+ fluxes and the mitochondrial membrane potential. *FEBS Lett,* 60, pp.176–179.

Dohm, C. P., et al., 2006. Bax inhibitor-1 protects neurons from oxygen-glucose deprivation. *J Mol Neurosci,* 29, pp.1–8.

Douglas, R. M., et al., 2006. The calcium-sensitive large-conductance potassium channel (BK/MAXI K) is present in the inner mitochondrial membrane of rat brain. *Neuroscience,* 139, pp.1249–1261.

Drago, I., Pizzo, P. and Pozzan, T., 2011. After half a century mitochondrial calcium in- and efflux machineries reveal themselves. *EMBO J,* 30, pp.4119–4125.

Drahota, Z., et al., 1969. A comparison of the effects of different divalent cations on a number of mitochondrial reactions linked to ion translocation. *Arch Biochem Biophys,* 130, pp.267–273.

Duchen, M. R., 2012. Mitochondria, calcium-dependent neuronal death and neurodegenerative disease. *Pflugers Arch,* 464, pp.111–121.

Elustondo, P. A., et al., 2013. Polyhydroxybutyrate targets mammalian mitochondria and increases permeability of plasmalemmal and mitochondrial membranes. *PLoS One,* 8, p.e75812.

Endo, T. and Yamano, K., 2010. Transport of proteins across or into the mitochondrial outer membrane. *Biochim Biophys Acta,* 1803, pp.706–714.

Feissner, R. F., et al., 2009. Crosstalk signaling between mitochondrial Ca2+ and ROS. *Front Biosci,* 14, pp.1197–1218.

Fernandez-Salas, E., et al., 2002. mtCLIC/CLIC4, an organellular chloride channel protein, is increased by DNA damage and participates in the apoptotic response to p53. *Mol Cell Biol,* 22, 3610–3620.

Furuichi, T., et al., 1989. Primary structure and functional expression of the inositol 1,4,5-trisphosphate-binding protein P400. *Nature,* 342, pp.32–38.

Gebert, N., et al., 2011. Dual function of Sdh3 in the respiratory chain and TIM22 protein translocase of the mitochondrial inner membrane. *Mol Cell,* 44, pp.811–818.

Ghezzi, D., et al., 2010. Severe X-linked mitochondrial encephalomyopathy associated with a mutation in apoptosis-inducing factor. *Am J Hum Genet,* 86, pp.639–649.

Ghosh, T., et al., 2007. A role for voltage-dependent anion channel Vdac1 in polyglutamine-mediated neuronal cell death. *PLoS One,* 2, p.e1170.

Giorgio, V., et al., 2013. Dimers of mitochondrial ATP synthase form the permeability transition pore. *Proc Natl Acad Sci U S A,* 110, pp.5887–5892.

Giovanni, A., et al., 2000. E2F1 mediates death of B-amyloid-treated cortical neurons in a manner independent of p53 and dependent on Bax and caspase 3. *J Biol Chem,* 275, pp.11553–11560.

Gould, T. W., et al., 2006. Complete dissociation of motor neuron death from motor dysfunction by Bax deletion in a mouse model of ALS. *J Neurosci,* 26, pp.8774–8786.

Gross, A., McDonnell, J. M. and Korsmeyer, S.J., 1999. BCL-2 family members and the mitochondria in apoptosis. *Genes Dev,* 13, pp.1899–1911.

Grosskreutz, J., Van Den Bosch, L. and Keller, B. U., 2010. Calcium dysregulation in amyotrophic lateral sclerosis. *Cell Calcium,* 47, pp.165–174.

Guillet, V., et al., 2011. Bioenergetic defect associated with mKATP channel opening in a mouse model carrying a mitofusin 2 mutation. *FASEB J,* 25, pp.1618–1627.

Gunter, T. E., et al., 2004. Calcium and mitochondria. *FEBS Lett,* 567, pp.96–102.

Gunter, T. E. and Sheu, S.S., 2009. Characteristics and possible functions of mitochondrial Ca(2+) transport mechanisms. *Biochim Biophys Acta,* 1787, pp.1291–1308.

Guo, X. W., et al., 1995. Molecular design of the voltage-dependent, anion-selective channel in the mitochondrial outer membrane. *J Struct Biol,* 114, pp.41–59.

Habib, S. J., et al., 2005. Assembly of the TOB complex of mitochondria. *J Biol Chem,* 280, pp.6434–6440.

Hajnoczky, G., et al., 2006. Mitochondrial calcium signalling and cell death: approaches for assessing the role of mitochondrial Ca2+ uptake in apoptosis. *Cell Calcium,* 40, pp.553–560.

Halestrap, A. P., 2009. What is the mitochondrial permeability transition pore? *J Mol Cell Cardiol,* 46, pp.821–831.

Hansford, R. G., 1991. Dehydrogenase activation by Ca2+ in cells and tissues. *J Bioenerg Biomembr,* 23, pp.823–854.

Hashimoto, T., et al., 2008. Magnesium exerts both preventive and ameliorating effects in an in vitro rat Parkinson disease model involving 1-methyl-4-phenylpyridinium (MPP+) toxicity in dopaminergic neurons. *Brain Res,* 1197, pp.143–151.

Hassouna, I., et al., 1996. Increase in bax expression in substantia nigra following 1-methyl-4-phenyl-1,2,3,6-tetrahydropyridine (MPTP) treatment of mice. *Neurosci Lett,* 204, pp.85–88.

Herrmann, J. M. and Hell, K., 2005. Chopped, trapped or tacked—protein translocation into the IMS of mitochondria. *Trends Biochem Sci,* 30, pp.205–211.

Hetz, C., et al., 2005. Bax channel inhibitors prevent mitochondrion-mediated apoptosis and protect neurons in a model of global brain ischemia. *J Biol Chem,* 280, pp.42960–42970.

Hill, K., et al., 1998. Tom40 forms the hydrophilic channel of the mitochondrial import pore for preproteins [see comment]. *Nature,* 395, pp.516–521.

Hiller, S., et al., 2008. Solution structure of the integral human membrane protein VDAC-1 in detergent micelles. *Science,* 321, pp.1206–1210.

Holden, M. J. and Colombini, M., 1993. The outer mitochondrial membrane channel, VDAC, is modulated by a protein localized in the intermembrane space. *Biochim Biophys Acta,* 1144, pp.396–402.

Hoogenboom, B. W., et al., 2007. The supramolecular assemblies of voltage-dependent anion channels in the native membrane. *J Mol Biol,* 370, pp.246–255.

Hur, J. Y., et al., 2012. Identification of novel gamma-secretase-associated proteins in detergent-resistant membranes from brain. *J Biol Chem,* 287, pp.11991–12005.

Inoue, I., et al., 1991. ATP-sensitive K+ channel in the mitochondrial inner membrane. *Nature,* 352, pp.244–247.

Ionov, I. D., 2007. Survey of ALS-associated factors potentially promoting Ca2+ overload of motor neurons. *Amyotroph Lateral Scler,* 8, pp.260–265.

Israelson, A., et al., 2010. Misfolded mutant SOD1 directly inhibits VDAC1 conductance in a mouse model of inherited ALS. *Neuron,* 67, pp.575–587.

Javadov, S. and Kuznetsov, A., 2013. Mitochondrial permeability transition and cell death: the role of cyclophilin d. *Front Physiol,* 4, p.76.

Jeon, K., et al., 2012. Bax inhibitor-1 enhances survival and neuronal differentiation of embryonic stem cells via differential regulation of mitogen-activated protein kinases activities. *Biochim Biophys Acta,* 1823, pp.2190–2200.

Jin, H., et al., 1996. A novel X-linked gene, DDP, shows mutations in families with deafness (DFN-1), dystonia, mental deficiency and blindness. *Nat Genet,* 14, pp.177–180.

Jonas, E. A., Buchanan, J. and Kaczmarek, L. K., 1999. Prolonged activation of mitochondrial conductances during synaptic transmission. *Science,* 286, pp.1347–1350.

Jonas, E. A., et al., 2004. Proapoptotic N-truncated BCL-xL protein activates endogenous mitochondrial channels in living synaptic terminals. *Proc Natl Acad Sci U S A,* 101, pp.13590–13595.

Jonas, E. A., 2009. Molecular participants in mitochondrial cell death channel formation during neuronal ischemia. *Exp Neurol,* 218, pp.203–212.

Jouaville, L. S., et al., 1999. Regulation of mitochondrial ATP synthesis by calcium: evidence for a long-term metabolic priming. *Proc Natl Acad Sci U S A,* 96, 13807–13812.

Jung, D. W. and Brierley, G. P., 1999. Matrix free Mg(2+) and the regulation of mitochondrial volume. *Am J Physiol,* 277, pp.C1194–C1201.

Kane, D. A. and Pavlov, E. V., 2013. Calculation of ion currents across the inner membrane of functionally intact mitochondria. *Channels (Austin),* 7, pp.426–431.

Kawamata, H. and Manfredi, G., 2010. Mitochondrial dysfunction and intracellular calcium dysregulation in ALS. *Mech Ageing Dev,* 131, pp.517–526.

Keinan, N., et al., 2013. The role of calcium in VDAC1 oligomerization and mitochondria-mediated apoptosis. *Biochim Biophys Acta,* 1833, pp.1745–1754.

Kerscher, O., et al., 1997. The Tim54p-Tim22p complex mediates insertion of proteins into the mitochondrial inner membrane. *J Cell Biol,* 139, pp.1663–1675.

Kim, J. S., He, L., and Lemasters, J. J., 2003. Mitochondrial permeability transition: a common pathway to necrosis and apoptosis. *Biochem Biophys Res Commun,* 304, pp.463–470.

Kinnally, K. W., Campo, M. L. and Tedeschi, H., 1989. Mitochondrial channel activity studied by patch-clamping mitoplasts. *J Bioenerg Biomembr,* 21, pp.497–506.

Kinnally, K. W., et al., 1991. Calcium modulation of mitochondrial inner membrane channel activity. *Biochem Biophys Res Commun,* 176, pp.1183–1188.

Kinnally, K. W., Antonenko, Y. N. and Zorov, D. B., 1992. Modulation of inner mitochondrial membrane channel activity. *J Bioenerg Biomembr,* 24, pp.99–110.

Kinnally, K. W., et al., 1996. Perspectives on the mitochondrial multiple conductance channel. *J Bioenerg Biomembr,* 28, pp.115–123.

Kinnally, K. W. and Antonsson, B., 2007. A tale of two mitochondrial channels, MAC and PTP, in apoptosis. *Apoptosis,* 12, pp.857–868.

Kinnally, K. W., et al., 2011. Is mPTP the gatekeeper for necrosis, apoptosis, or both? *Biochim Biophys Acta,* 1813, pp.616–622.

Kirichok, Y., Krapivinsky, G. and Clapham, D. E., 2004. The mitochondrial calcium uniporter is a highly selective ion channel. *Nature,* 427, pp.360–364.

Klitsch, T. and Siemen, D., 1991. Inner mitochondrial membrane anion channel is present in brown adipocytes but is not identical with the uncoupling protein. *J Membr Biol,* 122, pp.69–75.

Kokoszka, J. E., et al., 2004. The ADP/ATP translocator is not essential for the mitochondrial permeability transition pore. *Nature,* 427, pp.461–465.

Kokotas, H., Petersen, M. B. and Willems, P. J., 2007. Mitochondrial deafness. *Clin Genet,* 71, pp.379–391.

Kondo-Okamoto, N., Shaw, J. M. and Okamoto, K., 2008. Tetratricopeptide repeat proteins Tom70 and Tom71 mediate yeast mitochondrial morphogenesis. *EMBO Rep,* 9, pp.63–69.

Kostic, V., et al., 1997. Bcl-2: prolonging life in a transgenic mouse model of familial amyotrophic lateral sclerosis. *Science,* 277, pp.559–562.

Kovermann, P., et al., 2002. Tim22, the essential core of the mitochondrial protein insertion complex, forms a voltage-activated and signal-gated channel. *Mol Cell,* 9, pp.363–373.

Kozjak-Pavlovic, V., et al., 2007. Conserved roles of Sam50 and metaxins in VDAC biogenesis. *EMBO Rep,* 8, pp.576–582.

Kubota, T., et al., 2005. Mitochondria are intracellular magnesium stores: investigation by simultaneous fluorescent imagings in PC12 cells. *Biochim Biophys Acta,* 1744, pp.19–28.

Kudo, W., et al., 2012. Inhibition of Bax protects neuronal cells from oligomeric Abeta neurotoxicity. *Cell Death Dis,* 3, p.e309.

Kunkele, K. P., et al., 1998. The preprotein translocation channel of the outer membrane of mitochondria. *Cell,* 93, pp.1009–1019.

Kuramoto, T., et al., 2011. A mutation in the gene encoding mitochondrial Mg(2)+ channel MRS2 results in demyelination in the rat. *PLoS Genet,* 7, p.e1001262.

Kutik, S., et al., 2008. Dissecting membrane insertion of mitochondrial beta-barrel proteins. *Cell,* 132, pp.1011–1024.

Kwong, J. Q., et al., 2014. Genetic deletion of the mitochondrial phosphate carrier desensitizes the mitochondrial permeability transition pore and causes cardiomyopathy. *Cell Death Differ,* 21, pp.1209–1217.

Lemasters, J. J. and Holmuhamedov, E., 2006. Voltage-dependent anion channel (VDAC) as mitochondrial governator—thinking outside the box. *Biochim Biophys Acta,* 1762, pp.181–190.

Li, Q., et al., 2010. ALS-linked mutant superoxide dismutase 1 (SOD1) alters mitochondrial protein composition and decreases protein import. *Proc Natl Acad Sci U S A,* 107, pp.21146–21151.

Lindsten, T., et al., 2000. The combined functions of proapoptotic Bcl-2 family members bak and bax are essential for normal development of multiple tissues. *Mol Cell,* 6, pp.1389–1399.

Liu, D., et al., 2010. The KATP channel activator diazoxide ameliorates amyloid-beta and tau pathologies and improves memory in the 3xTgAD mouse model of Alzheimer's disease. *J Alzheimers Dis,* 22, pp.443–457.

Lohret, T. A., Jensen, R. E. and Kinnally, K. W., 1997. Tim23, a protein import component of the mitochondrial inner membrane, is required for normal activity of the multiple conductance channel, MCC. *J Cell Biol,* 137, 377–386.

Lu, L., et al., 2013. Voltage-dependent anion channel involved in the alpha-synuclein-induced dopaminergic neuron toxicity in rats. *Acta Biochim Biophys Sin (Shanghai),* 45, pp.170–178.

Ma, G., et al., 2009. Diazoxide reverses the enhanced expression of KATP subunits in cholinergic neurons caused by exposure to Abeta(1)(-)(4)(2). *Neurochem Res,* 34, pp.2133–2140.

Maack, C. and O'Rourke, B., 2007. Excitation-contraction coupling and mitochondrial energetics. *Basic Res Cardiol,* 102, pp.369–392.

Macaskill, A. F., et al., 2009. Miro1 is a calcium sensor for glutamate receptor-dependent localization of mitochondria at synapses. *Neuron,* 61, pp.541–55.

Magrane, J., et al., 2012. Mitochondrial dynamics and bioenergetic dysfunction is associated with synaptic alterations in mutant SOD1 motor neurons. *J Neurosci,* 32, pp.229–242.

Mammucari, C., et al., 2011. Molecules and roles of mitochondrial calcium signaling. *Biofactors,* 37, 219–227.

Manczak, M. and Reddy, P.H., 2012. Abnormal interaction of VDAC1 with amyloid beta and phosphorylated tau causes mitochondrial dysfunction in Alzheimer's disease. *Hum Mol Genet,* 21, pp.5131–5146.

Martinez-Caballero, S., et al., 2007. Tim17p regulates the twin pore structure and voltage gating of the mitochondrial protein import complex TIM23. *J Biol Chem,* 282, pp.3584–3593.

Meisinger, C., et al., 2001. Protein import channel of the outer mitochondrial membrane: a highly stable Tom40-Tom22 core structure differentially interacts with preproteins, small tom proteins, and import receptors. *Mol Cell Biol,* 21, pp.2337–2348.

Michels, G., et al., 2009. Regulation of the human cardiac mitochondrial Ca2+ uptake by 2 different voltage-gated Ca2+ channels. *Circulation,* 119, pp.2435–2443.

Mikoshiba, K., Huchet, M. and Changeux, J. P., 1979. Biochemical and immunological studies on the P400 protein, a protein characteristic of the Purkinje cell from mouse and rat cerebellum. *Dev Neurosci,* 2, 254–275.

Mitchell, P., 1961. Coupling of phosphorylation to electron and hydrogen transfer by a chemi-osmotic type of mechanism. *Nature,* 191, pp.144–148.

Mokranjac, D., et al., 2005. Role of Tim21 in mitochondrial translocation contact sites. *J Biol Chem*, 280, 23437–23440.

Moran, O., et al., 1990. Electrophysiological characterization of contact sites in brain mitochondria. *J Biol Chem,* 265, pp.908–913.

Murphy, R. C., et al., 1998. Two high conductance channels of the mitochondrial inner membrane are independent of the human mitochondrial genome. *FEBS Lett,* 425, 259–262.

Nakagawa, I., et al., 2005. MitoKATP-channel opener protects against neuronal death in rat venous ischemia. *Neurosurgery,* 57, pp.334–340; discussion 334–340.

Neumann, M., et al., 2006. Ubiquitinated TDP-43 in frontotemporal lobar degeneration and amyotrophic lateral sclerosis. *Science,* 314, pp.130–133.

Neupert, W. and Herrmann, J. M., 2007. Translocation of proteins into mitochondria. *Annu Rev Biochem,* 76, pp.723–749.

Okatsu, K., et al., 2012. Mitochondrial hexokinase HKI is a novel substrate of the Parkin ubiquitin ligase. *Biochem Biophys Res Commun,* 428, pp.197–202.

Otera, H., et al., 2007. A novel insertion pathway of mitochondrial outer membrane proteins with multiple transmembrane segments. *J Cell Biol,* 179, pp.1355–1363.

Ott, M., et al., 2007. The mitochondrial TOM complex is required for tBid/Bax-induced cytochrome c release. *J Biol Chem,* 282, pp.27633–27639.

Oyanagi, K., et al., 2006. Magnesium deficiency over generations in rats with special references to the pathogenesis of the Parkinsonism-dementia complex and amyotrophic lateral sclerosis of Guam. *Neuropathology,* 26, pp.115–128.

Palombo, I., Daley, D. O. and Rapp, M., 2013. Why is the GMN motif conserved in the CorA/Mrs2/Alr1 superfamily of magnesium transport proteins? *Biochemistry,* 52, pp.4842–4847.

Pan, X., et al., 2013. The physiological role of mitochondrial calcium revealed by mice lacking the mitochondrial calcium uniporter. *Nat Cell Biol,* 15, pp.1464–1472.

Parone, P. A., et al., 2013. Enhancing mitochondrial calcium buffering capacity reduces aggregation of misfolded SOD1 and motor neuron cell death without extending survival in mouse models of inherited amyotrophic lateral sclerosis. *J Neurosci,* 33, 4657–4671.

Pavlov, E., et al., 2005a. The mitochondrial channel VDAC has a cation-selective open state. *Biochim Biophys Acta,* 1710, pp.96–102.

Pavlov, E., et al., 2005b. A large, voltage-dependent channel, isolated from mitochondria by water-free chloroform extraction. *Biophys J,* 88, pp.2614–2625.

Pavlov, E. V., et al., 2001. A novel, high conductance channel of mitochondria linked to apoptosis in mammalian cells and Bax expression in yeast. *J Cell Biol,* 155, pp.725–731.

Peixoto, P. M., et al., 2007. Awaking TIM22, a dynamic ligand-gated channel for protein insertion in the mitochondrial inner membrane. *J Biol Chem,* 282, pp.18694–18701.

Peixoto, P. M., et al., 2009. MAC inhibitors suppress mitochondrial apoptosis. *Biochem J,* 423, 381–387.

Peixoto, P. M., Ryu, S. Y. and Kinnally, K. W., 2010. Mitochondrial ion channels as therapeutic targets. *FEBS Lett,* 584, pp.2142–2152.

Perera, M. N., et al., 2012. Bax and Bcl-xL exert their regulation on different sites of the ceramide channel. *Biochem J,* 445, pp.81–91.

Perez Velazquez, J. L., et al., 2000. Mitochondrial porin required for ischemia-induced mitochondrial dysfunction and neuronal damage. *Neuroscience,* 97, pp.363–369.

Perez Velazquez, J. L., et al., 2003. Anti-porin antibodies prevent excitotoxic and ischemic damage to brain tissue. *J Neurotrauma,* 20, 633–647.

Perkins, G., et al., 1997. Electron tomography of neuronal mitochondria: three-dimensional structure and organization of cristae and membrane contacts. *J Struct Biol,* 119, pp.260–272.

Perocchi, F., et al., 2010. MICU1 encodes a mitochondrial EF hand protein required for Ca(2+) uptake. *Nature,* 467, pp.291–296.

Petronilli, V., Szabo, I. and Zoratti, M., 1989. The inner mitochondrial membrane contains ion-conducting channels similar to those found in bacteria. *FEBS Lett,* 259, pp.137–143.

Plovanich, M., et al., 2013. MICU2, a paralog of MICU1, resides within the mitochondrial uniporter complex to regulate calcium handling. *PLoS One,* 8, p.e55785.

Popov-Celeketic, D., et al., 2008. Active remodelling of the TIM23 complex during translocation of preproteins into mitochondria. *Embo J,* 27, pp.1469–1480.

Poynor, M., Eckert, R. and Nussberger, S., 2008. Dynamics of the preprotein translocation channel of the outer membrane of mitochondria. *Biophys J,* 95, pp.1511–1522.

Rao, V. K., Carlson, E. A. and Yan, S. S., 2014. Mitochondrial permeability transition pore is a potential drug target for neurodegeneration. *Biochim Biophys Acta,* 1842, pp.1267–1272.

Reyes, N. A., et al., 2010. Blocking the mitochondrial apoptotic pathway preserves motor neuron viability and function in a mouse model of amyotrophic lateral sclerosis. *J Clin Invest,* 120, pp.3673–3679.

Richter, C. and Ghafourifar, P., 1999. Ceramide induces cytochrome c release from isolated mitochondria. *Biochem Soc Symp,* 66, pp.27–31.

Robin, E., et al., 2011. Postconditioning in focal cerebral ischemia: role of the mitochondrial ATP-dependent potassium channel. *Brain Res,* 1375, pp.137–146.

Rodriguez, M. J., et al., 2013. Targeting microglial K(ATP) channels to treat neurodegenerative diseases: a mitochondrial issue. *Oxid Med Cell Longev,* 2013, 194546.

Roise, D. and Schatz, G., 1988. Mitochondrial presequences. *J Biol Chem,* 263, pp.4509–4511.

Roman, I., et al., 2006. Hunting interactomes of a membrane protein: obtaining the largest set of voltage-dependent anion channel-interacting protein epitopes. *Mol Cell Proteomics,* 5, pp.1667–1680.

Ross, K., Rudel, T. and Kozjak-Pavlovic, V., 2009. TOM-independent complex formation of Bax and Bak in mammalian mitochondria during TNFalpha-induced apoptosis. *Cell Death Differ,* 16, pp.697–707.

Rostovtseva, T. and Colombini, M., 1996. ATP flux is controlled by a voltage-gated channel from the mitochondrial outer membrane. *J Biol Chem,* 271, pp.28006–28008.

Rostovtseva, T. and Colombini, M., 1997. VDAC channels mediate and gate the flow of ATP: implications for the regulation of mitochondrial function. *Biophys J,* 72, pp.1954–1962.

Rostovtseva, T. K., et al., 2004. Bid, but not Bax, regulates VDAC channels. *J Biol Chem,* 279, pp.13575–13583.

Runke, G., et al., 2006. Deletion variants of Neurospora mitochondrial porin: electrophysiological and spectroscopic analysis. *Biophys J,* 90, pp.3155–3164.

Ryu, S. Y., et al., 2010. Mitochondrial ryanodine receptors and other mitochondrial Ca2+ permeable channels. *FEBS Lett,* 584, pp.1948–1955.

Ryu, S. Y., et al., 2011. Single channel characterization of the mitochondrial ryanodine receptor in heart mitoplasts. *J Biol Chem,* 286, 21324–21329.

Saitoh, T., et al., 2007. Tom20 recognizes mitochondrial presequences through dynamic equilibrium among multiple bound states. *Embo J,* 26, pp.4777–4787.

Sanjuan Szklarz, L. K., et al., 2007. Preprotein transport machineries of yeast mitochondrial outer membrane are not required for Bax-induced release of intermembrane space proteins. *J Mol Biol,* 368, pp.44–54.

Saotome, M., et al., 2008. Bidirectional Ca2+-dependent control of mitochondrial dynamics by the Miro GTPase. *Proc Natl Acad Sci U S A,* 105, pp.20728–20733.

Sato, T., et al., 2000. Selective pharmacological agents implicate mitochondrial but not sarcolemmal K(ATP) channels in ischemic cardioprotection. *Circulation,* 101, pp.2418–2423.

Schein, S. J., Colombini, M. and Finkelstein, A., 1976. Reconstitution in planar lipid bilayers of a voltage-dependent anion-selective channel obtained from paramecium mitochondria. *J Membr Biol,* 30, pp.99–120.

Schindl, R., et al., 2007. Mrs2p forms a high conductance Mg2+ selective channel in mitochondria. *Biophys J,* 93, pp.3872–3883.

Schmitt, S., et al., 2005. Role of Tom5 in maintaining the structural stability of the TOM complex of mitochondria. *J Biol Chem,* 280, pp.14499–14506.

Seidlmayer, L. K., et al., 2012a. Inorganic polyphosphate—an unusual suspect of the mitochondrial permeability transition mystery. *Channels (Austin),* 6, pp.463–467.

Seidlmayer, L. K., et al., 2012b. Inorganic polyphosphate is a potent activator of the mitochondrial permeability transition pore in cardiac myocytes. *J Gen Physiol,* 139, pp.321–331.

Sherman, E. L., Go, N. E. and Nargang, F. E., 2005. Functions of the small proteins in the TOM complex of Neurospora crasssa. *Mol Biol Cell,* 16, pp.4172–4182.

Shindo, Y., et al., 2011. Newly developed Mg2+-selective fluorescent probe enables visualization of Mg2+ dynamics in mitochondria. *PLoS One,* 6, p.e23684.

Siemen, D., et al., 1999. Ca2+-activated K channel of the BK-type in the inner mitochondrial membrane of a human glioma cell line. *Biochem Biophys Res Commun,* 257, pp.549–554.

Sirrenberg, C., et al., 1996. Import of carrier proteins into the mitochondrial inner membrane mediated by Tim22. *Nature,* 384, pp.582–585.

Siskind, L. J. and Colombini, M., 2000. The lipids C2- and C16-ceramide form large stable channels. Implications for apoptosis. *J Biol Chem,* 275, pp.38640–38644.

Siskind, L. J., Kolesnick, R. N. and Colombini, M., 2002. Ceramide channels increase the permeability of the mitochondrial outer membrane to small proteins. *J Biol Chem,* 277, pp.26796–26803.

Smithen, M., et al., 2013. Role of polyhydroxybutyrate in mitochondrial calcium uptake. *Cell Calcium,* 54, pp.86–94.

Sorgato, M. C., Keller, B. U. and Stuhmer, W., 1987. Patch-clamping of the inner mitochondrial membrane reveals a voltage-dependent ion channel. *Nature,* 330, pp.498–500.

Sorgato, M. C., et al., 1989. Further investigation on the high-conductance ion channel of the inner membrane of mitochondria. *J Bioenerg Biomembr,* 21, pp.485–496.

Streb, H., et al., 1983. Release of Ca2+ from a nonmitochondrial intracellular store in pancreatic acinar cells by inositol-1,4,5-trisphosphate. *Nature,* 306, pp.67–69.

Suh, K. S., et al., 2004. The organellular chloride channel protein CLIC4/mtCLIC translocates to the nucleus in response to cellular stress and accelerates apoptosis. *J Biol Chem,* 279, pp.4632–4641.

Sun, Y., et al., 2012. Voltage-dependent anion channels (VDACs) recruit Parkin to defective mitochondria to promote mitochondrial autophagy. *J Biol Chem,* 287, pp.40652–40660.

Szabo, I. and Zoratti, M., 1991. The giant channel of the inner mitochondrial membrane is inhibited by cyclosporin A. *J Biol Chem,* 266, pp.3376–3379.

Szabo, I., Bernardi, P. and Zoratti, M., 1992. Modulation of the mitochondrial megachannel by divalent cations and protons. *J Biol Chem,* 267, pp.2940–2946.

Szabo, I. and Zoratti, M., 1992. The mitochondrial megachannel is the permeability transition pore. *J Bioenerg Biomembr,* 24, pp.111–117.

Szabo, I. and Zoratti, M., 1993. The mitochondrial permeability transition pore may comprise VDAC molecules. I. Binary structure and voltage dependence of the pore. *FEBS Lett,* 330, pp.201–205.

Szabo, I., et al., 2005. A novel potassium channel in lymphocyte mitochondria. *J Biol Chem,* 280, 12790–12798.

Szabo, I., et al., 2008. Mitochondrial potassium channel Kv1.3 mediates Bax-induced apoptosis in lymphocytes. *Proc Natl Acad Sci U S A,* 105, pp.14861–14866.

Szabo, I. and Zoratti, M., 2014. Mitochondrial channels: ion fluxes and more. *Physiol Rev,* 94, pp.519–608.

Szewczyk, A., Jarmuszkiewicz, W. and Kunz, W. S., 2009. Mitochondrial potassium channels. *IUBMB Life,* 61, 134–143.

Tai, K. K., McCrossan, Z. A. and Abbott, G.W., 2003. Activation of mitochondrial ATP-sensitive potassium channels increases cell viability against rotenone-induced cell death. *J Neurochem,* 84, pp.1193–1200.

Teshima, Y., et al., 2003. Mitochondrial ATP-sensitive potassium channel activation protects cerebellar granule neurons from apoptosis induced by oxidative stress. *Stroke,* 34, pp.1796–1802.

Thornton, N., et al., 2010. Two modular forms of the mitochondrial sorting and assembly machinery are involved in biogenesis of alpha-helical outer membrane proteins. *J Mol Biol,* 396, pp.540–549.

Ujwal, R., et al., 2008. The crystal structure of mouse VDAC1 at 2.3 A resolution reveals mechanistic insights into metabolite gating. *Proc Natl Acad Sci U S A,* 105, pp.17742–17747.

van der Laan, M., et al., 2005. Pam17 is required for architecture and translocation activity of the mitochondrial protein import motor. *Mol Cell Biol,* 25, pp.7449–7458.

van Wilpe, S., et al., 1999. Tom22 is a multifunctional organizer of the mitochondrial preprotein translocase. *Nature,* 401, pp.485–489.

Vande Velde, C., et al., 2011. Misfolded SOD1 associated with motor neuron mitochondria alters mitochondrial shape and distribution prior to clinical onset. *PLoS One,* 6, p.e22031.

Vander Heiden, M. G., et al., 2000. Outer mitochondrial membrane permeability can regulate coupled respiration and cell survival. *Proc Natl Acad Sci U S A,* 97, p.4666–4671.

Vasington, F. D. and Murphy, J. V., 1962. Ca ion uptake by rat kidney mitochondria and its dependence on respiration and phosphorylation. *J Biol Chem,* 237, pp.2670–2677.

Verschoor, A., Tivol, W. F. and Mannella, C. A., 2001. Single-particle approaches in the analysis of small 2D crystals of the mitochondrial channel VDAC. *J Struct Biol,* 133, pp.254–265.

Vila, M., et al., 2001. Bax ablation prevents dopaminergic neurodegeneration in the 1-methyl-4-phenyl-1,2,3,6-tetrahydropyridine mouse model of Parkinson's disease. *Proc Natl Acad Sci U S A,* 98, pp.2837–2842.

Virgili, N., et al., 2013. K(ATP) channel opener diazoxide prevents neurodegeneration: a new mechanism of action via antioxidative pathway activation. *PLoS One,* 8, p.e75189.

von Lewinski, F. and Keller, B. U., 2005. Ca2+, mitochondria and selective motoneuron vulnerability: implications for ALS. *Trends Neurosci,* 28, pp.494–500.

Vukosavic, S., et al., 1999. Bax and Bcl-2 interaction in a transgenic mouse model of familial amyotrophic lateral sclerosis. *J Neurochem,* 73, pp.2460–2468.

Wang, H. Q., et al., 2005. Interaction of presenilins with FKBP38 promotes apoptosis by reducing mitochondrial Bcl-2. *Hum Mol Genet,* 14, pp.1889–1902.

Wang, L., et al., 2011. The protective roles of mitochondrial ATP-sensitive potassium channels during hypoxia-ischemia-reperfusion in brain. *Neurosci Lett,* 491, pp.63–67.

Wang, X. and Schwarz, T. L., 2009. The mechanism of Ca2+ -dependent regulation of kinesin-mediated mitochondrial motility. *Cell,* 136, pp.163–174.

Wang, Y., et al., 2012. The newly identified K+ channel blocker talatisamine attenuates beta-amyloid oligomers induced neurotoxicity in cultured cortical neurons. *Neurosci Lett,* 518, pp.122–127.

Wei, M. C., et al., 2001. Proapoptotic BAX and BAK: a requisite gateway to mitochondrial dysfunction and death. *Science,* 292, pp.727–730.

Xie, J., et al., 2010. K(ATP) channel openers protect mesencephalic neurons against MPP+-induced cytotoxicity via inhibition of ROS production. *J Neurosci Res,* 88, pp.428–437.

Xu, W., et al., 2002. Cytoprotective role of Ca2+- activated K+ channels in the cardiac inner mitochondrial membrane. *Science,* 298, pp.1029–1033.

Yano H., et al., 2014. Inhibition of mitochondrial protein import by mutant huntingtin. *Nature neuroscience,* 17, pp.822–831.

Yasui, M., Kihira, T. and Ota, K., 1992a. Calcium, magnesium and aluminum concentrations in Parkinson's disease. *Neurotoxicology,* 13, 593–600.

Yasui, M., et al., 1992b. Magnesium and calcium contents in CNS tissues of amyotrophic lateral sclerosis patients from the Kii peninsula, *Japan. Eur Neurol,* 32, pp.95–98.

Yi, M., Weaver, D. and Hajnoczky, G., 2004. Control of mitochondrial motility and distribution by the calcium signal: a homeostatic circuit. *J Cell Biol,* 167, pp.661–672.

Young, J. C., Hoogenraad, N. J. and Hartl, F. U., 2003. Molecular chaperones Hsp90 and Hsp70 deliver preproteins to the mitochondrial import receptor Tom70. *Cell,* 112, pp.41–50.

Young, K. W., et al., 2010. Mitochondrial fragmentation and neuronal cell death in response to the Bcl-2/Bcl-x(L)/Bcl-w antagonist ABT-737. *Neuropharmacology,* 58, pp.1258–1267.

Zeng, J., Wang, G. and Chen, S. D., 2007. ATP-sensitive potassium channels: novel potential roles in Parkinson's disease. *Neurosci Bull,* 23, 370–376.

Zizi, M., et al., 1994. NADH regulates the gating of VDAC, the mitochondrial outer membrane channel. *J Biol Chem,* 269, pp.1614–1616.

Zoratti, M. and Szabo, I., 1995. The mitochondrial permeability transition. *Biochim Biophys Acta,* 1241, pp.139–176.

Zsurka, G., Gregan, J. and Schweyen, R.J., 2001. The human mitochondrial Mrs2 protein functionally substitutes for its yeast homologue, a candidate magnesium transporter. *Genomics,* 72, pp.158–168.

SECTION II

CONTROL OF MITOCHONDRIAL SIGNALING NETWORKS

4

MITOCHONDRIAL Ca^{2+} TRANSPORT IN THE CONTROL OF NEURONAL FUNCTIONS: MOLECULAR AND CELLULAR MECHANISMS

YURIY M. USACHEV

Department of Pharmacology, University of Iowa College of Medicine, Iowa City, Iowa

4.1 INTRODUCTION

The ability of mitochondria to transport Ca^{2+} has been recognized for several decades (Carafoli, 2003; Carafoli and Lehninger, 1971; Deluca and Engstrom, 1961; Drago, Pizzo, and Pozzan, 2011; Vasington and Murphy, 1962). The fact that mitochondrial Ca^{2+} transport remains a focus of intensive investigation to this day reflects the importance of this process in the control of many determinants of cell life and death (Babcock and Hille, 1998; Friel, 2000; Newmeyer and Ferguson-Miller, 2003; Nicholls and Budd, 2000; Orrenius, Zhivotovsky, and Nicotera, 2003; Pozzan and Rizzuto, 2000; Thayer et al., 2002) (Fig. 4.1). In neurons, mitochondria efficiently buffer Ca^{2+} influx during excitation, limiting the amplitude of the Ca^{2+} signal. Rapid Ca^{2+} buffering by mitochondria is followed by a much slower Ca^{2+} release back to the cytosol (Bernardi, 1999; Nicholls, 2005; Thayer, Usachev, and Pottorf, 2002), and such Ca^{2+} cycling across the inner mitochondrial membrane markedly prolongs the cytosolic Ca^{2+} response by producing a so-called $[Ca^{2+}]_i$ plateau (Figs. 4.1 and 4.2). This phenomenon occurs both in central (Medler and Gleason, 2002; White and Reynolds, 1996) and peripheral (Colegrove, Albrecht, and Friel, 2000a; Dedov and

The Functions, Disease-Related Dysfunctions, and Therapeutic Targeting of Neuronal Mitochondria, First Edition. Edited by Valentin K. Gribkoff, Elizabeth A. Jonas and J. Marie Hardwick.

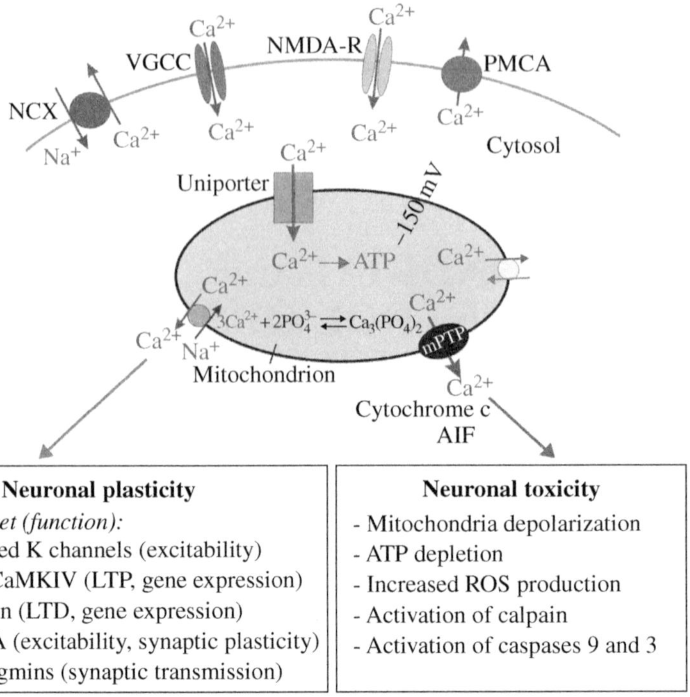

FIGURE 4.1 Mitochondrial Ca^{2+} signaling in neurons. Ca^{2+} entering neurons via voltage-gated Ca^{2+} channels (VGCC), glutamate NMDA receptors (NMDA-R), or other Ca^{2+}-permeable channels (e.g., TRP channels) is rapidly taken up by mitochondria. This Ca^{2+} uptake is mediated by the mitochondrial Ca^{2+} uniporter. Within mitochondria, Ca^{2+} stimulates ATP synthesis through activation of several Ca^{2+}-activated dehydrogenases (Denton, 2009; Glancy and Balaban, 2012). Free Ca^{2+} within the matrix is buffered into Ca-phosphate complexes. Ca^{2+} is released from mitochondria back into the cytosol via at least three mechanisms, Na^+/Ca^{2+} exchanger, H^+/Ca^{2+} exchanger, and mitochondrial permeability transition pore (mPTP). Under physiological conditions, Ca^{2+} mobilization from mitochondria is mediated primarily by the mitochondrial Na^+/Ca^{2+} exchanger. The Ca^{2+} released from mitochondria is ultimately extruded from neurons via the plasma membrane Ca^{2+}-ATPases (PMCA) and Na^+/Ca^{2+} exchangers (NCX). Ca^{2+} uptake and release by mitochondria reduces the amplitude of increase in cytosolic Ca^{2+} concentration ($[Ca^{2+}]_i$), and prolongs the duration of $[Ca^{2+}]_i$ elevation. By shaping $[Ca^{2+}]_i$, mitochondrial Ca^{2+} cycling modulates activity of many neuronal Ca^{2+} targets and influences numerous Ca^{2+}-dependent functions such as excitability, synaptic transmission, and gene regulation. Under pathological conditions, mitochondrial overload with Ca^{2+} triggers opening of the mPTP, a large channel permeable for Ca^{2+} and small proteins including the pro-apoptotic factors cytochrome c and apoptosis-inducing factor (AIF). Opening of the mPTP triggers a cascade of toxic events (e.g., mitochondrial depolarization, collapse of ATP synthesis, increase in reactive oxygen species (ROS) production, Ca^{2+} deregulation, activation of caspases, and the Ca^{2+}-dependent protease calpain), ultimately leading to neuronal demise. The deregulation of mitochondrial Ca^{2+} handling is implicated in neuronal toxicity in stroke, Alzheimer's disease, Parkinson's disease, Huntington's disease, amyotrophic lateral sclerosis (ALS), and other neurodegenerative disorders (Bezprozvanny, 2009; Mattson, Gleichmann, and Cheng, 2008; Pivovarova and Andrews, 2010). (*See insert for color representation of the figure.*)

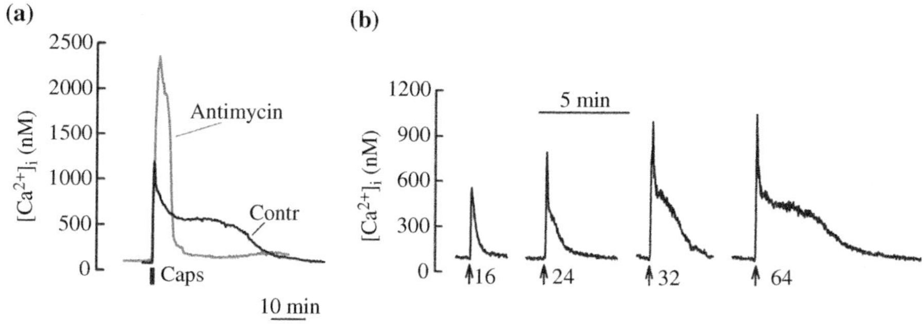

FIGURE 4.2 Mitochondrial Ca^{2+} cycling regulates amplitude and duration of $[Ca^{2+}]_i$ responses in neurons. Changes in cytosolic Ca^{2+} concentration ($[Ca^{2+}]_i$) were studied in cultured dorsal root ganglion (DRG) sensory neurons using the Ca^{2+}-sensitive indicator Fura-2 as previously described (Kim and Usachev, 2009; Kim et al., 2014; Schnizler et al., 2008; Usachev et al., 2002). **a**, Superimposition of $[Ca^{2+}]_i$ responses to the TRPV1 agonist capsaicin (1 μM; 30 s) recorded in control *(black)* or in the presence of the electron transport inhibitor antimycin (1 μM) and ATP-synthase inhibitor oligomycin (2 μM) *(grey trace)*. Both traces were obtained from the same cell. Inhibition of mitochondrial Ca^{2+} transport by the antimycin with oligomycin treatment increased the amplitude and reduced the duration of capsaicin-evoked $[Ca^{2+}]_i$ elevation (compare grey and black traces). **b**, $[Ca^{2+}]_i$ transients were evoked by trains of action potentials (8 Hz; *arrows*) of increased duration using extracellular field stimulation. The number of action potentials is shown under the traces. Increasing the strength of the stimulation led to a gradual appearance and prolongation of the $[Ca^{2+}]_i$ plateau phase mediated by Ca^{2+} release from mitochondria.

Roufogalis, 2000; Friel and Tsien, 1994; Shishkin et al., 2002; Thayer and Miller, 1990) neurons but is more pronounced in the latter.

Ca^{2+} transported into the mitochondrial matrix stimulates Ca^{2+}-dependent dehydrogenases and boosts ATP production to meet increased energy demand during excitation (Denton, 2009; Glancy and Balaban, 2012; Hajnoczky et al., 1995). However, excessive load of mitochondria with Ca^{2+} triggers neurotoxic processes in stroke and in Alzheimer's, Parkinson's, and Huntington's diseases (Bezprozvanny, 2009; Moskowitz, Lo, and Iadecola, 2010; Murphy, Fiskum, and Beal, 1999; Nicholls and Budd, 2000; Panov et al., 2002; Reynolds, 1999). In addition, mitochondrial-dependent alteration of the amplitude and duration of cytosolic Ca^{2+} signals influences the activation and function of numerous cytosolic Ca^{2+} targets, including Ca^{2+}-activated ion channels (e.g., small- and large-conductance Ca^{2+}-activated K^+ channels, SK and BK), Ca^{2+} sensors at the synapse (e.g., synaptotagmins and Doc2) and numerous Ca^{2+}-dependent enzymes (e.g., calmodulin-dependent protein kinases, protein kinases A and C, protein phosphatase 2B/calcineurin, and the Ca^{2+}-dependent protease calpain) (Berridge, Bootman, and Lipp, 1998; Clapham, 2007; Groffen et al., 2010; Kim and Usachev, 2009; Sah, 1996; Sugita et al., 2002; Xu, Mashimo, and Sudhof, 2007; Yao et al., 2011). Accordingly, mitochondrial Ca^{2+} transport regulates many Ca^{2+}-dependent functions including neuronal excitability, synaptic transmission, gene expression, and neuronal survival (see Fig. 4.1).

The fact that mitochondrial Ca^{2+} signaling is key to controlling neuronal life and death makes the components of mitochondrial Ca^{2+} transport attractive candidates for therapeutic targeting in the treatment of neurodegenerative and psychiatric disorders. Although progress in this area was long hindered by a lack of information on the identities of the molecules that mediate Ca^{2+} transport in and out of mitochondria, the recent discovery of several genes encoding key mitochondrial Ca^{2+} transporters (Marchi and Pinton, 2014; Pendin, Greotti, and Pozzan, 2014) is expected to accelerate our progress toward understanding and therapeutically exploiting this machinery.

In this chapter I will review recent advancements in the molecular biology of mitochondrial Ca^{2+} signaling in neurons and discuss how Ca^{2+} transport into and out of neuronal mitochondria contributes to the control of neuron excitability, synaptic transmission, and gene expression (see Fig. 4.1). The role of mitochondrial Ca^{2+} overload in neuronal toxicity, which is another important aspect of mitochondrial Ca^{2+} transport in neurons, is reviewed in the contribution by Brustovetsky (see Chapter 1).

4.2 PHYSIOLOGICAL AND PHARMACOLOGICAL CHARACTERISTICS OF MITOCHONDRIAL Ca^{2+} TRANSPORT IN NEURONS

In neurons at rest, total Ca^{2+} concentration in the matrix ($[Ca]_{mt}$) is low, usually <100 nM (Pivovarova et al., 1999; Stanika et al., 2012). During electrical or synaptic stimulation, $[Ca]_{mt}$ can increase to millimolar levels within seconds (Pivovarova et al., 1999, 2002). In the case of modest to strong stimulation, the rate of mitochondrial Ca^{2+} uptake in intact neurons can be as high as 200 μM/s relative to the mitochondrial volume (Colegrove et al., 2000a; Pivovarova et al., 1999). Ca^{2+} transport from the cytosol into the mitochondrial matrix requires crossing the outer mitochondrial membrane (OMM) and inner mitochondrial membrane (IMM). The OMM is permeable to molecules smaller than 5 kDa, including Ca^{2+}. This permeability is ascribed to voltage-dependent anion channels (VDACs) also known as mitochondrial porins that are highly expressed in the OMM (Rizzuto et al., 2012; Shoshan-Barmatz et al., 2010; Spat et al., 2008). It is generally assumed that $[Ca^{2+}]$ in the intermembrane space is the same as in the cytosol. Ca^{2+} transport across the IMM is mediated by a designated transporter, mitochondrial Ca^{2+} uniporter, and is driven by a steep electrical potential across the IMM ($\Delta\Psi_{mt} \sim -150$ mV) (Friel, 2000; Nicholls, 2005; Thayer et al., 2002). Recent electrophysiological and molecular studies strongly suggest that the mitochondrial Ca^{2+} uniporter functions as a Ca^{2+}-selective ion channel, which is consistent with the high rate of Ca^{2+} uptake by mitochondria (Chaudhuri et al., 2013; De Stefani et al., 2011; Fieni et al., 2012; Kirichok, Krapivinsky, and Clapham, 2004). The Ca^{2+} uniporter is potently blocked by its specific inhibitor Ru360 (IC_{50}~0.2-2 nM) (Kirichok et al., 2004; Matlib et al., 1998), as well as by the less selective parent compound Ruthenium Red (IC_{50}~7-9 nM), a highly charged compound that also blocks other Ca^{2+} channels including TRP channels and ryanodine receptors (Kirichok et al., 2004; Kostyuk and Verkhratsky, 1994; Matlib et al., 1998; Patapoutian et al., 2003; Wu Sweet, and Clapham, 2010). Mitochondrial Ca^{2+} uptake can also be inhibited by mitochondria-depolarizing agents, such as the protonophore p-trifluoromethoxyphenylhydrazone (FCCP) and the electron transport inhibitors rotenone and antimycin (Nicholls et al., 2003; Thayer et al., 2002).

The mitochondrial Ca^{2+} uniporter channel is gated by Ca^{2+}, and the rate of mitochondrial Ca^{2+} uptake depends on the cytosolic Ca^{2+} concentration ($[Ca^{2+}]_i$) as a power function of ~2 to 2.5 (Colegrove et al., 2000a; Gunter and Pfeiffer, 1990; Nicholls, 2012; Shutov et al., 2013). Previous studies in isolated mitochondria suggested that the EC_{50} of the $[Ca^{2+}]_i$ for uniporter activation is ~10 to 20 μM (Gunter and Pfeiffer, 1990). Such a requirement implied that mitochondria are involved in Ca^{2+} transport only in a setting of intense stimulation or pathological $[Ca^{2+}]_i$ elevations caused by toxic conditions. However, most recent studies using intact neurons indicate that mitochondrial Ca^{2+} uptake can be induced by $[Ca^{2+}]_i$ elevations as low as 200 to 300 nM, and that mitochondrial Ca^{2+} transport contributes to many aspects of Ca^{2+} signaling in neurons, both under physiological and pathological conditions (Chouhan et al., 2010; Colegrove et al., 2000a; David, 1999; Kim and Usachev, 2009; Medvedeva, Kim, and Usachev, 2008; Pivovarova et al., 1999, 2004; Shutov et al., 2013; Stout et al., 1998; Werth and Thayer, 1994). For example, using electron microscopy and electron probe X-ray microanalysis (EPMA) to study bullfrog sympathetic neurons, Pivovarova et al. (1999) found that depolarization induced by applying 30 mM KCl for 2 min led to an elevation of total mitochondrial $[Ca^{2+}]$ from a ~0 to 8.6 mM (12.8 mmol/kg dry weight), which corresponds to a rate of mitochondrial Ca^{2+} uptake of ~72 μM/s relative to the mitochondrial volume. Parallel measurements using the Ca^{2+}-sensitive indicator Fura-2 showed that for most of this stimulation period the $[Ca^{2+}]_i$ was at ~200 to 300 nM. In our recent study of rat sensory neurons, quantification of the rate of mitochondrial Ca^{2+} uptake as a function of the presynaptic $[Ca^{2+}]_i$ elevation gave the $K_{0.5}$ value of ~0.6 μM (Shutov et al., 2013), which is consistent with the high affinity of mitochondrial Ca^{2+} uptake mechanisms ($K_{0.5}$~0.4 to 0.5 μM) reported for nonmammalian synapses (Chouhan et al., 2010; David, 1999). Thus, mitochondrial Ca^{2+} uptake in neurons is sensitive to modest elevations in $[Ca^{2+}]_i$, is activated by cooperative Ca^{2+} binding and is rapid.

Within the matrix, free Ca^{2+} is rapidly buffered by its incorporation into Ca-phosphate complexes, primarily in the forms of $Ca_3(PO_4)_2$ and $CaHPO_4$ (Chalmers and Nicholls, 2003; Kristian et al., 2007). Heavy Ca^{2+} buffering in the matrix helps to explain why $[Ca^{2+}]_{mt}$ typically does not exceed 1 to 5 μM in neurons even after strong stimulation (Billups and Forsythe, 2002; Chalmers and Nicholls, 2003; David, 1999; David, Talbot, and Barrett, 2003). The bound-to-free Ca^{2+} ratio is a dynamic value that increases as the Ca^{2+} load of mitochondria increases (Chalmers and Nicholls, 2003). In neurons at rest or with a low mitochondrial Ca^{2+} load, the ratio is estimated to be ~4,000:1, and as mitochondrial Ca^{2+} approaches toxic levels the ratio increases to as much as 150,000:1 (Babcock et al., 1997; Chalmers and Nicholls, 2003; Pivovarova et al., 2004; Stanika et al., 2012). Such an adjustable Ca^{2+} buffering system is enabled by at least two additional aspects of mitochondrial Ca^{2+} uptake. One is the transport of phosphate into the matrix that accompanies Ca^{2+} uptake by mitochondria (Nicholls, 2005; Nicholls and Ferguson, 2013). The second is an efflux of H^+ that takes place during mitochondrial Ca^{2+} uptake. This is highly relevant because of the pH-dependence of the solubility of Ca^{2+}-phosphate complexes (Nicholls, 2005; Nicholls and Ferguson, 2013; Rossi and Lehninger, 1964). Indeed, alkalinization of the matrix is predicted to increase the formation of Ca^{2+}-phosphate complexes (Chalmers and Nicholls, 2003). Just the same, acidification of the matrix by protonophores such as FCCP causes Ca^{2+} to dissociate from phosphates, which in turn contributes to a massive release of Ca^{2+} from mitochondria in response to FCCP-induced depolarization of the organelle (Brocard, Tassetto, and Reynolds, 2001; Chalmers and Nicholls, 2003; Thayer and Miller, 1990).

Ca^{2+} accumulated in mitochondria can be released via at least three mechanisms, a Na^+/Ca^{2+} exchanger, a H^+/Ca^{2+} exchanger, and the mitochondrial permeability transition pore (mPTP) (Bernardi, 1999; Nicholls, 2005; Thayer et al., 2002). It is generally agreed that under physiological conditions the mitochondrial Na^+/Ca^{2+} exchanger (mtNCX) is the main pathway mediating Ca^{2+} efflux from mitochondria in neurons (Friel, 2000; Nicholls, 2005; Thayer et al., 2002), (but see Jonas et al., 1999). This conclusion is based on the sensitivity of Ca^{2+} transport to a selective inhibitor of mtNCX, CGP37157, as well as its dependence on intracellular Na^+ (Baron and Thayer, 1997; Colegrove et al., 2000a; Garcia-Chacon et al., 2006; Medvedeva et al., 2008; Thayer and Miller, 1990). Because the rate of Ca^{2+} efflux from mitochondria is significantly slower than that of Ca^{2+} uptake, Ca^{2+} rapidly accumulates in mitochondria following its entry into the cell during stimulation. Using an approach combining pharmacological tools and mathematical modeling, Friel and colleagues estimated that Ca^{2+} is removed from neuronal mitochondria at 35 to 75 nM/s, a rate ~1000-fold slower than the rate of mitochondrial Ca^{2+} uptake (Colegrove et al., 2000a; Colegrove, Albrecht, and Friel, 2000b). However, EPMA-based examination of sympathetic neurons showed that total matrix Ca^{2+} recovered from a depolarization-induced peak value of 22 mM to 0.9 mM within 15 minutes, which yields a much faster rate value of ~23 μM/s (Pivovarova et al., 1999). Removal of 1 Ca^{2+} by mtNCX is coupled to mitochondrial uptake of 3 Na^+ (Bernardi, 1999; Palty and Sekler, 2012). Thus, mtNCX is electrogenic, and its activity is predicted to be reduced by mitochondrial depolarization.

By rapidly sequestering the large amounts of Ca^{2+} that enter the cell during stimulation and then gradually releasing Ca^{2+} back into the cytosol, mitochondria can profoundly affect the amplitude and time course of cytosolic Ca^{2+} signals in neurons. This ability to modulate activity-induced $[Ca^{2+}]_i$ transients has been well documented in a wide variety of peripheral and central neurons (reviewed in Barrett, 2001; Friel, 2000; Miller, 1988; Nicholls, 2005; Pivovarova and Andrews, 2010; Thayer et al., 2002), and it is perhaps best illustrated in sensory and sympathetic neurons (Colegrove et al., 2000a; Friel and Tsien, 1994; Kim and Usachev, 2009; Lu, Zhang, and Gold, 2006; Medvedeva et al., 2008; Thayer and Miller, 1990; Werth and Thayer, 1994). Figure 4.2**a** compares $[Ca^{2+}]_i$ transients elicited by the TRPV1 agonist capsaicin in a rat dorsal root ganglion (DRG) neuron under control conditions (black trace) and following blockade of mitochondrial Ca^{2+} uptake by the electron transport inhibitor antimycin (red trace). In the control, $[Ca^{2+}]_i$ response showed a characteristic $[Ca^{2+}]_i$ plateau caused by Ca^{2+} release from mitochondria. Inhibition of mitochondrial Ca^{2+} uptake resulted in an about twofold increase in the amplitude of $[Ca^{2+}]_i$ elevation and elimination of the $[Ca^{2+}]_i$ plateau phase (Fig. 4.2**a**).

Another example (Fig. 4.2**b**) illustrates the function of mitochondria during electrical stimulation of a DRG neuron using trains of action potentials of incremental duration. In the case of mild stimulation (16 to 32 action potentials in this example), as the number of action potentials increased, the amplitude of the $[Ca^{2+}]_i$ response gradually increased as well. However, with the activation of mitochondrial Ca^{2+} uptake by larger $[Ca^{2+}]_i$ elevations, the rise in amplitude of the $[Ca^{2+}]_i$ response approached an upper limit because of robust buffering by mitochondria (compare $[Ca^{2+}]_i$ responses elicited by 32 and 64 action potentials in Fig. 4.2**a**). At the same time, the $[Ca^{2+}]_i$ transients began to exhibit a characteristic $[Ca^{2+}]_i$ plateau during the recovery phase, resulting from a Ca^{2+} release from mitochondria, and the duration of this plateau increased with stimulus strength. Indeed, the more Ca^{2+} is taken up by mitochondria during the stimulus, the longer it takes to unload the accumulated Ca^{2+} back into the cytosol, which is reflected by progressive prolongation of the plateau phase (Fig. 4.2**b**). Thus, the mitochondrion is not merely a Ca^{2+} buffer but also serves the more complex role of translating the amplitude of the input into the duration of the response. In other words, the $[Ca^{2+}]_i$ plateau carries memory of the type and intensity of the stimulus. As has been shown, this mitochondria-dependent modulation of Ca^{2+} signaling plays an important role in regulating neuronal excitability, synaptic plasticity, and gene expression (Billups and Forsythe, 2002; David and Barrett, 2003; Garcia-Chacon et al., 2006; Kim and Usachev, 2009; Medvedeva et al., 2008; Tang and Zucker, 1997; Vanden Berghe, Kenyon, and Smith, 2002; Wheeler et al., 2012).

Although mitochondria effectively buffer Ca^{2+} in central neurons, the $[Ca^{2+}]_i$ plateau is these cells is not as pronounced as that in sensory and sympathetic neurons (Brustovetsky et al., 2011; Castilho et al., 1998; Stanika et al., 2009; Wang and Thayer, 2002). This is illustrated in Figure 4.3 that compares mitochondrial Ca^{2+} signaling between DRG (**a**) and hippocampal (**b**) neurons, as assessed by simultaneously monitoring Ca^{2+} concentration in the cytosol ($[Ca^{2+}]_i$; black) and mitochondria ($[Ca^{2+}]_{mt}$; red). As shown in Figure 4.3**a**, in DRG neurons the $[Ca^{2+}]_i$ and $[Ca^{2+}]_{mt}$ displayed similar kinetics, whereas in hippocampal neurons $[Ca^{2+}]_i$ rapidly recovered to the baseline but $[Ca^{2+}]_{mt}$ remained elevated for more than 15 minutes after depolarization was terminated (Fig. 4.3**b**). One potential explanation for this difference is based on the geometry of the neurons; higher surface-to-volume ratio in the smaller and flatter hippocampal neurons may improve their ability to return $[Ca^{2+}]_i$ to baseline levels (via extrusion through plasma membrane Ca^{2+}-ATPase and Na^+/Ca^{2+} exchangers) despite Ca^{2+} release from mitochondria. In addition, the observed protracted recovery of $[Ca^{2+}]_{mt}$ in hippocampal neurons suggests that the rate of mitochondrial Ca^{2+} efflux is slower in these cells than in DRG neurons (see Fig. 4.3). Future molecular studies will determine whether or not this reflects the differences in the molecules that mediate mitochondrial Ca^{2+} release in DRG and hippocampal neurons.

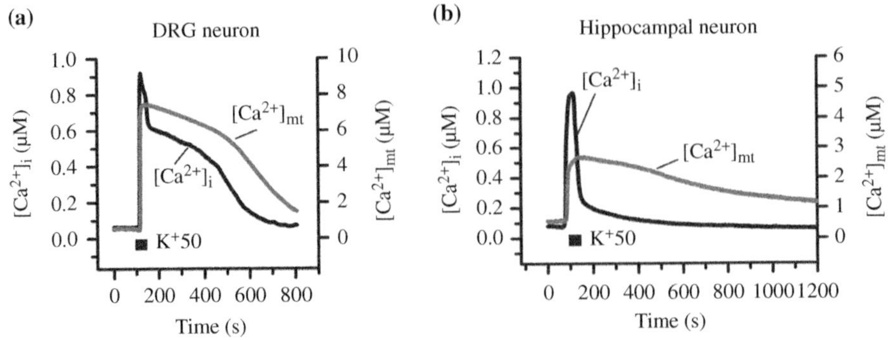

FIGURE 4.3 Comparison of depolarization-induced mitochondrial Ca^{2+} signals between dorsal root ganglion (DRG) and hippocampal neurons. Ca^{2+} concentration was simultaneously monitored in the cytosol ($[Ca^{2+}]_i$) and mitochondria ($[Ca^{2+}]_{mt}$) of DRG (**a**) and hippocampal (**b**) neurons using Fura2 *(black)* and genetic mitochondrial Ca^{2+} indicator mito-LAR-GECO1.2 *(grey)*, respectively, as recently described (Wu et al., 2014). Depolarization was induced by applying extracellular solution containing 50 mM KCl (30 s).

4.3 MOLECULAR COMPONENTS OF MITOCHONDRIAL Ca^{2+} TRANSPORT IN NEURONS

Although the physiological characteristics of mitochondrial Ca^{2+} transport have been known for several decades, the molecular nature of the transporters within the IMM remained enigmatic until 2010, when Mootha et al. (Perocchi et al., 2010) identified mitochondrial calcium uptake 1 (MICU1, also known as CBARA1) as an essential component of the mitochondrial Ca^{2+} uniporter complex. A year later, both the Mootha and Rizzuto groups identified mitochondrial calcium uniporter (MCU, also known as CCDC109A) as a pore-forming subunit of the complex (Baughman et al., 2011; De Stefani et al., 2011), and since then, a number of additional components essential to the function of the complex have been discovered (Fig. 4.4). These include an MCU paralog, MCUb (also known as CCDC109B) (Raffaello et al., 2013), two MICU1 paralogs, MICU2 and MICU3 (also known as EFHA1 and EFHA2, respectively) (Plovanich et al., 2013), mitochondrial calcium uniporter regulator 1 (MCUR1, also known as CCDC90A) (Mallilankaraman et al., 2012), and essential MCU regulator, EMRE (Sancak et al., 2013).

Based on molecular studies and computational modeling, it is proposed that MCU and MCUb represent pore-forming subunits of the mitochondrial Ca^{2+} uniporter, and that they can exist either as homo- or hetero-oligomers (most likely tetramers) (Baughman et al., 2011; Marchi and Pinton, 2014; Pendin et al., 2014; Raffaello et al., 2013). These proteins share ~50% homology and have a similar topology characterized by two transmembrane domains that span the IMM, with both the N- and C-termini facing the mitochondrial matrix. Also, in both proteins the two transmembrane domains are connected by a short linker region that is highly conserved among species and critical for Ca^{2+} selectivity (Baughman et al., 2011; De Stefani et al., 2011;

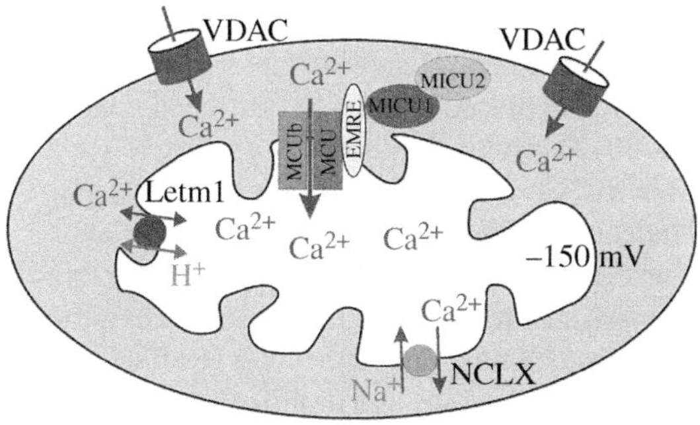

Expression and function of mitochondrial Ca^{2+} transporters in neurons

Protein	Alternative name	Function	Nervous system expression (method)	KD or KO phenotype	References
MCU	CCDC109A	Ca^{2+} uptake	Cortex, cerebellum, hippocacampus, hypothalamus, spinal cord, DRG (mouse gene expression atlas and PCR) Cortex and hippocampus (western blot)	Reduced mitochondrial Ca^{2+} uptake, protection from excitotoxicity	Plovanich et al., 2013 Qiu et al., 2013 Sankak et al., 2013 Shutov et al., 2013
MCUb	CCDC109B	Ca^{2+} uptake	Cortex, cerebellum, hypothalamus, spinal cord (mouse gene expression atlas and qPCR)		Sankak et al., 2013 Raffaello et al., 2013
MICU1	CBARA1	Ca^{2+} uptake regulator	Cortex, cerebellum, hypothalamus, spinal cord, DRG (mouse gene expression atlas and PCR) Cerebellum, ventral tegmental nuclei (immunohistochemistry)	Increased mitochondrial Ca^{2+} uptake, learning impairment, ataxia, neuropathies **loss-of-function mutations	Plovanich et al., 2013 Logan et al., 2014 Sankak et al., 2013 Shutov et al., 2013
MICU2	EFHA1	Ca^{2+} uptake regulator	Hypothalamus, spinal cord (mouse gene expression atlas)		Plovanich et al., 2013 Sankak et al., 2013
MICU3	EFHA2	Ca^{2+} uptake regulator	Cortex, cerebellum, hypothalamus, spinal cord (mouse gene expression atlas)		Plovanich et al., 2013 Sankak et al., 2013
EMRE		Ca^{2+} uptake regulator	Cortex, cerebellum, hypothalamus, spinal cord (mouse gene expression atlas)		Plovanich et al., 2013 Sankak et al., 2013
Letm1		Ca^{2+} uptake and release	Brain (Western blot) DRG (PCR)	Reduced mitochondrial Ca^{2+} uptake, impaired control of ATP synthesis in the brain, increased seizures	Tamai et al., 2008 Jiang et al., 2013 Shutov et al., 2013
NCLX	SLC8B1	Ca^{2+} release	Brain (western blot and immunohistochemistry with EM) DRG (PCR)		Palty et al., 2010 Shutov et al., 2013

FIGURE 4.4 Molecular components of mitochondrial Ca^{2+} transport in neurons. The upper panel shows the main components of the mitochondrial Ca^{2+} transport machinery in the inner mitochondrial membrane. The table summarizes the information about the expression and function of the mitochondrial Ca^{2+} transporters in neurons, including the data from knockdown (KD) and knockout (KO) studies, as well as a recent report on the loss-of-function MICU1 mutations in humans. (*See insert for color representation of the figure.*)

Marchi and Pinton, 2014; Pendin et al., 2014; Raffaello et al., 2013). This linker region is nearly identical between MCU and MCUb, with an exception of the Asp at position 256 in MCU being replaced with a Val at the corresponding position in MCUb (Raffaello et al., 2013). This substitution of a negatively charged Asp with an electrically neutral Val in MCUb was hypothesized to interfere with the Ca^{2+} conductance of the channel. Indeed, insertion of MCU into a lipid bilayer resulted in a Ca^{2+} channel conductance, whereas insertion of MCUb did not have this effect. Moreover, silencing of MCUb in HeLa cells led to an increase in mitochondrial Ca^{2+} uptake, and overexpression of this protein had the opposite effect (Raffaello et al., 2013). It was therefore proposed that MCUb functions as a dominant-negative subunit of the mitochondrial Ca^{2+} uniporter complex, and that the overall Ca^{2+} permeability of the complex is determined by the MCU/MCUb ratio, which appears to range from ~2:1 to 20:1 among various tissues (Chaudhuri and Clapham, 2014; Marchi and Pinton, 2014; Pendin et al., 2014; Raffaello et al., 2013). Consistent with this idea, conductance of the mitochondrial Ca^{2+} uniporter from tissues with a relatively high MCU/MCUb ratio (10 to 20 in skeletal muscle, kidney and liver) is 15- to 30-fold greater than in others characterized by a very low MCU/MCUb ratio (~2 to 3 in cardiac muscle) (Fieni et al., 2012; Raffaello et al., 2013).

Besides MCU and MCUb, a quantitative mass spectrometry analysis identified MICU1, MICU2, and EMRE as additional components of the mitochondrial Ca^{2+} uniporter complex (Sancak et al., 2013) (see Fig. 4.4). MICU1 and MICU2 contain two canonical Ca^{2+}-binding EF-hand domains, and therefore it was proposed that they control Ca^{2+}-dependent gating of the uniporter channel (Perocchi et al., 2010; Plovanich et al., 2013; Wang et al., 2014). Recent studies have shown that these proteins play nonredundant and essential roles in defining two important characteristics of the Ca^{2+} uniporter: the $[Ca^{2+}]_i$ activation threshold and Ca^{2+} cooperativity (Csordas et al., 2013; Kamer and Mootha, 2014; Patron et al., 2014). Specifically, MICU2 (and possibly MICU1) is responsible for keeping the channel closed when $[Ca^{2+}]_i$ is below the threshold, and MICU1 is required for cooperative (super-linear) Ca^{2+}-dependent activation of the uniporter when $[Ca^{2+}]_i$ is elevated (Kamer and Mootha, 2014; Patron et al., 2014). Both proteins are physically linked to MCU via EMRE, a small protein of 10 kDa. Immunoprecipitation analysis of mitochondrial proteins from cells lacking EMRE or MICU1 suggest that MICU1 interacts with EMRE directly, but that MICU2 is linked to EMRE via MICU1 (Sancak et al., 2013). MICU1 and MICU2 are likely located in the intermembrane space of mitochondria, consistent with the importance of cytosolic Ca^{2+} in regulating Ca^{2+} uptake by the uniporter (Csordas et al., 2013; Hung et al., 2014; Perocchi et al., 2010; Rhee et al., 2013), but see Hoffman et al. (2013) who propose that MICU1 is localized to the matrix. It should be mentioned that the third MICU paralog, MICU3, is expressed almost exclusively in the brain but its function is unknown (Plovanich et al., 2013).

Most of the information about the molecular organization of the mitochondrial Ca^{2+} uniporter complex comes from studies that were carried out in cell lines. As such, our knowledge of the expression and function of the MCU and MICU isoforms within various organ systems is limited. Here I will focus on the data available for the nervous system. At the mRNA level, expression of MCU, MCUb, MICU1, MICU2,

MICU3, and EMRE has been documented in the cerebral cortex, cerebellum, hippocampus, hypothalamus, and spinal cord (Plovanich et al., 2013; Qiu et al., 2013; Raffaello et al., 2013; Sancak et al., 2013). At the protein level, Western blotting has demonstrated MCU expression in cultured hippocampal and cortical neurons (Qiu et al., 2013), and immunohistochemistry has revealed MICU1 expression in the cerebellum and ventral tegmental nuclei (Logan et al., 2014). In addition, we have reported expression of MCU and MICU1 mRNAs in dorsal root ganglia of the rat (Shutov et al., 2013) (see Fig. 4.4).

Perhaps the strongest evidence available so far for the importance of the mitochondrial Ca^{2+} uniporter in the brain comes from a recent study describing loss-of-function mutations in *MICU1* in humans (Logan et al., 2014). These mutations led to intronic insertions that caused frame-shifts, and this resulted in a marked reduction in expression of the MICU1 mRNA. The individuals with these mutations suffer from learning difficulties, ataxia, and peripheral neuropathies (Logan et al., 2014). Cellular studies using fibroblasts from affected individuals demonstrated an elevation in Ca^{2+} uptake by mitochondria, both at rest and in response to low-intensity stimulation. The latter finding is consistent with the proposed role for MICU1 in gating mitochondrial Ca^{2+} entry via MCU. In another recent paper, Qiu et al. (2013) demonstrated that silencing of MCU reduced the mitochondrial Ca^{2+} load and provided protection against NMDA-induced toxicity in hippocampal and cortical neurons. Both of these discoveries highlight the importance of MCU and its regulatory proteins in neuronal functions and resistance to excitotoxic conditions.

Progress toward understanding mitochondrial Ca^{2+} transport had not been limited to identification of the key molecules of the mitochondrial Ca^{2+} uptake complex. Specifically, our understanding has been further advanced by the discovery of two genes that encode mitochondrial Ca^{2+} antiporters, *Letm1* and *NCLX* (Jiang, Zhao, and Clapham, 2009; Palty et al., 2010) (see Fig. 4.4). Leucine zipper-EF hand-containing transmembrane 1 (Letm1) has been proposed to function as a Ca^{2+}/H^+ exchanger in the IMM (Doonan et al., 2014; Jiang et al., 2009, 2013). The Letm1 stoichiometry is $1Ca^{2+}/2H^+$, a ratio that renders the antiport electroneutral (Tsai et al., 2014). The direction of Ca^{2+} transport by Letm1 depends on the Ca^{2+} and H^+ gradients across the inner mitochondrial membrane; for small $[Ca^{2+}]_i$ elevations, Letm1 contributes to Ca^{2+} accumulation by mitochondria, but in the case of high $[Ca^{2+}]_{mt}$ and low cytoplasmic pH, Letm1 transports Ca^{2+} out of the matrix (Jiang et al., 2009). There is also an alternative view that proposes that Letm1 functions as a K^+/H^+, but not Ca^{2+}/H^+ exchanger (Nowikovsky et al., 2004, 2012). Despite these conflicting views, it is generally agreed that Letm1 is critical for mitochondrial function. Indeed, Letm1 knockdown has been shown to result in mitochondrial swelling, depolarization, reduced ATP production, and increased generation of reactive oxygen species (ROS) (Doonan et al., 2014; Nowikovsky et al., 2004). In mice, Letm1 knockout results in embryonic lethality, and heterozygotes show a significant reduction of adenosine triphosphate (ATP) production and pyruvate dehydrogenase activity in the brain (Jiang et al., 2013). Given that pyruvate dehydrogenase is a Ca^{2+}-sensitive enzyme, the reduction in it activity in Letm1 deficient mice may reflect an impairment of mitochondrial Ca^{2+} transport (Denton, 2009; Glancy and

Balaban, 2012; Jiang et al., 2013). Letm1 is expressed in the brain and in the peripheral nervous system (Shutov et al., 2013; Tamai et al., 2008). Abnormalities in mitochondrial function associated with the loss of Letm1 have been implicated in the human disease, Wolf-Hirschhorn syndrome (WHS). This is a complex genetic disorder that is characterized by congenital malformation, mental retardation, and epilepsy and is caused by large deletions of the short arm of chromosome 4 (Andersen et al., 2014; Bergemann, Cole, and Hirschhorn, 2005). Importantly, mice heterozygous for Letm1 exhibit increased seizure activity, a prominent WHS symptom (Jiang et al., 2013). Collectively, these studies highlight the importance of Letm1 for brain function. Distribution of Letm1 in the nervous system and its role in neuronal Ca^{2+} signaling remain to be determined.

NCLX ($Na^+/Li^+/Ca^{2+}$ exchanger), also known as SLC8B1, is thought to function as a mitochondrial Na^+/Ca^{2+} exchanger and to be responsible for Ca^{2+} efflux from mitochondria (Palty et al., 2010). It belongs to the family of Ca^{2+}/cation exchangers (CCX) within the CaCa superfamily of Ca^{2+} transporters (Khananshvili, 2013; Lytton, 2007). Two other gene families within this superfamily encode the Na^+/Ca^{2+} exchangers (NCX1-3, also known as SLC8A1-3) and the $Na^+/K^+/Ca^{2+}$ exchangers (NCKX1-5, also known as SLC24A1-5), both of which are responsible for Ca^{2+} extrusion across the plasma membrane. A unique feature of NCLX relative to its NCX and NCKX counterparts is that it can operate in the absence of Na^+, exchanging Ca^{2+} for Li^+ (Palty et al., 2004, 2010). This interesting characteristic is also a property of the originally described mitochondrial Na^+/Ca^{2+} exchanger (Crompton, Kunzi, and Carafoli, 1977), and it can be exploited for pharmacological testing of the role of NCLX in intact cells. Although NCLX is expressed in the brain and in peripheral neurons (Palty et al., 2010; Shutov et al., 2013), its function in neurons remains unknown. Recent studies also reported that NCX1, NCX2, and NCX3 are located in the mitochondria of neurons, with NCX3 detected specifically at the OMM (Gobbi et al., 2007; Scorziello et al., 2013b). The latter observation combined with the other reports on the specialized and opposite roles of VDAC1 and VDAC2 in transporting Ca^{2+} across the OMM (De Stefani et al., 2012; Rapizzi et al., 2002), challenges the traditional view that Ca^{2+} can freely diffuse across the OMM (Rizzuto et al., 2012; Scorziello et al., 2013a; Shoshan-Barmatz et al., 2010; Spat et al., 2008). Future studies will determine the specific roles of NCLX and NCX in Ca^{2+} transport across the OMM and IMM in central and peripheral neurons.

4.4 MITOCHONDRIAL Ca^{2+} SIGNALING AND NEURONAL EXCITABILITY

One of the essential roles of Ca^{2+} as a second messenger is to regulate neuronal excitability. This is achieved through the gating of three types of ion channels: Ca^{2+}-activated K^+, Ca^{2+}-activated Cl^-, and Ca^{2+}-activated cation channels (Hartzell, Putzier, and Arreola, 2005; Marty, 1989; Partridge and Swandulla, 1988; Sah, 1996; Scott et al., 1995). The small- and large-conductance Ca^{2+}-activated K^+ channels, SK and BK, respectively, are widely expressed in neurons and play important roles in regulating their excitability (Sah, 1996; Wei et al., 2005). SK channels are opened when $[Ca^{2+}]_i$ is elevated to ~0.6 to 0.7 μM, and they regulate membrane afterhyperpolarization. BK channels require larger $[Ca^{2+}]_i$ elevations (to ~1 to 10 μM) and neuronal depolarization for their opening, and their activity contributes to the repolarization phase of the action potential (Berkefeld et al., 2006; Sah, 1996; Wei et al., 2005). Ca^{2+}-activated cation and Cl^- channels require $[Ca^{2+}]_i$ elevations in the 0.1- to 1-μM range to open, and the former channels contribute to neuronal depolarization, whereas the latter can either depolarize or hyperpolarize the membrane depending on the transmembrane Cl^- gradient and membrane potential (Hartzell et al., 2005; Partridge and Swandulla, 1988; Scott et al., 1995). Ca^{2+} also modulates many other channels and receptors including voltage-gated Ca^{2+} channels, TRP channels, and NMDA glutamate receptors (Budde, Meuth, and Pape, 2002; Caterina and Julius, 2001; Catterall, 2000; Catterall, Leal, and Nanou, 2013; Chuang, Neuhausser, and Julius, 2004; Clapham, 2003; McBain and Mayer, 1994).

Mitochondrial buffering and the release of Ca^{2+} have complex effects on Ca^{2+}-activated and -modulated conductances and neuronal excitability. For example, blocking mitochondrial Ca^{2+} uptake by applying protonophores or inhibitors of electron transport resulted in hyperpolarization mediated by Ca^{2+}-activated K^+ channels in hippocampal and myenteric neurons (Nowicky and Duchen, 1998; Vanden Berghe et al., 2002). Several reports also showed that blocking mitochondrial Ca^{2+} uptake reduced voltage-gated Ca^{2+} currents, likely the result of an increase in Ca^{2+}-dependent inactivation of the Ca^{2+} channels (Hernandez-Guijo et al., 2001; Medler and Gleason, 2002; Shutov et al., 2013). Thus, under normal conditions, mitochondrial Ca^{2+} uptake limits Ca^{2+}-dependent activation of BK and SK channels, and also minimizes Ca^{2+}-dependent inactivation of voltage-gated Ca^{2+} channels. The latter effect is especially critical for the synapse, where Ca^{2+} flux through these channels triggers transmitter release.

4.5 MITOCHONDRIAL Ca^{2+} CYCLING IN THE REGULATION OF SYNAPTIC TRANSMISSION

Presynaptic terminals are enriched for mitochondria, an observation that reflects high-energy demand at the site of transmitter release (De Camilli et al., 2001; Hollenbeck, 2005; Wimmer et al., 2006). The ATP produced by presynaptic mitochondria is required for the transport, release, and endocytosis of synaptic vesicles, as well as for the restoration of transmembrane ionic gradients by Na^+/K^+- and Ca^{2+}-ATPases (Harris, Jolivet, and Attwell, 2012; Ly and Verstreken, 2006). In addition to synthesizing ATP, mitochondria are involved in the control of presynaptic Ca^{2+} signaling. This mitochondrial role is also crucial for synaptic transmission because Ca^{2+} regulates almost every aspect of the synaptic vesicle life cycle including exocytosis, endocytosis, docking, and priming (Neher and Sakaba, 2008; Smith and Augustine, 1988; Sudhof and Scheller, 2001). Furthermore, Ca^{2+} modulates presynaptic voltage-gated Ca^{2+} channels (Budde et al., 2002; Catterall, 2000; Catterall et al., 2013). Therefore, Ca^{2+} uptake by presynaptic mitochondria and its delayed release from them can have profound effects on presynaptic Ca^{2+} signaling, synaptic transmission and plasticity.

Ca^{2+} buffering by mitochondria limits the amplitude of the activity-driven elevation of presynaptic $[Ca^{2+}]_i$ and regulates synaptic strength at many different synapses, including central synapses, nerve-muscular junction, and the first sensory synapse (Billups and Forsythe, 2002; David, 1999; David and Barrett, 2000; David and Barrett, 2003; Lee et al., 2007; Shutov et al., 2013; Tang and Zucker, 1997). Indeed, inhibition of mitochondrial Ca^{2+} uptake resulted in a larger presynaptic $[Ca^{2+}]_i$ elevation, an increase in the frequency of asynchronous release and accelerated synaptic depression at the neuro-muscular junction (David, 1999; David and Barrett, 2000; David and Barrett, 2003). Similar effects were described at the central synapses (Billups and Forsythe, 2002; Lee et al., 2007). In the case of GABAergic synaptic transmission between retinal amacrine cells, mitochondrial depolarization led to a reduction in evoked transmitter release that was attributed to an increase in Ca^{2+}-dependent inactivation of presynaptic voltage-gated Ca^{2+} channels in the absence of Ca^{2+} buffering by mitochondria (Medler and Gleason, 2002).

Mitochondrial Ca^{2+} transport also contributes to posttetanic potentiation (PTP) at neuromuscular and central synapses (Garcia-Chacon et al., 2006; Lee et al., 2007; Storozhuk et al., 2005; Tang and Zucker, 1997; Zhong, Beaumont, and Zucker, 2001). PTP is a form of short-term synaptic plasticity that is induced by high-frequency (tetanic) stimulation of the presynaptic terminal and is characterized by an enhancement of synaptic strength that lasts for many minutes (Zucker and Regehr, 2002). This phenomenon relies on a prolonged presynaptic $[Ca^{2+}]_i$ elevation (residual $[Ca^{2+}]_i$ or $[Ca^{2+}]_i$ plateau) following the initial $[Ca^{2+}]_i$ rise to the micromolar levels in response to intense stimulation (Neher and Sakaba, 2008; Zucker and Regehr, 2002). In many instances prolonged Ca^{2+} release from presynaptic mitochondria accounts for the residual $[Ca^{2+}]_i$ (Garcia-Chacon et al., 2006; Lee et al., 2007; Medvedeva et al., 2008; Tang and Zucker, 1997; Zhong et al., 2001). For example, high-frequency presynaptic stimulation produced a prolonged residual presynaptic $[Ca^{2+}]_i$ elevation and PTP in nerve muscular junctions of crayfish and mouse, as well at mossy fiber synapses of the rat hippocampus that were blocked by inhibitors of mitochondrial Ca^{2+} transport (Garcia-Chacon et al., 2006; Lee et al., 2007; Tang and Zucker, 1997; Zhong et al., 2001). Mitochondria-dependent PTP was also reported for GABAergic synaptic transmission between neocortical neurons, although presynaptic $[Ca^{2+}]_i$ was not measured in this study (Storozhuk et al., 2005). It is most likely that Ca^{2+} release from presynaptic mitochondria is mediated by the mitochondrial Na^+/Ca^{2+} exchanger (Garcia-Chacon et al., 2006; Lee et al., 2007; Medvedeva et al., 2008). However alternative mechanisms may also be involved at some synapses (Zhong et al., 2001).

One group found that mitochondria regulate presynaptic $[Ca^{2+}]_i$ and synaptic transmission at the first sensory synapse (Medvedeva et al., 2008; Shutov et al., 2013). This synapse is formed by primary sensory neurons and dorsal horn neurons and gates the flow of sensory information from the periphery to the central nervous system (Kuner, 2010; Woolf and Salter, 2006). The role of mitochondria at this synapse is especially pronounced in the case of synaptic transmission induced by activation of transient receptor potential vanilloid 1 (TRPV1), a channel that is involved in pain transmission and is activated by noxious heat, acidic pH, lipid-derived endovanilloids and inflammatory mediators (Caterina and Julius, 2001; De Petrocellis and Di Marzo, 2005; Patapoutian, Tate, and Woolf, 2009). In addition to being expressed in the periphery, TRPV1 is found on the central terminals of primary nociceptive neurons, where it regulates sensory synaptic transmission and contributes to central pain sensitization (Christoph et al., 2006; Cui et al., 2006; Kanai et al., 2005). A brief (20 to 30 s) stimulation of presynaptic TRPV1 produces prolonged enhancement of glutamate release that continues for many minutes after TRPV1 activation is terminated (Medvedeva et al., 2008; Nakatsuka et al., 2002; Yang et al., 1998). The TRPV1 channel is highly permeable for Ca^{2+}, and robust Ca^{2+} influx is a hallmark of its activation (Caterina and Julius, 2001; Dhaka, Viswanath, and Patapoutian, 2006). By monitoring presynaptic Ca^{2+} concentration in the cytosol and mitochondria, a significant portion of the Ca^{2+} entering through presynaptic TRPV1 receptors was found to be rapidly sequestered by presynaptic mitochondria via the mitochondrial Ca^{2+} uniporter and then gradually transported back to the cytosol via the mitochondrial Na^+/Ca^{2+} exchanger (Medvedeva et al., 2008). This Ca^{2+} cycling by mitochondria is responsible for the long-lasting presynaptic $[Ca^{2+}]_i$ elevation ($[Ca^{2+}]_i$ plateau) and enhanced glutamate release in response to TRPV1 activation. The more Ca^{2+} is taken up by mitochondria during stimulation, the longer the duration of the presynaptic $[Ca^{2+}]_i$ plateau, glutamate release, and synaptic activity. Indeed, in comparing the effects of weak and strong TRPV1 stimulation (0.1 and 1 μM capsaicin, respectively) we found the two stimuli produced presynaptic $[Ca^{2+}]_i$ elevations of approximately the same amplitudes, but that the duration of the $[Ca^{2+}]_i$ response and the synaptic activity induced by the weak stimulus were markedly shorter than those produced by the strong stimulus (Medvedeva et al., 2008). These findings suggest that mitochondria control transmission at the first sensory synapse by storing information about the strength of TRPV1 stimulation and translating it into the duration of the presynaptic $[Ca^{2+}]_i$ signal and synaptic activity (Medvedeva et al., 2008).

Findings from recent studies in *Drosophila* motor terminals suggest that mitochondrial Ca^{2+} uptake enhances energy metabolism by presynaptic mitochondria (Chouhan et al., 2012; Ivannikov and Macleod, 2013). These observations are consistent with the ability of mitochondrial Ca^{2+} to stimulate Ca^{2+}-dependent dehydrogenases, pyruvate dehydrogenase, isocitrate dehydrogenase, and α-ketoglutarate dehydrogenase, thereby increasing the input of critical substrates into the oxidative phosphorylation reaction and ATP synthesis (Denton, 2009; Glancy and Balaban, 2012). Given that $[Ca^{2+}]_{mt}$ rarely exceeds 5 μM, it is likely that pyruvate dehydrogenase and α-ketoglutarate dehydrogenase, each of which has Ca^{2+} K_d of ~0.8 μM (for comparison, isocitrate dehydrogenase has a Ca^{2+} K_d of~40 μM) (McCormack, Halestrap, and Denton, 1990), are the primary targets of matrix Ca^{2+}. Such a mechanism may be an important coordinator of ATP supply and demand at the synapse. Indeed, increased synaptic activity is associated with larger amounts of Ca^{2+} entering the presynaptic terminal, which in turn increases Ca^{2+} flux into the mitochondria and stimulates ATP synthesis. Thus, mitochondrial Ca^{2+} transport regulates synaptic transmission through at least two general mechanisms, with one contributing to the shaping of presynaptic $[Ca^{2+}]_i$ signals, and the other impacting the ATP synthesis at the synapse.

4.6 MITOCHONDRIAL Ca^{2+} TRANSPORT AND THE REGULATION OF GENE EXPRESSION IN NEURONS

Ca^{2+} is a crucial element of activity-dependent gene regulation in neurons, acting through the networks of Ca^{2+}-dependent transcription factors and regulators to drive specific gene expression programs (Deisseroth et al., 2003; Mellstrom and Naranjo, 2001; West, Griffith, and Greenberg, 2002). Ca^{2+}-dependent transcription factors vary in their sensitivities to the route of Ca^{2+} entry, as well as the amplitude and duration of the $[Ca^{2+}]_i$ elevation (Berridge, 1997; Deisseroth et al., 2003; Fields, Lee, and Cohen, 2005; Greer and Greenberg, 2008). Consequently, the spatial and temporal characteristics of Ca^{2+} signals and their magnitudes determine which specific transcription factors will preferentially be activated. As emphasized previously, mitochondrial Ca^{2+} transport can significantly reduce the amplitude and prolong the duration of the $[Ca^{2+}]_i$ response in neurons (Colegrove et al., 2000a; Garcia-Chacon et al., 2006; Kim and Usachev, 2009; Thayer and Miller, 1990). It is therefore predicted that the activity of transcription factors that are sensitive to the amplitude of $[Ca^{2+}]_i$ elevation will be suppressed by mitochondrial Ca^{2+} buffering, whereas the activity of those sensing the duration of Ca^{2+} signals will be enhanced by Ca^{2+} release from mitochondria. Indeed, recent reports show that mitochondrial Ca^{2+} transport can have the opposite effect on two ubiquitous Ca^{2+}-dependent transcription factors, CRE-binding protein (CREB) and nuclear factor of activated T-cells (NFAT), that are implicated in many aspects of neuronal development, experience-dependent plasticity, and neurological and psychiatric diseases (Arron et al., 2006; Cohen and Greenberg, 2008; Deisseroth et al., 2003; Nguyen and Di Giovanni, 2008; Quadrato et al., 2012; Robison and Nestler, 2011; Wu et al., 2007).

The transcription factor CREB is activated by the convergence of several Ca^{2+}-dependent signaling pathways that are highly sensitive to the route of Ca^{2+} entry and the amplitude of $[Ca^{2+}]_i$ elevation (Deisseroth et al., 2003; Greer and Greenberg, 2008; Wheeler et al., 2008, 2012; Wu, Deisseroth, and Tsien, 2001). CREB is preferentially stimulated by Ca^{2+} influx through L-type (Ca_v1) voltage-gated Ca^{2+} channels in neurons, but it can also be activated by Ca^{2+} entering through N- and P/Q-type (Ca_v2) voltage-gated Ca^{2+} channels (Brosenitsch and Katz, 2001; Dolmetsch et al., 2001; Wheeler et al., 2012). The more efficient coupling of Ca_v1 channels with CREB activation is likely explained by the close proximity (<40 nm; that is, nanodomain organization) of the channels to the downstream Ca^{2+} targets calmodulin (CaM) and CaM-dependent protein kinase II (CaMKII) (Hudmon et al., 2005; Wheeler et al., 2012). For comparison, the distance between Ca_v2 channels and CaMKII is estimated at >1 μm (Wheeler et al., 2012). Given that $[Ca^{2+}]_i$ decreases steeply as a function of distance from the mouth of Ca^{2+} channels (Deisseroth, Bito, and Tsien, 1996; Roberts, 1994), the influx of Ca^{2+} through Ca_v2 channels must be much higher than that through Ca_v1 channels to produce the same degree of CaMKII/CREB activation (Wheeler et al., 2012). Notably, mitochondria strongly buffer Ca^{2+} entering through Ca_v2 channels and prevent CREB activation. In turn, mitochondrial depolarization by FCCP causes a significant increase in both the amplitude of the $[Ca^{2+}]_i$ response and the magnitude of CREB activation (Wheeler et al., 2012). Thus, in the context of CREB activation, mitochondria act as a Ca^{2+} sink that prevents spreading of the Ca^{2+} signal from the channel to its cellular targets (e.g., CaMKII), and consequently suppresses CREB-dependent excitation-transcription coupling.

Another Ca^{2+}-dependent transcription factor, NFAT, is exquisitely sensitive to duration of Ca^{2+} signal, such that a prolonged $[Ca^{2+}]_i$ elevation is required for its activation (Dolmetsch et al., 1997; Feske et al., 2001). This requirement is explained by the NFAT activation mechanism. At rest, hyperphosphorylated NFAT resides in the cytosol. Upon excitation, $[Ca^{2+}]_i$ rise activates the Ca^{2+}-dependent protein phosphatase calcineurin (CaN) that dephosphorylates NFAT at several serine residues. This, in turn, unmasks the nuclear localization signal of NFAT and facilitates its import to the nucleus and thus initiates transcription. Several opposing protein kinases, including glycogen synthase kinase 3β (GSK3β) and c-Jun N-terminal kinase (JNK), phosphorylate NFAT, which facilitates binding of the nuclear exportin Crm-1 to NFAT and, consequently, deactivation and nuclear export of NFAT (Crabtree and Olson, 2002; Hogan et al., 2003). In other words, the activation status of NFAT depends on the balance between the Ca^{2+}-dependent activity of CaN and opposing constitutively active NFAT kinases. Consequently, sustained activation of NFAT requires a prolonged elevation of $[Ca^{2+}]_i$. We found that Ca^{2+} release from mitochondria and the resulting $[Ca^{2+}]_i$ plateau facilitate NFAT activation, as well as NFAT-mediated transcription in neurons in response to depolarization (Kim and Usachev, 2009). These observations are consistent with the height of $[Ca^{2+}]_i$ plateau (300 to 800 nM) and the $[Ca^{2+}]_i$ required for NFAT activation (>200 to 300 nM) (Kim and Usachev, 2009). Importantly, the release of Ca^{2+} from mitochondria also resulted in a prolonged elevation of $[Ca^{2+}]_i$ within the nucleus, which may be important in preventing the dissociation of activated CaN from nuclear NFAT and its displacement by Crm-1 (Zhu and McKeon, 1999). Mitochondrial depolarization by the electron transport inhibitor antimycin eliminated the $[Ca^{2+}]_i$ plateau and reduced NFAT activation, despite a significant increase in the amplitude of the electrically-evoked $[Ca^{2+}]_i$ response (Kim and Usachev, 2009). Moreover, preventing Ca^{2+} release from mitochondria by the mtNCX specific inhibitor CGP37157 also eliminated $[Ca^{2+}]_i$ plateau and inhibited NFAT activation. Collectively, these data show that by releasing Ca^{2+} into the cytosol and nucleus, mitochondria can significantly facilitate activation of the transcription factor NFAT in neurons.

Comparison of the inhibitory and facilitatory effects of mitochondrial Ca^{2+} transport on CREB and NFAT activation, respectively, suggests that mitochondrial Ca^{2+} signaling can function as a discriminatory mechanism, stimulating the execution of one transcriptional response versus others. The energetic status of mitochondria, their proximities to the Ca^{2+} sources and targets, as well as the expression and function of mitochondrial Ca^{2+} transporters, ultimately determine the extent to which the organelle contributes to activity-dependent gene regulation in neurons.

4.7 FUTURE DIRECTIONS

Recent discoveries of key molecular components of mitochondrial Ca^{2+} transport (see Fig. 4.4) have opened new avenues to neuroscientists studying mitochondrial signaling and function. The generation of animals in which components of the mitochondrial Ca^{2+} uniporter complex and other mitochondrial Ca^{2+} transporters have been deleted or disrupted is under way. These animal models will make it possible to address longstanding questions regarding the regulation and function of mitochondrial Ca^{2+} transport in neuronal excitability, plasticity, bioenergetics, and toxicity. Progress in the molecular and structural biology of mitochondrial Ca^{2+} transporters will also facilitate the development of new pharmacological tools that specifically and potently target mitochondrial Ca^{2+} transporters. This is crucial, as the inhibitors that are currently available have limited specificity, are unstable and have a limited (or none) capacity to penetrate through the plasma membrane (e.g., Ru360 and ruthenium red). Finally, the recent progress in mitochondrial Ca^{2+} research is expected to restore the interest of the pharmaceutical industry in mitochondrial Ca^{2+} transporters as potential drug targets for achieving neuroprotection in stroke, Alzheimer's disease, Parkinson's disease, ALS and other neurodegenerative conditions in which Ca^{2+} dysregulation has been implicated as a key underlying mechanism.

ACKNOWLEDGMENTS

This work was supported by National Institutes of Health (NIH) Grant NS087068 and the Fraternal Order of Eagles Diabetes Research Center (FOEDRC).

REFERENCES

Andersen, E. F., et al., 2014. Deletions involving genes WHSC1 and LETM1 may be necessary, but are not sufficient to cause Wolf-Hirschhorn Syndrome. *European journal of human genetics,* 22, pp.464–470.

Arron, J. R., et al., 2006. NFAT dysregulation by increased dosage of DSCR1 and DYRK1A on chromosome 21. *Nature,* 441, pp.595–600.

Babcock, D. F., et al., 1997. Mitochondrial participation in the intracellular Ca2+ network. *J Cell Biol,* 136, pp.833–844.

Babcock, D. F. and Hille, B., 1998. Mitochondrial oversight of cellular Ca2+ signaling. *Curr Opin Neurobiol,* 8, pp.398–404.

Baron, K. T. and Thayer, S. A., 1997. CGP37157 modulates mitochondrial Ca2+ homeostasis in cultured rat dorsal root ganglion neurons. *Eur J Pharmacol,* 340, pp.295–300.

Barrett, E. F., 2001. Contrasting contributions of endoplasmic reticulum and mitochondria to Ca2+ handling in neurons - Commentary. *Journal of General Physiology,* 118, pp.79–82.

Baughman, J. M., et al., 2011. Integrative genomics identifies MCU as an essential component of the mitochondrial calcium uniporter. *Nature,* 476, pp.341–345.

Bergemann, A. D., Cole, F. and Hirschhorn, K., 2005. The etiology of Wolf-Hirschhorn syndrome. *Trends in genetics,* 21, 188–195.

Berkefeld, H., et al., 2006. BKCa-Cav channel complexes mediate rapid and localized Ca2+-activated K+ signaling. *Science,* 314, pp.615–620.

Bernardi, P., 1999. Mitochondrial transport of cations: channels, exchangers, and permeability transition. *Physiol Rev,* 79, pp.1127–1155.

Berridge, M. J., 1997. The AM and FM of Calcium Signalling. *Nature,* 386, pp.759–760.

Berridge, M. J., Bootman, M. D. and Lipp, P., 1998. Calcium—a life and death signal. *Nature,* 395, pp.645–648.

Bezprozvanny, I., 2009. Calcium signaling and neurodegenerative diseases. *Trends in molecular medicine,* 15, pp.89–100.

Billups, B. and Forsythe, I. D., 2002. Presynaptic mitochondrial calcium sequestration influences transmission at mammalian central synapses. *J Neurosci,* 22, pp.5840–5847.

Brocard, J. B., Tassetto, M. and Reynolds, I. J., 2001. Quantitative evaluation of mitochondrial calcium content in rat cortical neurones following a glutamate stimulus. *Journal of Physiology,* 531, pp.793–805.

Brosenitsch, T. A. and Katz, D.M., 2001. Physiological patterns of electrical stimulation can induce neuronal gene expression by activating N-type calcium channels. *J Neurosci,* 21, pp.2571–2579.

Brustovetsky, T., et al., 2011. KB-R7943, an inhibitor of the reverse Na+ /Ca2+ exchanger, blocks N-methyl-D-aspartate receptor and inhibits mitochondrial complex I. *British Journal of Pharmacology,* 162, pp.255–270.

Budde, T., Meuth, S. and Pape, H. C., 2002. Calcium-dependent inactivation of neuronal calcium channels. *Nat Rev Neurosci,* 3, pp.873–883.

Carafoli, E. and Lehninger, A. L., 1971. A survey of the interaction of calcium ions with mitochondria from different tissues and species. *Biochemical Journal,* 122, pp.681–690.

Carafoli, E., 2003. Historical review: mitochondria and calcium: ups and downs of an unusual relationship. *Trends in Biochemical Sciences,* 28, pp.175–181.

Castilho, R. F., et al., 1998. Mitochondrial control of acute glutamate excitotoxicity in cultured cerebellar granule cells. *Journal of Neuroscience,* 18, pp.10277–10286.

Caterina, M. J. and Julius, D., 2001. The vanilloid receptor: a molecular gateway to the pain pathway. *Annu Rev Neurosci,* 24, pp.487–517.

Catterall, W. A., 2000. Structure and regulation of voltage-gated Ca2+ channels. *Annu Rev Cell Dev Biol,* 16, pp.521–555.

Catterall, W. A., Leal, K. and Nanou, E., 2013. Calcium channels and short-term synaptic plasticity. *Journal of Biological Chemistry,* 288, pp.10742–10749.

Chalmers, S. and Nicholls, D. G., 2003. The relationship between free and total calcium concentrations in the matrix of liver and brain mitochondria. *J Biol Chem,* 278, pp.19062–19070.

Chaudhuri, D., et al., 2013. MCU encodes the pore conducting mitochondrial calcium currents. *eLife,* 2, p.e00704.

Chaudhuri, D. and Clapham, D. E., 2014. Outstanding questions regarding the permeation, selectivity, and regulation of the mitochondrial calcium uniporter. *Biochemical and biophysical research communications,* 449, pp.367–369.

Chouhan, A. K., et al., 2010. Presynaptic mitochondria in functionally different motor neurons exhibit similar affinities for Ca2+ but exert little influence as Ca2+ buffers at nerve firing rates in situ. *J Neurosci,* 30, pp.1869–1881.

Chouhan, A. K., et al., 2012. Cytosolic calcium coordinates mitochondrial energy metabolism with presynaptic activity. *J Neurosci,* 32, pp.1233–1243.

Christoph, T., et al., 2006. Silencing of vanilloid receptor TRPV1 by RNAi reduces neuropathic and visceral pain in vivo. *Biochem Biophys Res Commun,* 350, pp.238–243.

Chuang, H. H., Neuhausser, W. M. and Julius, D., 2004. The super-cooling agent icilin reveals a mechanism of coincidence detection by a temperature-sensitive TRP channel. *Neuron,* 43, pp.859–869.

Clapham, D. E., 2003. TRP channels as cellular sensors. *Nature,* 426, pp.517–524.

Clapham, D. E., 2007. Calcium signaling. *Cell,* 131, pp.1047–1058.

Cohen, S. and Greenberg, M. E., 2008. Communication between the synapse and the nucleus in neuronal development, plasticity, and disease. *Annu Rev Cell Dev Biol,* 24, pp.183–209.

Colegrove, S. L., Albrecht, M. A. and Friel, D. D., 2000a. Dissection of mitochondrial Ca2+ uptake and release fluxes in situ after depolarization-evoked [Ca2+]i elevations in sympathetic neurons. *J Gen Physiol,* 115, pp.351–369.

Colegrove, S. L., Albrecht, M. A. and Friel, D. D., 2000b. Quantitative analysis of mitochondrial Ca2+ uptake and release pathways in sympathetic neurons: reconstruction of the recovery after depolarization-evoked [Ca2+](i) elevations. *Journal of General Physiology,* 115, pp.371–388.

Crabtree, G. R. and Olson, E. N., 2002. NFAT signaling: choreographing the social lives of cells. *Cell,* 109 Suppl, pp.S67–S79.

Crompton, M., Kunzi, M. and Carafoli, E., 1977. The calcium-induced and sodium-induced effluxes of calcium from heart mitochondria. Evidence for a sodium-calcium carrier. *Eur J Biochem,* 79, pp.549–558.

Csordas, G., et al., 2013. MICU1 controls both the threshold and cooperative activation of the mitochondrial Ca(2)(+) uniporter. *Cell metabolism,* 17, pp.976–987.

Cui, M., et al., 2006. TRPV1 receptors in the CNS play a key role in broad-spectrum analgesia of TRPV1 antagonists. *J Neurosci,* 26, pp.9385–9393.

David, G., 1999. Mitochondrial clearance of cytosolic Ca2+ in stimulated lizard motor nerve terminals proceeds without progressive elevation of mitochondrial matrix [Ca2+]. *J Neurosci,* 19, pp.7495–7506.

David, G. and Barrett, E. F., 2000. Stimulation-evoked increases in cytosolic [Ca2+] in mouse motor nerve terminals are limited by mitochondrial uptake and are temperature-dependent. *J Neurosci,* 20, pp.7290–7296.

David, G. and Barrett, E. F., 2003. Mitochondrial Ca2+ uptake prevents desynchronization of quantal release and minimizes depletion during repetitive stimulation of mouse motor nerve terminals. *J Physiol,* 548, pp.425–438.

David, G., Talbot, J. and Barrett, E. F., 2003. Quantitative estimate of mitochondrial [Ca2+] in stimulated motor nerve terminals. *Cell Calcium,* 33, pp.197–206.

De Camilli, P., et al., 2001. The structure of synapses. In: M. W. Cowan, T. C. Sudhof, C. F. Stevens, eds. *Synapses.* Baltimore, John Hopkins University Press, pp.89–133.

De Petrocellis, L. and Di Marzo, V., 2005. Lipids as regulators of the activity of transient receptor potential type V1 (TRPV1) channels. *Life Sci,* 77, pp.1651–1666.

De Stefani, D., et al., 2011. A forty-kilodalton protein of the inner membrane is the mitochondrial calcium uniporter. *Nature,* 476, pp.336–340.

De Stefani, D., et al., 2012. VDAC1 selectively transfers apoptotic Ca2+ signals to mitochondria. *Cell death and differentiation,* 19, pp.267–173.

Dedov, V. N. and Roufogalis, B. D., 2000. Mitochondrial calcium accumulation following activation of vanilloid (VR1) receptors by capsaicin in dorsal root ganglion neurons. *Neuroscience,* 95, pp.183–188.

Deisseroth, K., Bito, H. and Tsien, R. W., 1996. Signaling from synapse to nucleus—postsynaptic CREB phosphorylation during multiple forms of hippocampal synaptic plasticity. *Neuron,* 16, pp.89–101.

Deisseroth, K., et al., 2003. Signaling from synapse to nucleus: the logic behind the mechanisms. *Curr Opin Neurobiol,* 13, pp.354–365.

Deluca, H. F. and Engstrom, G. W., 1961. Calcium uptake by rat kidney mitochondria. *Proc Natl Acad Sci U S A,* 47, pp.1744–1750.

Denton, R. M., 2009. Regulation of mitochondrial dehydrogenases by calcium ions. *Biochimica et Biophysica Acta,* 1787, pp.1309–1316.

Dhaka, A., Viswanath, V. and Patapoutian, A., 2006. Trp ion channels and temperature sensation. *Annu Rev Neurosci,* 29, pp.135–161.

Dolmetsch, R. E., et al., 1997. Differential activation of transcription factors induced by Ca2+ response amplitude and duration. *Nature,* 386, pp.855–858.

Dolmetsch, R. E., et al., 2001. Signaling to the nucleus by an L-type calcium channel-calmodulin complex through the MAP kinase pathway. *Science,* 294, pp.333–339.

Doonan, P. J., et al., 2014. LETM1-dependent mitochondrial Ca2+ flux modulates cellular bioenergetics and proliferation. *FASEB J,* 28, pp.4936–4949.

Drago, I., Pizzo, P. and Pozzan, T., 2011. After half a century mitochondrial calcium in- and efflux machineries reveal themselves. *EMBO J,* 30, pp.4119–4125.

Feske, S., et al., 2001. Gene regulation mediated by calcium signals in T lymphocytes. *Nature Immunology,* 2, pp.316–324.

Fields, R .D., Lee, P. R. and Cohen, J.E., 2005. Temporal integration of intracellular Ca2+ signaling networks in regulating gene expression by action potentials. *Cell Calcium,* 37, pp.433–442.

Fieni, F., et al., 2012. Activity of the mitochondrial calcium uniporter varies greatly between tissues. *Nature Communications*, 3, p.1317.

Friel, D. D. and Tsien, R.W ., 1994. An FCCP-sensitive Ca2+ store in bullfrog sympathetic neurons and its participation in stimulus-evoked changes in [Ca2+]i. *J. Neurosci,* 14, pp.4007–4024.

Friel, D. D., 2000. Mitochondria as regulators of stimulus-evoked calcium signals in neurons. *Cell Calcium,* 28, pp.307–316.

Garcia-Chacon, L. E., et al., 2006. Extrusion of Ca2+ from mouse motor terminal mitochondria via a Na+-Ca2+ exchanger increases post-tetanic evoked release. *J Physiol (Lond),* 574, pp.663–675.

Glancy, B. and Balaban, R. S., 2012. Role of mitochondrial Ca2+ in the regulation of cellular energetics. *Biochemistry,* 51, pp.2959–2973.

Gobbi, P., et al., 2007. Mitochondrial localization of Na+/Ca2+ exchangers NCX1-3 in neurons and astrocytes of adult rat brain in situ. *Pharmacological Research,* 56, pp.556–565.

Greer, P. L. and Greenberg, M. E., 2008. From synapse to nucleus: calcium-dependent gene transcription in the control of synapse development and function. *Neuron,* 59, pp.846–860.

Groffen, A. J., et al., 2010. Doc2b is a high-affinity Ca2+ sensor for spontaneous neurotransmitter release. *Science.* 327, pp.1614–1618.

Gunter, T. E. and Pfeiffer, D. R., 1990. Mechanisms by which mitochondria transport calcium. *Am. J. Physiol,* 258, pp.C755–C786.

Hajnoczky, G., et al., 1995. Decoding of cytosolic calcium oscillations in the mitochondria. *Cell,* 82, pp.415–424.

Harris, J. J., Jolivet, R. and Attwell, D., 2012. Synaptic energy use and supply. *Neuron,* 75, pp.762–777.

Hartzell, C., Putzier, I. and Arreola, J., 2005. Calcium-activated chloride channels. *Annual Review of Physiology,* 67, pp.719–758.

Hernandez-Guijo, J. M., et al., 2001. Calcium-dependent inhibition of L, N, and P/Q Ca2+ channels in chromaffin cells: role of mitochondria. *J Neurosci,* 21, pp.2553–2560.

Hoffman, N. E., et al., 2013. MICU1 motifs define mitochondrial calcium uniporter binding and activity. *Cell Reports,* 5, pp.1576–1588.

Hogan, P. G., et al., 2003. Transcriptional regulation by calcium, calcineurin, and NFAT. *Genes Dev,* 17, pp.2205–2232.

Hollenbeck, P. J., 2005. Mitochondria and neurotransmission: evacuating the synapse. *Neuron,* 47, pp.331–333.

Hudmon, A., et al., 2005. CaMKII tethers to L-type Ca2+ channels, establishing a local and dedicated integrator of Ca2+ signals for facilitation. *J Cell Biol,* 171, pp.537–547.

Hung, V., et al., 2014. Proteomic mapping of the human mitochondrial intermembrane space in live cells via ratiometric APEX tagging. *Molecular Cell,* 55, pp.332–341.

Ivannikov, M. V. and Macleod, G.T., 2013. Mitochondrial free Ca(2)(+) levels and their effects on energy metabolism in Drosophila motor nerve terminals. *Biophysical Journal,* 104, pp.2353–2361.

Jiang, D., Zhao, L. and Clapham, D. E., 2009. Genome-wide RNAi screen identifies Letm1 as a mitochondrial Ca2+/H+ antiporter. *Science,* 326, pp.144–147.

Jiang, D., et al., 2013. Letm1, the mitochondrial Ca2+/H+ antiporter, is essential for normal glucose metabolism and alters brain function in Wolf-Hirschhorn syndrome. *Proc Natl Acad Sci U S A,* 110, pp.E2249–E2254.

Jonas, E. A., Buchanan, J. and Kaczmarek, L. K., 1999. Prolonged activation of mitochondrial conductances during synaptic transmission. *Science,* 286, 1347–1350.

Kamer, K. J. and Mootha, V. K., 2014. MICU1 and MICU2 play nonredundant roles in the regulation of the mitochondrial calcium uniporter. *EMBO Reports,* 15, pp.299–307.

Kanai, Y., et al., 2005. Involvement of an increased spinal TRPV1 sensitization through its up-regulation in mechanical allodynia of CCI rats. *Neuropharmacology,* 49, pp.977–984.

Khananshvili, D., 2013. The SLC8 gene family of sodium-calcium exchangers (NCX)—structure, function, and regulation in health and disease. *Molecular Aspects of Medicine,* 34, pp.220–235.

Kim, M. S. and Usachev, Y. M., 2009. Mitochondrial Ca2+ cycling facilitates activation of the transcription factor NFAT in sensory neurons. *J Neurosci,* 29, pp.12101–12114.

Kim, M. S., et al., 2014. Nerve growth hactor (NGF) regulates activity of the transcription factor NFAT in meurons via the phosphatidylinositol-3-kinase (PI3K)-akt-glycogen synthase kinase 3beta (GSK3beta) pathway. *J Biol Chem,* 289, pp.31349–31360.

Kirichok, Y., Krapivinsky, G. and Clapham, D. E., 2004. The mitochondrial calcium uniporter is a highly selective ion channel. *Nature,* 427, pp.360–364.

Kostyuk, P. and Verkhratsky, A., 1994. Calcium stores in neurons and glia. *Neuroscience,* 63, pp.381–404.

Kristian, T., et al., 2007. Calcium-induced precipitate formation in brain mitochondria: composition, calcium capacity, and retention. *J Neurochem,* 102, pp.1346–1356.

Kuner, R., 2010. Central mechanisms of pathological pain. *Nature Medicine,* 16, pp.1258–1266.

Lee, D., et al., 2007. Target cell-specific involvement of presynaptic mitochondria in post-tetanic potentiation at hippocampal mossy fiber synapses. *J Neurosci,* 27, pp.13603–13613.

Logan, C. V., et al., 2014. Loss-of-function mutations in MICU1 cause a brain and muscle disorder linked to primary alterations in mitochondrial calcium signaling. *Nature Genetics,* 46, pp.188–193.

Lu, S. G., Zhang, X. and Gold, M. S., 2006. Intracellular calcium regulation among subpopulations of rat dorsal root ganglion neurons. *J Physiol (Lond),* 577, pp.169–90.

Ly, C. V. and Verstreken, P., 2006. Mitochondria at the synapse. *Neuroscientist,* 12, pp.291–299.

Lytton, J., 2007. Na+/Ca2+ exchangers: three mammalian gene families control Ca2+ transport. *Biochem J.* 406, pp.365–382.

Mallilankaraman, K., et al., 2012. MCUR1 is an essential component of mitochondrial Ca2+ uptake that regulates cellular metabolism. *Nature Cell Biology,* 14, pp.1336–1343.

Marchi, S. and Pinton, P., 2014. The mitochondrial calcium uniporter complex: molecular components, structure and physiopathological implications. *J Physiol,* 592, pp.829–839.

Marty, A., 1989. The physiological role of calcium-dependent channels. *Trends Neurosci,* 12, pp.420–424.

Matlib, M. A., et al., 1998. Oxygen-bridged dinuclear ruthenium amine complex specifically inhibits Ca2+ uptake into mitochondria in vitro and in situ in single cardiac myocytes. *J Biol Chem,* 273, pp.10223–10231.

Mattson, M. P., Gleichmann, M. and Cheng, A., 2008. Mitochondria in neuroplasticity and neurological disorders. *Neuron,* 60, pp.748–766.

McBain, C. J. and Mayer, M. L., 1994. N-methyl-D-aspartic acid receptor structure and function. *Physiol Rev,* 74, pp.723–760.

McCormack, J. G., Halestrap, A. P. and Denton, R. M., 1990. Role of calcium ions in regulation of mammalian intramitochondrial metabolism. *Physiol Rev,* 70, pp.391–425.

Medler, K. and Gleason, E. L., 2002. Mitochondrial Ca2+ buffering regulates synaptic transmission between retinal amacrine cells. *J of Neurophysiol,* 87, pp.1426–1439.

Medvedeva, Y. V., Kim, M. S. and Usachev, Y. M., 2008. Mechanisms of prolonged presynaptic Ca2+ signaling and glutamate release induced by TRPV1 activation in rat sensory neurons. *J Neurosci,* 28, pp.5295–5311.

Mellstrom, B. and Naranjo, J.R., 2001. Mechanisms of Ca2+-dependent transcription [Review]. *Curr Opin Neurobiol,* 11, pp.312–319.

Miller, R. J., 1988. Calcium signalling in neurons. *Trends Neurosci,* 11, pp.415–419.

Moskowitz, M. A., Lo, E. H. and Iadecola, C., 2010. The science of stroke: mechanisms in search of treatments. *Neuron,* 67, pp.181–198.

Murphy, A. N., Fiskum, G. and Beal, M. F., 1999. Mitochondria in neurodegeneration: bioenergetic function in cell life and death. *J Cereb Blood Flow Metab,* 19, pp.231–245.

Nakatsuka, T., et al., 2002. Activation of central terminal vanilloid receptor-1 receptors and alpha beta-methylene-ATP-sensitive P2X receptors reveals a converged synaptic activity onto the deep dorsal horn neurons of the spinal cord. *J Neurosci,* 22, pp.1228–1237.

Neher, E. and Sakaba, T., 2008. Multiple roles of calcium ions in the regulation of neurotransmitter release. *Neuron,* 59, pp.861–872.

Newmeyer, D. D. and Ferguson-Miller, S., 2003. Mitochondria: releasing power for life and unleashing the machineries of death. *Cell,* 112, pp.481–490.

Nguyen, T. and Di Giovanni, S., 2008. NFAT signaling in neural development and axon growth. *Int J Dev Neurosci,* 26, pp.141–145.

Nicholls, D. G., 2005. Mitochondria and calcium signaling. *Cell Calcium,* 38, pp.311–317.

Nicholls, D. G., 2012. Mitochondria, sodium, and calcium in neuronal dysfunction. In: A. K. Reeve, K. J. Krishnan, M. R. Duchen and D. M. Turnbull, eds. *Mitochondrial dysfunction in neurodegenerative disorders.* London, Springer, pp.113–125.

Nicholls, D. G. and Budd, S. L., 2000. Mitochondria and neuronal survival. *Physiol Rev,* 80, pp.315–360.

Nicholls, D. G. and Ferguson, S. J., 2013. *Bionergetics.* San Diego, Academic Press.

Nicholls, D. G., et al., 2003. Interactions between mitochondrial bioenergetics and cytoplasmic calcium in cultured cerebellar granule cells. *Cell Calcium,* 34, pp.407–424.

Nowicky, A. V. and Duchen, M. R., 1998. Changes in [Ca2+]i and membrane currents during impaired mitochondrial metabolism in dissociated rat hippocampal neurons. *J Physiol* 507(Pt 1), pp.131–145.

Nowikovsky, K., et al., 2004. The LETM1/YOL027 gene family encodes a factor of the mitochondrial K+ homeostasis with a potential role in the Wolf-Hirschhorn syndrome. *J Biol Chem,* 279, pp.30307–30315.

Nowikovsky, K., et al., 2012. Perspectives on: SGP symposium on mitochondrial physiology and medicine: the pathophysiology of LETM1. *J Gen Physiol,* 139, pp.445–454.

Orrenius, S., Zhivotovsky, B. and Nicotera, P., 2003. Regulation of cell death: the calcium-apoptosis link. *Nat Rev Mol Cell Biol,* 4, pp.552–565.

Palty, R. and Sekler, I., 2012. The mitochondrial Na(+)/Ca(2+) exchanger. *Cell Calcium,* 52, pp.9–15.

Palty, R., et al., 2004. Lithium-calcium exchange is mediated by a distinct potassium-independent sodium-calcium exchanger. *J Biol Chem,* 279, pp.25234–25240.

Palty, R., et al., 2010. NCLX is an essential component of mitochondrial Na+/Ca2+ exchange. *Proc Natl Acad Sci U S A,* 107, pp.436–441.

Panov, A. V., et al., 2002. Early mitochondrial calcium defects in Huntington's disease are a direct effect of polyglutamines. *Nature Neuroscience,* 5, pp.731–736.

Partridge, L. D. and Swandulla, D., 1988. Calcium-activated non-specific cation channels. *Trends Neurosci,* 11, pp.69–72.

Patapoutian, A., Tate, S. and Woolf, C.J., 2009. Transient receptor potential channels: targeting pain at the source. *Nat Rev Drug Discov,* 8, pp.55–68.

Patapoutian, A., et al., 2003. ThermoTRP channels and beyond: mechanisms of temperature sensation. Nature reviews. *Neuroscience,* 4, pp.529–539.

Patron, M., et al., 2014. MICU1 and MICU2 finely tune the mitochondrial Ca2+ uniporter by exerting opposite effects on MCU activity. *Molecular Cell,* 53, pp.726–737.

Pendin, D., Greotti, E. and Pozzan, T., 2014. The elusive importance of being a mitochondrial Ca(2+) uniporter. *Cell Calcium,* 55, pp.139–145.

Perocchi, F., et al., 2010. MICU1 encodes a mitochondrial EF hand protein required for Ca(2+) uptake. *Nature,* 467, pp.291–296.

Pivovarova, N. B. and Andrews, S. B., 2010. Calcium-dependent mitochondrial function and dysfunction in neurons. *FEBS J,* 277, pp.3622–3636.

Pivovarova, N. B., et al., 1999. Depolarization-induced mitochondrial Ca accumulation in sympathetic neurons: spatial and temporal characteristics. *J Neurosci,* 19, 6372–6384.

Pivovarova, N. B., et al., 2002. correlated calcium uptake and release by mitochondria and endoplasmic reticulum of CA3 hippocampal dendrites after afferent synaptic stimulation. *J Neurosci,* 22, 10653–10661.

Pivovarova, N. B., et al., 2004. Excitotoxic calcium overload in a subpopulation of mitochondria triggers delayed death in hippocampal neurons. *J Neurosci,* 24, pp.5611–5622.

Plovanich, M., et al., 2013. MICU2, a paralog of MICU1, resides within the mitochondrial uniporter complex to regulate calcium handling. *PLoS One,* 8, p.e55785.

Pozzan, T. and Rizzuto, R., 2000. The renaissance of mitochondrial calcium transport. *Eur J Biochem,* 267, pp.5269–5273.

Qiu, J., et al., 2013. Mitochondrial calcium uniporter Mcu controls excitotoxicity and is transcriptionally repressed by neuroprotective nuclear calcium signals. *Nature Communications,* 4, p.2034.

Quadrato, G., et al., 2012. Nuclear factor of activated T cells (NFATc4) is required for BDNF-dependent survival of adult-born neurons and spatial memory formation in the hippocampus. *Proc Natl Acad Sci U S A,* 109, pp.E1499–E1508.

Raffaello, A., et al., 2013. The mitochondrial calcium uniporter is a multimer that can include a dominant-negative pore-forming subunit. *EMBO J,* 32, pp.2362–2376.

Rapizzi, E., et al., 2002. Recombinant expression of the voltage-dependent anion channel enhances the transfer of Ca2+ microdomains to mitochondria. *J Cell Biol,* 159, pp.613–624.

Reynolds, I. J., 1999. Mitochondrial membrane potential and the permeability transition in excitotoxicity. *Ann NY Acad Sci,* 893, pp.33–41.

Rhee, H. W., et al. 2013. Proteomic mapping of mitochondria in living cells via spatially restricted enzymatic tagging. *Science,* 339, pp.1328–1331.

Rizzuto, R., et al., 2012. Mitochondria as sensors and regulators of calcium signalling. Nature reviews. *Mole Cell Biol,* 13, pp.566–578.

Roberts, W. M., 1994. Localization of calcium signals by a mobile calcium buffer in frog saccular hair cells. *J. Neurosci,* 14, pp.3246–3262.

Robison, A. J. and Nestler, E. J., 2011. Transcriptional and epigenetic mechanisms of addiction. Nature reviews. *Neuroscience,* 12, pp.623–637.

Rossi, C. S. and Lehninger, A. L., 1964. Stoichiometry of respiratory stimulation, accumulation of Ca++ and phosphate, and oxidative phosphorylation in rat liver mitochondria. *J Biol Chem,* 239, pp.3971–3980.

Sah, P., 1996. Ca2+-activated K+ currents in neurones: types, physiological roles and modulation. *Trends Neurosci,* 19, pp.150–154.

Sancak, Y., et al., 2013. EMRE is an essential component of the mitochondrial calcium uniporter complex. *Science,* 342, pp.1379–1382.

Schnizler, K., et al., 2008. Protein kinase A anchoring via AKAP150 is essential for TRPV1 modulation by forskolin and prostaglandin E2 in mouse sensory neurons. *J Neurosci,* 28, pp.4904–4917.

Scorziello, A., et al., 2013a. New insights in mitochondrial calcium handling by sodium/calcium exchanger. *Adv Exp Med Biol,* 961, pp.203–209.

Scorziello, A., et al. 2013b. NCX3 regulates mitochondrial Ca(2+) handling through the AKAP121-anchored signaling complex and prevents hypoxia-induced neuronal death. *J Cell Sci,* 126, pp.5566–5577.

Scott, R. H., et al., 1995. Aspects of calcium-activated chloride currents: a neuronal perspective. *Pharmacol Therapeutic,* 66, pp.535–565.

Shishkin, V., et al., 2002. Role of mitochondria in intracellular calcium signaling in primary and secondary sensory neurones of rats. *Cell Calcium,* 32, pp.121–130.

Shoshan-Barmatz, V., et al., 2010. VDAC, a multi-functional mitochondrial protein regulating cell life and death. *Mole Aspects Med,* 31, pp.227–285.

Shutov, L. P., et al., 2013. Mitochondria and plasma membrane Ca2+-ATPase control presynaptic Ca2+ clearance in capsaicin-sensitive rat sensory neurons. *J Physiol,* 591, pp.2443–2462.

Smith, S. J. and Augustine, G. J., 1988. Calcium ions, active zones and synaptic transmitter release. *Trends Neurosci,* 11, pp.458–464.

Spat, A., et al., 2008. High- and low-calcium-dependent mechanisms of mitochondrial calcium signalling. *Cell Calcium,* 44, pp.51–63.

Stanika, R. I., et al., 2009. Coupling diverse routes of calcium entry to mitochondrial dysfunction and glutamate excitotoxicity. *Proc Natl Acad Sci U S A,* 106, pp.9854–9859.

Stanika, R. I., et al., 2012. Comparative impact of voltage-gated calcium channels and NMDA receptors on mitochondria-mediated neuronal injury. *J Neurosci,* 32, pp.6642–6650.

Storozhuk, M. V. et al., 2005. Possible role of mitochondria in posttetanic potentiation of GABAergic synaptic transmission in rat neocortical cell cultures. *Synapse,* 58, pp.45–52.

Stout, A. K., et al., 1998. Glutamate-induced neuron death requires mitochondrial calcium uptake. *Nature Neuroscience,* 1, pp.366–373.

Sudhof, T. C. and Scheller, R. H., 2001. Mechanism and regulation of neurotransmitter release. In: C. M. Cowan, T. C. Sudhof, C. F. Stevens, eds., Synapses. Baltimore: Johns Hopkins University Press, pp.177–215.

Sugita, S., et al., 2002. Synaptotagmins form a hierarchy of exocytotic Ca(2+) sensors with distinct Ca(2+) affinities. *EMBO Journal,* 21, pp.270–280.

Tamai, S., et al., 2008. Characterization of the mitochondrial protein LETM1, which maintains the mitochondrial tubular shapes and interacts with the AAA-ATPase BCS1L. *J Cell Sci,* 121, pp.2588–2600.

Tang, Y. G. and Zucker, R. S., 1997. Mitochondrial involvement in post-tetanic potentiation of synaptic transmission. *Neuron,* 18, pp.483–491.

Thayer, S. A. and Miller, R. J., 1990. Regulation of the free intracellular calcium concentration in rat dorsal root ganglion neurones in vitro. *J Physiol (Lond.),* 425, pp.85–115.

Thayer, S. A., Usachev, Y. M. and Pottorf, W. J., 2002. Modulating Ca2+ clearance from neurons. *Front Biosci,* 7, D1255–D1279.

Tsai, M. F., et al., 2014. Functional reconstitution of the mitochondrial Ca2+/H+ antiporter Letm1. *J Gen Physiol,* 143, pp.67–73.

Usachev, Y. M., et al., 2002. Bradykinin and ATP accelerate Ca2+ efflux from rat sensory neurons via protein kinase C and the plasma membrane Ca2+ pump isoform 4. *Neuron,* 33, pp.113–122.

Vanden Berghe, P., Kenyon, J. L. and Smith, T. K., 2002. Mitochondrial Ca2+ uptake regulates the excitability of myenteric neurons. *J Neurosci,* 22, pp.6962–6971.

Vasington, F. D. and Murphy, J. V., 1962. Ca ion uptake by rat kidney mitochondria and its dependence on respiration and phosphorylation. *J Biol Chem,* 237, pp.2670–2677.

Wang, G. J. and Thayer, S. A., 2002. NMDA-induced calcium loads recycle across the itochondrial inner membrane of hippocampal neurons in culture. *J Neurophysiol,* 87, pp.740–749.

Wang, L., et al., 2014. Structural and mechanistic insights into MICU1 regulation of mitochondrial calcium uptake. *EMBO J,* 33, pp.594–604.

Wei, A. D., et al., 2005. International Union of Pharmacology. LII. Nomenclature and molecular relationships of calcium-activated potassium channels. *Pharmacol Rev,* 57, pp.463–472.

Werth, J. L. and Thayer, S. A., 1994. Mitochondria buffer physiological calcium loads in cultured rat dorsal root ganglion neurons. *J of Neurosci,* 14, pp.348–356.

West, A. E., Griffith, E. C. and Greenberg, M. E., 2002. Regulation of transcription factors by neuronal activity. *Nat Rev Neurosci* 3, pp.921–931.

Wheeler, D. G., et al., 2008. CaMKII locally encodes L-type channel activity to signal to nuclear CREB in excitation-transcription coupling. *J Cell Biol,* 183, pp.849–863.

Wheeler, D. G., et al., 2012. Ca(V)1 and Ca(V)2 channels engage distinct modes of Ca(2+) signaling to control CREB-dependent gene expression. *Cell,* 149, pp.1112–1124.

White, R. J. and Reynolds, I. J., 1996. Mitochondrial depolarization in glutamate-stimulated neuronsan early signal specific to excitotoxin exposure. *J Neurosci,* 16, pp.5688–5697.

Wimmer, V. C., et al., 2006. Donut-like topology of synaptic vesicles with a central cluster of mitochondria wrapped into membrane protrusions: a novel structure-function module of the adult calyx of Held. *J Neurosci.* 26, pp.109–116.

Woolf, C. J. and Salter, M. W., 2006. Plasticity and pain: role of the dorsal horn. In: S. B. McMahon, K. Koltzenburg, eds., *Textbook of pain.* Cincinati, Elsevier, pp.91–105.

Wu, G. Y., Deisseroth, K. and Tsien, R. W., 2001. Activity-dependent CREB phosphorylation: convergence of a fast, sensitive calmodulin kinase pathway and a slow, less sensitive mitogen- activated protein kinase pathway. *Proc Natl Acad Sci U S A,* 98, pp.2808–2813.

Wu, H., et al., 2007. NFAT signaling and the invention of vertebrates. *Trends Cell Biol,* 17, pp.251–260.

Wu, J., et al., 2014. Red fluorescent genetically encoded Ca2+ indicators for use in mitochondria and endoplasmic reticulum. *Biochem J,* 464, 13–22.

Wu, L. J., Sweet, T. B. and Clapham, D. E., 2010. International Union of Basic and Clinical Pharmacology. LXXVI. Current progress in the mammalian TRP ion channel family. *Pharmacol Reviews,* 62, pp.381–404.

Xu, J., Mashimo, T. and Sudhof, T. C., 2007. Synaptotagmin-1, -2, and -9: Ca(2+) sensors for fast release that specify distinct presynaptic properties in subsets of neurons. *Neuron,* 54, pp.567–581.

Yang, K., et al., 1998. Capsaicin facilitates excitatory but not inhibitory synaptic transmission in substantia gelatinosa of the rat spinal cord. *Neurosci Lett,* 255, pp.135–138.

Yao, J., et al., 2011. Doc2 is a ca(2+) sensor required for asynchronous neurotransmitter release. *Cell,* 147, pp.666–677.

Zhong, N., Beaumont, V. and Zucker, R. S., 2001. Roles for mitochondrial and reverse mode Na+/Ca2+ exchange and the plasmalemma Ca2+ ATPase in post-tetanic potentiation at crayfish neuromuscular junctions. *J Neurosci,* 21, pp.9598–9607.

Zhu, J. and McKeon, F., 1999. NF-AT activation requires suppression of Crm1-dependent export by calcineurin. *Nature* 398, pp.256–260.

Zucker, R. S. and Regehr, W. G., 2002. Short-term synaptic plasticity. *Annu Rev Physiol,* 64, pp.355–405.

5

AMP-ACTIVATED PROTEIN KINASE (AMPK) AS A CELLULAR ENERGY SENSOR AND THERAPEUTIC TARGET FOR NEUROPROTECTION

PETRONELA WEISOVÁ

Max F. Perutz Laboratories, University of Vienna, Austria

SHONA PFEIFFER AND JOCHEN H. M. PREHN

Department of Physiology and Medical Physics, Centre for the Study of Neurological Disorders, Royal College of Surgeons in Ireland, Dublin, Ireland

5.1 INTRODUCTION

To drive biological processes, a marked amount of energy is stored in cells in the form of the molecule adenosine triphosphate (ATP). Neuronal cells in particular are highly dependent on large amounts of ATP to prevent ionic imbalance and to maintain their electrochemical gradients across the plasma membrane, important for neuronal function and firing. The ability of neurons to function in an efficient way is achieved by the existence of a "biological energy sensing" mechanism that provides cells with the ability to respond rapidly to depletion of intracellular ATP levels as a consequence of metabolic, traumatic, or toxic challenges.

AMP-activated protein kinase (AMPK) has consequently been evolved in eukaryotic cells as an ancient metabolic enzyme with the ability to sense compromised availability of cellular energy resulting from various physiological or pathological stimuli

The Functions, Disease-Related Dysfunctions, and Therapeutic Targeting of Neuronal Mitochondria, First Edition. Edited by Valentin K. Gribkoff, Elizabeth A. Jonas and J. Marie Hardwick.

(Carling, 2004; Hardie et al., 1999; Kahn et al., 2005). Decline in neuronal energy in response to different metabolic stress conditions is reflected by a corresponding immediate decrease in ATP levels; this unfavorable energetic status in neurons results in increased adenosine monophosphate (AMP) levels, serving as the key driver for the activation of AMPK. One of the most prominent roles of AMPK as a critical regulator in cellular energy function is the maintenance of neuronal energy homeostasis in response to an increased AMP/ATP ratio, by the activation and promotion of catabolic pathways and simultaneous inhibition of energy-consuming anabolic pathways. In doing so, AMPK increases cytosolic ATP levels. Interestingly, AMPK is also involved in regulation of a vast number of biological and physiological processes including life span regulation, autophagy, cell polarity, inflammation, apoptosis, and cell migration (Giri et al., 2004; Lee et al., 2010; Liang et al., 2007; Mirouse et al., 2007; Nakano et al., 2010). This chapter will highlight recent advances in understanding the role of AMPK activity in the brain and its impact on neuronal metabolism, its neuron-specific ("neuronlike") functions, its role in neuronal cell death and neurodegeneration, and its role as a potential therapeutic target.

5.1.1 AMPK Expression, Structure, and Activity Regulation in Brain

5.1.1a AMPK in the Control of Cellular Metabolic Function

AMPK, an enzyme with serine/threonine activity, continuously monitors cellular energetic status in the brain and is activated by elevated ratios of AMP/ATP and adenosine diphosphate (ADP)/ATP during conditions of compromised energy availability as well as by other signals such as hormones or an increase in Ca^{2+}. AMPK functions to maintain energy balance within the cells by controlling several intracellular systems leading to ATP production and restoration (Marsin et al., 2002; Wu and Wei, 2012); in this manner this evolutionary conserved signaling molecule regulates the functions of various metabolic processes. Upon activation, AMPK promotes molecular mechanisms of glucose metabolism, fatty acid oxidation, and mitochondrial biogenesis (Dzamko and Steinberg, 2009; Russell et al., 1999; Weisova et al., 2009; Zong et al., 2002) that generate ATP while simultaneously suppressing energy consuming pathways such as cell growth, cell cycle, protein synthesis, fatty acid synthesis, and ribosome biosynthesis (Carling, 2004; Hardie, Ross, and Hawley, 2012b; Weisova et al., 2011). In this regard, AMPK is well suited to respond rapidly to dynamic changes in metabolism during physiological (exercise, hormones) or pathological (ischemia, excitotoxicity) conditions, as well as to that of compromised nutrient availability (starvation, caloric restriction) (Stark, Ashley, and Andrews, 2013). It may also have long-term effects by regulation of gene transcription through the modulation of transcription factors (forkhead box class O3a [FOXO3a], nuclear factor kappa B; peroxisome proliferator-activated receptor gamma coactivator-1α [PGC-1α]) or by controlling the formation of transcription complexes by adapting rRNA synthesis (Weisova et al., 2011). Interestingly, activation of AMPK can be also mediated by cellular insults, hormones, or pharmacological agents without affecting the AMP/ATP ratio, demonstrating the ability to sense both basal and pathological conditions (Hardie, 2007a; Kahn et al., 2005; Ronnett et al., 2009).

5.1.1b Expression of AMPK in the Brain

As an evolutionarily conserved metabolic master, AMPK has been shown to be expressed in various types of tissues and cells, including brain (Culmsee et al., 2001; Poels et al., 2009; Ronnett et al., 2009). Most of the information about brain energy metabolism is usually considered predominantly to be about metabolism of neuronal cells (Belanger, Allaman, and Magistretti, 2011). In particular, neurons, in comparison to other types of cells, are highly dependent on a continuous supply of nutrients because they do not have sufficient mechanisms to store energy in the form of glycogen. Taking into consideration the high rate of metabolic activity displayed by neurons, neural activity and growth dictate their increased sensitivity to metabolic fluctuations (Poels et al., 2009). AMPK is detectable throughout early development to adulthood, present in both the embryonic and adult brain, with high levels of expression throughout embryonic hippocampal, cortical, and cerebellar granule neurons (Culmsee et al., 2001; Vingtdeux et al., 2010; Weisova et al., 2009). Notably, each of the AMPK subunits is expressed at high levels throughout the brain and central nervous system (CNS), predominantly confined to neurons with significant localization to the neuronal nucleus compared to weak cytoplasmic localization detected in astrocytes (Culmsee et al., 2001; Turnley et al., 1999; Vingtdeux et al., 2010). Moreover, there have been extensive studies examining the consequences of AMPK activity on hypothalamic neurons because of its known effect on feeding behavior (food intake and body weight), thus firmly positioning AMPK to regulate energy homeostasis at the whole-organismal level (Andersson et al., 2004; Claret et al., 2007; Minokoshi et al., 2004; Pimentel et al., 2013).

AMPK is present in brain cells as a heterotrimeric protein complex consisting of multiple subunits conserved in evolution. This complex is comprised of a catalytic α subunit and two regulatory subunits, β and γ (Kahn et al., 2005). Differential CNS expression patterns have revealed the existence of multiple subunit isoforms: two catalytic α (α1, α2), two regulatory β (β1, β2) and three γ (γ1, γ2, γ3) subunit isoforms, each with distinctive tissue-specific localization and function (Manwani and McCullough, 2013). This combinatory option of cell-type specific AMPK subunit expression throughout the brain tissue allows for heterogeneity and increased diversity of function in its activity as a response to various stimuli. All of these AMPK subunits have been reported to be expressed in brain, with the exception of γ3 subunit, which is considered a muscle specific isoform (Turnley et al., 1999; Yu et al., 2004). In addition, the expression of these different isoforms is dynamic and has been shown to change throughout brain development (Amato and Man, 2011a; Turnley et al., 1999). Therefore, spatial and temporal differences in the AMPK subunit expression pattern are not well characterized; notably with the AMPK complex demonstrating a constitutive level of activation in brain cells. However, the significance of this basal level of activation, particularly in neurons, warrants further investigation.

5.1.1c Molecular Activation of AMPK Signaling

The two α subunit isoforms contain a serine/threonine residue (Thr-172) that can be phosphorylated by upstream kinases (Hawley et al., 1995), the process required for activation of AMPK. Allosteric activation of AMPK is a complex mechanism. First, binding of AMP to AMPK γ subunits enables a conformational change and thus an increase in kinase activity of the AMPK α subunit. Thus, the binding of AMP (and possibly also ADP) to AMPK γ promotes phosphorylation of threonine 172 within the activation loop of AMPK α by upstream kinases referred to as AMPK kinases (AMPKKs) and at the same time inhibition of its dephosphorylation by protein phosphatase-2Cα (Sanders et al., 2007), leading to maximal AMPK activity. Regardless, if AMPK is in an active or inactive form, the AMPK β subunit acts as an important scaffold protein binding AMPK α and AMPK γ (Hardie and Hawley, 2001).

Up to the present day, three mammalian AMPK upstream kinases responsible for an increase in phosphorylation of the α subunit have been identified: the tumor suppressor LKB1 (liver kinase B1/serine threonine kinase 11); Ca^{2+}/Calmodulin–dependent protein kinase β (CaMKKβ), and transforming growth factor-β-activated kinase (TAK-1) (Hawley et al., 2003; Woods et al., 2003, 2005; Xie et al., 2006). LKB1 is ubiquitously expressed in most mammalian cells and is believed to be the predominant upstream kinase that activates AMPK in an AMP/ATP-sensitive manner (Hardie, 2007a). However, CaMKKβ is mainly expressed in brain tissue (Anderson et al., 1998) and triggers AMPK activation once intracellular Ca^{2+} levels are increased. In doing so CaMKKβ couples neuronal activity with neuronal energy metabolism. More comprehensive studies are required to determine the relevance of additional AMPK upstream kinase TAK-1 in brain tissue in particular.

5.1.1d Pharmacological Activation of AMPK

AMPK is the main metabolic sensor with the capability to modulate an ever-increasing number of signaling pathways and biological processes. As a broad and increasing spectrum of metabolic-related diseases with rising prevalence present themselves to date (metabolic syndrome, diabetes, ischemia, cardiovascular disease etc.), AMPK has become an attractive target for drug development in recent years. Already in 1999 AMPK activators have been suggested as potential drugs for the treatment of type 2 diabetes (Winder and Hardie, 1999). So far, a number of pharmacological agents have been used that can activate AMPK upon treatment of tissues/cells in vivo and in vitro. It has been clarified that agents with the potential to activate AMPK can act either by direct or indirect mechanisms; indeed, studies using the drugs activating AMPK have helped us to understand more precisely the role of this ancient energy master in modulating molecular signaling in different cells types.

Some of the most widely used pharmacological agents that indirectly activate AMPK in peripheral tissue and demonstrate antidiabetic properties are biguanide metformin and thiazolidinediones (TZDs) (Hardie, 2007b; Hawley et al., 2010; Zhang et al., 2009). Metformin activates AMPK by inhibition of mitochondrial function, thus increasing AMP/ATP ratio; however it also has AMPK-independent effects (Brunmair et al., 2004; Hardie, 2013; Hardie et al., 2012b). Moreover, TZDs are able to modulate AMPK activity by regulating the release of the hormone adiponectin, known to increase AMPK levels (Hardie et al., 2012b). From the group of indirect AMPK activators, an emerging role for several naturally occurring plant-derived products, including alkaloids, has been reported

(Hardie, 2007b, 2013; Hardie et al., 2012b; Russo, Russo, and Ungaro, 2013). Interestingly, most of these compounds are naturally present in food such as capsaicin from peppers (Hwang et al., 2005); curcumin in turmeric (Lim, Lim, and Wong, 2009); epigallocatechin-3-gallate (EGCG) present in green tea, theaflavin from black tea (Hardie, 2013; Hwang, Kwon, and Yoon, 2009); or have been derived from traditional medicines such as Berberine (Turner et al., 2008). An additional natural compound that deserves a special note and is currently often used in studies associated with neuronal research is Resveratrol, produced in grapes (Baur et al., 2006; Baur and Sinclair, 2006; Dasgupta and Milbrandt, 2007; Pasinetti et al., 2011; Vingtdeux et al., 2010), leading to the recent development of other novel small synthetic agents termed RSVAs (RSVA314 and RSVA405), analogues of resveratrol with higher potency in activating AMPK (Vingtdeux et al., 2011a).

Interestingly, despite the fact that these natural compounds have variable structures, they all can activate AMPK. Many of them have the capability to activate this enzyme indirectly by interference with mitochondrial ATP synthesis, thus increasing AMP and ADP levels (Hardie et al., 2012a; Hardie, 2013; Russo et al., 2013). Importantly, data obtained from studies using these natural compounds as AMPK activators should be interpreted with care because potential "off-target" effects as well as side effects may occur.

With this in mind, it is also necessary to develop alternative activators that act directly on AMPK to better understand the role of this principal energy sensor in various diseases. Direct activators of AMPK currently in use and derived from high throughput screening, A-769662 (Cool et al., 2006; Zhang et al., 2009) and a small molecular activator PT1, both have ability to activate AMPK allosterically without having an effect on AMP/ATP ratio (Pang et al., 2008). Another pharmacological agent acting as a direct activator of AMPK widely in circulation in research over the years also worthy of note is 5-aminoimidazole-4-carboxamid riboside (AICAR) (Corton et al., 1995; Guigas et al., 2009). The capability of AICAR to mimic effects of AMP binding on AMPK by its mono-phosphorylated derivative ZMP (5-aminoimidazole-4-carboxamide ribonucleoside) makes this drug a valuable and frequently used experimental tool (Weisova et al., 2011). Indeed, a large number of studies in the area of brain research have been using AICAR extensively as an agent to modulate AMPK activity both in vitro and in vivo (Anilkumar et al., 2013; Concannon et al., 2010; Davila et al., 2012; Weisova et al., 2009, 2011).

Another natural plant product of particular interest, considered to be one of the oldest medicinal agents known since ancient times, is salicylate, which has the capability to directly activate AMPK (Hardie, 2013; Hawley et al., 2012). This active component of willow bark, as well as a natural component of an everyday drug often in use (aspirin), has been recently shown to activate AMPK in human cell lines (Hawley et al., 2012). Salicylate activates AMPK allosterically by direct binding to the same site as the synthetic agent A-769662 (Hardie et al., 2012a; Hawley et al., 2012). Moreover, the structure of salicylate is similar to A-769662 (Hawley et al., 2012).

In a new study conducted in our laboratory we have described another small-molecule compound called Latrepirdine/Dimebon as an activator of AMPK (Weisova et al., 2013) and surprisingly, we have demonstrated in primary neuronal cultures that

Latrepirdine acts and increases AMPK levels at low, (sub-) nanomolar doses. This makes this agent currently one of the most effective pharmacological activators of AMPK (Weisova et al., 2013). As a result, we propose that further molecular and structural studies may present substantial potential in the identification of new compounds activating AMPK and may also elucidate new understanding of the mode of action of already known AMPK activators.

5.1.2 Other Roles for AMPK

5.1.2a Neuronal Development

AMPK has been described as an enzyme with wide expression in mammalian brain (Culmsee et al., 2001); however, additional functions in brain physiology have been only recently investigated. AMPK signaling has been ultimately linked to various aspects of neurodevelopment and function necessarily connected to processes such as neurogenesis, polarization, and axon formation (Amato et al., 2011; Amato and Man, 2011a, 2011b; Williams et al., 2011), communication between synapses and regulation of neuronal plasticity (Potter et al., 2010). In neurons, in addition to its energy sensing role, AMPK has been shown to be an important regulator of neuronal development (Dasgupta and Milbrandt, 2007). Increased phosphorylation and activation of AMPK using pharmacological approaches has been shown to promote neurite outgrowth in Neuro2a cells, whereas dephosphorylation suppresses this phenotype by inhibition of AMPK activity (Dasgupta and Milbrandt, 2007). In contrast, overactivation of AMPK has been shown to inhibit axon outgrowth by suppressing mTOR signaling in cortical neuron development (Williams et al., 2011). Moreover, enhancement of AMPK activity in early stages of neuronal development, through the pharmacological mimicking of energy-deficient conditions led to disrupted neuronal polarization and axon formation in hippocampal neurons and cortical slices (Amato et al., 2011); increased localization of PI3K at the neurite tip, a pivotal process important for neuronal polarization, has also been found to be suppressed with increases in AMPK activity (Amato et al., 2011).

In other studies, using genetic approaches to modulate the catalytic activity of AMPK, such as deletion of various AMPK subunits in different model organisms, has elucidated the role of this energetic enzyme in neurodevelopment (Dasgupta and Milbrandt, 2009; Tschape et al., 2002; Williams et al., 2011). In this regard, two recent studies employing *Drosophila* as a model organism as well as mice, reported that genetic loss of function of AMPK activity mediated by deletion of β1 subunit led to rapid and progressive neurodegeneration, combined with neurophysiological defects, whereas mutant mice lacking β1 subunit were observed to demonstrate severe neuronal loss throughout the CNS, accompanied by atrophy of the dentate gyrus and cerebellum, (Dasgupta and Milbrandt, 2009; Spasic, Callaerts, and Norga, 2008). Additionally, in the same model organism, a mutation in AMPK γ subunit also resulted in extensive neuronal degeneration, thus elucidating the significance of AMPK activity in growth factor signaling, maintenance of neuronal integrity and regulation of brain development (Tschape et al., 2002). A study exploiting neural stem-cell progenitors (NPCs) to test the role of AMPK has revealed the necessity of its function for normal brain development (Zang et al., 2009); however in another work, a genetic loss of function of AMPK as a result of the deletion of $\alpha1/\alpha2$ catalytic subunits in mammalian cortex has been shown to not have any effect on proper neurogenesis, radial migration, differentiation, or polarization during physiological conditions (Williams et al., 2011). Interestingly, this is in accordance with data from other multicellular organisms (*Drosophila, Caenorhabditis elegans*) in which AMPK α activity was not important for proper neuronal specification under basal conditions (Alessi, Sakamoto, and Bayascas, 2006; Mirouse et al., 2007; Narbonne and Roy, 2009). Further investigations are needed to resolve these discrepancies in the requirement of AMPK activity during CNS development.

Importantly, noticeable evidence exists in studies reporting the significance of a major upstream kinase of AMPK signaling, liver kinase B1 (LKB1), in axon specification and axon growth during neuronal polarization in vivo (Barnes et al., 2007; Shelly et al., 2007). However, it has been also illustrated that unlike in other cells, LKB1 may not be a predominant regulator of AMPK activity in immature neuronal tissue (Hawley et al., 2005). Under physiological conditions, cortical neurons with LKB1 deficiency exhibited neither change in phosphorylation status nor in total levels of AMPK, suggesting that LKB1 may act independently of AMPK function during neuronal development (Barnes et al., 2007; Shelly et al., 2007). However, in previous work, AMPK has been activated by intracellular Ca^{2+} elevation, whereas AMP levels remain unchanged (Hawley et al., 2005). An alternative AMPK upstream signaling kinase, termed *Ca^{2+}/calmodulin dependent protein kinase* (CaMKK), has also been suggested to have a function in modulating AMPK activity by Ca^{2+} (Green. Anderson, and Means, 2011; Hawley et al., 2005; Woods et al., 2005). Taking into consideration that Ca^{2+} ions have a central role in synaptic function and transmission as well as during excitotoxicity, activation of AMPK by Ca^{2+} may represent a molecular switch for coupling of neuronal metabolic regulation with synapse formation (Amato and Man, 2011a, 2011b).

5.1.2b Other Neuron-Specific Roles of AMPK

Several additional effects of the AMPK signaling cascade have been attributed exclusively to neurons (Hardie and Frenguelli, 2007; Kuramoto et al., 2007; Sopjani et al., 2010a, 2010b). AMPK has been described as a negative regulator of Na^+-coupled glutamate transport by down-regulation of the glutamate transporters EAAT3 and EAAT4 expressed in neuronal and glial cells (Sopjani et al., 2010a). Interestingly, activity of AMPK has been also linked to the modulation of GABA(B) receptor desensitization (Hardie and Frenguelli, 2007; Kuramoto et al., 2007, 2013), the process important for reduction of excitotoxicity, thus promoting neuronal survival. Furthermore, regulation of neuronal metabolism has also been linked to neuronal plasticity through demonstration of AMPK modulation of long-term potentiation (LTP) (Potter et al., 2010). In addition, pharmacological activation of AMPK induced a concentration-dependent decrease in spine density in cortical neurons in vitro (Mairet-Coello et al., 2013). We and others have recently reported the role of AMPK in regulation of neuronal excitability (Ikematsu et al., 2011; Weisova et al., 2013) through the use of the small molecule compound Latrepirdine, which activates

AMPK in the nanomolar concentration range, demonstrating reduced excitability of neurons by affecting plasma membrane potential (Weisova et al., 2013). In another study, introducing active AMPK into hippocampal neurons resulted in a massive reduction in the rate of action potential discharge (Ikematsu et al., 2011) and a decline in neuronal membrane excitability; this was mediated by direct phosphorylation of the voltage-gated potassium channel Kv2.1 by AMPK (Ikematsu et al., 2011). Importantly, a new role of AMPK in cognition and suppression of memory impairment has been recently reported (Kim et al., 2013).

Taken together, these observations indicate the existence of additional neuron-specific roles for AMPK signaling beyond its recognized metabolic functions.

5.1.3 AMPK in Neurological Diseases and Neurodegeneration

5.1.3a AMPK in Excitotoxicity and Ischemia

Ischemic or excitotoxic injury of the brain is accompanied by a cascade of molecular events leading to neuronal damage resulting in major part from cellular energy deficits. Previously it has been shown that AMPK activation in excitotoxic or ischemic tissue has the ability to enhance the metabolism of glucose through the increased expression of glucose transporter 3 on the plasma membrane, resulting in a potential increase in glycolytic flux, increased mitochondrial biogenesis or fatty acid oxidation to compensate for the energy deficit (Li and McCullough, 2010; Manwani and McCullough, 2013; Ramamurthy and Ronnett, 2012; Ronnett et al., 2009; Spasic, Callaerts, and Norga, 2009; Weisova et al., 2009). Despite many described reports of beneficial effects of AMPK on cellular energy levels in neurons, the outcome of its activation in ischemic or excitotoxic neuronal tissue is dependent on a variety of factors warranting consideration. The overall pro-survival or pro-death effects of AMPK activation are dependent on the duration and nature of neuronal insult as well as on duration and intensity of AMPK activity following metabolic stress, the extent of recovery of ATP and cellular bioenergetics and the type and metabolic capacity of the cell at the time of insult. These factors all play a role in determining whether the cell can tolerate the stimulus or will initiate cell death (Bolanos, Almeida, and Moncada, 2010; Davila et al., 2012; Manwani and McCullough, 2013; Ramamurthy and Ronnett, 2012; Weisova et al., 2011). In addition, when this multifunctional enzyme is activated in the “preconditioning paradigm,” it can exert protective responses in neurons on subsequent ischemic/excitotoxic insult (Anilkumar et al., 2013; Weisova et al., 2012, 2013). Taken as a whole, from in vitro studies, it has become apparent that activity of AMPK has a dual role in neuronal outcome, on one side having beneficial effects and promoting survival or on the other as the factor contributing to neuronal death (Concannon et al., 2010; Culmsee et al., 2001; Davila et al., 2012; Weisova et al., 2009, 2011; Zhang et al., 2010). Some of the recent work from our laboratory has demonstrated that the duration of AMPK activation had a direct influence on neuronal survival in neurons exposed to excitotoxicity (Davila et al., 2012). Transient activation of AMPK as a result of brief glutamate exposure in cerebellar granule neurons (CGNs) led to significant enhancement of glycolytic flux resulting from increased GLUT3 expression on neuronal plasma membranes, associated with hyperpolarization of mitochondrial membrane potential with subsequent recovery and overshoot of neuronal ATP levels (Weisova et al., 2009). However, sustained AMPK activation during excitotoxic injury or by pharmacological activation with AICAR resulted in activation of expression of the pro-apoptotic BH3-only protein Bim (Concannon et al., 2010). Moreover, it has been detected that this Bim activation requires not only translocation of Forkhead transcription factor FOXO3 to the nucleus, which is also observed during transient AMPK activation, but also phosphorylation by AMPK as a second activation step (Davila et al., 2012).

Based on more complex in vivo approaches using middle cerebral artery occlusion (MCAO) in a mouse model of ischemic stroke, it has been demonstrated that increased AMPK activation during ischemia exacerbated neuronal injury also in vivo (Manwani and McCullough, 2013; McCullough et al., 2005). In another study, in which activation of AMPK has been stimulated by the anti-diabetic drug metformin either acutely (3 days) or chronically (3 weeks) before stroke induction, again the duration of AMPK activity was the central factor directly affecting infarct volume in the brain (Li and McCullough, 2010; Manwani and McCullough, 2013). Other studies have revealed that inhibition of AMPK activity by administration of compound C or fatty acid synthase inhibitor C75, as well as by genetic deletion of the AMPK α2 catalytic subunit, results in reduction in brain damage induced by MCAO (Li et al., 2007; Li and McCullough, 2010; Manwani and McCullough, 2013; McCullough et al., 2005). Several other works using an AMPK subunit gene deletion approach have further elucidated the role of AMPK in stroke models (Manwani and McCullough, 2013).

Most of these in vivo studies have shown deleterious effects of AMPK in ischemic brain. However, a study by Kuramoto et al. (2007) has described AMPK as initiating a neuroprotective signaling cascade. In conditions of ischemic injury, activation of AMPK resulted in increased direct phosphorylation of $GABA_B$ receptor S783 in the R2 subunit. This led to enhanced receptor activity with a subsequent prolonged post-synaptic hyperpolarization, suggesting a reduction in neurotransmitter release from presynaptic neurons , thus promoting neuronal survival after chemical anoxia (Hardie and Frenguelli, 2007; Kuramoto et al., 2007; Li and McCullough, 2010). Notably, the effect of AMPK on phosphorylation of $GABA_{(B)}$ receptors is mainly detected in regions of the brain usually unaffected by ischemic injury (Li and McCullough, 2010; Manwani and McCullough, 2013). Therefore, a more precise link between AMPK and neuroprotection mediated by modulation of $GABA_{(B)}$ receptors in ischemia has to be established. Overall, additional investigations using selective activators of AMPK signaling, as well as testing various timing, insult severity and degree of AMPK activation in a tissue-specific manner might bring new insights to the task of clarifying the potential of AMPK as a promising therapeutic intervention for stroke.

5.1.3b AMPK in Neurodegeneration

Defects in AMPK signalling have been demonstrated in several neurodegenerative diseases such as Alzheimer's disease (AD), Huntington's disease (HD), Parkinson's disease (PD), and motor neuron disease, as well as in multiple sclerosis, peripheral diabetic neuropathy, and neuropathic pain (Cai et al., 2012; Chen et al., 2009; Ju et al., 2011; Lim et al., 2012; Melemedjian et al., 2011; Paintlia et al., 2013; Salminen and

Kaarniranta, 2012; Thornton et al., 2011; Vingtdeux et al., 2010, 2011b; Won et al., 2010). Among the few functional studies that explored the effect of AMPK in neurodegeneration, Ju and co-workers (2011) showed activation of the α1 isoform of AMPK in striatal neurons of humans and mice with HD. The authors then demonstrated that overactivation of AMPK in the striatum caused brain atrophy, facilitated neuronal loss, and increased formation of Htt aggregates in a transgenic mouse model (R6/2) of HD, suggesting a disease-accelerating function of AMPK in this model. The authors also proposed that nuclear accumulation of AMPK α1 was required for this toxicity. Similarly, reduced activity of AMPK has been suggested to protect against mutant SOD1-induced motoneuron degeneration (Lim et al., 2012). Because AMPK also activates autophagy, it is, however, also conceivable that AMPK activation exerts neuroprotective activities during neurodegeneration. In this context, AMPK activation has been shown to mitigate dopaminergic dysfunction in *Drosophila* models of PD (Ng et al., 2012). The small molecule activator of AMPK, Latrepirdine, has also been shown to stimulate autophagy and reduce the accumulation of alpha-synuclein in in vitro and in vivo PD models (Steele et al., 2013a), and to enhance autophagy and arrest the progression of neuropathology in an AD model in vivo (Steele and Gandy, 2013; Steele et al., 2013b). Given the role of AMPK as a central regulator of bioenergetics, autophagy, and apoptosis, further genetic studies in mammalian models of PD, as well as AD, are required to fully explore the role of AMPK in these neurodegenerative disorders.

5.2 CONCLUSION AND FUTURE PERSPECTIVES

Multiple pathways are able to activate AMPK in neurons. Moreover, AMPK has multiple effects on the cell, tissue, and whole-body level. As outlined previously, effects of AMPK activation and inhibition can also be highly context-dependent. Hence the biology of AMPK will remain an area of intense research in the neurosciences. Both an increase in our knowledge of the diverse substrates of AMPK, as well as a better temporal and spatial analysis of AMPK signaling effects will likely deliver new key insights into the role of AMPK in neurodevelopment, neuronal physiology, and neuronal injury. Novel systems-based approaches including computational modeling approaches may also provide new insights into the role of AMPK as a molecular "switch" in these processes. With this knowledge increase and with the development of agents selectively targeting the diverse activities of AMPK, AMPK may increasingly become an important molecular target for the treatment of ischemic and neurologic disorders.

REFERENCES

Alessi, D. R., Sakamoto, K. and Bayascas, J. R., 2006. LKB1-dependent signaling pathways. *Annu Rev Biochem*, 75, pp.137–163.

Amato, S., et al., 2011. AMP-activated protein kinase regulates neuronal polarization by interfering with PI 3-kinase localization. *Science*, 332, pp.247–251.

Amato, S. and Man, H. Y., 2011a. AMPK links cellular bioenergy status to the decision making of axon initiation in neurons. *Cell Logist*, 1, pp.103–105.

Amato, S. and Man, H. Y., 2011b. Bioenergy sensing in the brain: the role of AMP-activated protein kinase in neuronal metabolism, development and neurological diseases. *Cell Cycle*, 10, pp.3452–3460.

Anderson, K. A., et al., 1998. Components of a calmodulin-dependent protein kinase cascade. Molecular cloning, functional characterization and cellular localization of Ca2+/calmodulin-dependent protein kinase kinase beta. *J Biol Chem*, 273, pp.31880–31889.

Andersson, U., et al., 2004. AMP-activated protein kinase plays a role in the control of food intake. *J Biol Chem.* 279, pp.12005–12008.

Anilkumar, U., et al., 2013. AMP-activated protein kinase (AMPK)-induced preconditioning in primary cortical neurons involves activation of MCL-1. *J Neurochem*, 124, pp.721–734.

Barnes, A. P., et al., 2007. LKB1 and SAD kinases define a pathway required for the polarization of cortical neurons. *Cell*, 129, pp.549–563.

Baur, J. A., et al., 2006. Resveratrol improves health and survival of mice on a high-calorie diet. *Nature*, 444, pp.337–342.

Baur, J. A. and Sinclair, D. A., 2006. Therapeutic potential of resveratrol: The in vivo evidence. *Nat Rev Drug Discov*, 5, pp.493–506.

Belanger, M., Allaman, I. and Magistretti, P. J., 2011. Brain energy metabolism: focus on astrocyte-neuron metabolic cooperation. *Cell Metab*, 14, pp.724–738.

Bolanos, J. P., Almeida, A. and Moncada, S., 2010. Glycolysis: A bioenergetic or a survival pathway? *Trends Biochem Sci,* 35, pp.145–149.

Brunmair, B., et al., 2004. Thiazolidinediones, like metformin, inhibit respiratory complex I: a common mechanism contributing to their antidiabetic actions? *Diabetes*, 53, pp.1052–1059.

Cai, Z., et al., 2012. Roles of AMP-activated protein kinase in Alzheimer's disease. *Neuromolecular Med*, 14, pp.1–14.

Carling, D., 2004. The AMP-activated protein kinase cascade—a unifying system for energy control. *Trends Biochem Sci*, 29, pp.18–24.

Chen, Y., et al., 2009. Antidiabetic drug metformin (GlucophageR) increases biogenesis of Alzheimer's amyloid peptides via up-regulating BACE1 transcription. *Proc Natl Acad Sci U S A*, 106, 3907–3912.

Claret, M., et al., 2007. AMPK is essential for energy homeostasis regulation and glucose sensing by POMC and AgRP neurons. *J Clin Invest*, 117, pp.2325–2336.

Concannon, C. G., et al., 2010. AMP kinase-mediated activation of the BH3-only protein Bim couples energy depletion to stress-induced apoptosis. *J Cell Biol*, 189, pp.83–94.

Cool, B., et al., 2006. Identification and characterization of a small molecule AMPK activator that treats key components of type 2 diabetes and the metabolic syndrome. *Cell Metab*, 3, pp.403–416.

Corton, J. M., et al., 1995. 5-aminoimidazole-4-carboxamide ribonucleoside. A specific method for activating AMP-activated protein kinase in intact cells? *Eur J Biochem*, 229, pp.558–565.

Culmsee, C., et al., 2001. AMP-activated protein kinase is highly expressed in neurons in the developing rat brain and promotes neuronal survival following glucose deprivation. *J Mol Neurosci*, 17, pp.45–58.

Dasgupta, B. and Milbrandt, J., 2007. Resveratrol stimulates AMP kinase activity in neurons. *Proc Natl Acad Sci U S A*, 104, pp.7217–7722.

Dasgupta, B. and Milbrandt, J., 2009. AMP-activated protein kinase phosphorylates retinoblastoma protein to control mammalian brain development. *Dev Cell*, 16, pp.256–270.

Davila, D., et al., 2012. Two-step activation of FOXO3 by AMPK generates a coherent feed-forward loop determining excitotoxic cell fate. *Cell Death Differ*, 19, 1677–1688.

Dzamko, N. L. and Steinberg, G. R., 2009. AMPK-dependent hormonal regulation of whole-body energy metabolism. *Acta Physiol (Oxf)*, 196, pp.115–127.

Giri, S., et al., 2004. 5-aminoimidazole-4-carboxamide-1-beta-4-ribofuranoside inhibits proinflammatory response in glial cells: a possible role of AMP-activated protein kinase. *J Neurosci*, 24, pp.479–487.

Green, M. F., Anderson, K. A. and Means, A. R., 2011. Characterization of the CaMKK beta-AMPK signaling complex. *Cell Signal*, 23, 2005–2012.

Guigas, B., et al., 2009. Beyond AICA riboside: in search of new specific AMP-activated protein kinase activators. *IUBMB Life*, 61, pp.18–26.

Hardie, D. G., et al., 1999. AMP-activated protein kinase: An ultrasensitive system for monitoring cellular energy charge. *Biochem J*, 338 (Pt 3), pp.717–722.

Hardie, D. G. and Hawley, S. A., 2001. AMP-activated protein kinase: The energy charge hypothesis revisited. *Bioessays*, 23, pp.1112–1119.

Hardie, D. G., 2007a. AMP-activated/SNF1 protein kinases: Conserved guardians of cellular energy. *Nat Rev Mol Cell Biol*, 8, pp.774–785.

Hardie, D. G., 2007b. AMP-activated protein kinase as a drug target. *Annu Rev Pharmacol Toxicol*, 47, pp.185–210.

Hardie, D. G. and Frenguelli, B. G., 2007. A neural protection racket: AMPK and the GABA(B) receptor. *Neuron*, 53, pp.159–162.

Hardie, D. G., Ross, F. A. and Hawley, S. A., 2012a. AMP-activated protein kinase: A target for drugs both ancient and modern. *Chem Biol*, 19, pp.1222–1236.

Hardie, D. G., Ross, F. A. and Hawley, S. A., 2012b. AMPK: A nutrient and energy sensor that maintains energy homeostasis. *Nat Rev Mol Cell Biol*, 13, pp.251–262.

Hardie, D. G., 2013. AMPK: A target for drugs and natural products with effects on both diabetes and cancer. *Diabetes*, 62, pp.2164–2172.

Hawley, S. A., et al., 1995. 5'-AMP activates the AMP-activated protein kinase cascade, and Ca2+/calmodulin activates the calmodulin-dependent protein kinase I cascade, via three independent mechanisms. *J Biol Chem*, 270, pp.27186–27191.

Hawley, S. A., et al., 2003. Complexes between the LKB1 tumor suppressor, STRAD alpha/beta and MO25 alpha/beta are upstream kinases in the AMP-activated protein kinase cascade. *J Biol*, 2, p.28.

Hawley, S. A., et al., 2005. Calmodulin-dependent protein kinase kinase-beta is an alternative upstream kinase for AMP-activated protein kinase. *Cell Metab*, 2, pp.9–19.

Hawley, S. A., et al., 2010. Use of cells expressing gamma subunit variants to identify diverse mechanisms of AMPK activation. *Cell Metab*, 11, pp.554–565.

Hawley, S. A., et al., 2012. The ancient drug salicylate directly activates AMP-activated protein kinase. *Science*, 336, pp.918–22.

Hwang, J. T., et al., 2005. Genistein, EGCG, and capsaicin inhibit adipocyte differentiation process via activating AMP-activated protein kinase. *Biochem Biophys Res Commun*, 338, pp.694–699.

Hwang, J. T., Kwon, D. Y. and Yoon, S. H., 2009. AMP-activated protein kinase: a potential target for the diseases prevention by natural occurring polyphenols. *N Biotechnol*, 26, pp.17–22.

Ikematsu, N., et al., 2011. Phosphorylation of the voltage-gated potassium channel Kv2.1 by AMP-activated protein kinase regulates membrane excitability. *Proc Natl Acad Sci U S A*, 108, pp.18132–18137.

Ju, T. C., et al., 2011. Nuclear translocation of AMPK-alpha1 potentiates striatal neurodegeneration in Huntington's disease. *J Cell Biol*, 194, pp.209–227.

Kahn, B. B., et al., 2005. AMP-activated protein kinase: Ancient energy gauge provides clues to modern understanding of metabolism. *Cell Metab*, 1, pp.15–25.

Kim, S. J., et al., 2013. Neuroprotective effects of AMP-activated protein kinase on scopolamine induced memory impairment. *Korean J Physiol Pharmacol*, 17, pp.331–338.

Kuramoto, N., et al., 2007. Phospho-dependent functional modulation of GABA(B) receptors by the metabolic sensor AMP-dependent protein kinase. *Neuron*, 53, pp.233–247.

Kuramoto, N., et al., 2013. Dephosphorylation of endogenous GABA(B) receptor R2 subunit and AMPK alpha subunits which were measured by in vitro method using transfer membrane. *Neurochem Int*, 62, pp.137–144.

Lee, J. W., et al., 2010. The association of AMPK with ULK1 regulates autophagy. *PLoS One*, 5, p.e15394.

Li, J., et al., 2007. Neuroprotective effects of adenosine monophosphate-activated protein kinase inhibition and gene deletion in stroke. *Stroke*, 38, pp.2992–2999.

Li, J. and McCullough, L. D., 2010. Effects of AMP-activated protein kinase in cerebral ischemia. *J Cereb Blood Flow Metab*, 30, pp.480–492.

Liang, J., et al., 2007. The energy sensing LKB1-AMPK pathway regulates p27(kip1) phosphorylation mediating the decision to enter autophagy or apoptosis. *Nat Cell Biol*, 9, pp.218–224.

Lim, H. W., Lim, H. Y. and Wong, K. P., 2009. Uncoupling of oxidative phosphorylation by curcumin: Implication of its cellular mechanism of action. *Biochemical and Biophysical Research Communications*, 389, pp.187–192.

Lim, M. A., et al., 2012. Reduced activity of AMP-activated protein kinase protects against genetic models of motor neuron disease. *J Neurosci*, 32, pp.1123–1141.

Mairet-Coello, G., et al., 2013. The CAMKK2-AMPK kinase pathway mediates the synaptotoxic effects of Abeta oligomers through Tau phosphorylation. *Neuron*, 78, pp.94–108.

Manwani, B. and McCullough, L. D., 2013. Function of the master energy regulator adenosine monophosphate-activated protein kinase in stroke. *J Neurosci Res*, 91, pp.1018–1029.

Marsin, A. S., et al., 2002. The stimulation of glycolysis by hypoxia in activated monocytes is mediated by AMP-activated protein kinase and inducible 6-phosphofructo-2-kinase. *J Biol Chem*, 277, pp.30778–30783.

McCullough, L. D., et al., 2005. Pharmacological inhibition of AMP-activated protein kinase provides neuroprotection in stroke. *J Biol Chem*, 280, pp.20493–20502.

Melemedjian, O. K., et al., 2011. Targeting adenosine monophosphate-activated protein kinase (AMPK) in preclinical models reveals a potential mechanism for the treatment of neuropathic pain. *Mol Pain*, 7, p.70.

Minokoshi, Y., et al., 2004. AMP-kinase regulates food intake by responding to hormonal and nutrient signals in the hypothalamus. *Nature*, 428, pp.569–574.

Mirouse, V., et al., 2007. LKB1 and AMPK maintain epithelial cell polarity under energetic stress. *J Cell Biol*, 177, pp.387–392.

Nakano, A., et al., 2010. AMPK controls the speed of microtubule polymerization and directional cell migration through CLIP-170 phosphorylation. *Nat Cell Biol*, 12, pp.583–590.

Narbonne, P. and Roy, R., 2009. Caenorhabditis elegans dauers need LKB1/AMPK to ration lipid reserves and ensure long-term survival. *Nature*, 457, pp.210–214.

Ng, C. H., et al., 2012. AMP kinase activation mitigates dopaminergic dysfunction and mitochondrial abnormalities in Drosophila models of Parkinson's disease. *J Neurosci*, 32, pp.14311–14317.

Paintlia, A. S., et al., 2013. AMP-activated protein kinase signaling protects oligodendrocytes that restore central nervous system functions in an experimental autoimmune encephalomyelitis model. *Am J Pathol*, 183, pp.526–541.

Pang, T., et al., 2008. Small molecule antagonizes autoinhibition and activates AMP-activated protein kinase in cells. *J Biol Chem*, 283, pp.16051–16060.

Pasinetti, G. M., et al., 2011. Neuroprotective and metabolic effects of resveratrol: Therapeutic implications for Huntington's disease and other neurodegenerative disorders. *Exp Neurol*, 232, pp.1–6.

Pimentel, G. D., et al., 2013. The role of neuronal AMPK as a mediator of nutritional regulation of food intake and energy homeostasis. *Metabolism*, 62, pp.171–178.

Poels, J., et al., 2009. Expanding roles for AMP-activated protein kinase in neuronal survival and autophagy. *Bioessays*, 31, pp.944–952.

Potter, W. B., et al., 2010. Metabolic regulation of neuronal plasticity by the energy sensor AMPK. *PLoS One*, 5, p.e8996.

Ramamurthy, S. and Ronnett, G., 2012. AMP-activated protein kinase (AMPK) and energy-sensing in the brain. *Exp Neurobiol*, 21, pp.52–60.

Ronnett, G. V., et al., 2009. AMPK in the brain: its roles in energy balance and neuroprotection. *J Neurochem*, 109 Suppl 1, pp.17–23.

Russell, R. R., 3rd, et al., 1999. Translocation of myocardial GLUT-4 and increased glucose uptake through activation of AMPK by AICAR. *Am J Physiol*, 277, pp.H643–H649.

Russo, G. L., Russo, M. and Ungaro, P., 2013. AMP-activated protein kinase: a target for old drugs against diabetes and cancer. *Biochem Pharmacol*, 86, pp.339–350.

Salminen, A. and Kaarniranta, K., 2012. AMP-activated protein kinase (AMPK) controls the aging process via an integrated signaling network. *Ageing Res Rev*, 11, pp.230–241.

Sanders, M. J., et al., 2007. Investigating the mechanism for AMP activation of the AMP-activated protein kinase cascade. *Biochem J*, 403, pp.139–148.

Shelly, M., et al., 2007. LKB1/STRAD promotes axon initiation during neuronal polarization. *Cell*, 129, pp.565–577.

Sopjani, M., et al., 2010a. Down-regulation of Na+-coupled glutamate transporter EAAT3 and EAAT4 by AMP-activated protein kinase. *J Neurochem*, 113, pp.1426–1435.

Sopjani, M., et al., 2010b. Regulation of Na+-coupled glucose carrier SGLT1 by AMP-activated protein kinase. *Mol Membr Biol*, 27, pp.137–144.

Spasic, M. R., Callaerts, P. and Norga, K. K., 2008. Drosophila alicorn is a neuronal maintenance factor protecting against activity-induced retinal degeneration. *J Neurosci*, 28, pp.6419–6429.

Spasic, M. R., Callaerts, P. and Norga, K. K., 2009. AMP-activated protein kinase (AMPK) molecular crossroad for metabolic control and survival of neurons. *Neuroscientist*, 15, pp.309–316.

Stark, R., Ashley, S. E. and Andrews, Z. B., 2013. AMPK and the neuroendocrine regulation of appetite and energy expenditure. *Mol Cell Endocrinol*, 366, pp.215–223.

Steele, J. W. and Gandy, S., 2013. Latrepirdine (Dimebon(R)), a potential Alzheimer therapeutic, regulates autophagy and neuropathology in an Alzheimer mouse model. *Autophagy*, 9, pp.617–618.

Steele, J. W., et al., 2013a. Latrepirdine stimulates autophagy and reduces accumulation of alpha-synuclein in cells and in mouse brain. *Mol Psychiatry*, 18, pp,882–888.

Steele, J. W., et al., 2013b. Latrepirdine improves cognition and arrests progression of neuropathology in an Alzheimer's mouse model. *Mol Psychiatry*, 18, pp.889–897.

Thornton, C., et al., 2011. AMP-activated protein kinase (AMPK) is a tau kinase, activated in response to amyloid beta-peptide exposure. *Biochem J*, 434, pp.503–512.

Tschape, J. A., et al., 2002. The neurodegeneration mutant lochrig interferes with cholesterol homeostasis and Appl processing. *EMBO J*, 21, pp.6367–6376.

Turner, N., et al., 2008. Berberine and its more biologically available derivative, dihydroberberine, inhibit mitochondrial respiratory complex I: A mechanism for the action of berberine to activate AMP-activated protein kinase and improve insulin action. *Diabetes*, 57, pp.1414–1418.

Turnley, A. M., et al., 1999. Cellular distribution and developmental expression of AMP-activated protein kinase isoforms in mouse central nervous system. *J Neurochem*, 72, pp.1707–1716.

Vingtdeux, V., et al., 2010. AMP-activated protein kinase signaling activation by resveratrol modulates amyloid-beta peptide metabolism. *J Biol Chem*, 285, pp.9100–9113.

Vingtdeux, V., et al., 2011a. Novel synthetic small-molecule activators of AMPK as enhancers of autophagy and amyloid-beta peptide degradation. *FASEB J*, 25, pp.219–231.

Vingtdeux, V., et al., 2011b. AMPK is abnormally activated in tangle- and pre-tangle-bearing neurons in Alzheimer's disease and other tauopathies. *Acta Neuropathol*, 121, pp.337–349.

Weisova, P., et al., 2009. Regulation of glucose transporter 3 surface expression by the AMP-activated protein kinase mediates tolerance to glutamate excitation in neurons. *J Neurosci*, 29, pp.2997–3008.

Weisova, P., et al., 2011. Role of 5'-adenosine monophosphate-activated protein kinase in cell survival and death responses in neurons. *Antioxid Redox Signal*, 14, pp.1863–1876.

Weisova, P., et al., 2012. "Mild mitochondrial uncoupling" induced protection against neuronal excitotoxicity requires AMPK activity. *Biochim Biophys Acta*, 1817, pp.744–753.

Weisova, P., et al., 2013. Latrepirdine is a potent activator of AMP-activated protein kinase and reduces neuronal excitability. *Transl Psychiatry*, 3, p.e317.

Williams, T., et al., 2011. AMP-activated protein kinase (AMPK) activity is not required for neuronal development but regulates axogenesis during metabolic stress. *Proc Natl Acad Sci U S A*, 108, pp.5849–5854.

Winder, W. W. and Hardie, D. G., 1999. AMP-activated protein kinase, a metabolic master switch: possible roles in type 2 diabetes. *Am J Physiol*, 277, pp.E1–E10.

Won, J. S., et al., 2010. Involvement of AMP-activated-protein-kinase (AMPK) in neuronal amyloidogenesis. *Biochem Biophys Res Commun*, 399, pp.487–491.

Woods, A., et al., 2003. LKB1 is the upstream kinase in the AMP-activated protein kinase cascade. *Curr Biol*, 13, pp.2004–2008.

Woods, A., et al., 2005. Ca2+/calmodulin-dependent protein kinase kinase-beta acts upstream of AMP-activated protein kinase in mammalian cells. *Cell Metab*, 2, pp.21–33.

Wu, S. B. and Wei, Y. H., 2012. AMPK-mediated increase of glycolysis as an adaptive response to oxidative stress in human cells: implication of the cell survival in mitochondrial diseases. *Biochim Biophys Acta*, 1822, pp.233–247.

Xie, M., et al., 2006. A pivotal role for endogenous TGF-beta-activated kinase-1 in the LKB1/AMP-activated protein kinase energy-sensor pathway. *Proc Natl Acad Sci U S A*, 103, pp.17378–17383.

Yu, H., et al., 2004. Cloning and characterization of mouse 5'-AMP-activated protein kinase gamma3 subunit. *Am J Physiol Cell Physiol*, 286, pp.C283–C292.

Zang, Y., et al., 2009. AMP-activated protein kinase is involved in neural stem cell growth suppression and cell cycle arrest by 5-aminoimidazole-4-carboxamide-1-beta-D-ribofuranoside and glucose deprivation by down-regulating phospho-retinoblastoma protein and cyclin D. *J Biol Chem*, 284, 6175–6184.

Zhang, B. B., Zhou, G. and Li, C., 2009. AMPK: an emerging drug target for diabetes and the metabolic syndrome. *Cell Metab*, 9, pp.407–416.

Zhang, X., et al., 2010. A pharmacological activator of AMP-activated protein kinase protects hypoxic neurons in a concentration-dependent manner. *Neurochem Res*, 35, pp.1281–1289.

Zong, H., et al., 2002. AMP kinase is required for mitochondrial biogenesis in skeletal muscle in response to chronic energy deprivation. *Proc Natl Acad Sci U S A*, 99, pp.15983–15987.

6

HDAC6: A MOLECULE WITH MULTIPLE FUNCTIONS IN NEURODEGENERATIVE DISEASES

YAN YAN AND RENJIE JIAO

State Key Laboratory of Brain and Cognitive Sciences, Institute of Biophysics, Chinese Academy of Sciences, Beijing, China

6.1 INTRODUCTION

Neurodegenerative diseases are a group of multifarious neurological disorders featured by age-related and progressive loss of pathological neurons. Alzheimer's disease, Huntington's disease, Parkinson's disease, and amyotrophic lateral sclerosis are the major examples among these disorders. Despite that there are diverse molecular mechanisms of pathogenesis, selective neuronal death is being considered now as the main physiological feature of neurodegenerative disease, which is usually caused by cellular toxicity such as congenital or acquired mutated protein aggregates, deficiency of inclusion clearance, mitochondrial dysfunction, or excessive reactive oxygen species (ROS) formation. In recent years, many lines of evidence have shown that histone deacetylase 6 (HDAC6), a unique member in the HDAC family, can prevent neuronal cell death through promoting potentially protective inclusion formation and mediating the autophagic degradation of protein aggregates and dysfunctional mitochondria. In this chapter, we first describe the structural basis of HDAC6 as a unique member of the HDAC family. We then summarize the current understanding of how HDAC6 functions as a "scavenger" to facilitate the formation of protective inclusions and the clearance of toxic protein aggregates or

The Functions, Disease-Related Dysfunctions, and Therapeutic Targeting of Neuronal Mitochondria, First Edition. Edited by Valentin K. Gribkoff, Elizabeth A. Jonas and J. Marie Hardwick.

dysfunctional mitochondria. It will also be discussed that as an integral factor of stress surveillance, HDAC6 may be treated as a promising drug target to relieve symptoms or for the cure of neurodegenerative disorders.

6.2 MOLECULAR PROPERTIES OF HDAC6

Lysine acetylation on proteins is a dynamic process controlled by antagonistic actions of two superfamilies of enzymes: histone acetyltransferases (HATs) and HDACs. Protein acetylation and deacetylation catalyzed by these two families are two of the most frequent modifications that are involved in numerous developmental processes that, when deregulated, lead to the occurrence of diseases (Haberland, Montgomery, and Olson, 2009). HDACs deacetylate the acetylated lysines of not only histones but also nonhistone substrates, such as E2F and SP1, regulating cellular processes such as gene transcription and cell cycle progression (Bouwman and Philipsen, 2002; Marzio et al., 2000). Altered activities of different HDACs have been reported to contribute to cancer development and other diseases such as neurodegenerative diseases (Morrison, Majdzadeh, and D'Mello, 2007). Our focus will be on the molecular properties of HDAC6 and how it functions in the clearance of abnormal protein aggregates and dysfunctional mitochondria, before which a brief introduction to the entire HDAC family and its functions in neurodegeneration is also described.

6.2.1 Classes of the HDAC Family

To date, 18 HDACs have been identified in mammals (Haberland et al., 2009) and 6 in *Drosophila* (Cho et al., 2005). These 18 mammalian HDACs can be classified into four classes, I, II, III, and IV, according to their similarities to the HDAC members in budding yeasts (Fig. 6.1). Class I HDACs (including HDAC1, 2, 3, and 8) show high sequence similarity to the yeast transcriptional regulator RPD3 (Taunton, Hassig, and Schreiber, 1996). Class II HDACs are highly homologous to the yeast HDA1 (Yang and Seto, 2008) and this group can be further subdivided into class IIa (including HDAC4, 5, 7, and 9) and class IIb (including HDAC6 and 10). Class IV HDAC consists of the sole member of HDAC11, whose deacetylase domain is similar to classes I and II HDACs, but with short N- and C- termini. Class III HDACs, also known as Sirtuins, including 1 to 7, in yeasts (Longo and Kennedy, 2006; Schwer and Verdin, 2008), are usually referred to as a group of nonclassical HDACs with NAD^+-dependent enzymatic domains, whereas the other three classes of HDACs contain Zn^+-dependent enzymatic domains.

6.2.1a Class I HDACs

Human HDAC1 and HDAC2 (with sequence identity of 80%) are found to localize to the nucleus (Johnstone, 2002). HDAC1 can repress or activate gene transcription by being recruited to the promoter region of specific genes (Cunliffe, 2004; Nicolas et al., 2000). Redundant roles of HDAC1 and HDAC2 have been observed in mouse heart development (Montgomery et al., 2007). HDAC3 has been reported to localize

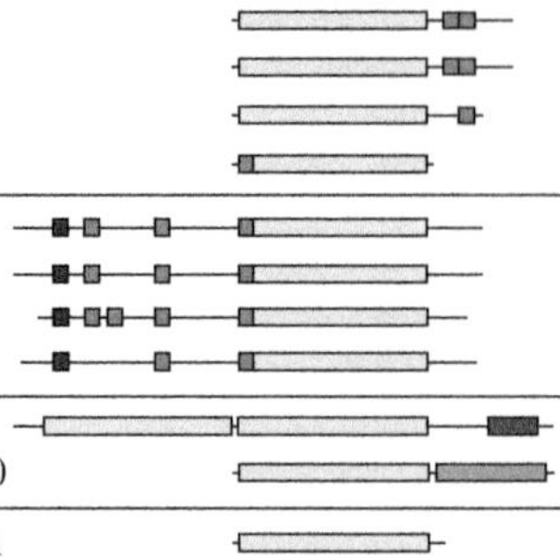

Classes		Protein domains	Protein length (aa)	Subcellular localization	Functions in neurodegenerative diseases
I	HDAC1		482	Nuclear	HD
	HDAC2		488	Nuclear	—
	HDAC3		428	Nuclear/cytoplasmic	HD
	HDAC8		377	Nuclear	—
IIa	HDAC4		1084	Nuclear/cytoplasmic	PD
	HDAC5		1122	Nuclear/cytoplasmic	HD
	HDAC7		855	Nuclear/cytoplasmic	—
	HDAC9		1011	Nuclear	ND*
IIb	HDAC6		1215	Nuclear/cytoplasmic	PD, HD, AD, ALS
	HDAC10		669	Nuclear/cytoplasmic	ND*
IV	HDAC11		347	Nuclear	—

FIGURE 6.1 Classes of human histone deacetylase (HDAC) family members and their protein structures, cellular localizations, and possible connections to neurodegenerative diseases. HDACs consist of four classes (I, II, III, and VII) according to their sequence identities and similarities to the yeast orthologues. Members of Class III HDACs are not included here because of their unique enzyme kinetics. The light blue-filled rectangle boxes indicate deacetylase domains. The 14-3-3 chaperone binding motifs and the myocyte enhancer factor 2 (MEF2) binding motifs are shown in grey and dark squares, respectively. Class IIb members possess an additional ubiquitin-binding domain *(dark grey)* or a leucine rich domain *(light grey)*. Parkinson's disease (PD), Alzheimer's disease (AD), Huntington's disease (HD), and amyotrophic lateral sclerosis (ALS) indicate that the corresponding HDACs may function in these neurodegenerative diseases. – shows that no relevant studies are available; ND* (neurodegeneration), known functions in neurodegeneration but not specified.

both to the nucleus and to the cytoplasm likely because of its nuclear export signal (NES) contained in the C-terminus (Yang et al., 2002). Conditional knockout of mouse HDAC3 in both the liver and the heart leads to an accumulation of lipids in these tissues (Knutson et al., 2008). Studies of HDAC1 and HDAC3 in *Caenorhabditis elegans* show that ectopic expression of the two HDACs can exacerbate the polyglutamine expansion toxicity in a Huntington's disease model (Bates et al., 2006). HDAC8 shares less sequence identity (30% compared to HDAC1) with other class I HDACs (de Ruijter et al., 2003). It is one of the best structurally and mechanistically characterized HDACs (Wolfson, Pitcairn, and Fierke, 2013). It is reported that HDAC8 can recognize and deacetylate the substrate of estrogen-related regulatorα (EERα), thus inhibiting its binding with DNA (Villena and Kralli, 2008).

6.2.1b Class II HDACs

All the human Class II HDACs contain both NLS and NES sequences, and a nondeacetylase region that can provide for protein-protein interactions (Haberland et al., 2009). HDAC4 and HDAC5, two of the Class IIa HDACs, both have been shown to be highly expressed in the heart and the brain of mice (Chang et al., 2004; Chawla et al., 2003; Vega et al., 2004). The translocation of HDAC4 from the nucleus to the cytoplasm is regulated by neuronal activity in cultured hippocampal neurons

(Chawla et al., 2003). Accumulation of HDAC5 is also observed in patients of Huntington's disease (Hoshino et al., 2003), and ectopic expression of HDAC5 can enhance apoptosis in cerebellar granule neurons (Linseman et al., 2003). HDAC7 has been proven to translocate from the inner mitochondrial membrane space to the cytoplasm during apoptosis (Bakin and Jung, 2004). HDAC9 contains two variants, a long isoform (usually called HDAC9), which possesses a deacetylase domain, and a short isoform (called HDRP), which lacks the catalytic domain (Petrie et al., 2003). Both of the two isoforms show functional association with HDAC3 (Mejat et al., 2005). HDRP is presumably considered as a protective factor in neurodegenerative diseases (Morrison et al., 2006).

HDAC6 is an unusual Class IIb HDAC, which has two deacetylase domains and a ubiquitin-binding domain (BUZ) (Haberland et al., 2009). The cytoplasmic localization and special substrates make HDAC6 a unique molecule in neurodegenerative diseases as compared with other HDACs (Du and Jiao, 2011). This chapter will discuss in detail the processes of how HDAC6 protects neurons from various toxicities of protein aggregates or dysfunctional mitochondria, and how it interacts with other proteins that are also essential for neurodegenerative diseases.

HDAC10, like HDAC6, also contains two deacetylase domains, one of which is non-functional (Guardiola and Yao, 2002). A recent study shows that HDAC10 protects neuroblastoma cells against drug toxicity through promoting autophagic activity (Oehme et al., 2013).

6.2.1c Classes III and IV HDACs

HDAC11 is the only member of class IV HDACs. Little is known about this HDAC except for its nuclear localization associated with HDAC6 (Gao et al., 2002). Class III HDACs consist of seven Sirtuins that are similar to yeast Sir2. As a result that the functional mechanisms of their enzyme kinetics are different from most of the HDACs (Longo and Kennedy, 2006; Schwer and Verdin, 2008), this class of HDACs will not be discussed in this chapter.

6.2.2 HDAC6

6.2.2a Molecular Structure and Substrates of HDAC6

As the largest member (1215 amino acids in human, see Fig. 6.1) of the HDAC family, HDAC6 has been first characterized in mice (Verdel and Khochbin, 1999), and later its homologs in *C. elegans* (Bertos et al., 2004), *Drosophila* (Barlow et al., 2001), and humans (Grozinger, Hassig, and Schreiber, 1999) were identified as well. The tandem deacetylase domains and a special BUZ finger domain are present in all homologues of HDAC6 in different species (Boyault et al., 2007a). Human HDAC6 also contains two NES motifs and a unique Ser-Glu-containing (SE14) region, which is reported to ensure its cytoplasmic localization (Bertos et al., 2004) (Fig. 6.2).

Double deacetylase domains render HDAC6 more complex in terms of enzymatic modulations and substrates compared with other HDACs; however, which of two domains (or both) are required for deacetylation of its substrates is controversial (Boyault et al., 2007a). α-tubulin is the first identified substrate of HDAC6

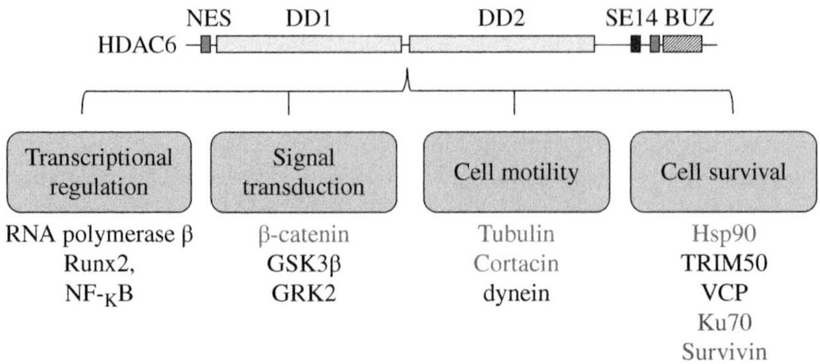

FIGURE 6.2 Protein domains and functions of HDAC6. HDAC6 possesses two deacetylase domains (*light grey rectangles*) and an ubiquitin-binding zinc (BUZ) finger domain (*cross rule rectangle*). There are also two nuclear export signal (NES) domains (*middle grey rectangles*) and a unique Ser-Glu-containing (SE14) region (*dark grey rectangle*). Functions of HDAC6 are summarized as indicated in the grey boxes: transcriptional regulation, signal transduction, cell motility, and cell survival. Grey terms indicate the substrates of HDAC6, whereas the proteins whose names are in black represent the interacting partners of HDAC6.

(Hubbert et al., 2002; Matsuyama et al., 2002). Some groups reported that both of the two domains contribute to the deacetylation of α-tubulin (Zhang et al., 2003b), whereas some others showed that the deacetylation of α-tubulin is mediated by the second deacetylase domain (Zou et al., 2006). There are also reports showing that the deacetylation of α-tubulin is not solely catalyzed by HDAC6 (Zhang et al., 2006). However, it is generally agreed that HDAC6 is the major catalytic deacetylase of α-tubulin.

The F-actin binding protein cortactin and heat shock protein 90 (HSP90) are two other substrates of HDAC6 (Valenzuela-Fernandez et al., 2008). Overexpression of HDAC6 leads to the hypo-acetylation of cortactin and further alters its association with F-actin to impair cell motility (Zhang et al., 2007). The well-known molecular chaperone HSP90 can be acetylated at multiple sites (Kovacs et al., 2005) and HDAC6 has been found to be one of its deacetylases (Kovacs et al., 2005). As a protein chaperone, deacetylated HSP90 by HDAC6 promotes the association of its client proteins such as glucocorticoid receptor (GR) (Murphy et al., 2005). Other substrates of HDAC6 include Ku70 (Subramanian et al., 2011), Survivin (Riolo et al., 2012), and β-catenin (Gao Hubbert and Yao, 2010), deacetylation of which regulates genome stability, apoptosis, and tumorgenesis (Li, Shin, and Kwon, 2013).

6.2.2b Interacting Proteins of HDAC6

Since cytoskeleton-related proteins α-tubulin and cortactin have been identified as direct substrates of HDAC6, it has been subsequently demonstrated that HDAC6 is involved in microtubule- and F-actin-dependent cellular trafficking and in numerous functions that depend on the cellular trafficking system (Boyault et al., 2007a). In addition to functions that require its deacetylase activity, other functions of

HDAC6 also regulate a variety of other cellular and physiological processes through physical interaction with its partners such as VCP/p97 (Boyault et al., 2006) and Aurora A (Pugacheva et al., 2007) (see Fig. 6.2). The BUZ at the C-terminus of HDAC6 has been reported to bind specifically with poly-ubiquitinated proteins to mediate the autophagic degradation of these proteins (Kawaguchi et al., 2003). Together, through its BUZ domain and the deacetylase domain, HDAC6 is supposed to be an integral factor that modulates multiple cellular processes such as gene transcription (Zhang and Kone, 2002), signal transduction (Pugacheva et al., 2007), cell morphology and migration (Gao et al., 2007; Palazzo, Ackerman, and Gundersen, 2003), misregulation of which often leads to diseases such as cancers and neurodegenerative diseases (Aldana-Masangkay and Sakamoto, 2011; Pandey et al., 2007). The relationship between HDAC6 and neurodegenerative diseases will be the focus of the following sections. Here we briefly summarize the functions of HDAC6 that are not related to the nervous system to give a holistic sketch to this promising drug target.

Several studies have shown that the maintenance of cell polarity and migration require the deacetylase activity of HDAC6 (Boyault et al., 2007a; Valenzuela-Fernandez et al., 2008). Selective inhibition of HDAC6 with tubacin, a specific deacetylase inhibitor, increases the acetylation level of cortactin and thus inhibits the binding of cortactin to F-actin, which is required for cell motility (Zhang et al., 2007). Despite that HDAC6 has been mainly reported to be localized in the cytoplasm in *Drosophila*, mice, and humans (de Ruijter et al., 2003), several studies also show that HDAC6 can regulate activities of transcriptional regulators in the nucleus. HDAC6 has the ability to directly bind to sumoylated p300, a histone acetyltransferase, to mediate transcriptional repression (Han et al., 2009). Inhibition of HDAC6 disrupts β-catenin nuclear localization and consequently inhibits Wnt signaling (Li et al., 2008).

6.3 HDAC6 AND NEURODEGENERATIVE DISEASES

Neurodegenerative diseases are a broad range of neurological disorders, characterized by specific neuronal loss and microscopically visible plaques or inclusion bodies (Sastry and Rao, 2000). The initiating pathogenesis of these disorders is multifactorial in Alzheimer's disease (AD), Huntington's disease (HD), Parkinson's disease (PD), or amyotrophic lateral sclerosis (ALS). Selective neuronal regions such as that of dopaminergic neurons in the substantia nigra of patients with PD are vulnerable to cellular toxicity that is caused by genetic factors, aging-related factors, and environmental factors (Lansbury and Lashuel, 2006). Discovering the causes of neuronal cell toxicity and factors or pathways that are involved in coping with the toxicity will give us better understanding of the pathogenesis of these neurodegenerative diseases and may provide insights for therapeutic drug design.

Recent studies have demonstrated that proteinaceous toxicity and mitochondrial dysfunction are the primary nosogenic elements contributing to neuron damage in neurodegenerative diseases (Gil and Rego, 2008; Lin and Beal, 2006; Robinson, 2008). The cellular stress response systems are known to monitor noxious stresses and deal with stressful situations (Morimoto, 2011). HDAC6 is one of such kind of stress sensing factors that can detect and eliminate the deleterious sources (Matthias, Yoshida, and Khochbin, 2008). Here we describe the protective roles of HDAC6 in coping with proteotoxic stresses and the clearance of dysfunctional mitochondria.

6.3.1 HDAC6 and Elimination of Proteotoxicity in Neurodegenerative Diseases

Aggregation and deposition of misfolded proteins are the common features in almost all kinds of neurodegenerative diseases (Sastry and Rao, 2000). Pathogenic mutations in misfolded proteins contribute to the failure of the elaborate degradation process known as the ubiquitin-proteasome system (UPS) (Lansbury and Lashuel, 2006). Mutated misfolded proteins that cannot be detected and cleared by the UPS system will then tend to aggregate into first small molecular weight oligomers and subsequently high molecular weight fibrils (Bemporad and Chiti, 2012) before pathologically visible inclusions or plaques form during the development of neurodegenerative disease (Sastry and Rao, 2000). Disease-related proteins usually undergo multistep aggregation before delivery to inclusion bodies (Brown, 2010; Glabe, 2008), which raises two basic questions as to (1) what is the toxic form of these misfolded proteins; and (2) how does the cell cope with this toxicity? The stress surveillance factor HDAC6 offers a platform to answer these questions because HDAC6 functions as a multilevel regulator in removing proteotoxicity during the development of neurodegenerative diseases.

6.3.1a Proteotoxicity in Neurodegenerative Diseases

Genetically or environmentally mutated proteins expressed in cultured cells or animal models have been reported to be often misfolded to phenocopy the pathogenic features of neurodegenerative diseases—neuronal cell death and formation of inclusions. Numerous studies in these models have shown that conversion of disease-related proteins from native monomer to misfolded oligomers is implicated as the primary pathogenic event in AD, PD, and other neurodegenerative diseases (Taylor, Hardy, and Fischbeck, 2002).

AD, the most common neurodegenerative disease, is characterized by the formation of amyloid deposits that are associated with three causative genes: an amyloid precursor protein (APP) gene (Goate et al., 1991), which produces Aβ peptides and two presenilin genes, PS1 and PS2 (Levy-Lahad et al., 1995; Moussaoui et al., 1996). Aβ peptides are prone to form conformational oligomer species and larger deposits (Lansbury and Lashuel, 2006). These metastable, transient, and heterogenic misfolded oligomers are considered to be responsible for neurotoxicity and neuronal death (Benilova, Karran, and De Strooper, 2012). Cultured cells exposed to medium supplemented with Aβ oligomers show decreased cell survival and proliferation as assayed by MTT reduction (Kayed et al., 2003). Other studies using markers of apoptosis (Wogulis et al., 2005) have also shown evidence that oligomer species of Aβ peptides are noxious to neuronal survival. Toxic Aβ oligomers have been found to dysregulate neuronal activities such as Ca^{2+} homeostasis (Malchiodi-Albedi et al., 2011). Another misfolded protein, α-Synuclein, is considered as a causative factor in PD (Sastry and Rao, 2000). It is reported to form annular oligomers that can capture lipids from cellular membranes, consequently resulting in the disruption of cellular membrane functions (Reynolds et al., 2011).

In summary, oligomer species of disease-related proteins have been proven to be the predominant cause of proteotoxicity that leads to neuronal damage in neurodegenerative diseases.

6.3.1b HDAC6 and Elimination of Proteotoxicity

Age-related and progressive features in neurodegenerative diseases imply that neuronal loss is a chronic process as a consequence of the imbalance of proteotoxic production and elimination. Clearance of toxic misfolded oligomers requires two primary processes: conversion of toxic oligomer species into protective protein aggregates and degradation of these aggregates. HDAC6 has been found to participate in both processes of proteotoxic elimination. Here we present several lines of evidence that HDAC6 functions as a protective factor to promote the conversion of harmful oligomers into nontoxic inclusions and to facilitate clearance of these inclusions.

6.3.1bi HDAC6 and Inclusion Formation

Inclusions or protein deposits in the selective brain regions of patients have long been recognized as a common epiphenomenon of neurodegenerative diseases (Taylor et al., 2002). Studies in cultured cell lines expressing pathologenic proteins such as α-Synuclein show that the misfolded proteins are prone to accumulate from oligomer species to high molecular weight protein aggregates that are often located at the microtubule organizing center (MTOC) (Richter-Landsberg and Leyk, 2013). These cellular aggregates (also named aggresomes) have been shown to be similar to physiological inclusions in terms of morphology and composition (Du and Jiao, 2011). Both aggresomes and inclusions are ubiquitin-positive and non-toxic (Taylor et al., 2002). Exploring mechanisms of aggresome formation from toxic oligomers may be the key to the prevention or the cure of neurodegenerative diseases. HDAC6 has been shown to be one of such important mediators in the process of aggresome formation (Pandey et al., 2007).

The first line of evidence from Kawaguchi et al. (2003) reported that HDAC6 locates at ubiquitin-positive protein aggresomes and mediates the aggresome formation at the region of MTOC in cultured cells. Both the BUZ domain and the dynein motor binding (DMB) domain are required for the misfolded proteins to be transported to the MTOC (Kawaguchi et al., 2003). Importantly, the authors also found that HDAC6 is accumulated in the inclusions of patients who suffer from PD. Further studies showed that aggresome formation in the cellular model expressing disease-related proteins such as Tau or polyglutanime also requires the deacetylase activity of HDAC6 (Ding, Dolan, and Johnson, 2008; Dompierre et al., 2007; Perez et al., 2009). These observations suggest that HDAC6 plays an important role in the formation of aggresomes in a microtubule-dependent manner.

Two important questions about HDAC6-mediated aggregation of misfolded proteins are what signals trigger this event and how HDAC6 recognizes the signals. Boyault et al. (2006, 2007b) found that the BUZ domain of HDAC6 directly binds to polyubiquitin chains and controls the turnover of these polyubiquitin chains. The HDAC6 interacting partner p97/VCP, a chaperone-like ATPase, antagonistically binds to polyubiquitin chains with HDAC6 and determines the fate of polyubiquitinated proteins (UPS degradation or protein aggregate formation) depending on its relative protein concentration compared with HDAC6 (Boyault et al., 2006, 2007a). These findings imply that the ubiquitination of misfolded proteins might be the signal to recruit HDAC6 and promote the aggregation process. In support of this view, Olzmann et al. (2007) showed that HDAC6 recognizes the K63-linked ubiquitination of misfolded protein DJ-1, a pathological protein that has been found in early-onset PD. This K63-linked ubiquitin chain, which is regulated by E3 ligase Parkin and E2 enzyme Uev1/UbcH3, is required for pericentriolar aggresome formation of misfolded DJ-1 (Olzmann et al., 2007). Ubiquitination of misfolded proteins by another E3 ligase TRIM50 has also been shown to be associated with HDAC6 to promote protein aggregate formation (Fusco et al., 2012), further supporting that the ubiquitination of misfolded proteins triggers the activation of HDAC6-dependent aggresome formation.

A recent study from our own lab provides in vivo evidence that HDAC6 promotes aggresome formation in dopaminergic neurons in a well-established *Drosophila* model of PD (Du et al., 2010). Loss of HDAC6 decreases the number of dopaminergic neurons and exacerbates retinal degeneration. The oligomeric species of overexpressed α-Synuclein interact directly with HDAC6 and accumulate in the brain of HDAC6 mutant flies (Du et al., 2010). These findings suggest that HDAC6 may recognize specific oligomer species that are harmful to cells and facilitate the formation of protective inclusions. It would be reasonable to predict that these misfolded α-Synuclein oligomers are also polyubiquitinated, which has yet to be characterized. In addition, HDAC6 can bind to a ubiquitin-like modifier FAT10 to promote FAT10- positive aggresome formation (Kalveram, Schmidtke, and Groettrup, 2008).

In summary, as shown in Figure 6.3, HDAC6 recognizes the toxic oligomers of misfolded proteins through binding to the polyubiquitin chain with its BUZ domain, mediating trafficking of the misfolded proteins along the microtubules through its deacetylase domain, and eventually promoting the formation of protective inclusions at the MTOC region of the cells.

6.3.1bii HDAC6 and Inclusion Clearance

Inclusion formation from toxic oligomer species of misfolded proteins is not the final, but the first, step to eliminate proteotoxicity. Autophagic degradation of aggresomes is considered as the second protective process to cope with cellular toxic stress (Rubinsztein, 2006, 2007). HDAC6 has been proven to play an important role as well in the process of autophagic clearance of aggresomes (Ding and Yin, 2008).

First, Iwata and colleagues (2005) showed that HDAC6 regulates lysosome enrichment at the MTOC region to accomplish autophagic degradation of Huntington aggregates in a microtubule dependent manner. Another study reported that HDAC6 is required for the fusion of autophagosomes with lysosomes and eventually for the clearance of ubiquitin-positive aggregates (Lee et al., 2010a). Further investigations indicate that HDAC6-mediated autophagosome maturation requires the F-actin

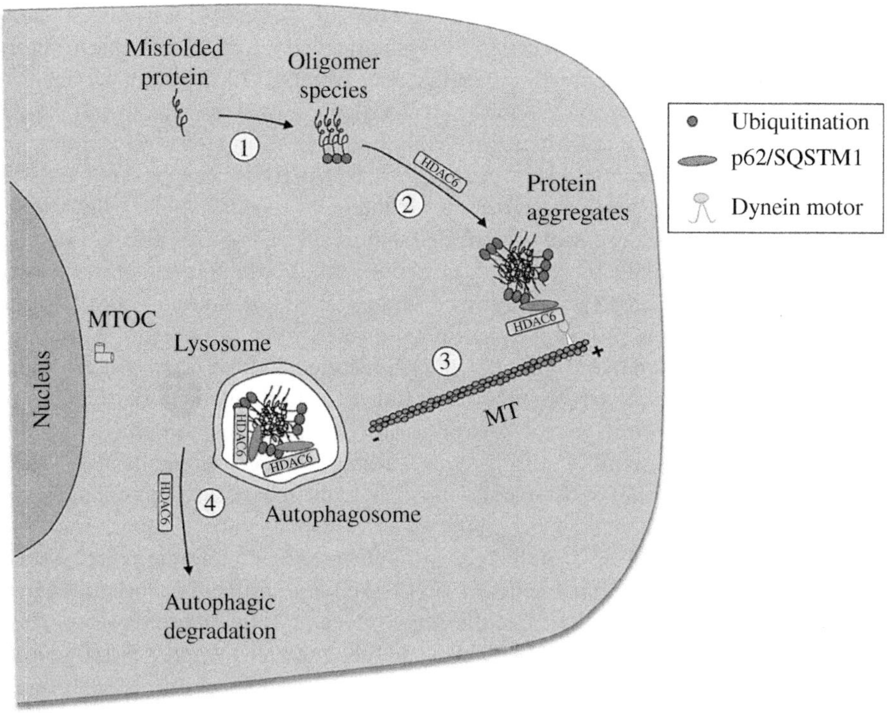

FIGURE 6.3 Schematic processes of HDAC6-mediated autophagic elimination of proteotoxicity. Proteotoxicity elimination requires HDAC6 at multi-levels. Step 1, genetic or environmental causes lead to accumulated misfolded proteins that is prone to aggregate into oligomer species. These oligomer species, if not eliminated, are toxic to the cells leading to eventually cell deaths. Step 2, the toxic oligomers with markers such as specific ubiquitinations can be recognized by the ubiquitin-binding domain of HDAC6. Recruitment of HDAC6 to the oligomers facilitates the formation of high weight protein aggregates. Step 3, the ubiquitin-positive protein aggregates can be transported to the microtubule organizing center (MTOC) by HDAC6-mediated microtubule (MT) network. Step 4, HDAC6 promotes the fusion of protein aggregates-containing autophagosomes and lysosomes, which is essential for the eventual clearance of protein aggregates. (*See insert for color representation of the figure.*)

network, which is independent of starvation-induced autophagy (Lee et al., 2010a). This process has been defined as quality control (QC) autophagy (Lee and Yao, 2010). These findings suggest that HDAC6 regulates late stage autophagic degradation of aggresomes in a cytoskeleton dependent manner. However, the differences between protein aggregate-induced and starvation-induced autophagosomes are highlighted since HDAC6 is only required for QC autophagosome maturation but not for starvation-induced autophagosome maturation (Lee et al., 2010a).

The next question is how HDAC6 selectively recognizes and mediates aggresome clearance by autophagic degradation to clear proteotoxicity, which raises the question of how HDAC6 is connected with autophagic components. To address this point, another protein p62/SQSTM1 also takes the stage here. p62/SQSTM1 is a polyubiquitin-binding protein with a light chain 3 (LC3)-interacting region (LIR) that directly binds autophagic component LC3 (Geetha and Wooten, 2002). Several lines of evidence have shown that p62/SQSTM1 colocalizes with both ubiquitin-positive aggregates and autophagic structures, promoting the formation and autophagic degradation of aggresomes (Bjorkoy et al., 2005). HDAC6 has been reported to accumulate in the ubiquitin-positive aggresomes together with p62/SQSTM1 (Into et al., 2010). A recent study successfully identified that p62/SQSTM1 physically interacts with HDAC6 to regulate its deacetylase activity, thus affecting F-actin network assembly (Yan et al., 2013). Taking into account all these findings points to the conclusion that, like the process of protein aggregate formation, ubiquitination may serve as an original signal to trigger the autophagic clearance of protein aggresomes as well. Instead of directly recognizing the ubiquitination signal in the process of aggregate/inclusion formation, HDAC6 physically binds to the autophagic adaptor p62/SQSTM1 that is required for polyubiquitinated cargo recognition of early autophagic flux and is then subsequently involved in the late event of autophagosome maturation (also Fig. 6.3).

However, this hypothesis is challenged by another interesting finding that HDAC6 can physically interact with unanchored K63-linked ubiquitin chains produced by a deubiquitinating enzyme Poh1 after the impairment of cellular proteasomes (Hao et al., 2013). This may imply that HDAC6 can recognize unanchored polyubiquitin chains as the initiating signal to accelerate the autophagic clearance of protein aggresomes. However, it is not clear how HDAC6 functions precisely as an integrator to connect the aggregate degradation with autophagic flux. Further efforts are needed to answer this question.

6.3.2 HDAC6 and Autophagic Clearance of Dysfunctional Mitochondria

Aging, the greatest risk factor in neurodegenerative diseases, has been suggested to be associated with the dysfunction of mitochondria. Damaged mitochondria with either mutations in mitochondrial DNA (mtDNA) or retention of ROS can cause pathological features of neurodegenerative diseases (Lin and Beal, 2006). Elimination of the damaged mitochondria and its toxic products may provide an effective strategy to prevent the neuronal death of degenerative diseases. HDAC6 has been shown to be involved in the elimination of dysfunctional mitochondria (Kirkin et al., 2009). Here we mainly discuss two questions of (1) how damaged mitochondria result in neurodegeneration, and (2) how HDAC6 recognizes these dysfunctional mitochondria and mediates their clearance.

6.3.2a Mitochondrial Dysfunction and Neurodegenerative Diseases

Dysfunction of mitochondria was first reported to be associated with PD when addition of rotenone, a inhibitor of mitochondrial complex-I of the electron transport chain, led to pathological parkinsonism manifested as the formation of ubiquitin-positive inclusions and neuronal loss (Betarbet et al., 2000; Fornai et al., 2005). Further studies showed that mutations in mtDNA (Chinnery et al., 2001; Zhang et al., 2003a) or other causative genes such as *parkin* (Palacino et al., 2004; Pesah et al., 2004), *DJ-1* (Bonifati et al., 2003), and *PINK1* (Valente et al., 2004) result in mitochondrial impairment and are associated with neurodegeneration. The damaged mitochondria release some intermembrane space proteins to the cytoplasm and subsequently activate the caspase cascade and apoptosis (Corral-Debrinski et al., 1992). It is now generally accepted that dysfunctional mitochondria caused by either genetic mutations or environmental stresses will result in pathological neuronal death through activating the cellular apoptotic pathway (Lin and Beal, 2006). The increased ROS arising from the damaged mitochondria may also activate or inactivate signaling pathways to further affect the processing of causative proteins, such as Aβ or Tau, to subsequently exacerbate protein aggregate formation (Tamagno et al., 2005). Many studies have focused on the relationship between dysfunctional mitochondria and neurodegenerative diseases and have suggested that dysfunction of mitochondria may play a central role in causing age-related progressive neurodegenerative changes. Understanding how to clear the damaged mitochondria will offer a promising strategy to develop drugs that can prevent toxic stresses and ease or cure degenerative diseases.

6.3.2b HDAC6 and Clearance of Damaged Mitochondria

It is known that dysfunctional mitochondria are eliminated by autophagy, a process also called mitophagy (Kim, Rodriguez-Enriquez, and Lemasters, 2007). Like autophagic degradation of protein aggresomes by QC autophagy, HDAC6 also plays an important role in the mitophagy process (Kirkin et al., 2009). The question is how HDAC6 recognizes the damaged mitochondria and modulates the autophagic clearance; therefore, it will be essential for us to understand the HDAC6-mediated stress surveillance as a whole.

Early studies have shown that clearance of effete or damaged mitochondria requires autophagic activity (Elmore et al., 2001). Further findings indicate that several ubiquitin E3 ligases such as NIX (Schweers et al., 2007) and LISTERIN (Chu et al., 2009) reside in the mitochondria and are required for autophagic elimination of dysfunctional mitochondria, giving rise to the assumption that ubiquitination may be the primary signal for autophagic degradation of damaged mitochondria. Thence, a general hypothesis has been appreciated that ubiquitination of factors causing cellular stresses such as protein aggregates, damaged mitochondria, peroxisomes, ribosomes and bacteria, is a common marker for triggering autophagic clearance of these stressors (Kirkin et al., 2009). In this interesting model, both HDAC6 and p62/SQSTM1 play a central role in mediating functional autophagy (Kirkin et al., 2009). In support of this model, Lee et al. (2010b) showed that HDAC6 promotes autophagic clearance of ubiquitinated mitochondria in an E3 ligase parkin dependent manner. In summary, HDAC6 functions as a general stress sensor to facilitate autophagic clearance of dysfunctional mitochondria that are considered to be a risk factor in age-related neurodegeneration.

However, it is still largely unclear what the authentic markers are that make damaged mitochondria the selective cargo of HDAC6-dependent autophagy and how HDAC6 is regulated during this process (see Fig. 6.4). More efforts are needed to explore these questions, which certainly will shed new light on finding strategies for the treatment of neurodegenerative diseases.

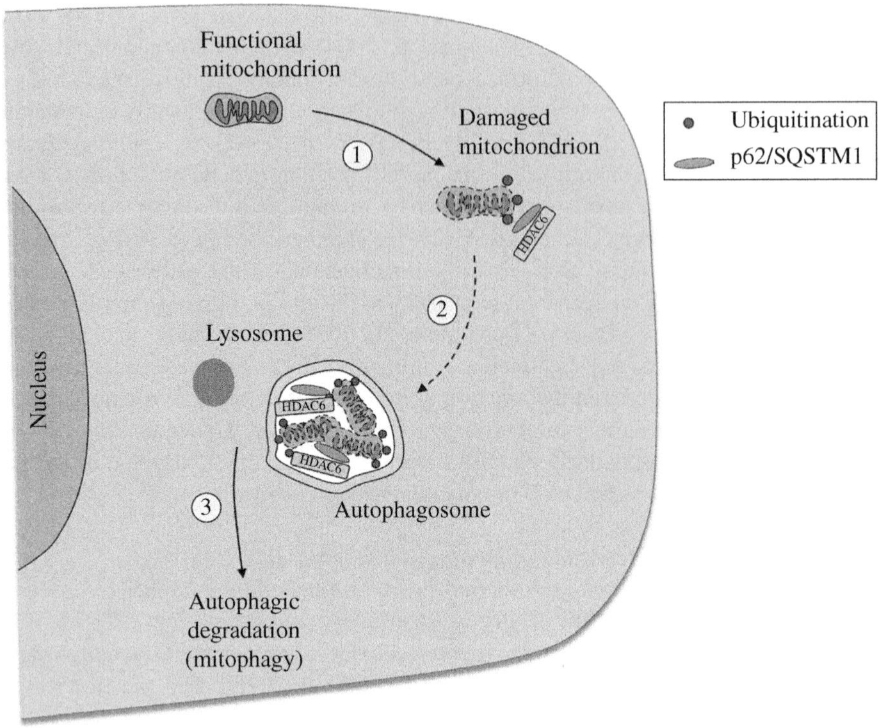

FIGURE 6.4 Schematic processes of HDAC6-mediated clearance of damaged mitochondria. HDAC6 is required for the autophagic clearance of dysfunctional mitochondria. Step 1, the damaged mitochondria that are resulted from causative mutations or environmental factors are accumulated in the cytoplasm, leading to cellular toxicity. Step 2, HDAC6 is recruited to the damaged mitochondria by specific ubiquitin signals and the recruitment of HDAC6 to the dysfunctional mitochondria enhances the autophagic clearance (mitophagy). Step 3, autophagic activities are necessary for the degradation of damaged mitochondria. It is unknown whether HDAC6 is also involved in the fusion of the damaged mitochondria-containing autophagosomes with the lysosomes. (*See insert for color representation of the figure.*)

6.4 PERSPECTIVES

HDAC6, as a member of the HDACs family, is involved in various nuclear and cytoplasmic processes (Boyault et al., 2007a). Here we have summarized multiple functions of HDAC6 in coping with cellular stresses such as toxic oligomer species of misfolded proteins and damaged mitochondria. With its two deacetylase domains and an ubiquitin-binding domain, HDAC6 can recognize the ubiquitinated toxic factors and bring them for selective autophagic clearance, thus protecting cells from undergoing the process of apoptosis and the animals, including humans, from suffering neurodegeneration.

Patients with neurodegenerative diseases usually suffer from chronic toxic stresses and degeneration in selective regions of the brain, leading to pathological symptoms in the later period (Sastry and Rao, 2000). Therefore, suppression or elimination of causative toxicity should be an effective approach to prevent or relieve the symptoms of neurodegenerative diseases (Du and Jiao, 2011). As a positive regulator of such stressor elimination, HDAC6 has become more and more accepted as a potential drug target for the treatment of neurodegeneration. However, limitations remain and need to be overcome. First, it is technically difficult to search for deacetylase activators of HDAC6. Since the deacetylase activity of HDAC6 is required for the autophagic elimination of protein aggregates or dysfunctional mitochondria, signaling pathways or modulators that can regulate the deacetylase activity of HDAC6 should be an alternative choice to improve HDAC6-dependent autophagic activity. It will be beneficial to examine whether there are modifications such as phosphorylation or acetylation that are required for the deacetylase activity of HDAC6. Secondly, it is essential for organisms or cells to maintain stable protein level or catalytic activity of HDAC6. Loss of HDAC6 causes deterioration of neurons in the fly brain, whereas excessive HDAC6 also leads to developmental defects (Du et al., 2010), indicating that the protein level of HDAC6 should be finely tuned. It remains to be elucidated whether the protein level of HDAC6 is controlled by the UPS, in which a specific ubiquitin E3 ligase may be responsible for maintaining the protein level of HDAC6. Finally, HDAC6 has been reported to be associated with cancer metastasis (Rodriguez-Gonzalez et al., 2008), and thus selective inhibitors of HDAC6 are considered as potential drugs for the treatment of cancers (McConkey, White, and Yan, 2012). Therefore, enhancement of HDAC6 activity for the treatment of neurodegeneration may automatically increase the risk of cancer. The use of HDAC6 as a drug target has to be investigated in more detail so that more specific drugs may be developed for a particular purpose.

REFERENCES

Aldana-Masangkay, G. I. and Sakamoto, K. M., 2011. The role of HDAC6 in cancer. *J Biomed Biotechnol*, 2011, p.875824.

Bakin, R. E. and Jung, M. O., 2004. Cytoplasmic sequestration of HDAC7 from mitochondrial and nuclear compartments upon initiation of apoptosis. *J Biol Chem*, 279, pp.51218–51225.

Barlow, A. L., et al., 2001. dSIR2 and dHDAC6: Two novel, inhibitor-resistant deacetylases in *Drosophila melanogaster*. *Exp Cell Res*, 265, pp.90–103.

Bates, E. A., et al., 2006. Differential contributions of *Caenorhabditis elegans* histone deacetylases to huntingtin polyglutamine toxicity. *J Neurosci*, 26, pp.2830–2838.

Bemporad, F. and Chiti, F., 2012. Protein misfolded oligomers: Experimental approaches, mechanism of formation, and structure-toxicity relationships. *Chem Biol*, 19, pp.315–327.

Benilova, I., Karran, E. and De Strooper, B., 2012. The toxic Abeta oligomer and Alzheimer's disease: An emperor in need of clothes. *Nat Neurosci*, 15, pp.349–357.

Bertos, N. R., et al., 2004. Role of the tetradecapeptide repeat domain of human histone deacetylase 6 in cytoplasmic retention. *J Biol Chem*, 279, pp.48246–48254.

Betarbet, R., et al., 2000. Chronic systemic pesticide exposure reproduces features of Parkinson's disease. *Nat Neurosci*, 3, 1301–1306.

Bjorkoy, G., et al., 2005. p62/SQSTM1 forms protein aggregates degraded by autophagy and has a protective effect on huntingtin-induced cell death. *J Cell Biol*, 171, pp.603–614.

Bonifati, V., et al., 2003. Mutations in the DJ-1 gene associated with autosomal recessive early-onset parkinsonism. *Science*, 299, pp.256–259.

Bouwman, P. and Philipsen, S., 2002. Regulation of the activity of Sp1-related transcription factors. *Mol Cell Endocrinol*, 195, pp.27–38.

Boyault, C., et al., 2006. HDAC6-p97/VCP controlled polyubiquitin chain turnover. *EMBO J*, 25, pp.3357–3366.

Boyault, C., et al., 2007a. HDAC6, at the crossroads between cytoskeleton and cell signaling by acetylation and ubiquitination. *Oncogene*, 26, pp.5468–5476.

Boyault, C., et al., 2007b. HDAC6 controls major cell response pathways to cytotoxic accumulation of protein aggregates. *Genes Dev*, 21, pp.2172–2181.

Brown, D. R., 2010. Oligomeric alpha-synuclein and its role in neuronal death. *IUBMB Life*, 62, pp.334–339.

Chang, S., et al., 2004. Histone deacetylases 5 and 9 govern responsiveness of the heart to a subset of stress signals and play redundant roles in heart development. *Mol Cell Biol*, 24, pp.8467–8476.

Chawla, S., et al., 2003. Neuronal activity-dependent nucleocytoplasmic shuttling of HDAC4 and HDAC5. *J Neurochem*, 85, pp.151–159.

Chinnery, P. F., et al., 2001. Point mutations of the mtDNA control region in normal and neurodegenerative human brains. *Am J Hum Genet*, 68, pp.529–532.

Cho, Y., et al., 2005. Individual histone deacetylases in Drosophila modulate transcription of distinct genes. *Genomics*, 86, pp.606–617.

Chu, J., et al., 2009. A mouse forward genetics screen identifies LISTERIN as an E3 ubiquitin ligase involved in neurodegeneration. *Proc Natl Acad Sci U S A*, 106, pp.2097–2103.

Corral-Debrinski, M., et al., 1992. Mitochondrial DNA deletions in human brain: Regional variability and increase with advanced age. *Nat Genet*, 2, pp.324–329.

Cunliffe, V. T., 2004. Histone deacetylase 1 is required to repress Notch target gene expression during zebrafish neurogenesis and to maintain the production of motoneurones in response to hedgehog signalling. *Development*, 131, pp.2983–2995.

de Ruijter, A. J., et al., 2003. Histone deacetylases (HDACs): Characterization of the classical HDAC family. *Biochem J*, 370, pp.737–749.

Ding, H., Dolan, P. J. and Johnson, G. V., 2008. Histone deacetylase 6 interacts with the microtubule-associated protein tau. *J Neurochem*, 106, pp.2119–2130.

Ding, W. X. and Yin, X. M., 2008. Sorting, recognition and activation of the misfolded protein degradation pathways through macroautophagy and the proteasome. *Autophagy*, 4, pp.141–150.

Dompierre, J. P., et al., 2007. Histone deacetylase 6 inhibition compensates for the transport deficit in Huntington's disease by increasing tubulin acetylation. *J Neurosci*, 27, pp.3571–3583.

Du, G., et al., 2010. Drosophila histone deacetylase 6 protects dopaminergic neurons against {alpha}-synuclein toxicity by promoting inclusion formation. *Mol Biol Cell*, 21, pp.2128–2137.

Du, G. and Jiao, R., 2011. To prevent neurodegeneration: HDAC6 uses different strategies for different challenges. *Commun Integr Biol*, 4, 139–142.

Elmore, S. P., et al., 2001. The mitochondrial permeability transition initiates autophagy in rat hepatocytes. *FASEB J*, 15, pp.2286–2287.

Fornai, F., et al., 2005. Parkinson-like syndrome induced by continuous MPTP infusion: Convergent roles of the ubiquitin-proteasome system and alpha-synuclein. *Proc Natl Acad Sci U S A*, 102, pp.3413–3418.

Fusco, C., et al., 2012. The E3-ubiquitin ligase TRIM50 interacts with HDAC6 and p62, and promotes the sequestration and clearance of ubiquitinated proteins into the aggresome. *PLoS One*, 7, p.e40440.

Gao, L., et al., 2002. Cloning and functional characterization of HDAC11, a novel member of the human histone deacetylase family. *J Biol Chem*, 277, pp.25748–25755.

Gao, Y. S., et al., 2007. Histone deacetylase 6 regulates growth factor-induced actin remodeling and endocytosis. *Mol Cell Biol*, 27, 8637–8647.

Gao, Y. S., Hubbert, C. C. and Yao, T. P., 2010. The microtubule-associated histone deacetylase 6 (HDAC6) regulates epidermal growth factor receptor (EGFR) endocytic trafficking and degradation. *J Biol Chem*, 285, pp.11219–11226.

Geetha, T. and Wooten, M. W., 2002. Structure and functional properties of the ubiquitin binding protein p62. *FEBS Lett*, 512, pp.19–24.

Gil, J. M. and Rego, A. C., 2008. Mechanisms of neurodegeneration in Huntington's disease. *Eur J Neurosci*, 27, pp.2803–2820.

Glabe, C. G., 2008. Structural classification of toxic amyloid oligomers. *J Biol Chem*, 283, pp.29639–29643.

Goate, A., et al., 1991. Segregation of a missense mutation in the amyloid precursor protein gene with familial Alzheimer's disease. *Nature*, 349, pp.704–706.

Grozinger, C. M., Hassig, C. A. and Schreiber, S. L., 1999. Three proteins define a class of human histone deacetylases related to yeast Hda1p. *Proc Natl Acad Sci U S A*, 96, pp.4868–4873.

Guardiola, A. R. and Yao, T. P., 2002. Molecular cloning and characterization of a novel histone deacetylase HDAC10. *J Biol Chem*, 277, pp.3350–3356.

Haberland, M., Montgomery, R. L. and Olson, E. N., 2009. The many roles of histone deacetylases in development and physiology: Implications for disease and therapy. *Nat Rev Genet*, 10, pp.32–42.

Han, Y., et al., 2009. Acetylation of histone deacetylase 6 by p300 attenuates its deacetylase activity. *Biochem Biophys Res Commun*, 383, pp.88–92.

Hao, R., et al., 2013. Proteasomes activate aggresome disassembly and clearance by producing unanchored ubiquitin chains. *Mol Cell*, 51, pp.819–828.

Hoshino, M., et al., 2003. Histone deacetylase activity is retained in primary neurons expressing mutant huntingtin protein. *J Neurochem*, 87, pp.257–267.

Hubbert, C., et al., 2002. HDAC6 is a microtubule-associated deacetylase. *Nature*, 417, pp.455–458.

Into, T., et al., 2010. Regulation of MyD88 aggregation and the MyD88-dependent signaling pathway by sequestosome 1 and histone deacetylase 6. *J Biol Chem*, 285, pp.35759–35769.

Iwata, A., et al., 2005. HDAC6 and microtubules are required for autophagic degradation of aggregated huntingtin. *J Biol Chem*, 280, pp.40282–40292.

Johnstone, R. W., 2002. Histone-deacetylase inhibitors: Novel drugs for the treatment of cancer. *Nat Rev Drug Discov*, 1, pp.287–s99.

Kalveram, B., Schmidtke, G. and Groettrup, M., 2008. The ubiquitin-like modifier FAT10 interacts with HDAC6 and localizes to aggresomes under proteasome inhibition. *J Cell Sci*, 121, 4079–4088.

Kawaguchi, Y., et al., 2003. The deacetylase HDAC6 regulates aggresome formation and cell viability in response to misfolded protein stress. *Cell*, 115, pp.727–738.

Kayed, R., et al., 2003. Common structure of soluble amyloid oligomers implies common mechanism of pathogenesis. *Science*, 300, pp.486–489.

Kim, I., Rodriguez-Enriquez, S. and Lemasters, J. J., 2007. Selective degradation of mitochondria by mitophagy. *Arch Biochem Biophys*, 462, pp.245–253.

Kirkin, V., et al., 2009. A role for ubiquitin in selective autophagy. *Mol Cell*, 34, pp.259–269.

Knutson, S. K., et al., 2008. Liver-specific deletion of histone deacetylase 3 disrupts metabolic transcriptional networks. *EMBO J*, 27, 1017–1028.

Kovacs, J. J., et al., 2005. HDAC6 regulates Hsp90 acetylation and chaperone-dependent activation of glucocorticoid receptor. *Mol Cell*, 18, pp.601–607.

Lansbury, P. T. and Lashuel, H. A., 2006. A century-old debate on protein aggregation and neurodegeneration enters the clinic. *Nature*. 443, pp.774–779.

Lee, J. Y., et al., 2010a. HDAC6 controls autophagosome maturation essential for ubiquitin-selective quality-control autophagy. *EMBO J*, 29, pp.969–980.

Lee, J. Y., et al., 2010b. Disease-causing mutations in parkin impair mitochondrial ubiquitination, aggregation, and HDAC6-dependent mitophagy. *J Cell Biol*, 189, pp.671–679.

Lee, J. Y. and Yao, T. P., 2010. Quality control autophagy: A joint effort of ubiquitin, protein deacetylase and actin cytoskeleton. *Autophagy*, 6, pp.555–557.

Levy-Lahad, E., et al., 1995. A familial Alzheimer's disease locus on chromosome 1. *Science*, 269, pp.970–973.

Li, Y., et al., 2008. HDAC6 is required for epidermal growth factor-induced beta-catenin nuclear localization. *J Biol Chem*, 283, pp.12686–12690.

Li, Y., Shin, D. and Kwon, S. H., 2013. Histone deacetylase 6 plays a role as a distinct regulator of diverse cellular processes. *FEBS J*, 280, pp.775–793.

Lin, M. T. and Beal, M. F., 2006. Mitochondrial dysfunction and oxidative stress in neurodegenerative diseases. *Nature*, 443, pp.787–795.

Linseman, D. A., et al., 2003. Inactivation of the myocyte enhancer factor-2 repressor histone deacetylase-5 by endogenous Ca(2+) //calmodulin-dependent kinase II promotes depolarization-mediated cerebellar granule neuron survival. *J Biol Chem*, 278, pp.41472–41481.

Longo, V. D. and Kennedy, B. K., 2006. Sirtuins in aging and age-related disease. *Cell*, 126, pp.257–268.

Malchiodi-Albedi, F., et al., 2011. Amyloid oligomer neurotoxicity, calcium dysregulation, and lipid rafts. *Int J Alzheimers Dis*, 2011, p.906964.

Marzio, G., et al., 2000. E2F family members are differentially regulated by reversible acetylation. *J Biol Chem*, 275, pp.10887–10892.

Matsuyama, A., et al., 2002. In vivo destabilization of dynamic microtubules by HDAC6-mediated deacetylation. *EMBO J*, 21, pp.6820–6831.

Matthias, P., Yoshida, M. and Khochbin, S., 2008. HDAC6 a new cellular stress surveillance factor. *Cell Cycle* 7, pp.7–10.

McConkey, D. J., White, M. and Yan, W., 2012. HDAC inhibitor modulation of proteotoxicity as a therapeutic approach in cancer. *Adv Cancer Res*, 116, pp.131–163.

Mejat, A., et al., 2005. Histone deacetylase 9 couples neuronal activity to muscle chromatin acetylation and gene expression. *Nat Neurosci*, 8, pp.313–321.

Montgomery, R. L., et al., 2007. Histone deacetylases 1 and 2 redundantly regulate cardiac morphogenesis, growth, and contractility. *Genes Dev*, 21, pp.1790–1802.

Morimoto, R. I., 2011. The heat shock response: systems biology of proteotoxic stress in aging and disease. *Cold Spring Harb Symp Quant Biol*, 76, pp.91–99.

Morrison, B. E., et al., 2006. Neuroprotection by histone deacetylase-related protein. *Mol Cell Biol*, 26, pp.3550–3564.

Morrison, B. E., Majdzadeh, N. and D'Mello, S. R., 2007. Histone deacetylases: Focus on the nervous system. *Cell Mol Life Sci*, 64, pp.2258–2269.

Moussaoui, S., et al., 1996. Immunohistochemical analysis of presenilin-1 expression in the mouse brain. *FEBS Lett*, 383, pp.219–222.

Murphy, P. J., et al., 2005. Regulation of the dynamics of hsp90 action on the glucocorticoid receptor by acetylation/deacetylation of the chaperone. *J Biol Chem*, 280, pp.33792–33799.

Nicolas, E., et al., 2000. RbAp48 belongs to the histone deacetylase complex that associates with the retinoblastoma protein. *J Biol Chem*, 275, pp.9797–9804.

Oehme, I., et al., 2013. Histone deacetylase 10 promotes autophagy-mediated cell survival. *Proc Natl Acad Sci U S A*, 110, pp.E2592–E2601.

Olzmann, J. A., et al., 2007. Parkin-mediated K63-linked polyubiquitination targets misfolded DJ-1 to aggresomes via binding to HDAC6. *J Cell Biol*, 178, pp.1025–1038.

Palacino, J. J., et al., 2004. Mitochondrial dysfunction and oxidative damage in parkin-deficient mice. *J Biol Chem*, 279, pp.18614–18622.

Palazzo, A., Ackerman, B. and Gundersen, G. G., 2003. Cell biology: Tubulin acetylation and cell motility. *Nature*, 421, p.230.

Pandey, U. B., et al., 2007. HDAC6 at the intersection of autophagy, the ubiquitin-proteasome system and neurodegeneration. *Autophagy*, 3, pp.643–645.

Perez, M., et al., 2009. Tau—an inhibitor of deacetylase HDAC6 function. *J Neurochem*, 109, pp.1756–1766.

Pesah, Y., et al., 2004. Drosophila parkin mutants have decreased mass and cell size and increased sensitivity to oxygen radical stress. *Development*, 131, pp.2183–2194.

Petrie, K., et al., 2003. The histone deacetylase 9 gene encodes multiple protein isoforms. *J Biol Chem*, 278, pp.16059–16072.

Pugacheva, E. N., et al., 2007. HEF1-dependent Aurora A activation induces disassembly of the primary cilium. *Cell*, 129, pp.1351–1363.

Reynolds, N. P., et al., 2011. Mechanism of membrane interaction and disruption by alpha-synuclein. *J Am Chem Soc*, 133, pp.19366–19375.

Richter-Landsberg, C. and Leyk, J., 2013. Inclusion body formation, macroautophagy, and the role of HDAC6 in neurodegeneration. *Acta Neuropathol*, 126, pp.794–807.

Riolo, M. T., et al., 2012. Histone deacetylase 6 (HDAC6) deacetylates survivin for its nuclear export in breast cancer. *J Biol Chem*, 287, pp.10885–10893.

Robinson, P. A., 2008. Protein stability and aggregation in Parkinson's disease. *Biochem J*, 413, pp.1–13.

Rodriguez-Gonzalez, A., et al., 2008. Role of the aggresome pathway in cancer: Targeting histone deacetylase 6-dependent protein degradation. *Cancer Res*, 68, pp.2557–2560.

Rubinsztein, D. C., 2006. The roles of intracellular protein-degradation pathways in neurodegeneration. *Nature*, 443, pp.780–786.

Rubinsztein, D. C., 2007. Autophagy induction rescues toxicity mediated by proteasome inhibition. *Neuron*, 54, pp.854–856.

Sastry, P. S. and Rao, K. S., 2000. Apoptosis and the nervous system. *J Neurochem*, 74, pp.1–20.

Schweers, R. L., et al., 2007. NIX is required for programmed mitochondrial clearance during reticulocyte maturation. *Proc Natl Acad Sci U S A*, 104, pp.19500–19505.

Schwer, B. and Verdin, E., 2008. Conserved metabolic regulatory functions of sirtuins. *Cell Metab*, 7, pp.104–112.

Subramanian, C., et al., 2011. HDAC6 deacetylates Ku70 and regulates Ku70-Bax binding in neuroblastoma. *Neoplasia*, 13, pp.726–734.

Tamagno, E., et al., 2005. Beta-site APP cleaving enzyme up-regulation induced by 4-hydroxynonenal is mediated by stress-activated protein kinases pathways. *J Neurochem*, 92, pp.628–636.

Taunton, J., Hassig, C. A. and Schreiber, S. L., 1996. A mammalian histone deacetylase related to the yeast transcriptional regulator Rpd3p. *Science*, 272, pp.408–411.

Taylor, J. P., Hardy, J. and Fischbeck, K. H., 2002. Toxic proteins in neurodegenerative disease. *Science*, 296, pp.1991–1995.

Valente, E. M., et al., 2004. Hereditary early-onset Parkinson's disease caused by mutations in PINK1. *Science*, 304, pp.1158–1160.

Valenzuela-Fernandez, A., et al., 2008. HDAC6: A key regulator of cytoskeleton, cell migration and cell-cell interactions. *Trends Cell Biol*, 18, pp.291–297.

Vega, R. B., et al., 2004. Histone deacetylase 4 controls chondrocyte hypertrophy during skeletogenesis. *Cell*, 119, pp.555–566.

Verdel, A. and Khochbin, S., 1999. Identification of a new family of higher eukaryotic histone deacetylases. Coordinate expression of differentiation-dependent chromatin modifiers. *J Biol Chem*, 274, pp.2440–2445.

Villena, J. A. and Kralli, A., 2008. ERRalpha: A metabolic function for the oldest orphan. *Trends Endocrinol Metab*, 19, 269–276.

Wogulis, M., et al., 2005. Nucleation-dependent polymerization is an essential component of amyloid-mediated neuronal cell death. *J Neurosci*, 25, pp.1071–1080.

Wolfson, N. A., Pitcairn, C. A. and Fierke, C. A., 2013. HDAC8 substrates: Histones and beyond. *Biopolymers*, 99, pp.112–126.

Yan, J., et al., 2013. SQSTM1/p62 Interacts with HDAC6 and Regulates Deacetylase Activity. *PLoS One*, 8, p.e76016.

Yang, W. M., et al., 2002. Functional domains of histone deacetylase-3. *J Biol Chem*, 277, pp.9447–9454.

Yang, X. J. and Seto, E., 2008. The Rpd3/Hda1 family of lysine deacetylases: From bacteria and yeast to mice and men. *Nat Rev Mol Cell Biol*, 9, pp.206–218.

Zhang, W. and Kone, B. C., 2002. NF-kappaB inhibits transcription of the H(+)-K(+)-ATPase alpha(2)-subunit gene: role of histone deacetylases. *Am J Physiol Renal Physiol*, 283, pp.F904–F911.

Zhang, J., et al., 2003a. Strikingly higher frequency in centenarians and twins of mtDNA mutation causing remodeling of replication origin in leukocytes. *Proc Natl Acad Sci U S A*, 100, pp.1116–1121.

Zhang, Y., et al., 2003b. HDAC-6 interacts with and deacetylates tubulin and microtubules in vivo. *EMBO J*, 22, pp.1168–1179.

Zhang, Y., et al., 2006. Two catalytic domains are required for protein deacetylation. *J Biol Chem*, 281, pp.2401–2404.

Zhang, X., et al., 2007. HDAC6 modulates cell motility by altering the acetylation level of cortactin. *Mol Cell*, 27, pp.197–213.

Zou, H., et al., 2006. Characterization of the two catalytic domains in histone deacetylase 6. *Biochem Biophys Res Commun*, 341, pp.45–50.

7

NEURONAL MITOCHONDRIAL TRANSPORT

Adam L. Knight, Yanmin Chen, Tao Sun and Zu-Hang Sheng

Synaptic Functions Section, Porter Neuroscience Research Center, National Institute of Neurological Disorders and Stroke, National Institutes of Health, Bethesda, Maryland, USA

7.1 INTRODUCTION

Mitochondria provide cells with the energy necessary to support a variety of neuronal functions including axonal growth and branching, synapse assembly (Lee and Peng, 2008), action potential generation (Attwell and Laughlin, 2001), and synaptic transmission (Verstreken et al., 2005). Mitochondria also buffer Ca^{2+} in the synapse to maintain neurotransmission (David and Barrett, 2003; Medler and Gleason, 2002) and induce short-term synaptic plasticity (Billups and Forsythe, 2002; Kang et al., 2008; Levy et al., 2003; Tang and Zucker, 1997; Werth and Thayer, 1994). Reduced mitochondrial transport to the axon terminals impairs synaptic transmission, which is likely caused by altered Ca^{2+} transients or insufficient adenosine triphosphate (ATP) supplies (Guo et al., 2005; Ma et al., 2009; Stowers et al., 2002).

Neurons are polarized cells composed of a cell body, a long thin axon, and dendrites with multiple branches. These complex morphological features provide challenges for neurons in maintaining energy homeostasis. As mitochondria are primarily generated in the soma, one major challenge is efficiently transporting them to distal areas with high-energy demand (e.g., axonal branches, synaptic terminals, and active growth cones) (Fig. 7.1) (Morris and Hollenbeck, 1993; Ruthel and Hollenbeck, 2003). Axons undergo dynamic remodeling in response to synaptic

The Functions, Disease-Related Dysfunctions, and Therapeutic Targeting of Neuronal Mitochondria, First Edition. Edited by Valentin K. Gribkoff, Elizabeth A. Jonas and J. Marie Hardwick.

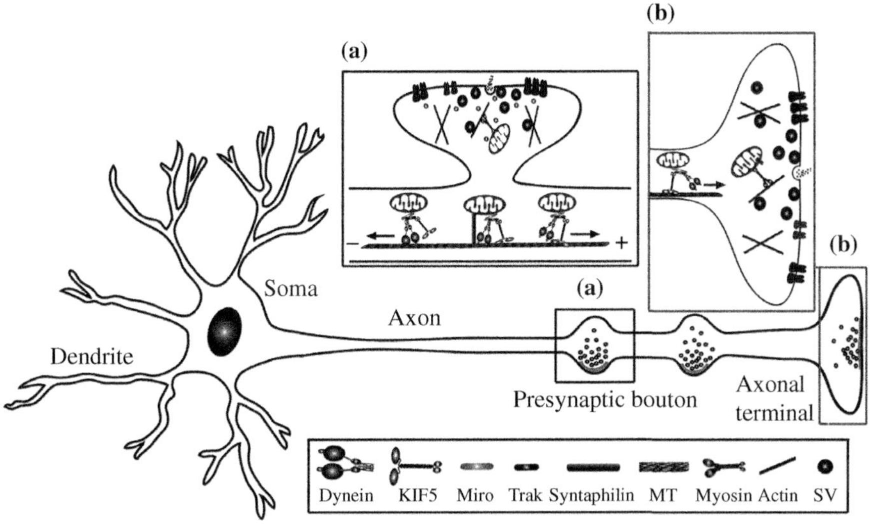

FIGURE 7.1 Mitochondrial trafficking and anchoring in neurons. Neurons require specialized mechanisms for delivery and distribution of mitochondria in distal regions to maintain energy and Ca^{2+} homeostasis. This figure highlights mitochondrial transport to the presynaptic bouton **(a)** and the axon terminal **(b)**. Long-range microtubule (MT)-based transport of mitochondria depends on MT polarity. In axons, MT plus-ends (+) are directed toward the terminal and minus-ends (–) are oriented toward the soma. KIF5 motors drive anterograde axonal transport to distal synapses while dynein motors drive retrograde transport to the soma. Trak motor adaptors mediate both anterograde and retrograde mitochondrial transport (van Spronsen et al., 2013). Myosin motors drives short-range mitochondrial movement along actin filaments within presynaptic terminals. Mitochondria can be anchored to MTs via a docking interaction with syntaphilin (Kang et al., 2008) or through an unknown actin-based docking receptor (Chada and Hollenbeck, 2004). Docking mechanisms ensure proper distribution of neuronal mitochondria along axons and at synapses. Adapted from Sheng, Z.-H., 2014. Mitochondrial trafficking and anchoring in neurons: New insight and implications. *J Cell Biol*, 204, pp.1087–1098, by permission.

activity and during neuronal development, therefore altering mitochondrial distribution and trafficking. As neurons are postmitotic cells, another major challenge is to maintain proper quality control of aged or dysfunctional mitochondria throughout the life span of an organism. Mitochondria alter their transport and distribution when their integrity is impaired (Cai et al., 2012; Chang and Reynolds, 2006; Miller and Sheetz, 2004), with aged and dysfunctional mitochondria being returned to the soma for more efficient degradation, where mature lysosomes are mainly localized (Cai et al., 2010; Lee, Sato, and Nixon, 2011; Overly and Hollenbeck, 1996).

Mitochondrial trafficking and anchoring must be efficiently regulated to recruit and redistribute healthy mitochondria to maintain local energy homeostasis, and to remove dysfunctional mitochondria from distal terminals for degradation. Research into regulation of axonal mitochondrial trafficking and anchoring is thus an important frontier in neurobiology. This chapter focuses on (1) mechanistic insight into how microtubule (MT)-based mitochondrial trafficking and anchoring is regulated in neurons, and (2) an updated overview of how synaptic function and mitochondrial quality control are influenced by axonal mitochondrial motility. Additional insight and overviews from different perspectives can be found in other in-depth reviews (Birsa et al., 2013; Court and Coleman, 2012; Frederick and Shaw, 2007; MacAskill and Kittler, 2010; Morfini et al., 2009; Saxton and Hollenbeck, 2012; Schon and Przedborski, 2011; Schwarz, 2013; Sheng, 2014; Sheng and Cai, 2012).

7.2 COMPLEX MOTILITY PATTERNS OF AXONAL MITOCHONDRIA

Mitochondrial motility can be monitored in live primary neurons by time-lapse imaging mitochondria labeled with fluorescent mitochondrial tags (e.g., DsRed-Mito) or dyes (e.g., Mitotracker Green or MitoTracker® Red CMXRos). Mitochondria display complex motility patterns, moving bidirectionally, and frequently pausing and changing direction. Motile mitochondria can become stationary and then remobilized in response to changes in metabolic status and synaptic activity. The relative velocity of mitochondrial transport can be quite variable (ranging from ~0.32 to 0.91 μm per second) because of these bidirectional movements and frequent pauses (Macaskill et al., 2009a; Morris and Hollenbeck, 1995). Approximately 30% of axonal mitochondria are motile in mature cultured neurons (Kang et al., 2008). By applying dual-channel live imaging of presynaptic terminals and axonal mitochondria in hippocampal neurons, we recently revealed five distinct mobility patterns of axonal mitochondria: mitochondria docked outside of the synapse (54.07% ± 2.53%) or at the synapse (16.29% ± 1.66%); passing through the synapse (14.77% ± 1.58%); pausing at the synapse briefly (<200 seconds, 7.01% ± 1.29%) or for extended periods of time (>200 seconds, 8.30% ± 1.52%) (Sun et al., 2013). These results are in accordance with a previous report in cultured cortical neurons (Chang, Honick, and Reynolds, 2006).

A key pathogenic change in several major age-associated neurodegenerative diseases is altered mitochondrial transport (see review by Sheng and Cai, 2012). Although studies with primary neurons have been useful toward elucidating mechanisms of mitochondrial transport and anchoring; this in vitro culture system using embryonic or newborn mice is rather limited for investigating age-associated diseases because the physiological conditions in vivo and the aging process cannot be replicated in a culture dish. An *ex vivo* system would therefore much better reflect physiological environment and age-associated changes in mitochondrial transport, as was shown in a previous study with acute nerve explants from transgenic mice in which neuronal mitochondria are labeled with mito-CFP or mito-YFP (Misgeld et al., 2007). To further investigate age-associated defects in mitochondrial trafficking and impaired removal and degradation of damaged mitochondria from distal terminals, it is critical to establish neuron cultures directly isolated from adult diseased mice or from patient-derived iPSCs.

7.3 MECHANISMS OF MITOCHONDRIAL TRANSPORT

Mitochondria display complex mobility patterns along neuronal processes, undergoing bidirectional transport and frequently pausing, or docking. This is a result of their coupling with anterograde and retrograde motors and anchoring machinery. Mitochondria bind to motors via adaptor proteins and receptors on the mitochondrial outer membrane. The uniform polarity of MTs within an axon makes it particularly well-suited for investigating mechanisms regulating bidirectional transport. Dynein drives retrograde transport toward the soma, and kinesin drives anterograde transport toward distal synapses.

7.3.1 Kinesin Motors and Anterograde Transport

The members of kinesin-1 (KIF5) family are the main kinesin motors involved in neuronal mitochondrial transport (Cai, Gerwin, and Sheng, 2005; Gorska-Andrzejak et al., 2003; Hurd and Saxton, 1996; Pilling et al., 2006; Tanaka et al., 1998). KIF5 heavy chain (KHC) is composed of a motor domain with an ATPase at the amino terminal, and a carboxyl terminal cargo-binding tail. There are three KIF5 isoforms in mammals (KIF5A, KIF5B, and KIF5C), with KIF5B being ubiquitously expressed and KIF5A and KIF5C being restricted to neurons (Kanai et al., 2000).

Adaptor proteins link the mitochondrion to KIF5 motors, and this motor-adaptor coupling enables effective regulation of mitochondrial transport. In *Drosophila*, Milton functions as a motor adaptor by linking KIF5 and Miro, a mitochondrial outer membrane receptor (Glater et al., 2006; Stowers et al., 2002). Milton mutation decreases anterograde mitochondrial trafficking to distal axonal terminals, without affecting the trafficking of other cargos. Trak1 and Trak2 are two mammalian Milton orthologs required for mitochondrial trafficking (Koutsopoulos et al., 2010; MacAskill et al., 2009b; Smith et al., 2006). Reduced Trak1 levels impair axonal mitochondrial trafficking (Brickley and Stephenson, 2011), whereas overexpression of Trak2 in hippocampal neurons increases axonal mitochondrial trafficking (Chen and Sheng, 2013). However, unlike in *Drosophila*, mammalian Trak proteins are also required for neuronal trafficking of endosomal compartments, GABAA receptors, and K^+ channels (Grishin et al., 2006; Kirk, Chin, and Li, 2006; Webber, Li, and Chin, 2008).

Miro is a Rho-GTPase containing two GTPase domains and two Ca^{2+}-binding EF hand motifs (Fransson, Ruusala, and Aspenstrom, 2006; Frederick et al., 2004). The recently resolved crystal structure of *Drosophila* Miro reveals two additional EF hands that each pair with a canonical EF hand (Klosowiak et al., 2013). Mutation of the *Drosophila* miro impairs anterograde mitochondrial transport to distal axonal terminals (Guo et al., 2005). In mammals, there are two Miro isoforms: Miro1 and Miro2. In hippocampal neurons, mitochondrial transport is regulated by the Miro-Trak complex (MacAskill et al., 2009b), and increased Miro1 expression enhances axonal mitochondrial transport (Chen and Sheng, 2013), likely via increased recruitment of Trak2 and motor proteins to the mitochondria. Thus, the mitochondrial motor-adaptor-receptor complexes of KIF5, Milton/Trak, and Miro work together to drive anterograde transport (Fig. 7.2**a**).

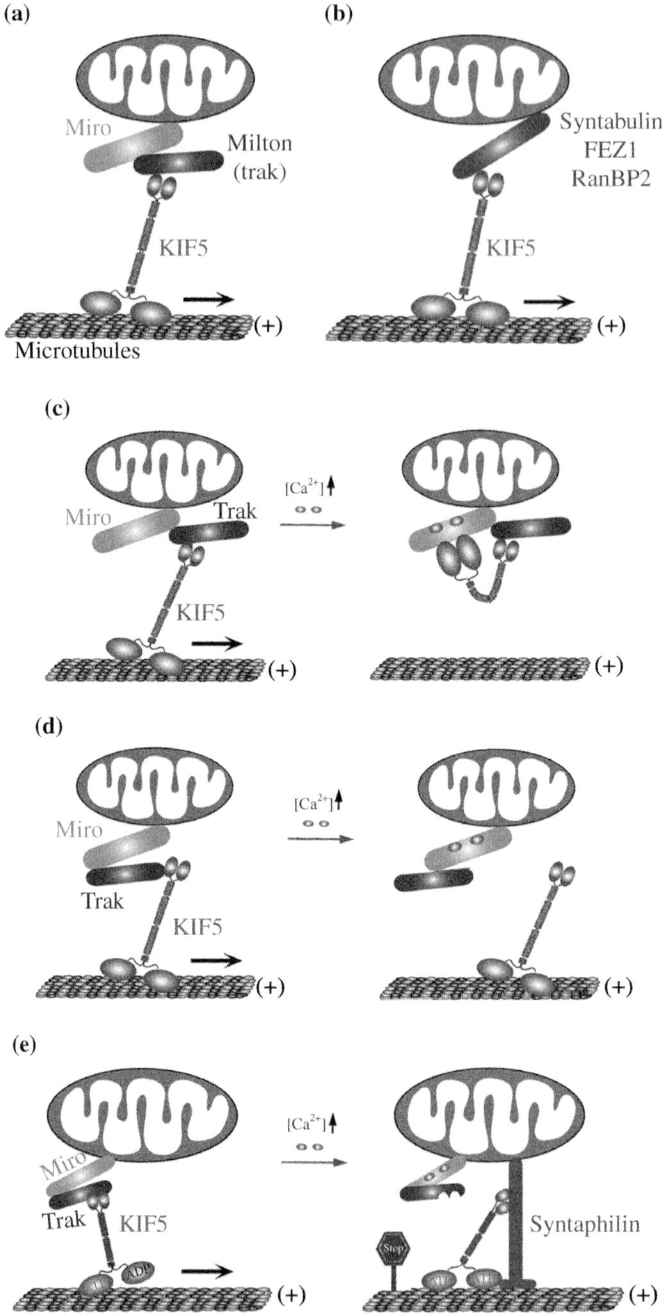

FIGURE 7.2 Regulation of mitochondrial transport by synaptic activity. **a**, KIF5-driven mitochondrial transport is mediated by the Miro-Milton/TRAK complexes. **b**, Syntabulin, RanBP2, or FEZ1 function as alternative KIF5 adaptors driving anterograde mitochondrial transport. **c** and **d**, Two alternative models of the role of the Miro-Ca^{2+} pathway in regulating

Syntabulin is another motor adaptor in neurons that links KIF5 and mitochondria (Cai et al., 2005) by binding to the KIF5 cargo-binding domain (Fig. 7.2**b**). Syntabulin targets the mitochondrial outer membrane via its carboxyl-terminal transmembrane domain. Depleting syntabulin or blocking the syntabulin–KIF5 coupling decreases anterograde axonal mitochondrial transport in hippocampal neurons. There are several other putative KIF5 motor adaptors that mediate mitochondrial transport, including FEZ1 and RanBP2 (Cho et al., 2007; Fujita et al., 2007). The presence of multiple mitochondrial motor adaptors highlights the complexity of neuronal mitochondrial transport. One potential explanation is that KIF5 motors use different adaptors to bind mitochondria, thus allowing neurons to regulate mitochondrial transport by integrating various physiological and pathological signals. It will therefore be important to investigate how assembly and activity of these motor-adaptor complexes is regulated by the sensing of cellular levels of adenosine diphosphate (ADP) and calcium, to reveal how neurons coordinate transport and distribution with metabolic demands and synaptic function.

7.3.2 Dynein Motors and Retrograde Transport

In contrast to anterograde KIF5 motor-adaptor complexes, little is known about dynein motor-adaptor complexes driving retrograde mitochondrial transport. Dynein motors consist of motor domain-containing heavy chains (DHC) and several other chains for regulating motility or associating with their cargoes including intermediate chains (DIC), light intermediate chains (DLIC), and light chains (DLC). Dynein motors are loaded on mitochondria and mediate retrograde trafficking in *Drosophila* axons (Pilling et al., 2006). The *Drosophila* Miro (dMiro), a mitochondrial receptor for KIF5, may also bind dynein motors for retrograde transport (Guo et al., 2005; Russo et al., 2009). dMiro mutants display reduced bidirectional transport and dMiro overexpression also alters mitochondrial transport in both directions. Supporting this is a recent finding that mammalian Trak1 and Trak2 have both KIF5B- and dynein/dynactin-binding sites (van Spronsen et al., 2013), indicating that Trak can mediate bidirectional axonal transport through interaction

---

mitochondrial transport. Miro has 4 Ca^{2+}-binding EF-hand motifs. The KIF5 C-terminal cargo-binding domain binds to the Miro-Trak adaptor complex. The binding of Ca^{2+} to the EF-hands of Miro causes the motor domain to detach from MTs, preventing motor-MT engagement (**c**) (Wang and Schwarz, 2009). Alternatively, binding of Ca^{2+} releases KIF5 from mitochondria (**d**) (MacAskill et al., 2009a). The influx of Ca^{2+} triggered by synaptic activity stops motile mitochondria at activated synapses. **e**, The syntaphilin-mediated "engine-switch and brake" model. A Miro-Ca^{2+} sensing pathway initiates the switch of KIF5 binding from Miro-Trak to syntaphilin, immobilizing axonal mitochondria by inhibiting the motor ATPase activity. Thus, syntaphilin switches off the "KIF5 motor engine" by sensing a "Stop Sign" (increased Ca^{2+} levels) and places a brake on mitochondria, anchoring them to the MTs. Stationary KIF5-loaded mitochondria remain bound to MT tracks, whereas the KIF5 ATPase is in an inactive state (Chen and Sheng, 2013). Adapted from Sheng, Z.-H., 2014. Mitochondrial trafficking and anchoring in neurons: New insight and implications. *J Cell Biol*, 204, pp.1087–1098, by permission.

with KIF5 and dynein motors (see Fig. 7.1). These data highlight a new model of cargo-adaptor proteins in regulating the activity of opposing motors to control bidirectional axonal transport.

7.3.3 Interplay of Opposing Motor Proteins

The finding that mitochondria move bidirectionally along axons suggests that retrograde and anterograde motors are both loaded on the same mitochondrion. For example, dynein-loaded mitochondria undergo both anterograde and retrograde transport (Hirokawa et al., 1990). Disrupting kinesin-driven transport complexes is not sufficient for dynein to dominate mitochondrial retrograde transport (MacAskill et al., 2009a; Wang and Schwarz, 2009). Instead of simply competing with each other, kinesin and dynein motors likely coordinate bidirectional transport (Ligon et al., 2004; Welte, 2004). Although dynein primarily mediates transport toward the minus-end of MTs, it can also move bidirectionally (Mallik et al., 2005). One candidate molecule for coordinating the activity of these opposing motors is dynactin. Mutation of dynactin in *Drosophila* impairs both retrograde and anterograde transport (Haghnia et al., 2007). DHC and dynactin $p150^{Glued}$ mutations also disrupt fast bidirectional transport and lead to axonal swellings filled with mitochondria and other cargos (Martin et al., 1999). Furthermore, kinesin-1 inhibition decreases bidirectional mitochondrial transport, revealing that it may be required for dynein-driven retrograde movement (Pilling et al., 2006). An important question is how transport direction is determined. One potential mechanism is a tug-of-war between opposing motor complexes. Changes in the activity of one motor would be inversely related to changes with the opposing motor, with the more powerful motor ultimately dictating transport direction. An alternative mechanism is the relative processivity of the opposing motors, with changes altering the bidirectional bias and the amount of time spent in the stationary phase.

Cumulating evidence supports a link between dynein motor activity and kinesin adaptors. First, mutation of the kinesin-binding JIP-1/JNK scaffold protein, APLIP1, also impairs retrograde mitochondrial transport in *Drosophila* (Horiuchi et al., 2005). Second, dMiro mutation affected both the distance and duration of mitochondrial transport (Russo et al., 2009), suggesting that dMiro may integrate cellular signals with anterograde/retrograde motors to determine direction of transport. Third, depleting the mitochondrial fusion protein Mfn2 in neurons decreases anterograde/retrograde trafficking and increases the length of pauses through a mechanism involving Miro (Misko et al., 2010). It is likely that Miro facilitates crosstalk between kinesin and dynein motors. It would be necessary to elucidate the role of Miro in coordinating bidirectional transport of mitochondria in response to changes in neuronal physiology.

7.4 MECHANISMS OF AXONAL MITOCHONDRIAL ANCHORING

To maintain axonal and synaptic function, neurons must recruit mitochondria to areas with high-energy demand and Ca^{2+} buffering capacity. Because ATP diffusion is limited in the cell (Hubley, Locke, and Moerland, 1996), particularly within extended long axons (Sun et al., 2013), anchored stationary mitochondria are necessary to sustain synaptic transmission by maintaining energy and Ca^{2+} homeostasis in axons and at synapses (Billups and Forsythe, 2002; Kang et al., 2008). Neurons require special mechanisms for mitochondrial docking because two-thirds of axonal mitochondria are stationary. Syntaphilin is an intriguing anchoring protein identified in our laboratory that specifically targets axonal mitochondria and functions as a "static anchor" for docking mitochondria in axons (Chen, Gerwin, and Sheng, 2009; Kang et al., 2008). Deletion of *syntaphilin* robustly increases axonal mitochondrial motility, whereas overexpressing syntaphilin immobilizes axonal mitochondria. We further provided a docking mechanism: syntaphilin recruits motile mitochondria into stationary pools by anchoring them to MTs in axons. Syntaphilin is therefore an attractive molecular target for investigating the regulation of mitochondrial trafficking in axons and anchoring at active synapses; the *syntaphilin* KO mouse provides an ideal genetic model for studies of the role of enhanced axonal mitochondrial motility on presynaptic function and mitochondrial quality control.

7.5 REGULATION OF MITOCHONDRIAL TRANSPORT BY SYNAPTIC ACTIVITY

Synaptic activity is tightly correlated with axonal mitochondrial transport and distribution. Under sustained synaptic activity, mitochondria are recruited to the synapse by sensing elevated levels of intracellular Ca^{2+}, which are triggered by activating NMDA receptors in dendrites or voltage-dependent Ca^{2+} channels in axons (Chang et al., 2006; Rintoul et al., 2003; Szabadkai et al., 2006; Yi, Weaver, and Hajnóczky, 2004). These studies raise the key question of how mitochondria get recruited to active synapses in response to Ca^{2+} influx. Three groups recently shed light onto this by identifying Miro as a Ca^{2+} sensor that regulates mitochondrial trafficking (MacAskill et al., 2009a; Saotome et al., 2008; Wang and Schwarz, 2009). Elevated levels of intracellular Ca^{2+} recruit and immobilize mitochondria at active synapses via the Miro-Ca^{2+}-sensing pathway by deactivating the KIF5-Trak-Miro transport complex. In this Miro-Ca^{2+} sensing model, KIF5 motors drive mitochondrial transport by forming a complex with Miro-Milton/Trak at resting Ca^{2+} levels. When mitochondria pass by active synapses, elevated Ca^{2+} binds the EF-hands in Miro inducing a conformational change, which uncouples the transport complex and immobilizes mitochondria at activated synapses (Figs. 7.2**c** and 7.2**d**) (MacAskill et al., 2009a; Wang and Schwarz, 2009). However, these studies disagreed on whether kinesin motors remain associated with mitochondria after immobilization. It is important to note that the Miro-Ca^{2+} pathway arrests both anterograde and retrograde mitochondrial trafficking.

Our recent work supports the hypothesis that the Miro-Ca^{2+}-sensing pathway immobilizes mitochondria through an anchoring mechanism. We found that activating this pathway does not prevent axonal mitochondrial trafficking in *syntaphilin*-null hippocampal neurons, even though dendritic mitochondria are effectively immobilized (Chen and Sheng, 2013). This suggests that syntaphilin plays a key role in activity-dependent immobilization of mitochondria in axons. Syntaphilin competes against Trak2 for KIF5 binding and prevents KIF5 ATPase activity in vitro, suggesting that it prevents the KIF5 motor moving along the MTs. Elevated Ca^{2+} levels and synaptic activity favor the syntaphilin-KIF5 interaction by disrupting the kinesin-Trak-Miro complex. Furthermore, syntaphilin is recruited to axonal mitochondria under conditions of sustained neuronal activity, although it is unknown how this process occurs. Overexpression of syntaphilin lacking the axon-sorting domain induces translocation to all mitochondria in neurons (Kang et al., 2008). These reports converge to suggest that syntaphilin is recruited to mitochondria after being delivered to axons. An intriguing question still unanswered is how mitochondria anchored by syntaphilin are remobilized. One possibility is that syntaphilin is degraded via ubiquitin-mediated pathways because it was found to be a cellular target of the E3 ubiquitin ligase Cullin 1 (Yen and Elledge, 2008). It will be important to elucidate mechanisms regulating activity-dependent recruitment and subsequent removal of syntaphilin from mitochondria.

We propose a new "engine-switch and brake" model (Fig. 7.2**e**) (Chen and Sheng 2013), in which syntaphilin anchors mitochondria onto MTs in response to elevated Ca^{2+} ("stop" sign), switching off kinesin and putting the brake on mitochondria. KIF5-loaded mitochondria therefore remain associated with MT tracks in axons during stationary periods. Importantly, our new model may explain how the Miro-Ca^{2+}-sensing mechanism immobilizes neuronal mitochondria, reconciling the disagreements in the field. Although MacAskill et al. (2009a) suggest that Ca^{2+} binds Miro and disconnects motors from dendritic mitochondria (Fig. 7.2**d**), Wang and Schwarz (2009) propose that KIF5 remains associated with immobilized axonal mitochondria under conditions of increased Ca^{2+} (Fig. 7.2**c**). One potential explanation for this discrepancy is the fact that these two studies selectively observed mitochondrial trafficking in axons versus dendrites, which is especially important considering that syntaphilin is absent from dendrites. The presence of syntaphilin may enable axonal mitochondria to be more responsive to alterations in synaptic activity. Once Ca^{2+} levels return to baseline, motor complexes can be quickly reformed and transport mitochondria to new active synapses. An important point in this new model is that loading of the motor complex onto mitochondria is not sufficient for transport—the anchoring machinery must also be released. In support of this, a recent report demonstrated that dynein and KIF5 motors remain bound to both motile and stationary prion protein vesicles (Encalada et al., 2011).

7.6 MITOCHONDRIAL TRANSPORT AND SYNAPTIC TRANSMISSION

Mitochondria are transported to synapses, where they provide the ATP supply and Ca^{2+} buffering capacity necessary to maintain synaptic transmission (Kang et al., 2008; Tang and Zucker, 1997; Werth and Thayer, 1994). Reduced mitochondrial transport depletes the number of mitochondria in synaptic terminals. Reduced local ATP supply impairs a variety of ATP-dependent processes such as vesicular neurotransmitter loading, Ca^{2+} removal from synaptic terminals, and synaptic vesicle transport from the reserve pools to release sites. *Drosophila* milton mutants have fewer synaptic mitochondria and exhibit defective neurotransmission (Stowers et al., 2002). Drp1 mutants also have severe defects in synaptic localization of mitochondria and reduced recruitment of the synaptic vesicles from the reserve pool, which causes rapid synaptic vesicle depletion during prolonged trains of stimulation pulses (Verstreken et al., 2005). A recent study with cultured superior cervical ganglion neurons demonstrated that syntabulin, a mitochondrial adaptor for KIF5 motors, is critical for maintaining presynaptic function (Ma et al., 2009). Syntabulin loss-of-function reduces mitochondrial number in axonal processes, accompanied by reduced basal synaptic activity, accelerated synaptic depression during high-frequency firing, and slowed recovery rate of synapses after synaptic vesicle depletion. ATP application to presynaptic neurons rescues these defects, suggesting that efficient delivery of mitochondria to axonal terminals is critical for maintenance of synaptic function in mature neurons.

Neurotransmitter release is activated by Ca^{2+} influx, and factors that buffer Ca^{2+} levels at presynaptic terminals may affect the probability of synaptic vesicle release. Repetitive stimulation will accumulate Ca^{2+} within presynaptic boutons by prolonged activation of voltage-gated calcium channels. Using a combination of electrophysiology and calcium imaging at the calyx of Held, mitochondria were found to rapidly sequester large quantities of cytoplasmic Ca^{2+} and modulate neurotransmitter release (Billups and Forsythe, 2002). This reveals a potential role for mitochondria in inducing short-term presynaptic plasticity via modulating intracellular Ca^{2+} levels. However, presynaptic mitochondria at neuromuscular junctions may simply maintain Ca^{2+} homeostasis by sequestering excess Ca^{2+} during repetitive stimulation. dMiro mutants have defects in mitochondrial transport to the neuromuscular junctions, which subsequently decreases Ca^{2+} buffering capacity and alters neurotransmission during prolonged stimulation (Guo et al., 2005).

7.7 MITOCHONDRIAL TRANSPORT AND PRESYNAPTIC VARIABILITY

An important characteristic of synaptic physiology is the large variation of synaptic strength in response to repeated stimulations, and the effects of this on neural circuits are increasingly recognized (Murthy, Sejnowski, and Stevens, 1997). One fundamental question is how this variation arises. The consensus is that stochastic molecular and structural changes at synapses are responsible for the heterogeneity in synaptic transmission between neuron to neuron or between synapse to synapse (see reviews by Atwood and Karunanithi, 2002; Branco and Staras, 2009; Marder and Goaillard, 2006; Ribrault, Sekimoto, and Triller, 2011; Stein, Gossen, and Jones, 2005). Given the fact that ~30% of axonal mitochondria are highly motile, some of which dynamically pass through presynaptic boutons in hippocampal neurons, it raises a relevant question as to whether the dynamic movement of axonal mitochondria contributes to the pulse-to-pulse variability of presynaptic strength at single-bouton levels.

Mitochondria that are docked within presynaptic terminals provide stable ATP supply and maintain synaptic energy homeostasis. Motile mitochondria that pass by presynaptic boutons dynamically affect ATP levels and influence a variety of ATP-dependent synaptic functions including the establishment of a proton gradient for neurotransmitter loading, Ca^{2+} removal from terminals, and mobilization of synaptic vesicles to release sites. We recently found that axonal mitochondrial mobility is one of the primary mechanisms responsible for the variation in presynaptic strength (Sun et al., 2013). By applying dual-channel live imaging and electrophysiology in hippocampal slices and cultured neurons from *syntaphilin* knockout mice, we demonstrated that enhanced axonal mitochondrial transport increases pulse-to-pulse variability of EPSC amplitudes; blocking mitochondrial transport in axons diminished this variability. Our results reveal two important findings: (1) synaptic vesicle release at individual boutons can change over time with dynamic changes in mitochondrial transport and distribution; (2) because EPSC amplitudes are averaged across multiple synapses, at any given time, these individual boutons display different patterns of mitochondrial distribution and motility; mitochondria move either into or out of boutons or pass through boutons during stimulation, thus contributing to the pulse-to-pulse variability. Using dual-channel live imaging at the single-bouton level to monitor mitochondrial motility and synaptic vesicle release, we found that mitochondrial transport through presynaptic boutons significantly affects vesicle release via fluctuations in synaptic ATP levels. Motile mitochondria passing through a bouton lacking a stationary mitochondrion can provide a temporary supply of ATP, thus influencing ATP-dependent processes such as synaptic vesicle cycling, mobilization, and replenishment (Fig. 7.3). These data are supported by a recent report using a new quantitative reporter for presynaptic ATP. Synaptic vesicle cycling and neurotransmission consumes most presynaptic ATP (Rangaraju, Calloway, and Ryan, 2014). Collectively, multiple ATP-dependent processes likely contribute to presynaptic variability. Our study revealed that dynamic axonal mitochondrial transport underlies presynaptic variation (Sun et al., 2013).

7.8 MITOCHONDRIAL TRANSPORT AND AXONAL BRANCHING

Docked mitochondria function as local energy power plants. Elevated levels of ADP decrease mitochondrial motility, thus recruiting mitochondria to areas with ATP depletion (Mironov, 2007). However, it is unknown how mitochondrial motility and energy consumption are coordinated. Recent studies showed that axonal branching depends on stable ATP supply from docked mitochondria at branching points. LKB1 regulates axonal branching in cortical neurons in vitro and in vivo through the AMPK-like kinase, NUAK1 (Courchet et al., 2013). AMP-activated protein kinase (AMPK) is a master regulator of energy homeostasis and becomes activated on cellular stresses that reduce ATP supplies. Decreased NUAK1 or LKB1 enhances axonal mitochondrial transport, whereas increased expression reduces axonal mitochondrial transport. Courchet et al. (2013) further tested the link between mitochondrial motility and axonal branching by manipulating syntaphilin expression levels.

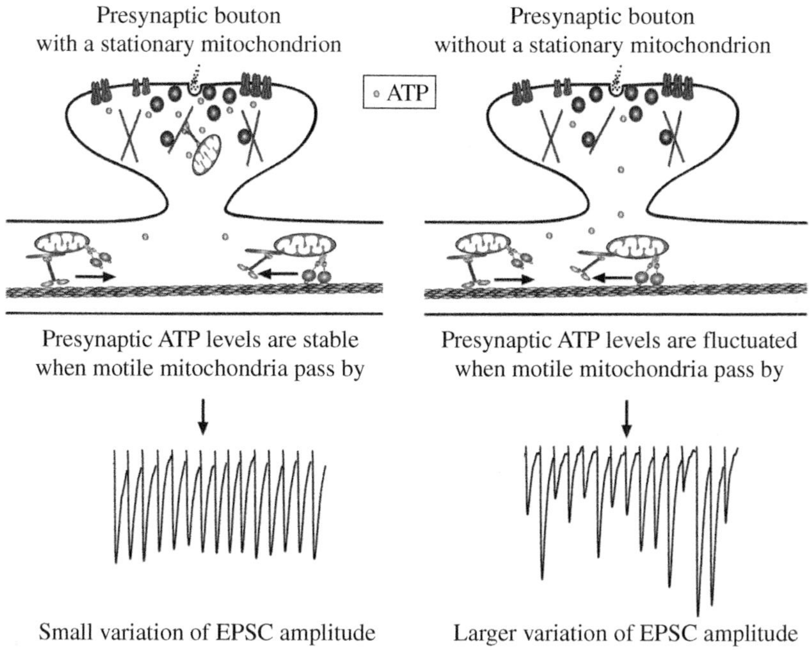

FIGURE 7.3 Motile axonal mitochondria contribute to presynaptic variation. Mitochondria docked within presynaptic terminals provide stable adenosine triphosphate (ATP) supplies that power neurotransmission *(left panel)*. In the absence of mitochondria within the presynaptic bouton *(right panel)*, there is no stable ATP supply; motile mitochondria passing through this bouton can temporarily supply ATP, altering synaptic energy levels and causing fluctuation of ATP-dependent synaptic functions. Mitochondrial transport is a primary source of the variability in synaptic vesicle release and amplitude of excitatory postsynaptic currents (EPSCs) (Sun et al., 2013). Adapted from Sheng, Z.-H., 2014. Mitochondrial trafficking and anchoring in neurons: New insight and implications. *J Cell Biol*, 204, pp.1087–1098, by permission.

Decreased syntaphilin expression enhanced mitochondrial transport and decreased axonal branching, whereas overexpression of syntaphilin inhibited axonal mitochondrial transport and increased axonal branching. This highlights a key role for syntaphilin-mediated mitochondrial anchoring in AMPK-induced branching in axons, and demonstrates that LKB1 and NUAK1 may regulate axonal branching via recruiting mitochondria to and anchoring them at branch sites.

AMPK serves as an energy sensor for intracellular ATP consumption, positively regulating signal transduction pathways to replenish cellular ATP levels. A new study provided further evidence demonstrating that sustained neuronal depolarization depletes the cellular ATP and activates AMPK (Tao, Matsuki, and Koyama, 2013). AMPK activation enhances anterograde axonal mitochondrial flux and induces ATP-dependent axonal branches. These findings establish roles for the AMPK pathway in regulating axonal branching by recruiting and anchoring axonal mitochondria at branching points. A potential mechanism for recruiting mitochondria to support axonal branching is through localized intra-axonal protein synthesis (Spillane et al., 2013). Inhibition of protein synthesis and mitochondrial respiration in axons impairs axonal branching. Docked mitochondria therefore likely coordinate axonal branching through supporting mRNA translation and protein synthesis in specific axon hot spots.

7.9 MITOCHONDRIAL TRANSPORT AND MITOPHAGY

Dysfunctional mitochondria have reduced capacity for ATP production and Ca^{2+} buffering and also release reactive oxygen species (ROS). Mitochondrial transport defects prevent the replenishment of healthy mitochondria in distal processes, disrupting synaptic energy homeostasis and Ca^{2+} transients. ROS may enhance synaptic stress and accelerate neurodegeneration. There are multiple cellular mechanisms for maintaining mitochondrial quality control, including fusion to repair dysfunctional mitochondria (see reviews by Chen and Chan 2009; Westermann, 2010) and mitophagy to eliminate irreversibly damaged mitochondria. The sequestration and subsequent degradation of damaged mitochondria via the autophagy-lysosomal system is a critical cellular pathway for maintaining mitochondrial integrity and function. Recent studies from many laboratories have identified the PINK1/Parkin pathway as being critical for mitochondrial quality control (Clark et al., 2006; Gautier Kitada and Shen, 2008; Narendra et al., 2008). Importantly, Parkin and PINK1 loss-of-function mutations are associated with recessive forms of Parkinson's disease. Mitochondrial depolarization triggers the accumulation of PINK1 on the outer membrane of mitochondria and recruits the cytosolic E3 ubiquitin ligase Parkin, which subsequently ubiquitinates mitochondrial proteins (Geisler et al., 2010; Poole et al., 2010; Yoshii et al., 2011; Ziviani, Tao, and Whitworth, 2010). This process enables dysfunctional mitochondria to be engulfed by the isolation membranes, followed by fusion with late endosomes/lysosomes for degradation.

Because mature acidic lysosomes are primarily localized in the soma (Cai et al., 2010; Lee et al., 2011; Overly and Hollenbeck, 1996), neurons face special challenges to remove aged and dysfunctional mitochondria in the distal regions. Recently, we reported unique features of spatial and dynamic Parkin-mediated mitophagy in mature cortical neurons on chronic dissipation of mitochondrial membrane potential ($\Delta\psi_m$) for 24 hours (Cai et al., 2012). Given the fact that mitochondrial depolarization quickly triggers apoptosis in vitro, it is critical to maintain high-quality cultures for the neurons to survive long enough to observe mitophagy. Chronic treatment with $\Delta\psi_m$ uncoupling reagents at low concentrations induces the accumulation of Parkin-tagged mitochondria in the proximal regions of processes and the soma. This regionally restricted pattern for Parkin-tagged mitochondria was also consistently observed for two $\Delta\psi_m$ uncoupling reagents CCCP (10 μM) and antimycin A (1 μM) or mitochondrial ATP synthase inhibitor oligomycin (1 μM). This distribution pattern is a result of altered transport of chronically depolarized mitochondria, reduced anterograde and enhanced retrograde transport, thus resulting in reduced net anterograde flux of dysfunctional mitochondria into distal processes. Altered transport may serve a protective role under conditions of chronic mitochondrial stress, allowing damaged mitochondria to be selectively returned to the soma for degradation and healthy mitochondria to be allowed to remain at distal locations (Fig. 7.4).

Decreased anterograde trafficking of depolarized mitochondria is consistent with reports showing Miro degradation on the surface of mitochondria upon mitophagy induction. Miro interacts with Parkin and PINK1 and is ubiquitinated by Parkin on mitochondrial depolarization and Parkin translocation (Chan et al., 2011; Liu et al., 2012; Sarraf et al., 2013; Wang et al., 2011; Weihofen et al., 2009). Miro degradation on the surface of mitochondria may selectively impair anterograde transport, thus

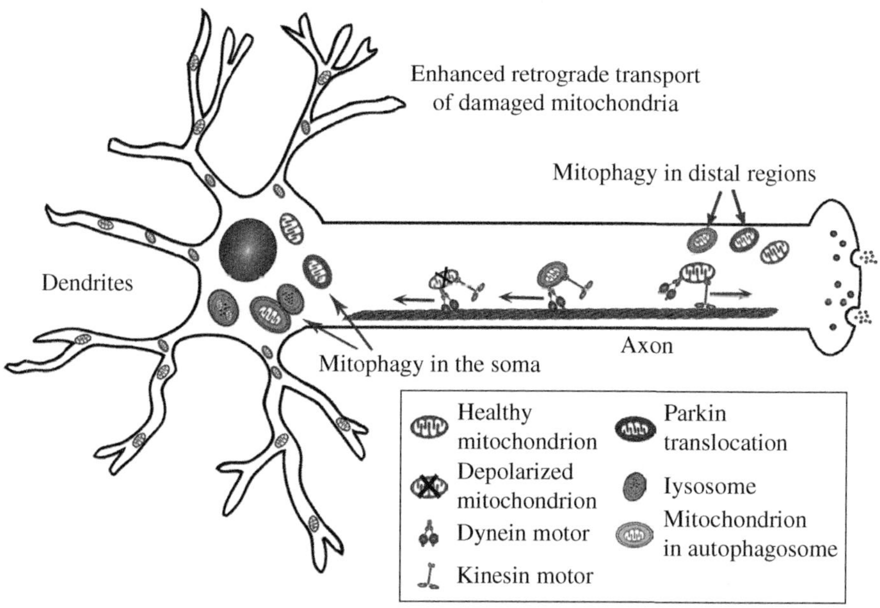

FIGURE 7.4 Coordination of mitochondrial transport and mitophagy. Chronically depolarizing mitochondrial membrane potential ($\Delta\psi_m$) with uncoupling reagents induces the accumulation of Parkin-tagged mitochondria in the soma and proximal regions of processes. The restriction of dysfunctional mitochondria to these regions is the result of enhanced retrograde and reduced anterograde transport (Cai et al., 2012). This spatial organization enables damaged mitochondria in distal axons to be eliminated through the autophagy-lysosomal pathway in the soma, where mature lysosomes are primarily localized. Dysfunctional mitochondria can also recruit Parkin in axonal terminals when anchored by syntaphilin (Cai et al., 2012) or immobilized by degradation of Miro (Chan et al., 2011; Liu et al., 2012; Sarraf et al., 2013; Wang et al., 2011; Weihofen et al., 2009; Yoshli et al., 2011). The predominant retrograde transport of autophagic vacuoles including those engulfing mitochondria from distal axons to the proximal regions supports the hypothesis that dysfunctional mitochondria immobilized in distal regions may be also degraded in the soma (Maday et al., 2012). Adapted from Sheng, Z.-H., 2014. Mitochondrial trafficking and anchoring in neurons: New insight and implications. *Journal of Cell Biology*, 204, pp.1087–1098, by permission.

relatively enhancing retrograde transport to the soma (see Fig. 7.4). Our study also showed that depolarized mitochondria at distal processes also recruit Parkin once the mitochondria are anchored by syntaphilin (Cai et al., 2012). The predominant retrograde transport of autophagic vacuoles including those engulfing mitochondria from distal axons to the soma supports the hypothesis that dysfunctional mitochondria immobilized in distal regions may be degraded in the soma (Maday, Wallace, and Holzbaur, 2012). Therefore, mitochondrial transport and quality control are tightly coordinated to ensure removal of aged and damaged mitochondria from distal processes. Two previous reports on the correlation between mitochondrial $\Delta\psi_m$ and transport in neurons are controversial because they used acute treatment with high concentrations of $\Delta\psi_m$ dissipating reagents (Miller and Sheetz, 2004; Verburg and Hollenbeck, 2008). We examined mitochondrial trafficking and mitophagy in cortical neurons after chronic treatment with low concentration of CCCP or antimycin with pan-caspase inhibitor Z-VAD for 24 hours (Cai et al., 2012). Under these conditions, most neurons survive and altered mitochondrial transport can be detected. Thus, in mature cultured neurons, prolonged $\Delta\psi_m$ dissipation may provide a more physiological model of chronic mitochondrial stress in vivo.

7.10 CONCLUSIONS AND NEW CHALLENGES

The identification of proteins that regulate mitochondrial trafficking and anchoring has advanced our understanding of the molecular and cellular pathways that enable neurons to efficiently respond to physiological and pathological signals. However, there are many key questions that need to be investigated: (1) How does the Miro-Ca^{2+}-sensing pathway immobilize mitochondria in both anterograde and retrograde movement at active synapses? (2) How do the KIF5 and dynein motor proteins and their adaptors coordinate bidirectional mitochondrial transport? (3) How are mitochondria recruited in response to changes in metabolic requirement during axonal growth/branching and regeneration? (4) How is mitochondrial motility and quality control regulated in sensing various cellular stress and pathological signals? (5) How does neuronal mitochondria coordinate their transport with integrity and function? In particular, how do neurons remove damaged mitochondria from axonal terminals while maintaining healthy mitochondria at distal synapses? Mitochondrial dysfunction is a key pathological hallmark of several major age-associated neurodegenerative diseases (see reviews by Chen and Chan, 2009; Court and Coleman, 2012; Itoh et al., 2013; Schon and Przedborski, 2011; Sheng and Cai, 2012). The findings from these studies will explain the challenges that neurons have in regulating mitochondrial motility during development, healthy aging, and age-associated neurodegenerative disorders. Research into the fundamental processes of mitochondrial trafficking and quality control will advance our understanding of neurological diseases.

ACKNOWLEDGMENTS

A. L. Knight and Y. Chen contributed equally to this publication.

REFERENCES

Attwell D. and Laughlin, S. B., 2001. An energy budget for signaling in the grey matter of the brain. *J Cereb Blood Flow Metab*, 21, pp.1133–1145.

Atwood H. L. and Karunanithi, S., 2002. Diversification of synaptic strength: presynaptic elements. *Nat Rev Neurosci*, 3, pp.497–516.

Billups, B. and Forsythe, I. D., 2002. Presynaptic mitochondrial calcium sequestration influences transmission at mammalian central synapses. *Journal of Neuroscience*, 22, pp.5840–5847.

Birsa, N., et al., 2013. Mitochondrial trafficking in neurons and the role of the Miro family of GTPase proteins. *Biochem Soc Trans*, 41, pp.1525–1531.

Branco, T. and Staras. K., 2009. The probability of neurotransmitter release: Variability and feedback control at single synapses. *Nat Rev Neurosci*, 10, pp.373–383.

Brickley, K. and Stephenson, F.A., 2011. Trafficking kinesin protein (TRAK)-mediated transport of mitochondria in axons of hippocampal neurons. *Journal of Biological Chemistry*, 286, pp.18079–18092.

Cai, Q., Gerwin, C. and Sheng, Z.-H., 2005. Syntabulin-mediated anterograde transport of mitochondria along neuronal processes. *Journal of Cell Biology*, 170, pp.959–969.

Cai, Q., et al., 2012. Spatial parkin translocation and degradation of damaged mitochondria via mitophagy in live cortical neurons. *Curr Biol*, 22, pp.545–552.

Cai, Q., et al., 2010. Snapin-regulated late endosomal transport is critical for efficient autophagy-lysosomal function in neurons. *Neuron*, 68, pp.73–86.

Chada, S. R. and Hollenbeck, P. J., 2004. Nerve growth factor signaling regulates motility and docking of axonal mitochondria. *Current Biology*, 14, pp.1272–1276.

Chan, N. C., et al., 2011. Broad activation of the ubiquitin-proteasome system by Parkin is critical for mitophagy. *Human Molecular Genetics*, 20, pp.1726–1737.

Chang, D. T. W., Honick, A. S. and Reynolds, I. J., 2006. Mitochondrial trafficking to synapses in cultured primary cortical neurons. *Journal of Neuroscience*, 26, pp.7035–7045.

Chang, D. T. W. and Reynolds, I. J., 2006. Mitochondrial trafficking and morphology in healthy and injured neurons. *Progress in Neurobiology*, 80, pp.241–268.

Chen, H. and Chan, D. C., 2009. Mitochondrial dynamics—fusion, fission, movement, and mitophagy—in neurodegenerative diseases. *Human Molecular Genetics*, 18, pp.R169–R176.

Chen, Y. and Sheng, Z.-H., 2013. Kinesin-1-syntaphilin coupling mediates activity-dependent regulation of axonal mitochondrial transport. *Journal of Cell Biology*, 202, pp.351–364.

Chen, Y. M., Gerwin, C. and Sheng, Z.-H., 2009. Dynein light chain LC8 regulates syntaphilin-mediated mitochondrial docking in axons. *Journal of Neuroscience*, 29, pp.9429–9438.

Cho, K.-I., et al., 2007. Association of the kinesin-binding domain of RanBP2 to KIF5B and KIF5C determines mitochondria localization and function. *Traffic*, 8, pp.1722–1735.

Clark, I. E., et al., 2006. Drosophila pink1 is required for mitochondrial function and interacts genetically with parkin. *Nature*, 441, pp.1162–1166.

Courchet, J., et al., 2013. Terminal axon branching is regulated by the LKB1-NUAK1 kinase pathway via presynaptic mitochondrial capture. *Cell*, 153, pp.1510–1525.

Court, F. A. and Coleman, M. P., 2012. Mitochondria as a central sensor for axonal degenerative stimuli. *Trends in Neurosciences*, 35, pp.364–372.

David, G. and Barrett, E. F., 2003. Mitochondrial Ca2+ uptake prevents desynchronization of quantal release and minimizes depletion during repetitive stimulation of mouse motor nerve terminals. *Journal of Physiology*, 548, pp.425–438.

Encalada, S. E., et al., 2011. Stable kinesin and dynein assemblies drive the axonal transport of mammalian prion protein vesicles. *Cell*, 144, pp.551–565.

Fransson, S., Ruusala, A. and Aspenstrom, P., 2006. The atypical Rho GTPases Miro-1 and Miro-2 have essential roles in mitochondrial trafficking. *Biochemical and Biophysical Research Communications*, 344, pp.500–510.

Frederick, R. L. and Shaw, J. M., 2007. Moving mitochondria: Establishing distribution of an essential organelle. *Traffic*, 8, pp.1668–1675.

Frederick, R. L., et al., 2004. Yeast Miro GTPase, Gem1p, regulates mitochondrial morphology via a novel pathway. *Journal of Cell Biology*, 167, pp.87–98.

Fujita, T., et al., 2007. Axonal guidance protein FEZ1 associates with tubulin and kinesin motor protein to transport mitochondria in neurites of NGF-stimulated PC12 cells. *Biochemical and Biophysical Research Communications*, 361, pp.605–610.

Gautier, C. A., Kitada, T. and Shen, J., 2008. Loss of PINK1 causes mitochondrial functional defects and increased sensitivity to oxidative stress. *Proc Natl Acad Sci U S A*, 105, pp.11364–11369.

Geisler, S., et al., 2010. PINK1/Parkin-mediated mitophagy is dependent on VDAC1 and p62/SQSTM1. *Nature Cell Biology*, 12, pp.119–131.

Glater, E. E., et al., 2006. Axonal transport of mitochondria requires milton to recruit kinesin heavy chain and is light chain independent. *Journal of Cell Biology*, 173, pp.545–557.

Górska-Andrzejak, J., et al., 2003. Mitochondria are redistributed in Drosophila photoreceptors lacking milton, a kinesin-associated protein. *J Comp Neurol*, 463, pp.372–388.

Grishin, A., et al., 2006. Identification of gamma-aminobutyric acid receptor-interacting factor 1 (TRAK2) as a trafficking factor for the K+ channel Kir2.1. *J Biol Chem*, 281, pp.30104–30111.

Guo, X., et al., 2005. The GTPase dMiro is required for axonal transport of mitochondria to Drosophila synapses. *Neuron*, 47, pp.379–393.

Haghnia, M., 2007. Dynactin is required for coordinated bidirectional motility, but not for dynein membrane attachment. *Mol Biol Cell*, 18, pp.2081–2089.

Hirokawa, N., et al., 1990. Brain dynein (MAP1C) localizes on both anterogradely and retrogradely transported membranous organelles *in vivo*. *J Cell Biol*, 111, pp.1027–1037.

Horiuchi, D., et al., 2005. APLIP1, a kinesin binding JIP-1/JNK scaffold protein, influences the axonal transport of both vesicles and mitochondria in Drosophila. *Curr Biol*, 15, pp.2137–2141.

Hubley, M. J., Locke, B. R. and Moerland, T. S., 1996. The effects of temperature, pH, and magnesium on the diffusion coefficient of ATP in solutions of physiological ionic strength. *Biochim Biophys Acta*, 1291, pp.115–121.

Hurd, D. D. and Saxton, W. M., 1996. Kinesin mutations cause motor neuron disease phenotypes by disrupting fast axonal transport in Drosophila. *Genetics*, 144, pp.1075–1085.

Itoh, K., et al., 2013. Mitochondrial dynamics in neurodegeneration. *Trends in Cell Biology* 23, pp.64–71.

Kanai, Y., et al., 2000. KIF5C, a novel neuronal kinesin enriched in motor neurons. *J Neurosci*, 20, pp.6374–6384.

Kang, J.-S., et al., 2008. Docking of axonal mitochondria by syntaphilin controls their mobility and affects short-term facilitation. *Cell*, 132, pp.137–148.

Kirk, E., Chin, L.-S. and Li. L., 2006. GRIF1 binds Hrs and is a new regulator of endosomal trafficking. *Journal of Cell Science*, 119, pp.4689–4701.

Klosowiak, J. L., et al., 2013. Structural coupling of the EF hand and C-terminal GTPase domains in the mitochondrial protein Miro. *EMBO Rep*, 14, pp.968–974.

Koutsopoulos, O. S., et al., 2010. Human Miltons associate with mitochondria and induce microtubule-dependent remodeling of mitochondrial networks. *Biochim Biophys Acta*, 1803, pp.564–574.

Lee, C. W. and Peng, H. B., 2008. The function of mitochondria in presynaptic development at the neuromuscular junction. *Molecular Biology of the Cell*, 19, pp.150–158.

Lee, S., Sato, Y. and Nixon, R. A., 2011. Lysosomal proteolysis inhibition selectively disrupts axonal transport of degradative organelles and causes an Alzheimer's-like axonal dystrophy. *Journal of Neuroscience*, 31, pp.7817–7830.

Levy, M., et al., 2003. Mitochondrial regulation of synaptic plasticity in the hippocampus. *J Biol Chem*, 278, pp.17727–17734.

Ligon, L. A., et al., 2004. A direct interaction between cytoplasmic dynein and kinesin I may coordinate motor activity. *J Biol Chem*, 279, pp.19201–19208.

Liu, S., et al., 2012. Parkinson's disease-associated kinase PINK1 regulates Miro protein level and axonal transport of mitochondria. *PLoS Genet*, 8, p.e1002537.

Ma, H., et al., 2009. KIF5B motor adaptor syntabulin maintains synaptic transmission in sympathetic neurons. *Journal of Neuroscience*, 29, pp.13019–13029.

MacAskill, A. F. and Kittler, J. T., 2010. Control of mitochondrial transport and localization in neurons. *Trends in Cell Biology*, 20, pp.102–112.

MacAskill, A. F., et al., 2009a. Miro1 is a calcium sensor for glutamate receptor-dependent localization of mitochondria at synapses. *Neuron*, 61, pp.541–555.

MacAskill, A. F., et al., 2009b. GTPase dependent recruitment of Grif-1 by Miro1 regulates mitochondrial trafficking in hippocampal neurons. *Mol Cell Neurosci*, 40, pp.301–312.

Maday, S., Wallace, K. E. and Holzbaur, E. L. F., 2012. Autophagosomes initiate distally and mature during transport toward the cell soma in primary neurons. *Journal of Cell Biology*, 196, pp.407–417.

Mallik, R., et al., 2005. Building complexity: An in vitro study of cytoplasmic dynein with in vivo implications. *Curr Biol*, 15, pp.2075–2085.

Marder, E. and Goaillard, J.-M., 2006. Variability, compensation and homeostasis in neuron and network function. *Nat Rev Neurosci*, 7, pp.563–574.

Martin, M., et al., 1999. Cytoplasmic dynein, the dynactin complex, and kinesin are interdependent and essential for fast axonal transport. *Molecular Biology of the Cell*, 10, pp.3717–3728.

Medler, K. and Gleason, E. L., 2002. Mitochondrial Ca(2+) buffering regulates synaptic transmission between retinal amacrine cells. *Journal of Neurophysiology*, 87, pp.1426–1439.

Miller, K. E. and Sheetz, M. P., 2004. Axonal mitochondrial transport and potential are correlated. *Journal of Cell Science*, 117, pp.2791–2804.

Mironov, S. L., 2007. ADP regulates movements of mitochondria in neurons. *Biophys J*, 92, pp.2944–2952.

Misgeld, T., et al., 2007. Imaging axonal transport of mitochondria in vivo. *Nat Meth*, 4, pp.559–561.

Misko, A., et al., 2010. Mitofusin 2 is necessary for transport of axonal mitochondria and interacts with the Miro/Milton complex. *Journal of Neuroscience*, 30, pp.4232–4240.

Morfini, G. A., et al., 2009. Axonal transport defects in neurodegenerative diseases. *Journal of Neuroscience*, 29, pp.12776–12786.

Morris, R. L. and Hollenbeck, P. J., 1993. The regulation of bidirectional mitochondrial transport is coordinated with axonal outgrowth. *Journal of Cell Science*, 104, pp.917–927.

Morris, R. L. and Hollenbeck, P. J., 1995. Axonal transport of mitochondria along microtubules and F-actin in living vertebrate neurons. *Journal of Cell Biology*, 131, pp.1315–1326.

Murthy, V. N., Sejnowski, T. J. and Stevens, C. F., 1997. Heterogeneous release properties of visualized individual hippocampal synapses. *Neuron*, 18, pp.599–612.

Narendra, D., et al., 2008. Parkin is recruited selectively to impaired mitochondria and promotes their autophagy. *J Cell Biol*, 183, pp.795–803.

Overly, C. C. and Hollenbeck, P. J., 1996. Dynamic organization of endocytic pathways in axons of cultured sympathetic neurons. *J Neurosci*, 16, pp.6056–6064.

Pilling, A. D., et al., 2006. Kinesin-1 and Dynein are the primary motors for fast transport of mitochondria in Drosophila motor axons. *Molecular Biology of the Cell*, 17, pp.2057–2068.

Poole, A. C., et al., 2010. The mitochondrial fusion-promoting factor mitofusin is a substrate of the PINK1/parkin pathway. *PLoS ONE*, 5, p.e10054.

Rangaraju, V., Calloway, N. and Ryan, T. A., 2014. Activity-driven local ATP synthesis is required for synaptic function. *Cell*, 156, pp.825–835.

Ribrault, C., Sekimoto, K. and Triller, A., 2011. From the stochasticity of molecular processes to the variability of synaptic transmission. *Nat Rev Neurosci*, 12, pp.375–387.

Rintoul, G. L., et al., 2003. Glutamate decreases mitochondrial size and movement in primary forebrain neurons. *Journal of Neuroscience*, 23, pp.7881–7888.

Russo, G. J., et al., 2009. Drosophila Miro is required for both anterograde and retrograde axonal mitochondrial transport. *Journal of Neuroscience*, 29, pp.5443–5455.

Ruthel, G. and Hollenbeck, P. J., 2003. Response of mitochondrial traffic to axon determination and differential branch growth. *Journal of Neuroscience*, 23, pp.8618–8624.

Saotome, M., et al., 2008. Bidirectional Ca2+-dependent control of mitochondrial dynamics by the Miro GTPase. *Proc Natl Acad Sci U S A*, 105, pp.20728–20733.

Sarraf, S. A., et al., 2013. Landscape of the PARKIN-dependent ubiquitylome in response to mitochondrial depolarization. *Nature*, 496, pp.372–376.

Saxton, W. M. and Hollenbeck, P. J., 2012. The axonal transport of mitochondria. *Journal of Cell Science*, 125, pp.2095–2104.

Schon, E. A. and Przedborski, S., 2011. Mitochondria: the next (neurode)generation. *Neuron*, 70, pp.1033–1053.

Schwarz, T. L., 2013. Mitochondrial trafficking in neurons. *Cold Spring Harbor Perspectives in Biology* 5, doi: 10.1101/cshperspect.a011304.

Sheng, Z.-H., 2014. Mitochondrial trafficking and anchoring in neurons: New insight and implications. *J Cell Biol*, 7, pp.1087–1098.

Sheng, Z.-H. and Cai, Q., 2012. Mitochondrial transport in neurons: Impact on synaptic homeostasis and neurodegeneration. *Nat Rev Neurosci*, 13, pp.77–93.

Smith, M. J., et al., 2006. Mapping the GRIF-1 binding domain of the kinesin, KIF5C, substantiates a role for GRIF-1 as an adaptor protein in the anterograde trafficking of cargoes. *J Biol Chem*, 281, pp.27216–27228.

Spillane, M., et al., 2013. Mitochondria coordinate sites of zxon branching through localized intra-axonal protein synthesis. *Cell Reports*, 5, pp.1564–1575.

Stein, R. B., Gossen, E. R. and Jones, K. E., 2005. Neuronal variability: Noise or part of the signal? *Nat Rev Neurosci*, 6, pp.389–397.

Stowers, R. S, et al., 2002. Axonal transport of mitochondria to synapses depends on milton, a novel Drosophila protein. *Neuron*, 36, pp.1063–1077.

Sun, T., et al., 2013. Motile axonal mitochondria contribute to the variability of presynaptic strength. *Cell Reports*, 4, pp.413–419.

Szabadkai, G., et al., 2006. Mitochondrial dynamics and Ca2+ signaling. *Biochim Biophys Acta*, 1763, pp.442–449.

Tanaka, Y., et al., 1998. Targeted disruption of mouse conventional kinesin heavy chain, kif5B, results in abnormal perinuclear clustering of mitochondria. *Cell*, 93, pp.1147–1158.

Tang, Y. and Zucker, R. S., 1997. Mitochondrial involvement in post-tetanic potentiation of synaptic transmission. *Neuron*, 18, pp.483–491.

Tao, K., Matsuki, N. and Koyama, R., 2013. AMP-activated protein kinase mediates activity-dependent axon branching by recruiting mitochondria to axon. *Dev Neurobiol*, 74, pp.557–573.

van Spronsen, M., et al., 2013. TRAK/Milton motor-adaptor proteins steer mitochondrial trafficking to axons and dendrites. *Neuron*, 77, pp.485–502.

Verburg, J. and Hollenbeck, P. J, 2008. Mitochondrial membrane potential in axons increases with local nerve growth factor or semaphorin signaling. *Journal of Neuroscience*, 28, pp.8306–8315.

Verstreken, P., et al., 2005. Synaptic mitochondria are critical for mobilization of reserve pool vesicles at Drosophila neuromuscular junctions. *Neuron*, 47, pp.365–378.

Wang, X. and Schwarz, T. L., 2009. The mechanism of Ca2+ -dependent regulation of kinesin-mediated mitochondrial motility. *Cell*, 136, pp.163–174.

Wang, X., et al., 2011. PINK1 and parkin target miro for phosphorylation and degradation to arrest mitochondrial motility. *Cell*, 147, pp.893–906.

Webber, E., Li, L. and Chin, L.-S., 2008. Hypertonia-associated protein Trak1 is a novel regulator of endosome-to-lysosome trafficking. *J Mol Biol*, 382, pp.638–651.

Welte, M. A., 2004. Bidirectional transport along microtubules. *Curr Biol*, 14, pp.R535–R537.

Weihofen, A., et al., 2009. Pink1 forms a multiprotein complex with Miro and Milton, linking Pink1 function to mitochondrial trafficking. *Biochemistry*, 48, pp.2045–2052.

Werth, J. L. and Thayer, S. A., 1994. Mitochondria buffer physiological calcium loads in cultured rat dorsal root ganglion neurons. *J Neurosci*, 14, pp.348–356.

Westermann, B., 2010. Mitochondrial fusion and fission in cell life and death. *Nat Rev Mol Cell Biol*, 11, pp.872–884.

Yen, H.-C. and Elledge, S. J., 2008. Identification of SCF ubiquitin ligase substrates by global protein stability profiling. *Science*, 322, pp.923–929.

Yi, M., Weaver, D. and Hajnóczky, G., 2004. Control of mitochondrial motility and distribution by the calcium signal: A homeostatic circuit. *Journal of Cell Biology*, 167, pp.661–672.

Yoshii, S. R., et al., 2011. Parkin mediates proteasome-dependent protein degradation and rupture of the outer mitochondrial membrane. *Journal of Biological Chemistry*, 286, pp.19630–19640.

Ziviani, E., Tao, R. N. and Whitworth, A. J., 2010. Drosophila parkin requires PINK1 for mitochondrial translocation and ubiquitinates mitofusin. *Proc Natl Acad Sci U S A*, 107, pp.5018–5023.

8

MITOCHONDRIA IN CONTROL OF HYPOTHALAMIC METABOLIC CIRCUITS

CAROLE M. NASRALLAH AND TAMAS L. HORVATH
Interdepartmental Neuroscience Program, Program in Integrative Cell Signaling and Neurobiology of Metabolism, Section of Comparative Medicine, Yale University School of Medicine, New Haven, Connecticut, USA

8.1 INTRODUCTION

The fundamental aspect of any living organism is its ability to convert organic molecules from the environment via chemical reactions using oxygen for energy production, cellular structure development, differentiation, and growth. Given the fundamental role of mitochondria in metabolic pathways relating to oxygen, regulation of mitochondria must be critical for adequate energy balance of the organism. In complex organisms, such as mammals, it is also fundamental that the various tissues are coordinated in such a manner that the energetic needs of the whole body are most effectively met. The central nervous system plays a crucial organizational role in which the hypothalamus is a key brain center. A large number of critical players in these central regulatory elements of metabolism reside within the hypothalamus, and recent studies have revealed that mitochondrial dynamics within specific circuits of the hypothalamus are important for the proper adaptive responses of the whole body to the changing metabolic milieu. In this chapter, we build our view on mitochondrial contributions to central metabolic principles from the perspective of this

The Functions, Disease-Related Dysfunctions, and Therapeutic Targeting of Neuronal Mitochondria, First Edition. Edited by Valentin K. Gribkoff, Elizabeth A. Jonas and J. Marie Hardwick.

hypothalamic point of view. Here we aim to provide an overview of the mechanisms by which mitochondria are involved in the central regulation of energy balance and how mitochondrial dysfunction can lead to metabolic disorders.

8.2 YIN-YANG RELATIONSHIP BETWEEN COMPONENTS OF HYPOTHALAMIC FEEDING AND SATIETY CIRCUITS

Changes in environment are unpredictable, forcing the body to constantly adapt. Although many circuits and signals, including gut signaling, are involved in the regulation of energy balance, the hypothalamus is a primary site that processes and integrates information on the body's nutritional status. The hypothalamic melanocortin model of energy regulation has been the most influential model explaining neuroendocrine control of energy homeostasis (Cone, 2005). In this model, the hypothalamic arcuate nucleus is considered a critical region for various reasons. First, neurons within the arcuate nucleus are anatomically placed in close proximity to fenestrated capillaries at the base of the hypothalamus, giving them better access to a variety of humoral signals that have restricted entry to other portions of the brain (Broadwell et al., 1983; Norsted, Gomuc, and Meister, 2008). These cells respond rapidly to fluctuations in nutritional signals such as glucose, insulin, or fatty acids, a feature that makes arcuate nucleus cells resemble first-order sensory neurons. However, the presence of receptors for nutritional signals in other hypothalamic nuclei support the notion that cells within the arcuate nucleus are not the only ones having a first hand on metabolic signals.

Within the arcuate nucleus two sets of cells have been identified in terms of their effects on food intake and energy expenditure (Fig. 8.1). Cells whose increased activation leads to decreases in food intake and increased energy expenditure produce the opioid precursor pro-opiomelanocortin (POMC) and the cocaine-amphetamine regulated transcript (CART) (Elias et al., 1998). The POMC precursor peptide is cleaved into α-melanocyte stimulating hormone (α-MSH), β-MSH, γ-MSH, adrenocorticotrophic hormone (ACTH), and β-endorphin (Castro and Morrison, 1997; Mountjoy, 2010). Of these, α- and β-MSH reduce body weight and food intake, and increase energy expenditure in animals including humans (Biebermann et al., 2006; McMinn et al., 2000; Poggioli, Vergoni, and Bertolini, 1986). Both α- and β-MSH exert their effects by acting upon melanocortin receptor subtypes 3 and 4, MC3 and MC4, respectively (Roselli-Rehfuss et al., 1993). These are located throughout the brain but are particularly abundant in most of the hypothalamic nuclei including the ARC, paraventricular nucleus (PVN), lateral hypothalamus (LH), dorsomedial hypothalamus (DMH), and various anterior hypothalamic regions (Mountjoy et al., 1994). In contrast, cells whose increased activity leads to orexigenic responses and decreased energy expenditure secrete neuropeptide Y (NPY) and the agouti related protein (AgRP). A member of the pancreatic polypeptide family, NPY was found to be a potent stimulator of food intake and to reduce energy expenditure (Clark et al., 1984; Stanley and Leibowitz, 1984). Also in support for a role of NPY in energy regulation, genetically obese

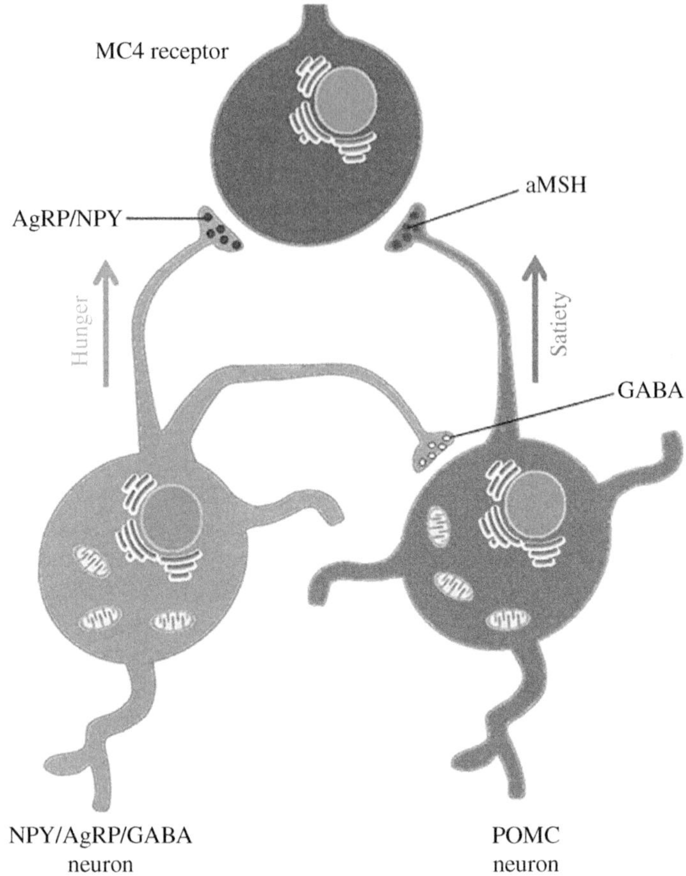

FIGURE 8.1 Melanocortin system in the arcuate nucleus of the hypothalamus: control of energy homeostasis. The primary neurons of the melanocortin system are the orexigenic NPY/AgRP/GABA neurons and the anorexigenic pro-opiomelanocortin (POMC) neurons. Both sets of neurons send projections to the paraventricular nucleus (PVN). Additionally, the NPY/AgRP neurons send inhibitory projections to the neighboring POMC cells.

animals (i.e., ob/ob and db/db mice, zucker rats, etc.) as well animals in negative energetic state (i.e., fasted animals, lactating animals) showed increased levels of ARC NPY synthesis and protein content (Chua et al., 1991; Dryden et al., 1995; Wilding et al., 1993). Although NPY is secreted by neurons throughout the central nervous system, the arcuate nucleus contains the largest concentration of NPY secreting cells. However, the key feature of NPY cells in relation to the melanocortin system is that they also secrete AgRP (Hahn et al., 1998). This peptide, like NPY, produces potent long lasting feeding responses, and these are generated because AgRP acts as a natural antagonist for the MC3 and MC4 melanocortin receptor, thereby reducing the anorectic effect of α-MSH (Ollmann et al., 1997). Furthermore, ectopic expression of Agouti, an AgRP like peptide, results in the obese and hyperinsulinemic phenotype of the *yA* agouti mouse (Yen et al., 1994). Regarding acute feeding regulation, it is now also apparent that the inhibitory amino acid neurotransmitter, gamma amino butyric acid (GABA) plays a crucial role. First, GABA was localized to arcuate nucleus NPY/AgRP neurons (Horvath et al., 1997). This observation helped resolve the paradox when genetic knockdown of either NPY (Palmiter et al., 1998) or AgRP or NPY and AgRP (Qian et al., 2002) together failed to dramatically impair feeding of mice. When adult AgRP neurons were ablated causing aphagia (Gropp et al., 2005; Luquet et al., 2005) and ultimate death of mice, it was GABA reinstatement to the appropriate projection areas of NPY/AgRP neurons that rescued the feeding and survival of mice (Wu, Boyle, and Palmiter, 2009). Also, selective removal of vesicular GABA transporter impacted feeding and energy metabolism of mice (Tong et al., 2008). Thus, many of the recently uncovered rapid action of NPY/AgRP neurons evoked with remote activation (Aponte, Atasoy, and Sternson, 2011; Krashes et al., 2011) is likely mediated by GABA release of these orexigenic neurons.

Both NPY/AgRP and POMC/CART cells in the arcuate nucleus have leptin and ghrelin receptors (Cheung, Clifton, and Steiner et al., 1997; Willesen, Kristensen, and Romer, 1999). Peripheral administration of either leptin or ghrelin modulates gene expression in the arcuate nucleus, with leptin increasing the activity of POMC/CART cells (Cowley et al., 2001; Elias et al., 1999; Schwartz et al., 1997), and ghrelin increasing the activity of NPY/AgRP cells (Cowley et al., 2003; Dickson and Luckman, 1997; Nakazato et al., 2001). Although leptin also targets and inhibits NPY/AgRP cells (Elias et al., 1999; Spanswick et al., 1997), the degree to which ghrelin modulates the activity of POMC cells has not been well determined. Nevertheless, what brings about the opposite outcome in AgRP and POMC neurons regarding electric activity of these cells is currently unknown. Recent studies on the bioenergetics of these cells with particular emphasis on mitochondrial functions may bring new insights on these processes.

8.3 MITOCHONDRIA AND THEIR DYNAMICS

The name *mitochondrion* was introduced in 1898 by Carl Benda, and studies covering the past 100 years have demonstrated the fundamental role of these organelles in providing energy to the cells. Mitochondria are the primary energy source for cellular function by converting nutrient flux into adenosine triphosphate (ATP). Products of glucose and fatty acids are actively transported across the inner mitochondrial membrane and into the matrix where they enter the tricarboxylic acid (TCA) cycle and are oxidized to carbon dioxide. In the process, there is production of GTP and reduced cofactors, nicotinamide adenine dinucleotide (NADH) and flavin adenine dinucleotide ($FADH_2$). These cofactors are then oxidized by the electron transport chain. This process is accompanied by a transfer of protons (H^+ ions) across the inner membrane that creates an electrochemical proton gradient, providing energy to power ATP synthase and the generation of ATP. However protons can reenter the mitochondrial matrix through facilitated diffusion, a process called *mitochondrial uncoupling*. This proton leak is in part mediated by inner membrane proteins encoded by the nuclear DNA, the uncoupling proteins (UCPs). There is also a leak of electrons from the respiratory chain leading to the generation of reactive oxygen species (ROS).

However, the role of mitochondria is known to extend far beyond the sole role of supplying energy for cellular function. Over the past few decades, mitochondria have been shown to be involved in signaling through reactive oxygen species, regulation of the membrane potential, apoptosis-programmed cell death, calcium signaling (including calcium-evoked apoptosis), and regulation of cellular metabolism. In 1914, Lewis and Lewis were already describing mitochondrial dynamics in living cells "The mitochondria are almost never at rest, but are continually changing their position and also their shape" (Lewis and Lewis, 1914, p.332). Mitochondria continually adapt their shapes through fusion, fission, and motility in response to mitochondrial life cycle and changes in energy demand and supply balance. Mitochondrial dynamics are essential for the proper function of the cell including viability and apoptosis but also bioenergetic adaptation. Fission is thought to occur when mitochondria are damaged or accumulate deleterious components and in response to high levels of cell stress. Fission is thought to segregate damaged mitochondria and can lead to autophagy (Gomes and Scorrano, 2008; Twig et al., 2008). However, fission may also correlate with the changing availability of fuels. Specifically, fission appears to be triggered in hypothalamic cells when fat utilization is increasing while carbohydrates are diminishing (Andrews et al., 2008; Coppola et al., 2007; Dietrich, Liu, and Horvath, 2013; Ozcan, 2013). Fission is mediated by various proteins, including Drp1, a member of the dynamin superfamily (Hoppins, Lackner, and Nunnari, 2007; Okamoto and Shaw, 2005).

Fusion of mitochondria can compensate for one another's defects by sharing RNA or proteins (Youle and van der Bliek, 2012). Fusion also enhances communication with the endoplasmic reticulum. Fusion between inner membranes of the mitochondria can be mediated by Opal 1, while fusion between outer membranes can be controlled by other members of the dynamin superfamily Mfn1 and Mfn2 (Hoppins et al., 2007; Okamoto and Shaw, 2005). From our work on the hypothalamus, it seems reasonable to suggest that fusion of mitochondria correlates with shifts in fuel availability toward carbohydrates (Ozcan, 2013; Schneeberger et al., 2013). Fission and fusion proteins are encoded in the nucleus. Chen, Chomyn, and Chan (2005) showed in fibroblasts that complete inhibition of mitochondrial fusion through repression of Mfn1 and Mfn2 expression induces severe cellular defects, including poor growth, change in mitochondrial membrane potential, and decreased respiration. Parone et al. (2008) showed in HeLa cells that down-regulation of Drp1 expression inhibits fission and leads a decrease in mitochondrial respiration. In neuronal cells like the Purkinje cells in the cerebellum and the dopamine cells in the midbrain, the lack of Mfn1 or Mfn2 led to the degeneration of the neurons or axon fibers (Chen, McCaffery, and Chan, 2007; Lee et al., 2012; Pham et al., 2012).

Mitochondria are also in constant communication with other cellular organelles such as the endoplasmic reticulum (ER) (Boncompagni et al., 2009; Csordas et al., 2006). Recent data demonstrate that Mfn2 is localized in mitochondria-ER sites and its key role in anchoring mitochondria to the ER in mouse embryonic fibroblasts (de Brito and Scorrano, 2008), thus suggesting a link between mitochondria-ER contact and mitochondrial fusion and function. A major role of mitochondria-ER interaction is thought to be the regulation of mitochondrial lipid biosynthesis and calcium signaling (Hayashi et al., 2009). The close contact with the ER facilitates mitochondrial uptake of free calcium released from the ER (Rizzuto et al., 2009). This calcium transfer provides a high local concentration for mitochondrial functions that are calcium dependent, including the activation of the TCA cycle. Calcium influx also plays a role in the regulation of mitochondrial morphology and dynamics. And lastly calcium plays a role in the regulation of apoptosis (Rowland and Voeltz, 2012).

Mitochondrial dysfunctions have been associated with the development of metabolic alterations and implicated in various human disorders such as diabetes, obesity, neurodegenerative diseases, and aging (Albers and Beal, 2000; Bournat and Brown, 2010; Morais and De Strooper, 2010; Trifunovic and Larsson, 2008). Whether these principles of mitochondrial function and dynamics are relevant for adaptive responses to the central melanocortin circuitry was not well established until recently.

8.4 METABOLIC PRINCIPLES OF HUNGER AND SATIETY PROMOTION: MITOCHONDRIA IN SUPPORT OF FAT VERSUS GLUCOSE UTILIZATION

For more than 20 years, we have been studying the arcuate nucleus NPY/AgRP and POMC neurons (Horvath et al., 1992a; Horvath, Naftolin and Leranth, 1992b). One intriguing aspect of their activity is that when AgRP neurons are active, POMC cells are quiescent, and vice versa. Of course it makes perfect sense from the perspective of their functional relevance (hunger versus satiety), but how it is regulated at the cellular level had been elusive. When glucose levels decrease and circulating ghrelin increases before meals (or after prolonged absence of food), AgRP neurons are activated and POMC neurons are silent promoting hunger and lipid metabolism in the periphery. In the past 10 years the idea emerged that during negative energy balance, pathways promoting lipid metabolism are also activated in the brain, specifically in the hypothalamus, and indeed, it is lipid utilization and metabolism that appear to drive AgRP neurons during negative energy balance (Andrews et al., 2008). This process is associated with AMP-activated protein kinase (AMPK) activation, decreased malonyl CoA levels, and enhanced activity of carnitine palmitoyltransferase I (Andrews et al., 2008; Lopez et al., 2008). It is accepted that in the periphery AMPK, controlled by AMP/ATP ration, regulates mitochondrial biogenesis during negative energy balance (Reznick and Shulman, 2006), while its activity is also elevated in the hypothalamus under similar metabolic conditions (McCrimmon et al., 2004; Minokoshi et al., 2004). It is also accepted that AMPK induces the uptake of long chain fatty acyl CoAs by the mitochondria through CPT1 (Andrews et al., 2008) to generate ATP via fatty acid beta oxidation (Hardie and Carling, 1997). Hypothalamic CPT1 was also identified as a crucial regulator of feeding (Obici et al., 2003) indicating that an intracellular lipid-metabolizing pathway characterized in detail in peripheral tissues is present in the hypothalamus. Specifically we identified this pathway as an important regulator of AgRP neuronal activity during fasting and in response to ghrelin (Andrews et al., 2008). We showed that hypothalamic AMPK action to promote feeding hinges on mitochondrial oxygen consumption mediated by the mitochondrial uncoupling protein, UCP2 (Andrews et al., 2008). Around the same time, AMPK activation has also been shown to increase the expression of mitochondrial uncoupling protein UCP2 in other tissues (Xie et al., 2008). These observations raised the notion that UCP2 enables long chain fatty acid oxidation in the mitochondria (Andrews et al., 2008). UCPs are transmembrane proteins that mediate the reentry of protons into the mitochondrial matrix, or proton leak. In 1978 Nicholls and his group discovered the first uncoupling protein UCP1 (Heaton et al., 1978). UCP1 is highly expressed in brown adipose tissue and regulates thermogenesis in mammals (Cannon and Nedergaard, 2004; Enerback et al., 1997). Its homologue, with partial and regulated uncoupling activity, UCP2 is expressed in a wide variety of tissues including the central nervous system (Alan et al., 2009; Andrews, Diano, and Horvath, 2005a; Krauss, Zhang, and Lowell, 2005) where they control ROS generation (Affourtit et al., 2007; Andrews et al., 2005b) associated with fatty acid beta oxidation in the mitochondria (Andrews et al., 2008). The absence of UCP2 is associated with decreased fatty acid utilization and an increase in glucose oxidation (Parton et al., 2007). UCP2 increases mitochondria number in AgRP neurons during fasting and in response to ghrelin (Andrews et al., 2008; Coppola et al., 2007). Whether these processes resembling mitochondrial fission are critical or simply correlative to these changes was not addressed in these studies. However, information on metabolic principles of satiety in the same regions suggested that mitochondrial switches are important contributors to changing neuronal functions.

When glucose and leptin levels rise in the circulation following food intake, POMC neurons are activated and promote cessation of feeding (Fig. 8.2). Glucose is critical for POMC neuronal activation (Ibrahim et al., 2003). In turn, glucose uptake and metabolism (glucose sensing) in POMC neurons have an important role in regulating systemic glucose homeostasis (Parton et al., 2007). It has been argued that glucose-dependent POMC activity is promoted, in part, by the activation of ATP sensitive potassium (KATP) channels, which leads to the depolarization of the cells (Ashford, Boden, and Treherne, 1990; Ibrahim et al., 2003; Miki et al., 2001). However, in conditions of elevated glucose and leptin levels, increasing activity of POMC neurons was also associated with an increase in the by-product of glucose metabolism, ROS levels in these cells (Andrews et al., 2008; Horvath, Andrews, and Diano, 2009). Selective sequestration of ROS in the hypothalamus promoted hunger (Andrews et al., 2008), and increasing hypothalamic ROS levels POMC neuronal activation and satiety seem to promote the activation of these cells and satiety (Diano et al., 2011). We also revealed that impairment of POMC neuronal firing on diet-induced obesity can be reversed by interference ROS buffering in the hypothalamus by suppression of PPARγ signaling (Diano et al., 2011). In addition, it is likely that UCP2 inhibits glucose-induced activation of POMC neurons in diet-induced obesity in part by buffering ROS and also by decreasing the yield of ATP (Zhang et al., 2001).

The preceding information shows that fundamentally different bioenergetics support the needs for hunger versus satiety promotion. Because hunger and satiety

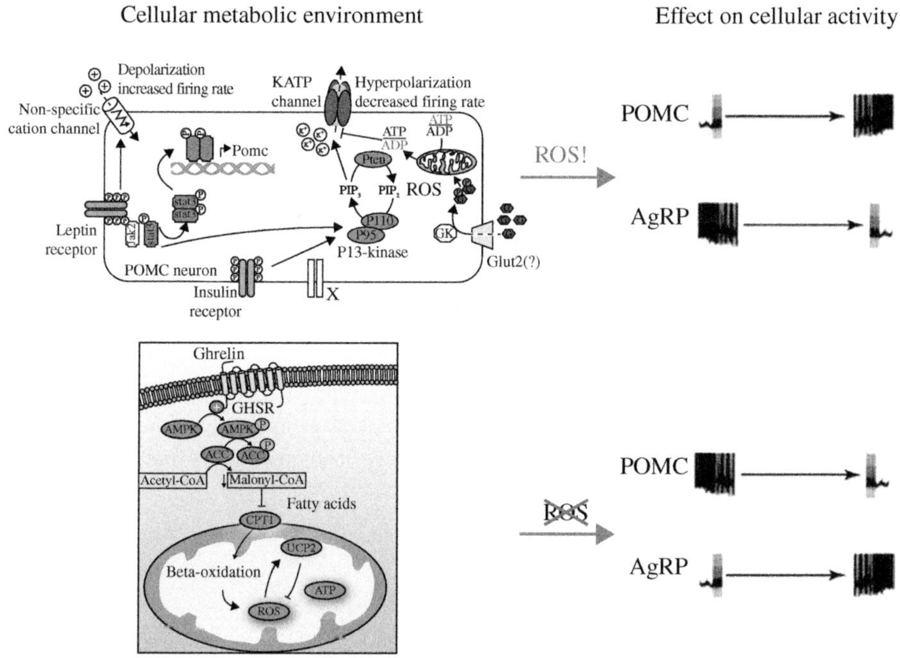

FIGURE 8.2 Regulation of pro-opiomelanocortin (POMC) and NPY/AgRP neuron activation. The firing of POMC neurons during elevated glucose levels (satiety) relies on cellular glucose metabolism and signaling pathways activated by leptin and insulin receptors *(top)*. It appears that a fundamental signaling event that affects action potential generation in these cells is glucose-derived reactive oxygen species (ROS). The same event, however, hyperpolarizes NPY/AgRP neurons and suppresses their firing. During low glucose levels, fatty acid metabolism is increased *(bottom)*. Active phosphorylated AMP-activated protein kinase (AMPK) phosphorylates ACC, leading to a decrease in its enzymatic activity. This latter event decreases levels of malonyl-CoA, releasing CPT1 to uptake fatty acids inside the mitochondria. Fatty acid metabolism, promoted by UCP2 activity, reduces intracellular ROS levels leading to the decreased activity of POMC neurons but elevated firing of NPY/AgRP neurons. (*See insert for color representation of the figure.*)

follow each other within hours, mitochondrial dynamics might be crucial in these processes to make proper switches in fuel utilization to support the proper activity of orexigenic and anorexigenic neurons.

8.5 MITOCHONDRIA DYNAMICS AND CELLULAR ENERGETICS

Recent studies have demonstrated a link between cellular energetics and mitochondrial architecture. Mitochondrial dynamics are altered in response to various combinations of nutrients in beta-cells (Molina et al., 2009). The rates of both fusion and fission vary in response to changes in metabolism. Mitochondrial fusion and fission regulation responds to the state of energy supply to Cells (Gomes, Di Benedetto, and Scorrano, 2011). Aberrant mitochondrial fission causes mitochondrial dysfunction and insulin resistance in skeletal muscle (Jheng et al., 2012). In myotubes, repression of Mfn2 causes the morphological fragmentation of the mitochondrial network and alters mitochondrial metabolism by reducing mitochondrial membrane potential and glucose oxidation (Bach et al., 2003). Mfn2 deficiency is associated with mitochondrial dysfunction and insulin resistance in skeletal muscle and liver (Sebastian et al., 2012). Mfn2 expression has been shown to be reduced in obese Zucker rats, patients who are obese and have type 2 diabetes (Bach et al., 2003; Hernandez-Alvarez et al., 2010), thus implicating fusion in the pathophysiology of these disorders.

8.5.1 Fission and Fusion of Mitochondria in Hypothalamic Feeding Circuits (Fig. 8.3)

Unlike in other neuronal populations, the repression of Mfn1 or Mfn2 in neurons of the arcuate nucleus does not lead to degeneration of neurons or axon fibers. Recently, we assessed the role of Mfn1 and Mfn2 in Agrp neurons using electron microscopy,

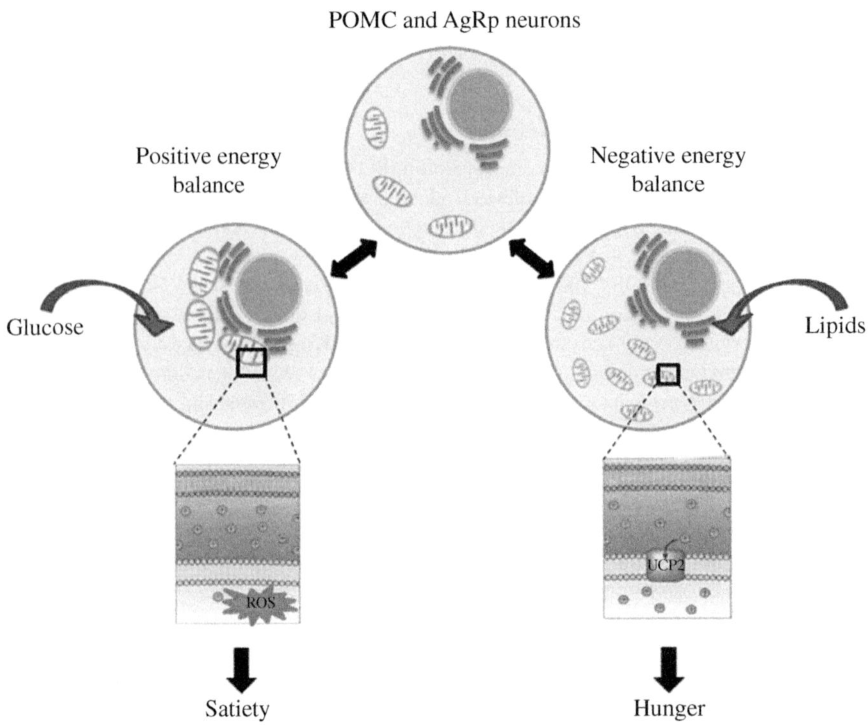

FIGURE 8.3 Feeding states and mitochondrial dynamics in AgRP and pro-opiomelanocortin (POMC) neurons. The activity of POMC neurons promotes satiety associated with glucose utilization, elevated reactive oxygen species (ROS) levels, increase in mitochondrial fusion and contact with the ER. By contrast, the activity of NPY/AgRP neurons during fasting or caloric restriction promotes fat utilization and hunger associated with low ROS levels and increase mitochondrial fission. (*See insert for color representation of the figure.*)

electrophysiology, and metabolic phenotyping (Dietrich et al., 2013). We showed that, in line with our previous studies (Andrews et al., 2008; Coppola et al., 2007), negative energy balance increases mitochondrial fission, whereas feeding and overfeeding promoted mitochondrial fusion (Dietrich et al., 2013). These fusion processes were regulated by the mitofusins, Mfn1 and Mfn2. In response to high fat diet exposure, there was a robust fusion of mitochondria in AgRP neurons leading a paradoxical elevation in electric activity of these orexigenic neurons at a time of elevated glucose and leptin levels (Dietrich et al., 2013). These cells are thought to be turned off when glucose and leptin levels are elevated, such as emerging on high fat diet. Both Mfn1- and Mfn2-regulated fusion mechanisms seemed to be important for these, because knocking either of them out in AgRP neurons prevented mitochondrial fusion, electric activation of these cells on high fat diet, and the emergence of diet induced obesity. Mitochondrial fusion in AgRP neurons was critical to provide sufficient ATP levels for AgRP neuronal firing in the obese, because if intracellular ATP levels were clamped in the impaired AgRP cells of transgenic mice, their activity reverted to that of controls (Dietrich et al., 2013). These observations offer new insights into the cell biological and bioenergetical principles of AgRP neuronal control by various metabolic states. They also raise some intriguing questions. What are the fundamental differences in activation of AgRP neurons during fasting, which is supported by mitochondrial fission (likely enhanced by UCP2 [Andrews et al., 2008; Coppola et al., 2007] and Drp1 [Carneiro et al., 2012]), and during overfeeding, which relies on fusion of AgRP mitochondria. Is it possible that from a neurobiological perspective, on high fat diet, AgRP neurons transform to a POMC-like phenotype? From the perspective of mitochondrial dynamics, this may not be the case as POMC neurons manifest differential responses. Most notably, the same mitofusins have different impacts on these satiety-promoting neurons. First, Schneeberger et al. (2013) found no observable effect of Mfn1 in POMC neurons and related physiological functions. Most critically, Mfn2 was less involved with fusion of mitochondria in these cells, but more so with mitochondria-endoplasmic reticulum (ER) interactions (Schneeberger et al., 2013). When Mfn2 was selectively ablated from POMC neurons, animals became morbidly obese. We observed that mitochondria-ER interactions were reduced in POMC neurons of obese, high-fat diet-fed mice. These events were accompanied by down-regulation of Mfn2. Knocking down Mfn2 in POMC neurons decreased of the number of mitochondria-ER contacts, reduced α-MSH levels, and resulted in an obese phenotype. The administration of a single dose of α-MSH normalized food intake and reduced body weight in these animals. Importantly, the lack of Mfn2 in POMC neurons lead to ER stress in POMC neurons and ER stress-induced leptin resistance. Inhibition of hypothalamic ER stress with chemical Chaperones was sufficient to reverse the metabolic phenotype of these transgenic mice. The dual role of Mfn2 in mitochondrial fusion and mitochondrial tethering to the ER suggests a key role in coordinating ER and mitochondrial functions to promote proper glucose utilization by these neurons. As noted previously, POMC neurons are driven by glucose uptake and metabolism. Agrp neurons did not have altered mitochondria-ER interactions in the absence of Mfn2 emphasizing cell-specific action for these molecules (Dietrich et al., 2013).

8.6 MITOCHONDRIAL DYSFUNCTION AND METABOLIC DISORDERS

Given the key role of mitochondria in energy balance and fuel utilization, it is not surprising that mitochondrial dysfunction impacts whole-body metabolic homeostasis and leads to metabolic disorders. Type 2 diabetes and obesity are major public health concerns and have reached epidemic proportions world-wide. In the United States, 8% of the population is diabetic and more than one third of adults are obese. Most approaches to treat obesity attempt to suppress appetite and food craving by targeting pathways that promote satiety (Dietrich and Horvath, 2012). Because of their non-selectivity, they have side effects affecting the central nervous system and peripheral tissues, most notably cardiomyocytes (Dietrich and Horvath, 2012). Furthermore, conceptually, as described previously, satiety is promoted by sustained ROS elevation. Thus, successful satiety promotion, by default, could result in oxidative stress-induced damages to central and peripheral tissues (Katz, 2007; Ritchie and Delbridge, 2006).

Obesity is associated with an increased risk of developing insulin resistance and type 2 diabetes (Kahn, Hull, and Utzschneider, 2006). Patients with diabetes suffer from insulin resistance, pancreatic beta cell dysfunction (improper insulin secretion), hyperglycemia, and altered regulation of food intake/energy expenditure pathways. Obesity-associated insulin resistance and diabetes are associated with the inability to switch from lipid utilization to carbohydrate utilization under positive energy balance and an overall reduced lipid oxidation (Kelley et al., 2002) arguing for a crucial role for mitochondrial dynamics and their impairment in these pathologies. Excess fuel availability in conditions of overnutrition and obesity increases ROS levels and alterations in mitochondrial dynamics and insulin resistance emphasizing the importance of mitochondrial redox dynamics in the events (Bonnard et al., 2008; Evans et al., 2003; Samocha-Bonet et al., 2012). In line with these, patients who are obese and with diabetes show reduced expression of peripheral Mfn2 (Bach et al., 2003; Hernandez-Alvarez et al., 2010; Sebastian et al., 2012). These studies point to mitochondrial dynamics as key in the pathogenesis of obesity and diabetes. Medications used to treat diabetes focus on decreasing blood glucose levels. Along with calorie restriction and exercise, the most commonly used drug to treat diabetes type 2 is Metformin. It inhibits glucogenesis in the liver through different pathways, one of which is AMPK dependent (Kim et al., 2008; Zhou et al., 2001). Metformin increases the AMP/ATP ratio, leading to an increase in the activation of AMPK (Zhang, He, and Balschi, 2007). Metformin has also been shown to decrease ROS levels through the inhibition of complex 1 of the respiratory chain (Owen, Doran, and Halestrap, 2000). Protection against oxidative stress occurs through an AMPK-dependent stimulation of UCP2 expression, again emphasizing the involvement of mitochondrial processes in these mechanisms.

ER stress is present in both obesity and type 2 diabetes. The use of chemicals aiming to decrease ER stress shows beneficial effects in both conditions. The metabolic overload that occurs during obesity, leading to ER stress in the hypothalamus (Ozcan et al., 2009), correlates with increased peroxisome proliferation in POMC neurons.

This increased peroxisome proliferation during diet-induced obesity is an adaptive response to control the ROS levels in POMC cells (Diano et al., 2011). As described previously in this review, ER stress observed in metabolic disorders is associated with a down-regulation of Mfn2. The lack of Mfn2 inhibits mitochondria-ER tethering and lead to uncontrolled ER stress. Thus, development of approaches targeting mitochondria-ER interactions via fission and fusion mechanism may prove to be a promising direction to control ER stress in metabolic disorders.

8.7 CONCLUSIONS

Mitochondria have been implicated in the development of chronic diseases such as obesity and diabetes but also in cancer and neurodegenerative disorders. It is why targeting mitochondria is an emerging and promising therapeutic approach. Recent studies unmasked important roles for mitochondrial fission and fusion events in central regulation of whole body metabolism and glucose control. Because different fission and fusion events occur in parallel in various cell and tissue types, targeting any specific mitochondrial mechanism for the treatment of chronic disorders in a systemic manner may be challenging. The outcome of a strategy may be different and potentially undesired in one cell population or tissue compared to another, leading to the failure of the therapy at the whole-body level. Thus, to better use the emerging concepts of mitochondrial dynamics for drug development, fundamental aspects of these intracellular processes need further explorations.

REFERENCES

Affourtit, C., et al., 2007. Novel uncoupling proteins. *Novartis Found Symp* 287, pp.70–80, 80–91.

Alan, L., et al., 2009. Absolute levels of transcripts for mitochondrial uncoupling proteins UCP2, UCP3, UCP4, and UCP5 show different patterns in rat and mice tissues. *J Bioenerg Biomembr*, 41, pp.71–78.

Albers, D. S. and Beal, M. F., 2000. Mitochondrial dysfunction and oxidative stress in aging and neurodegenerative disease. *J Neural Transm Suppl*, 59, pp.133–154.

Andrews, Z. B., Diano, S. and Horvath, T. L., 2005a. Mitochondrial uncoupling proteins in the CNS: in support of function and survival. *Nat Rev Neurosci*, 6, 829–840.

Andrews, Z. B., et al., 2005b. Uncoupling protein-2 is critical for nigral dopamine cell survival in a mouse model of Parkinson's disease. *J Neurosci*, 25, pp.184–191.

Andrews, Z. B., et al., 2008. UCP2 mediates ghrelin's action on NPY/AgRP neurons by lowering free radicals. *Nature*, 454, pp.846–851.

Aponte, Y., Atasoy, D. and Sternson, S. M., 2011. AGRP neurons are sufficient to orchestrate feeding behavior rapidly and without training. *Nat Neurosci*, 14, pp.351–355.

Ashford, M. L., Boden, P. R. and Treherne, J. M., 1990. Glucose-induced excitation of hypothalamic neurones is mediated by ATP-sensitive K+ channels. *Pflugers Arch*, 415, pp.479–483.

Bach, D., et al., 2003. Mitofusin-2 determines mitochondrial network architecture and mitochondrial metabolism. A novel regulatory mechanism altered in obesity. *J Biol Chem*, 278, pp.17190–17197.

Biebermann, H., et al., 2006. A role for beta-melanocyte-stimulating hormone in human body-weight regulation. *Cell Metab*, 3, pp.141–146.

Boncompagni, S., et al., 2009. Mitochondria are linked to calcium stores in striated muscle by developmentally regulated tethering structures. *Mol Biol Cell*, 20, pp.1058–1067.

Bonnard, C., et al., 2008. Mitochondrial dysfunction results from oxidative stress in the skeletal muscle of diet-induced insulin-resistant mice. *J Clin Invest*, 118, pp.789–800.

Bournat, J. C. and Brown, C. W., 2010. Mitochondrial dysfunction in obesity. *Curr Opin Endocrinol Diabetes Obes*, 17, pp.446–452.

Broadwell, R. D., et al., 1983. Brain-blood barrier? Yes and no. *Proc Natl Acad Sci U S A*, 80, pp.7352–7356.

Cannon, B. and Nedergaard, J., 2004. Brown adipose tissue: Function and physiological significance. *Physiol Rev*, 84, pp.277–359.

Carneiro, L., et al., 2012. Importance of mitochondrial dynamin-related protein 1 in hypothalamic glucose sensitivity in rats. *Antioxid Redox Signal*, 17, pp.433–444.

Castro, M. G. and Morrison, E., 1997. Post-translational processing of proopiomelanocortin in the pituitary and in the brain. *Crit Rev Neurobiol*, 11, pp.35–57.

Chen, H., Chomyn, A. and Chan, D. C., 2005. Disruption of fusion results in mitochondrial heterogeneity and dysfunction. *J Biol Chem*, 280, pp.26185–26192.

Chen, H., McCaffery, J. M. and Chan, D. C., 2007. Mitochondrial fusion protects against neurodegeneration in the cerebellum. *Cell*, 130, pp.548–562.

Cheung, C. C., Clifton, D. K. and Steiner, R. A., 1997. Proopiomelanocortin neurons are direct targets for leptin in the hypothalamus. *Endocrinology*, 138, pp.4489–4492.

Chua, S. C., Jr., et al., 1991. Food deprivation and hypothalamic neuropeptide gene expression: Effects of strain background and the diabetes mutation. *Brain Res Mol Brain Res*, 11, pp.291–299.

Clark, J. T., et al., 1984. Neuropeptide Y and human pancreatic polypeptide stimulate feeding behavior in rats. *Endocrinology*, 115, pp.427–429.

Cone, R. D., 2005. Anatomy and regulation of the central melanocortin system. *Nat Neurosci*, 8, pp.571–578.

Coppola, A., et al., 2007. A central thermogenic-like mechanism in feeding regulation: An interplay between arcuate nucleus T3 and UCP2. *Cell Metab*, 5, pp.21–33.

Cowley, M. A., et al., 2001. Leptin activates anorexigenic POMC neurons through a neural network in the arcuate nucleus. *Nature*, 411, pp.480–484.

Cowley, M. A., et al., 2003. The distribution and mechanism of action of ghrelin in the CNS demonstrates a novel hypothalamic circuit regulating energy homeostasis. *Neuron*, 37, pp.649–661.

Csordas, G., et al., 2006. Structural and functional features and significance of the physical linkage between ER and mitochondria. *J Cell Biol*, 174, pp.915–921.

de Brito, O. M. and Scorrano, L., 2008. Mitofusin 2 tethers endoplasmic reticulum to mitochondria. *Nature*, 456, pp.605–610.

Diano, S., et al., 2011. Peroxisome proliferation-associated control of reactive oxygen species sets melanocortin tone and feeding in diet-induced obesity. *Nat Med*, 17, pp.1121–1127.

Dickson, S. L. and Luckman, S. M., 1997. Induction of c-fos messenger ribonucleic acid in neuropeptide Y and growth hormone (GH)-releasing factor neurons in the rat arcuate nucleus following systemic injection of the GH secretagogue, GH-releasing peptide-6. *Endocrinology*, 138, pp.771–777.

Dietrich, M. O. and Horvath, T. L., 2012. Limitations in anti-obesity drug development: The critical role of hunger-promoting neurons. *Nat Rev Drug Discov*, 11, pp.675–691.

Dietrich, M. O., Liu, Z. W. and Horvath, T. L., 2013. Mitochondrial dynamics controlled by mitofusins regulate Agrp neuronal activity and diet-induced obesity. *Cell*, 155, pp.188–199.

Dryden, S., et al., 1995. Increased neuropeptide Y secretion in the hypothalamic paraventricular nucleus of obese (fa/fa) Zucker rats. *Brain Res*, 690, pp.185–188.

Elias, C. F., et al., 1998. Leptin activates hypothalamic CART neurons projecting to the spinal cord. *Neuron*, 21, pp.1375–1385.

Elias, C. F., et al., 1999. Leptin differentially regulates NPY and POMC neurons projecting to the lateral hypothalamic area. *Neuron*, 23, pp.775–786.

Enerback, S., et al., 1997. Mice lacking mitochondrial uncoupling protein are cold-sensitive but not obese. *Nature*, 387, pp.90–94.

Evans, J. L., et al., 2003. Are oxidative stress-activated signaling pathways mediators of insulin resistance and beta-cell dysfunction? *Diabetes*, 52, pp.1–8.

Gomes, L.C. and Scorrano, L., 2008. High levels of Fis1, a pro-fission mitochondrial protein, trigger autophagy. *Biochim Biophys Acta*, 1777, pp.860–866.

Gomes, L. C., Di Benedetto, G. and Scorrano, L., 2011. Essential amino acids and glutamine regulate induction of mitochondrial elongation during autophagy. *Cell Cycle*, 10, pp.2635–2639.

Gropp, E., et al., 2005. Agouti-related peptide-expressing neurons are mandatory for feeding. *Nat Neurosci*, 8, pp.1289–1291.

Hahn, T. M., et al., 1998. Coexpression of Agrp and NPY in fasting-activated hypothalamic neurons. *Nat Neurosci*, 1, pp.271–272.

Hardie, D. G. and Carling, D., 1997. The AMP-activated protein kinase--fuel gauge of the mammalian cell? *Eur J Biochem*, 246, pp.259–273.

Hayashi, T., et al., 2009. MAM: more than just a housekeeper. *Trends Cell Biol*, 19, pp.81–88.

Heaton, G. M., et al., 1978. Brown-adipose-tissue mitochondria: Photoaffinity labelling of the regulatory site of energy dissipation. *Eur J Biochem*, 82, pp.515–521.

Hernandez-Alvarez, M. I., et al., 2010. Subjects with early-onset type 2 diabetes show defective activation of the skeletal muscle PGC-1{alpha}/Mitofusin-2 regulatory pathway in response to physical activity. *Diabetes Care*, 33, pp.645–651.

Hoppins, S., Lackner, L. and Nunnari, J., 2007. The machines that divide and fuse mitochondria. *Annu Rev Biochem*, 76, pp.751–780.

Horvath, T. L., et al., 1992a. Neuropeptide-Y innervation of beta-endorphin-containing cells in the rat mediobasal hypothalamus: a light and electron microscopic double immunostaining analysis. *Endocrinology*, 131, pp.2461–2467.

Horvath, T. L., Naftolin, F. and Leranth, C., 1992b. GABAergic and catecholaminergic innervation of mediobasal hypothalamic beta-endorphin cells projecting to the medial preoptic area. *Neuroscience*, 51, pp.391–399.

Horvath, T. L., et al., 1997. Heterogeneity in the neuropeptide Y-containing neurons of the rat arcuate nucleus: GABAergic and non-GABAergic subpopulations. *Brain Res*, 756, pp.283–286.

Horvath, T. L., Andrews, Z. B. and Diano, S., 2009. Fuel utilization by hypothalamic neurons: Roles for ROS. *Trends Endocrinol Metab*, 20, pp.78–87.

Ibrahim, N., et al., 2003. Hypothalamic proopiomelanocortin neurons are glucose responsive and express K(ATP) channels. *Endocrinology*, 144, pp.1331–1340.

Jheng, H. F., et al., 2012. Mitochondrial fission contributes to mitochondrial dysfunction and insulin resistance in skeletal muscle. *Mol Cell Biol*, 32, pp.309–319.

Kahn, S. E., Hull, R. L. and Utzschneider, K. M., 2006. Mechanisms linking obesity to insulin resistance and type 2 diabetes. *Nature*, 444, pp.840–846.

Katz, A., 2007. Modulation of glucose transport in skeletal muscle by reactive oxygen species. *J Appl Physiol* (1985), 102, pp.1671–1676.

Kelley, D. E., et al., 2002. Dysfunction of mitochondria in human skeletal muscle in type 2 diabetes. *Diabetes*, 51, pp.2944–2950.

Kim, Y. D., et al., 2008. Metformin inhibits hepatic gluconeogenesis through AMP-activated protein kinase-dependent regulation of the orphan nuclear receptor SHP. *Diabetes*, 57, pp.306–314.

Krashes, M. J., et al., 2011. Rapid, reversible activation of AgRP neurons drives feeding behavior in mice. *J Clin Invest*, 121, pp.1424–1428.

Krauss, S., Zhang, C. Y. and Lowell, B. B., 2005. The mitochondrial uncoupling-protein homologues. *Nat Rev Mol Cell Biol*, 6, pp.248–261.

Lee, S., et al., 2012. Mitofusin 2 is necessary for striatal axonal projections of midbrain dopamine neurons. *Hum Mol Genet*, 21, pp.4827–4835.

Lewis, M. R. and Lewis, W. H., 1914. Mitochondria in tissue culture. *Science*. 39, pp.330–333.

Lopez, M., et al., 2008. Hypothalamic fatty acid metabolism mediates the orexigenic action of ghrelin. *Cell Metab*, 7, pp.389–399.

Luquet, S., et al., 2005. NPY/AgRP neurons are essential for feeding in adult mice but can be ablated in neonates. *Science*, 310, pp.683–685.

McCrimmon, R. J., et al., 2004. Potential role for AMP-activated protein kinase in hypoglycemia sensing in the ventromedial hypothalamus. *Diabetes*, 53, pp.1953–1958.

McMinn, J. E., et al., 2000. Effect of intracerebroventricular alpha-MSH on food intake, adiposity, c-Fos induction, and neuropeptide expression. *Am J Physiol Regul Integr Comp Physiol*, 279, pp.R695–R703.

Miki, T., et al., 2001. ATP-sensitive K+ channels in the hypothalamus are essential for the maintenance of glucose homeostasis. *Nat Neurosci*, 4, pp.507–512.

Minokoshi, Y., et al., 2004. AMP-kinase regulates food intake by responding to hormonal and nutrient signals in the hypothalamus. *Nature*, 428, pp.569–574.

Molina, A. J., et al., 2009. Mitochondrial networking protects beta-cells from nutrient-induced apoptosis. *Diabetes*, 58, pp.2303–2315.

Morais, V. A. and De Strooper, B., 2010. Mitochondria dysfunction and neurodegenerative disorders: cause or consequence. *J Alzheimers Dis*, 20 Suppl 2, pp.S255–S263.

Mountjoy, K. G., et al., 1994. Localization of the melanocortin-4 receptor (MC4-R) in neuroendocrine and autonomic control circuits in the brain. *Mol Endocrinol*, 8, pp.1298–1308.

Mountjoy, K. G., 2010. Functions for pro-opiomelanocortin-derived peptides in obesity and diabetes. *Biochem J*, 428, pp.305–324.

Nakazato, M., et al., 2001. A role for ghrelin in the central regulation of feeding. *Nature*, 409, pp.194–198.

Norsted, E., Gomuc, B. and Meister, B., 2008. Protein components of the blood-brain barrier (BBB) in the mediobasal hypothalamus. *J Chem Neuroanat*, 36, pp.107–121.

Obici, S., et al., 2003. Inhibition of hypothalamic carnitine palmitoyltransferase-1 decreases food intake and glucose production. *Nat Med*, 9, pp.756–761.

Okamoto, K. and Shaw, J. M., 2005. Mitochondrial morphology and dynamics in yeast and multicellular eukaryotes. *Annu Rev Genet*, 39, pp.503–536.

Ollmann, M. M., et al., 1997. Antagonism of central melanocortin receptors in vitro and in vivo by agouti-related protein. *Science*, 278, pp.135–138.

Owen, M. R., Doran, E. and Halestrap, A. P., 2000. Evidence that metformin exerts its anti-diabetic effects through inhibition of complex 1 of the mitochondrial respiratory chain. *Biochem J*, 348 Pt 3, pp.607–614.

Ozcan, L., et al., 2009. Endoplasmic reticulum stress plays a central role in development of leptin resistance. *Cell Metab*, 9, pp.35–51.

Ozcan, U., 2013. Mitofusins: Mighty regulators of metabolism. *Cell*, 155, pp.17–18.

Palmiter, R. D., et al., 1998. Life without neuropeptide Y. *Recent Prog Horm Res*, 53, pp.163–199.

Parone, P. A., et al., 2008. Preventing mitochondrial fission impairs mitochondrial function and leads to loss of mitochondrial DNA. *PLoS One*, 3, p.e3257.

Parton, L. E., et al., 2007. Glucose sensing by POMC neurons regulates glucose homeostasis and is impaired in obesity. *Nature*, 449, pp.228–232.

Pham, A. H., et al., 2012. Loss of Mfn2 results in progressive, retrograde degeneration of dopaminergic neurons in the nigrostriatal circuit. *Hum Mol Genet*, 21, pp.4817–4826.

Poggioli, R., Vergoni, A. V. and Bertolini, A., 1986. ACTH-(1-24) and alpha-MSH antagonize feeding behavior stimulated by kappa opiate agonists. *Peptides*, 7, pp.843–848.

Qian, S., et al., 2002. Neither agouti-related protein nor neuropeptide Y is critically required for the regulation of energy homeostasis in mice. *Mol Cell Biol*, 22, pp.5027–5035.

Reznick, R. M. and Shulman, G. I., 2006. The role of AMP-activated protein kinase in mitochondrial biogenesis. *J Physiol*, 574, pp.33–39.

Ritchie, R. H. and Delbridge, L. M., 2006. Cardiac hypertrophy, substrate utilization and metabolic remodelling: cause or effect? *Clin Exp Pharmacol Physiol*, 33, pp.159–166.

Rizzuto, R., et al., 2009. Ca(2+) transfer from the ER to mitochondria: when, how and why. *Biochim Biophys Acta*, 1787, pp.1342–1351.

Roselli-Rehfuss, L., et al., 1993. Identification of a receptor for gamma melanotropin and other proopiomelanocortin peptides in the hypothalamus and limbic system. *Proc Natl Acad Sci U S A*, 90, pp.8856–8860.

Rowland, A. A. and Voeltz, G. K., 2012. Endoplasmic reticulum-mitochondria contacts: function of the junction. *Nat Rev Mol Cell Biol*, 13, pp.607–625.

Samocha-Bonet, D., et al., 2012. Overfeeding reduces insulin sensitivity and increases oxidative stress, without altering markers of mitochondrial content and function in humans. *PLoS One*, 7, p.e36320.

Schneeberger, M., et al., 2013. Mitofusin 2 in POMC neurons connects ER stress with leptin resistance and energy imbalance. *Cell*, 155, pp.172–187.

Schwartz, M. W., et al., 1997. Leptin increases hypothalamic pro-opiomelanocortin mRNA expression in the rostral arcuate nucleus. *Diabetes*, 46, pp.2119–2123.

Sebastian, D., et al., 2012. Mitofusin 2 (Mfn2) links mitochondrial and endoplasmic reticulum function with insulin signaling and is essential for normal glucose homeostasis. *Proc Natl Acad Sci U S A*, 109, pp.5523–5528.

Spanswick, D., et al., 1997. Leptin inhibits hypothalamic neurons by activation of ATP-sensitive potassium channels. *Nature*, 390, pp.521–525.

Stanley, B. G. and Leibowitz, S. F., 1984. Neuropeptide Y: Stimulation of feeding and drinking by injection into the paraventricular nucleus. *Life Sci*, 35, pp.2635–2642.

Tong, Q., et al., 2008. Synaptic release of GABA by AgRP neurons is required for normal regulation of energy balance. *Nat Neurosci*, 11, pp.998–1000.

Trifunovic, A. and Larsson, N. G., 2008. Mitochondrial dysfunction as a cause of ageing. *J Intern Med*, 263, pp.167–178.

Twig, G., et al., 2008. Fission and selective fusion govern mitochondrial segregation and elimination by autophagy. *EMBO J*, 27, pp.433–446.

Wilding, J. P., et al., 1993. Increased neuropeptide-Y messenger ribonucleic acid (mRNA) and decreased neurotensin mRNA in the hypothalamus of the obese (ob/ob) mouse. *Endocrinology*, 132, pp.1939–1944.

Willesen, M. G., Kristensen, P. and Romer, J., 1999. Co-localization of growth hormone secretagogue receptor and NPY mRNA in the arcuate nucleus of the rat. *Neuroendocrinology*, 70, pp.306–316.

Wu, Q., Boyle, M. P. and Palmiter, R. D., 2009. Loss of GABAergic signaling by AgRP neurons to the parabrachial nucleus leads to starvation. *Cell*, 137, pp.1225–1234.

Xie, Z., et al., 2008. Upregulation of mitochondrial uncoupling protein-2 by the AMP-activated protein kinase in endothelial cells attenuates oxidative stress in diabetes. *Diabetes*, 57, pp.3222–3230.

Yen, T. T., et al., 1994. Obesity, diabetes, and neoplasia in yellow A(vy)/- mice: ectopic expression of the agouti gene. *FASEB J*, 8, pp.479–488.

Youle, R. J. and van der Bliek, A. M., 2012. Mitochondrial fission, fusion, and stress. *Science*, 337, 1062–1065.

Zhang, C. Y., et al., 2001. Uncoupling protein-2 negatively regulates insulin secretion and is a major link between obesity, beta cell dysfunction, and type 2 diabetes. *Cell*, 105, pp.745–755.

Zhang, L., He, H. and Balschi, J. A., 2007. Metformin and phenformin activate AMP-activated protein kinase in the heart by increasing cytosolic AMP concentration. *Am J Physiol Heart Circ Physiol*, 293, pp.H457–H466.

Zhou, G., et al., 2001. Role of AMP-activated protein kinase in mechanism of metformin action. *J Clin Invest*, 108, pp.1167–1174.

9

MITOCHONDRIA ANCHORED AT THE SYNAPSE

George A. Spirou, Dakota Jackson and Guy A. Perkins

Center for Neuroscience, Departments of Physiology and Pharmacology, Robert C. Byrd Health Sciences Center, West Virginia University School of Medicine, Morgantown, West Virginia, USA

9.1 INTRODUCTION

Following the advent of new sectioning and staining techniques in the late 19th century, a high density of mitochondria was recognized as a defining feature of nerve terminals (Held, 1897). Subsequent application of the electron microscope to living tissue posed new challenges for identifying synaptic structures because micrographs were typically single sections on the order of 100 nm thickness and did not include associated axons. Indeed, the presence of mitochondria was considered to be a defining characteristic of nerve terminals and lent confidence to the association of newly discovered organelles and synaptic vesicles with the terminal (De Robertis and Bennett, 1955; Palay, 1956; Robertson, 1955). Subsequent decades of work revealed that mitochondria are found in all medium- and large-sized terminals but are not always present in bouton-sized terminals. Even among the initial ultrastructural studies of synapses it was reported that in visual cortex, mitochondria were found in only about one-half of nerve terminals (Gray, 1959). Proper evaluation of mitochondrion prevalence requires labor-intensive serial sectioning, which has been practiced by few laboratories in the nearly seven decades since the application of electron microscopy (EM) to neural tissue. Serial section reconstruction of a population of

The Functions, Disease-Related Dysfunctions, and Therapeutic Targeting of Neuronal Mitochondria, First Edition. Edited by Valentin K. Gribkoff, Elizabeth A. Jonas and J. Marie Hardwick.

about 200 hippocampal boutons revealed that just about one-half contain mitochondria and most of these contain a single mitochondrion (Shepherd and Harris, 1998). These data are a snapshot of mitochondrial positioning because we know from in vitro studies that mitochondria move along the axon and periodically visit all boutons (reviewed in other chapters). We expect that the recent introduction of several techniques for automated and high throughput serial sectioning (Briggman and Bock, 2011) will provide new sets of exhaustive, quantitative information on the number, sizes, and positioning of mitochondria in nerve terminals.

9.2 CALIBRATED POSITIONING OF MITOCHONDRIA

Much attention has been devoted to the movement and delivery of mitochondria to nerve terminals with relatively little consideration of their locations and spacing within terminals. (Note that we use the terminology *terminal* to indicate all synaptic swellings, including en passant structures, and not just terminal swellings at the end of an axon.) With meaningful frequency of illustration throughout the literature, subsets of mitochondria appear to take up calibrated positions relative to the presynaptic membrane and nearby synapses. The earliest observation of such positioning was made by Gray (1963) in cat spinal cord. He described a mitochondrion located nearby an adherens junction with an intervening dense zone of material. This dense zone was crossed by two striations, one located near the mitochondrion and the other about mid-way between the mitochondrion and the adherens junction. The cristae of the mitochondrion appeared to orient perpendicular to the cell membrane. This observation was repeated in a thorough review by Gray and Guillery (1966) of the first dozen years of studying synapse structure by electron microscopy. During this time period of investigation, considerable attention was devoted to identifying the functional sites of communication between neurons. Much attention was drawn to any region of apposed pre- and postsynaptic membrane that exhibited an electron dense thickened structure, whether it be symmetrically distributed on both membranes or asymmetrically distributed. Consensus arose that structures exhibiting asymmetric thicker bands postsynaptically, a 300 to 600 Å diameter vesicle collection presynaptically and often the presence of angular densities called dense projections (Gray, 1963) along the presynaptic membrane, constituted a functional contact (Palay, 1956). This functional zone was referred to as a *synaptic contact* or *synapse* (Gray and Guillery, 1966; Peters and Palay, 1966). In crisp images of the synapse and surrounding region, the intermembrane space, termed the *synaptic cleft*, expanded to a distance of about 200 Å (Palay, 1956). Roles for other regions of potential functional contact at membrane thickenings were debated. Farquhar and Palade (1963) published their ultrastructural description of adhering junctions characterized primarily by symmetric thickenings underlying apposed cytoplasmic membranes with an intermembrane cleft. These structures were termed *zonula adhaerens* or *macula adhaerens* and considered to provide adhesion between cells. In this context, the membranous region of the mitochondrion-associated structure observed by Gray (1963) was initially considered to be a *macula adhaerens* (Gray and Guillery, 1966).

Another type of adhering structure was first observed in contacts between synapse-containing axonal swellings and either dendrites or cell bodies of the dorsal lateral geniculate nucleus. This structure could exhibit symmetric or asymmetric membrane thickenings with an absence of synaptic vesicles in the vicinity of the presynaptic thickening and obvious neurofilaments extending from the postsynaptic thickening. Based on this latter feature, this composite structure was named the filamentous contact (Colonnier and Guillery, 1964). As at synapses, membrane separation increased but to an even larger extent in the filamentous contact (200 to 300 Å; Lieberman and Spacek, 1997) and intercellular material was evident. In single sections, filamentous contacts often extended along the membrane apposition for longer distances than synapses (up to 7 μm; Lieberman and Spacek, 1997), so are larger than synapses and could be adjacent to synapses. In oblique section, they appeared as a patchwork of globular subunits. Although mitochondria were not listed initially as an element of the filamentous contact, in one image (figure 11 of Colonnier and Guillery, 1964) a mitochondrion was positioned over a filamentous contact with an intervening dense zone and cross striations similar to the structure described by Gray (1963). A potential relationship between these organelle arrays was therefore indicated. Subsequently, positioning of mitochondria over filamentous contacts and an intervening layer of agranular reticulated membrane at both pre- and postsynaptic locations were considered to be additional features of this structure (Lieberman and Spacek, 1997). Similar arrangements were noted in hippocampal boutons contacting dendritic spines, although of smaller dimension due to spatial constraints (Spacek and Harris, 1998). In addition to thalamic relay nuclei (somatosensory and visual thalamus), similar structures were noted in published studies of mammalian lateral vestibular nucleus and spinal cord and also avian ciliary ganglion and midbrain of electric fish (see discussion in Lieberman and Spacek, 1997).

The mitochondrion organelle's structure first described by Gray (1963), and occasionally present or noted in descriptions of filamentous contacts (previous paragraph), was highlighted with undescribed frequency in large auditory nerve terminals in the mammalian cochlear nucleus (Cant and Morest, 1979; Tolbert and Morest, 1981). In their study, filaments extended between an adhering junction and mitochondrion. The two cross-filament striations described by Gray (1963) were further resolved as an electron-dense plaque positioned adjacent to the outer mitochondrial membrane and what appeared to be, in single sections, a linked chain of vesicles or tubules of varying diameter but smaller than synaptic vesicles (schematic from Tolbert and Morest [1981]). Investigation of large terminals of cochlear nucleus globular bushy cells in the lateral nucleus of the trapezoid body (LNTB) of the superior olive revealed similar structures (Spirou, Rowland, and Berrebi, 1998). The aforementioned features were present, along with confirmation that the cristae of the mitochondrion, which were better preserved than in early EM studies, appeared oriented parallel to each other and perpendicular to the presynaptic membrane. The presence of postsynaptically arrayed mitochondria was also noted. Based on these conglomerate structures, this organelle arrangement was named the mitochondria-associated adherens complex (MAC). MACs could have larger dimensions than synapses, extending the length of tangentially sectioned mitochondria and aligned with up to four mitochondria arrayed in parallel. Given the prevalence of

MACs in large nerve terminals of the auditory brain stem, their prevalence in the calyx of Held, the largest nerve terminal in the mammalian brain, was documented (Rowland, Irby, and Spirou, 2000). MACs were located within a cluster of synapses and the two structures were often contiguous, as can occur with the filamentous contacts (Lieberman and Spacek, 1997). Three-dimensional (3D) reconstruction of MAC-forming mitochondria revealed them to be more branched than non-MAC mitochondria and they could form two MACs (Rowland et al., 2000). Essentially, the key features of the MAC were established: (1) adhering junction, (2) presynaptically positioned mitochondrion (often more than one), (3) cristae oriented in parallel and perpendicular to the presynaptic membrane, (4) filaments extending between the mitochondrion and the adhering junction, (5) mitochondrial plaque adjacent to the mitochondrion, (6) reticulated membrane interposed between the mitochondrial plaque and adhering junction, (7) less precisely positioned postsynaptic mitochondrion (often more than one). In the following sections we describe each of the features of the MAC, their representation in high-resolution electron tomography (ET) images, and speculation about the molecular identities of some of its constituent structures.

9.3 MITOCHONDRIA AND CRISTA STRUCTURE

Even before the application of ET, in the mid-1990s, to image mitochondria, it was recognized that mitochondrial cristae display striking interorganism and intertissue variations in shape, size, connectivity, and packing arrangement (Munn, 1974). The shape of cristae was usually described as either tubular or lamellar. By imaging tissue sections at multiple angles and computing the 3D tomographic volume, ET added the third dimension of resolution through the tissue thickness with distances between reconstructed image planes as small as 2 nm. Technological improvements in tissue registration and calibrated rotation, image analysis algorithms applicable to large stacks of image files, software designed for efficient segmentation (tracing of membranes and compartments across image planes), high voltage microscopes, and fast computer hardware to store and manipulate large image files supported a renaissance in ET methodology about 20 years ago. It became possible to identify and classify small features of the tomographic volume in serial computational slices for the quantification of mitochondrial architecture and identification of novel structures and structural arrangements (Perkins et al., 1997b; Soto et al., 1994).

Application of ET to cristae structure over the past 15 years has revealed that the majority of cristae in tissues are composed of both tubular and lamellar segments with the tubes generally arranged more peripherally and the lamellae usually more centrally located. Cristae that are entirely tubular are common, yet those that are entirely lamellar are rare, except within brown-fat mitochondria and the MAC. These two examples illustrate that even though the basic cristae architecture is similar across tissues (see figure 3 of Vafai and Mootha, 2012), crista structure and heterogeneity of molecular composition is often specialized for particular functions within cells (Mannella 2008; Perkins and Ellisman, 2007). ET revealed a novel structure, previously obscured because of its small size, called the *crista junction* (Frey and

Mannella, 2000; Perkins et al., 1997a, 1997b). The crista junction was so named because it connects the intracristal space to the intermembrane space at the mitochondrial periphery through a narrow tubule with a circular opening having a diameter between 9 and 20 nm, with few exceptions, in healthy, unperturbed cells, and tissues. Lamellar cristae can have multiple crista junctions anchoring them to the inner boundary membrane and even on opposite sides of the volume for those cristae extending across the volume width.

The uniformity of the crista junction structure suggests two aspects of compartmentation of crista membrane from the inner boundary membrane. One way of compartmenting is to prevent the mixing of membrane proteins. The fluid movement of integral membrane proteins between inner boundary and cristae membranes, which are contiguous, may be impeded by the sharp curvature of the crista junction, resembling a T-intersection. A large protein scaffold, anchored in the mitochondrial inner membrane and spanning the intermembrane space to contact the outer membrane (defined as a contact site), may also impede the diffusion of membrane-associated proteins. This newly named "mitochondrial contact site and cristae organizing system" (MICOS) (Pfanner et al., 2014) is comprised of proteins thought to reside at crista junctions, such as ChChd3/Mic19 and mitofilin/Mic60 in addition to the mitochondrial fusion protein Opa1 and the electron transport chain complex V. Incorporating membrane geometry, including crista junction membrane curvature, the modeling of Sukhorukov and Bereiter-Hahn (2009) lends support to the proposition that the crista junction helps to govern the protein composition between inner membrane compartments. Consistent with this hypothesis is the finding that the electron transport chain complexes are enriched in the cristae membranes, whereas translocases are preferentially located in the inner boundary membrane (Davies et al., 2011; Gilkerson, Selker, and Capaldi, 2003; Vogel et al., 2006; Wurm and Jakobs, 2006). Another aspect of compartmentation applies to soluble molecules that can diffuse in the intracristal space. The crista junction could act as a diffusional bottleneck for solutes passing between the intracristal space and the intermembrane space. In support of this hypothesis, a model of the diffusion of ions and metabolites inside cristae suggested the possibility of concentration gradients of Ca^{2+} and phosphate between the intermembrane space and deep into the cristae (Mannella, Lederer, and Jafri, 2013).

The crista junction membranes may have a protein composition different from other structures within the mitochondrion and which can have profound effects on crista and crista junction structure, adenosine triphosphate (ATP) generation, glycolysis, and respiration. For example, RNAi knockdown of ChChd3/Mic19 whose binding partners are mitofilin/Mic60 and OPA1 and which are known to influence crista morphology, produced fragmented mitochondria, reduced oxygen consumption and glycolytic rates, vesiculation and loss of cristae, accompanied by crista junction shrinkage suggesting substantial cristae remodeling (Darshi et al., 2011). Similarly, knockdown of ChChd6/Mic25, whose binding partners are mitofilin and ChChd3, disrupted cristae structure, ATP production, and oxygen consumption (An et al., 2012). More direct evidence exists for the presence of mitofilin at the crista junction via immunogold electron microscopy (Rabl et al., 2009). Mutants of mitofilin or its knockdown display an almost complete loss of crista junctions and the cristae appear as large, extended stacks of lamellar membranes that are detached from the inner boundary membrane (Alkhaja et al., 2012; Harner et al., 2011; Hoppins et al., 2011; John et al., 2005; Mun et al., 2010; Rabl et al., 2009; von der Malsburg et al., 2011; Zerbes et al., 2012a). With these mutants and knockdown, respiratory activity was reduced. Even as the molecular scaffold of the crista junction, its formation and architecture is being elucidated, ongoing research continues to tease out its interactome, including apparent antagonism between mitofilin and the F_1F_O–ATP synthase to control crista junction and cristae curvature.

Bioenergetic states of mitochondria often correlate with distinctive cristae structure. Mitochondria in vitro can adopt many different structural conformations (Munn, 1974). However, the principal conformations in situ are termed *orthodox* and *condensed*. During low respiratory activity (state 4), mitochondria are characterized by a relatively large matrix volume, concomitantly small intracristal volume, and with the inner boundary membrane closely apposed to the outer membrane—the orthodox conformation. In contrast, during high respiratory activity (state 3; Martino et al., 1979), mitochondria assume the condensed conformation characterized by a relatively small (condensed) matrix volume and enlarged cristae. In this state, cristae remodeling has been observed to increase their interconnectivity (Renken et al., 2002). Another way to increase ATP synthesizing capacity is to increase the abundance of cristae, their membrane surface area and packing density because the oxidative phosphorylation complexes are enriched in the cristae membranes. One way to increase cristae membrane surface area is for cristae to form lamellae.

ET of the MAC in the calyx of Held revealed novel features of cristae and crista junctions (Perkins et al., 2010). Cristae were lamellar, which is unusual for brain mitochondria (Perkins et al., 1997a, 2001). Lamellae were regularly spaced, were not interconnected, and were relatively flat with slightly more curvature near the lateral boundary of the mitochondrion. Not surprisingly, the cristae membrane surface area was found to be large in MAC mitochondria relative to other tissues, indicating a high capacity for ATP generation. The cristae, then, imparted a polarized structure to the mitochondria with their parallel arrays oriented perpendicular to the presynaptic membrane. They were further polarized in that the crista junctions were located under the outer mitochondrial membrane (OMM) proximal and distal to the presynaptic membrane and intriguingly occurred at nearly two times higher density facing (proximal to) the presynaptic membrane (Perkins et al., 2010). Non-MAC mitochondria in the calyx of Held also had lamellar cristae but did not exhibit a difference in the density of crista junctions at each end of the lamellar stack of cristae.

9.4 ADHERING JUNCTIONS AND LINKAGES TO THE CYTOSKELETON

Junctional contacts between cells were first noticed using light microscopy and became a topic ripe for investigation following application of EM to biological tissue. Initial ultrastructural surveys of cell-cell contacts in epithelial tissues revealed a series of symmetric densities aligned along the cytoplasmic side of apposed cell membranes (Odland, 1958; Selby, 1955). An emerging picture of variability among these junctional contacts found organization in the careful definition of junctional subtypes by Farquhar and Palade (1963). They defined three subtypes in epithelial cells: *zonula occludens*, *zonula adhaerens*, and *macula adhaerens*.

Zonula occludens defined regions where the outer membranes of two cells fuse to form a continuous, beltlike trilaminar structure along the extent of the adhesion. In common parlance these are referred to as tight junctions and form diffusional barriers even for ions and small molecules. Note that implementation of new EM staining and freeze fracture techniques revealed that gap junctions could perforate these regions of tight membrane apposition (Peters, Palay, and Webster, 1991). Zonula adhaerens, or intermediary junctions, exhibit a nearly 200 A intercellular space filled with an amorphous condensation of protein and parallel, symmetric cytoplasmic densities commonly referred to as dense plaques. These structures also ringed the epithelial cell but were shifted basally from the tight junctions. The dense plaques could be attached to thin filaments that extended into the cytoplasm. Variations on the geometry of cytoplasmic densities in other tissues led to the terms *fascia* (beltlike) and *punctum* (spotlike) adherens. Macula adhaerens were shifted further basally than intermediate junctions and named for their button-like geometry. They were similar in ultrastructural characteristics to intermediate junctions but with a slightly widened intercellular space (240 to 280 Å) relative to the intermediary junction and more organized intercellular material. The dense plaques were attached to intermediate filaments extending into the cytoplasm (Farquhar and Palade 1963; Odland, 1958). Zonula and macula adhaerens are most similar to the adhering junctions of the MAC and filamentous contact. During the last 20 years, considerable effort has been devoted to provide a molecular definition of adhering junctions.

Following identification of the cadherin superfamily (Sano et al., 1993; Yoshida-Noro, Suzuki, and Takeichi, 1984), it was determined these proteins form the primary adhesional contacts between cells. Cadherins exhibit five Ca^{2+} binding domains that confer rigidity and the ability to form homophilic interactions with cadherins on juxtaposed cells. Cadherin bridges physically span the intercellular space in a repeating pattern of linkages in both zonula and macula adhaerens (Al-Amoudi et al., 2007; He, Cowin, and Stokes, 2003; Pieperhoff et al., 2010). The macula adhaerens primarily use the cadherin proteins desmoglein and desmocollin. These proteins are not expressed in neurons or glia, possibly because of the lower levels of mechanical stress generally placed on neurons in the brain (Green and Simpson, 2007). This molecular definition of adhering junctions in the brain classifies them as variations of zonula adhaerens and likely not macula adhaerens (Franke, 2009; He et al., 2003). In the remainder of this section we will refer to the nonsynaptic adhesive structures in the brain using the general term *adhering junctions*. Adhering junctions use at least one additional adhering membrane spanning protein, called *nectin*. Like the cadherins, nectin molecules can form homophilic linkages that span adhering junctions, but unlike cadherins, these linkages are not Ca^{2+} dependent. The dense plaques of adhering junctions in the brain are comprised of these membrane spanning proteins in addition to proteins that link them to the cytoskeleton (Oda and Takeichi, 2011). Cadherins are linked to the actin cytoskeleton by α- and β-catenin

(Hirohashi and Kanai, 2003; Ozawa, Baribault, and Kemler, 1989). In parallel, nectins are linked to actin by afadin (Franke et al., 2009). The dense plaques also incorporate a diverse set of other actin cytoskeletal linking proteins (Nakanishi and Takai, 2004) including neurojungin (Paffenholz et al., 1999), plakoglobin, a member of the catenin family found also in *macula adhaerens* (Kofron et al., 2002), and the tight junction proteins ZO-(1-3) (Inagaki et al., 2003).

Adhering junctions can exist in close relation to synaptic sites, including those using classical cadherin/catenin complexes and, as revealed by immunoelectron and serial electron microscopy, ring the edges of synapses (Spacek and Harris, 1998; Uchida et al., 1996). Synapses also incorporate adhesion molecules and utilize both cadherin and nectins (Yamagata, Sanes, and Weiner, 2003; Yamamoto et al., 2013). Perkins and colleagues (2010) revealed that the MAC adhering junction tethered its mitochondria via a cytoskeletal superstructure comprised of microfilaments and larger filaments (and devoid of synaptic vesicles). The size of the larger filaments (~16 nm) are suggestive of possible coil-coiled assemblies of smaller filaments, perhaps including intermediate filaments (8 to 12 nm; Albers and Fuchs, 1992). Because adhering junctions were defined as binding to microfilaments, the MAC (and filamentous contact) may represent a new variant of a cell-cell adhesion.

9.5 LINKAGES OF THE OMM TO THE MITOCHONDRIAL PLAQUE AND RETICULATED MEMBRANE

In standard TEM sections, the mitochondrial plaque appears as a dense bar oriented parallel to the presynaptic membrane (Spirou et al., 1998; Tolbert and Morest, 1981). In tomographic sections, it is revealed to be a network of fibers and globular electron dense elements. ET also revealed microfilaments that link the OMM to the filamentous and globular elements that comprise the mitochondrial plaque (Perkins et al., 2010). This study also indicated that reticulated membrane is present but less prevalent between the mitochondrion and adherens junction than suggested by TEM. Reticulated membrane may be more prominent in filamentous contacts, but this organelle complex has not been investigated using ET. Recently, greater attention has been given to mitochondrial association with other organelles including reticulated membrane such as endoplasmic reticulum (ER), microtubules for mitochondrial transport and other mitochondria (Kornmann, 2013; Marchi, Patergnani, and Pinton, 2014; Vance, 2014). These linkages must derive in part from proteins that insert or attach to the OMM and interact with the aforementioned structures directly or via adaptor proteins. Mitochondria have not been localized adjacent to adhering junctions, so there has been no attention, to our knowledge, to identifying related linking proteins.

The molecular identities of OMM linking proteins are actively under investigation (Saxton and Hollenbeck, 2012). Most attention has been focused on mitochondrial transport along microtubules. Anterograde transport is mediated by kinesin motors. Disruption of syntabulin, which colocalizes and comigrates with mitochondria, limits anterograde transport, so this molecule may provide a link between kinesin and the OMM and may directly insert into the OMM (Cai, Gerwin, and Sheng, 2005).

Syntaphilin also colocalizes with mitochondria and can bind microtubules directly and may help to lock nonmotile mitochondria in place (Kang et al., 2008). The picture for linking the OMM to retrograde transport, mediated by dynein motors, is less clear. Dynein and dynactin interacting protein form part of a protein complex linking dynein to the OMM, but it is unknown whether these molecules insert into the OMM or operate through a receptor protein (Haghnia et al., 2007; Pilling et al., 2006). Mitochondria may also move on actin filaments through microtubule-poor cellular regions or for fine positioning within the cell. A myosin-based mechanism may position mitochondria during yeast budding (Lackner, 2013). This mechanism may also occur in vertebrates because myosin V and VI depletion leads to increased motility, so myosin-based mechanisms may help to immobilize mitochondria (Lackner, 2013; Pathak, Sepp, and Hollenbeck, 2010).

The physical coupling of mitochondria with the ER is achieved by protein associations that are part of the ER–mitochondria encounter structure (ERMES) best characterized in yeast (reviewed in Kornmann, 2013). The mammalian ERMES, most relevant to consideration here, has not been fully elucidated despite recent advances. The miro GTPases form part of the mammalian ERMES (Kornmann, Osman, and Walter, 2011) and serve an additional role in regulating mitochondrial movement. Additional tethers include mitofusin-2 anchored in the ER and interacting with both mitochondrial mitofusin-2 and mitofusin-1, chaperone Grp75 that links the ER Ca^{2+} channel IP3R with VDAC, a large-opening pore on the mitochondrial outer membrane, the ER protein VAPB in association with the mitochondrial protein PTPIP51 (De Vos et al., 2012) and the ER protein Bap31 connected with mitochondrial Fis1, a fission protein (Iwasawa et al., 2011). IP3R-VDAC coupling has been hypothesized to position the IP3R opening opposite to VDAC's pore to facilitate optimal Ca^{2+} transfer across organelles (Szabadkai et al., 2006). In summary, multiple proteins could link mitochondria to the mitochondrial plaque, reticulated membrane, microfilaments, intermediate filaments, and microtubules to hold mitochondria in place and maintain ionic and biochemical communication with neighboring structures.

9.6 FUNCTIONS OF THE ORGANELLE COMPLEX

The conglomerate organelle composition of MACs and filamentous contacts implicates them in multiple functions at axonal swellings and terminals. The adherens junction structural element is consistent with mechanical stability as an important function of the structure (Colonnier and Guillery, 1964), but this role was not considered primary because of their presence deep in the brain where mechanical stress is minimized (Lieberman and Spacek, 1997). However, neural structures likely alter their arrangements during vesicle fusion and endocytosis and are subjected to mechanical stress during traumatic events such as concussion. Adhesion structures were also speculated to consist of molecular constituents that could specify topographic connections between synaptic partners, especially in sensory systems (Colonnier and Guillery, 1964). A general role throughout the brain of specifying neural connections was questioned because of the seemingly regional expression of the filamentous contact

(Liberman and Spacek, 1997). Nonetheless, the increasing number of neural territories and species that exhibit features of MACs and filamentous contacts is growing, so they may play rather ubiquitous roles in nervous system structure. The complexity of the cadherin superfamily provides a molecular substrate for targeted neural connections (Takeichi, 2007). Suggestions that adhering junctions function generally during neural development, then decrease in prevalence with maturation (Lieberman and Spacek 1997), require further investigation. Adhering junctions may form before attraction of mitochondria, as MACs did not appear at the calyx of Held until postnatal day 3, after the onset of rapid growth (Hoffpauir et al., 2006).

The presence of a presynaptic mitochondrion implies roles for MACs and filamentous contacts in ATP generation and Ca^{2+} sensing to support synaptic transmission (Lieberman and Spacek, 1997; Rowland et al., 2000; Spirou et al., 1998). Both endocytosis and exocytosis require ATP (Attwell and Laughlin, 2001). MACs are ringed by synapses and mitochondria at the edge of filamentous contacts that can be adjacent to synapses, so their role in exocytosis is plausible. Endocytic regions of the nerve terminal are not well specified, but in many electron micrographs are within or not far from active zones and likely reachable by ATP diffusing from these mitochondria. Mitochondria synthesize glutamate, the major excitatory neurotransmitter of the vertebrate central nervous sysem (Maechler and Wollheim, 1999). Most of the accumulated observations of MACs and filamentous contacts are in synaptic regions of excitatory neurons, so glutamate production may be an important function of the mitochondrial component of these structures. It may be for these reasons that MACs are so prevalent in the large excitatory terminals of the auditory brainstem (Cant and Morest, 1979; Rowland et al., 2000; Spirou et al., 1998; Tolbert and Morest, 1981), which function at high rates and with high temporal precision (Lorteije et al., 2009). Mitochondria buffer Ca^{2+} ions and provide an energetically favorable sink or store via their negative membrane potential (Duchen, 2012; Raffaello, De Stefani, and Rizzuto, 2012). They have been implicated in short-term plasticity and support of sustained synaptic transmission at medium and high rates (Tang and Zucker, 1997; Verstreken et al., 2005). Through their anchored location adjacent to many active zones in MACs and perhaps in filamentous contacts, they become sentinel mitochondria for cellular events surrounding membrane currents and vesicle fusion and reloading. They would be the first mitochondria to detect local increases in Ca^{2+} concentration and bind or buffer Ca^{2+}. They would also be the first to detect local decreases in Ca^{2+} concentration and could release Ca^{2+} to maintain synaptic transmission (Billups and Forsythe, 2002). Note that maintenance of synaptic transmission involves continued vesicle fusion, local vesicle reloading and repriming, and mobilization from reserve to readily releasable pools, each of which may have different set points for Ca^{2+} concentration. A single sentinel mitochondrion is positioned through its branched structure to coordinate and modulate Ca^{2+} levels in each of these locales. Ca^{2+} level is a more efficient stimulator of ATP synthesis than ADP (Denton and McCormack, 1980). Sentinel mitochondria can begin this process in anticipation of future demand while signaling to their neighbors closer to the core of the axonal process to ramp up their ATP production. Indeed, at low and moderate levels of activity, Ca^{2+} entry at the active zone may never be sensed directly by most mitochondria in the nerve terminal.

The internal structure of the sentinel mitochondria in MACs and perhaps filamentous contacts (see Figure 3 in Lieberman and Spacek 1997) is unique. Their lamellar cristae and corresponding lamellar matrix structure, along with the high density of crista junctions at their presynaptic face, could facilitate Ca^{2+} diffusion into and out of the mitochondrion (Perkins et al., 2010). Consistent with this hypothesis, modeling studies support Ca^{2+} concentration gradients inside cristae (Mannella et al., 2013). The high ratio of crista surface area to OMM surface area indicates efficient ATP production. Rapid movement of Ca^{2+} along crista conduits, then, can serve the twin needs of Ca^{2+} buffering and efficient ATP production. The polarized locations of crista junctions may direct entry and exit of biochemicals such as ATP into their regions of greatest requirement. The finding that disruption of crista junctions reduces respiration rate and ATP production supports this hypothesis. The branched structure of MAC mitochondria, together with the observation that the region of the MAC superstructure excludes synaptic vesicles, provides a physical means to partition vesicle pools and their associated active zones. Active zones interspersed or at the edges of MAC clusters may have different pool sizes and see differential control of Ca^{2+} in ATP dynamics. Therefore, the geometry of MACs could generate active zones with heterogeneous release probabilities and dynamics.

The MAC superstructure remains enigmatic in its molecular composition and function. Because it links to an adherens junction, it may be involved in bidirectional intercellular signaling through cadherin and catenin molecules (Takeichi, 2007). This system could provide a biochemical report of presynaptic activity to the postsynaptic cell independent of adaptive and rapidly varying presynaptic Ca^{2+} levels and postsynaptic receptor desensitization. Likewise, postsynaptic status may be signaled to dial up or down presynaptic rates of vesicle fusion. MAC superstructure proteins may have phosphorylation sites so the localized density of crista junctions could efficiently modulate phosphorylation and transmission of other signaling molecules into and out of the mitochondrion. MAC superstructure proteins may also sense the local presynaptic biochemical milieu for sentinel mitochondria independent of intercellular communication.

9.7 MACs AND FILAMENTOUS CONTACTS: A CONTINUUM OF STRUCTURE?

Adhering junctions are common elements of nerve terminals through the nervous system. They reach higher levels of complexity in the filamentous contact and MAC. These structures may exist adjacently and in varying representation of a rigid superstructure or reticulated membrane in the space between the mitochondrion and adhering junction. These variations in structural expression may reflect the particular metabolic and signaling demands among the different neural circuits and regions of the nervous system. The presence of adhering junction proteins at the edges of synaptic contacts and as part of these organelle complexes in a seemingly increasing number of brain regions implicates them generally in nonsynaptic signaling in nerve terminals that complements communication via the contents of synaptic vesicles.

ACKNOWLEDGMENTS

This work was supported by NIH Grants R01 DC007695, P20 GM103503 and P41 GM103412.

REFERENCES

Al-Amoudi, A., et al., 2007. The molecular architecture of cadherins in native epidermal desmosomes. *Nature*, 450, pp.832–837.

Albers, K. and Fuchs, E., 1992. The molecular biology of intermediate filament proteins. *Int Rev Cytol*, 134, pp.243–279.

Alkhaja, A. K., et al., 2012. MINOS1 is a conserved component of mitofilin complexes and required for mitochondrial function and cristae organization. *Mol Biol Cell*, 23, pp.247–257.

An, J., et al., 2012. CHCM1/CHCHD6, novel mitochondrial protein linked to regulation of mitofilin and mitochondrial cristae morphology. *J Biol Chem*, 287, pp.7411–7426.

Attwell, D. and Laughlin, S. B., 2001. An energy budget for signaling in the grey matter of the brain. *J Cereb Blood Flow Metab*, 21, pp.1133–1145.

Billups, B. and Forsythe, I. D., 2002. Presynaptic mitochondrial calcium sequestration influences transmission at mammalian central synapses. *J Neurosci*, 22, pp.5840–5847.

Briggman, K. L. and Bock, D. D., 2012. Volume electron microscopy for neuronal circuit reconstruction. *Curr Opin Neurobiol*, 22, pp.154–161.

Cai, Q., Gerwin, C. and Sheng, Z. H., 2005. Syntabulin-mediated anterograde transport of mitochondria along neuronal processes. *J Cell Biol*, 170, pp.959–969.

Cant, N. B. and Morest, D. K., 1979. The bushy cells in the anteroventral cochlear nucleus of the cat. A study with the electron microscope. *Neuroscience*, 4, pp.1925–1945.

Colonnier, M. and Guillery, R. W., 1964. Synaptic organization in the lateral geniculate nucleus of the monkey. *Zeitschrift fur Zellforshung*, 62, pp.333–355.

Darshi, M., et al., 2011. ChChd3, an inner mitochondrial membrane protein, is essential for maintaining crista integrity and mitochondrial function. *J Biol Chem*, 286, pp.2918–2932.

Davies, K. M., et al., 2011. Macromolecular organization of ATP synthase and complex I in whole mitochondria. *Proc Natl Acad Sci U S A*, 108, pp.14121–14126.

Denton R. M. and McCormack J. G. 1980. The role of calcium in the regulation of mitochondrial metabolism. *Biochem Soc Trans*, 8, pp.266–268.

De Martino, C., et al., 1979. Morphological, histochemical and biochemical studies on germ cell mitochondria of normal rats. *Cell Tissue Res*, 196, pp.1–22.

De Robertis, E. D. and Bennet, S., 2003. Some features of the submicroscopic morphology of synapses in the frog and earthworm. *Vertex*, 14(51), pp.66–71.

De Vos, K. J., et al., 2012. VAPB interacts with the mitochondrial protein PTPIP51 to regulate calcium homeostasis. *Hum Mol Genet*, 21, pp.1299–1311.

Duchen, M. R., 2012. Mitochondria, calcium-dependent neuronal death and neurodegenerative disease. *Pflugers Arch*, 464, pp.1111121.

Farquhar, M. G. and Palade, G. E., 1963. Junctional complexes in various epithelia. *J Cell Biol*, 17, pp.375–412.

Franke, W. W., 2009. Discovering the molecular components of intercellular junctions—a historical view. *Cold Spring Harb Perspect Biol*, 1, p.a003061.

Franke, W. W., et al., 2009. The junctions that don't fit the scheme: Special symmetrical cell-cell junctions of their own kind. *Cell Tissue Res*, 338, pp.1–17.

Frey, T. G. and Mannella, C. A., 2000. The internal structure of mitochondria. *Trends Biochem Sci*, 25, pp.319–324.

Gilkerson, R. W., Selker, J. M. L. and Capaldi, R. A., 2003. The cristal membrane of mitochondria is the principal site of oxidative phosphorylation. *Fed Eur Biochem Soc*, 546, pp.355–358.

Gray, E. G., 1959. Axo-somatic and axo-dendritic synapses of the cerebral cortex: An electron microscope study. *J Anat*, 93, pp.420–433.

Gray, E. G., 1963. Electron microscopy of presynaptic organelles of the spinal cord. *J Anat*, 97, pp.101–106.

Gray, E. G. and Guillery, R. W., 1966. Synaptic morphology in the normal and degenerating nervous system. *Int Rev Cytol*, pp.111–182.

Green, K. J. and Simpson, C. L., 2007. Desmosomes: new perspectives on a classic. *J Invest Dermato*, 127, pp.2499–2515.

Haghnia, M., et al., 2007. Dynactin is required for coordinated bidirectional motility, but not for dynein membrane attachment. *Mol Biol Cell*, 18, pp.2081–2089.

Harner, M., et al., 2011. The mitochondrial contact site complex, a determinant of mitochondrial architecture. *EMBO J*, 30, pp.4356–4370.

He, W., Cowin, P. and Stokes, D. L., 2003. Untangling desmosomal knots with electron tomography. *Science*, 302, pp.109–113.

Held, H., 1893. Die centrale Gehörleitung. Arch Anat Physiol Anat Abt, 201–248.

Hirohashi, S. and Kanai, Y., 2003. Cell adhesion system and human cancer morphogenesis. *Cancer Sci*, 94, 575–581.

Hoffpauir, B. K., et al., 2006. Synaptogenesis of the calyx of Held: rapid onset of function and one-to-one morphological innervation. *J Neurosci*, 26, pp.5511–5523.

Hoppins, S., et al., 2011. A mitochondrial-focused genetic interaction map reveals a scaffold-like complex required for inner membrane organization in mitochondria. *J Cell Biol*, 195, pp.323–340.

Inagaki, M., et al., 2003. Nectin-dependent localization of ZO-1 at puncta adhaerentia junctions between the mossy fiber terminals and the dendrties of the pyramidal cells in the CA3 area of adult mouse hippocampus. *J Comp Neurol*, 460, pp.514–524.

Iwasawa, R., et al., 2011. Fis1 and Bap31 bridge the mitochondria-ER interface to establish a platform for apoptosis induction. *EMBO J*, 30, pp.556–568.

John, G. B., et al. (2005). The mitochondrial inner membrane protein mitofilin controls cristae morphology. *Mol Biol Cell*, 16, pp.1543–1554.

Kang, J. S., et al., 2008. Docking of axonal mitochondria by syntaphilin controls their mobility and affects short-term facilitation. *Cell*, 132, pp.137–148.

Kornmann, B., 2013. The molecular hug between the ER and the mitochondria. *Curr Opin Cell Biol*, 25, pp.443–448.

Kornmann, B., Osman, C. and Walter, P., 2011. The conserved GTPase Gem1 regulates endoplasmic reticulum-mitochondria connections. *Proc Natl Acad Sci U S A*, 108, pp.14151–14156.

Lackner, L. L., 2013. Determining the shape and cellular distribution of mitochondria: the integration of multiple activities. *Curr Opin Cell Biol*, 25, pp.471–476.

Lieberman, A. R. and Spacek, J., 1997. Filamentous contacts: the ultrastructure and three-dimensional organization of specialized non-synaptic interneuronal appositions in thalamic relay nuclei. *Cell Tissue Res*, 288, pp.43–57.

Lorteije, J. A. M., et al., 2009. Reliability and precision of the mouse calyx of Held synapse. *J Neurosci*, 29, pp.13770–13784.

Maechler, P. and Wollheim, C. B., 1999. Mitochondrial glutamate acts as a messenger in glucose-induced insulin exocytosis. *Nature*, 402, pp.685–689.

Mannella, C. A., 2008. Structural diversity of mitochondria: Functional implications. *Ann N Y Acad Sci*, 1147, pp.171–179.

Mannella, C. A., Lederer, W. J. and Jafri, M. S., 2013. The connection between inner membrane topology and mitochondrial function. *J Mol Cell Cardiol*, 62, pp.51–57.

Marchi, S., Patergnani, S. and Pinton, P., 2014. The endoplasmic reticulum-mitochondria connection: one touch, multiple functions. *Biochim Biophys Acta*, 1837, pp.461–469.

Martino, C. De, et al., 1979. Cell and tissue morphological, histochemical and biochemical studies on germ cell mitochondria of normal rats. *Cell Tissue Res*, 22, pp.1–22.

Mun, J. Y., et al., 2010. Caenorhabditis elegans mitofilin homologs control the morphology of mitochondrial cristae and influence reproduction and physiology. *J Cell Physiol*, 224, pp.748–756.

Munn, E. A., 1974. The structure of mitochondria. London: Academic Press.

Nakanishi H. and Takai, Y., 2004. Roles of nectins in cell adhesion, migration and polarization. *Biol Chem*, 385, pp.885–892.

Oda, H. and Takeichi, M., 2011. Evolution: Structural and functional diversity of cadherin at the adherens junction. *J Cell Biol*, 193, pp.1137–1146.

Odland, G. F., 1958. The fine structure of the interrelationship of cells in the human epidermis. *J Biophys Biochem Cytol*, 4, pp.529–538.

Ozawa, M., Baribault, H. and Kemler, R., 1989. The cytoplasmic domain of the cell adhesion molecule uvomorulin associates with three independent proteins structurally related in different species. *EMBO J*, 8, pp.1711–1717.

Paffenholz, R., et al., 1999. The arm-repeat protein NPRAP (neurojungin) is a constituent of the plaques of the outer limiting zone in the retina, defining a novel type of adhering junction. *Exp Cell Res*, 250, pp.452–464.

Palay, B. Y. S. L., 1956. Synapses in the central nervous system. *J Biophys Cytol*, 2, pp.193–202.

Pathak, D., Sepp, K. J. and Hollenbeck, P. J., 2010. Evidence that myosin activity opposes microtubule-based axonal transport of mitochondria. *J Neurosci*, 30, pp.8984–8992.

Perkins, G., et al., 1997a. Electron tomography of neuronal mitochondria: three-dimensional structure and organization of cristae and membrane contacts. *J Struct Biol*, 119, pp.260–272.

Perkins, G., et al., 1997b Electron tomography of large, multicomponent biological structures. *J Struct Biol*, 120, pp.219–227.

Perkins, G. A. and Ellisman, M. H., 2011. Mitochondrial configurations in peripheral nerve suggest differential ATP production. *J Structural Biology*, 173, pp.117–127.

Perkins, G. A., et al., 2001. Membrane architecture of mitochondria in neurons of the central nervous system. *J Neurosci Res*, 66, pp.857–865.

Perkins, G. A., et al., 2010. The micro-architecture of mitochondria at active zones: Electron tomography reveals novel anchoring scaffolds and cristae structured for high-rate metabolism. *J Neurosci*, 30, pp.1015–1026.

Peters, A. and Palay, S. L., 1966. The morphology of laminae A and A1 of the dorsal nucleus of the lateral geniculate body of the cat. *J Anat*, 100, pp.451–486.

Peters, A., Palay, S. L. and Webster, H., 1991. *The Fine Structure of the Nervous System: Neurons and Their Supporting Cells*, 3rd ed. New York: Oxford University Press.

Pfanner, N., et al., 2014. Uniform nomenclature for the mitochondrial contact site and cristae organizing system. *J Cell Biol*, 204, pp.1083–1086.

Pieperhoff, S., et al., 2010. Desmosomal molecules in and out of adhering junctions: normal and diseased States of epidermal, cardiac and mesenchymally derived cells. *Dermatol Res Pract*, 2010, p.139167.

Pilling, A. D., et al., 2006. Kinesin-1 and dynein are the primary motors for fast transport of mitochondria in drosophila motor axons. *Mol Biol Cell*, 17, pp.2057–2068.

Rabl, R., et al., 2009. Formation of cristae and crista junctions in mitochondria depends on antagonism between Fcj1 and Su e/g. *J Cell Biol*, 185, pp.1047–1063.

Raffaello, A., De Stefani, D. and Rizzuto, R., 2012. The mitochondrial Ca(2+) uniporter. *Cell Calcium*, 52, pp.16–21.

Renken, C., et al., 2002. A thermodynamic model describing the nature of the crista junction: a structural motif in the mitochondrion. *J Struct Biol*, 138, pp.137–144.

Robertson, J. D., 1956. The ultrastructure of a reptilian myoneural junction. *J Biophy Biochem Cytol*, 2, pp.381–394.

Rowland, K. C., Irby, N. K. and Spirou, G. A., 2000. Specialized synapse-associated structures within the calyx of Held. *J Neurosci*, 20, pp.9135–9144.

Sano, K., et al., 1993. Protocadherins: A large family of cadherin-related molecules in central nervous system. *EMBO J*, 12, pp.2249–2256.

Saxton, W. M. and Hollenbeck, P. J., 2012. The axonal transport of mitochondria. *J Cell Sci*, 125, pp.2095–2104.

Selby, C. C., 1955. An electron microscope study of the epidermis on mammalian skin in thin section. *J Biophys Biochem Cytol*, 1, pp.429–444.

Shepherd, G. M. and Harris, K. M., 1998. Three-dimensional structure and composition of CA3-->CA1 axons in rat hippocampal slices: implications for presynaptic connectivity and compartmentalization. *J Neurosci*, 18, pp.8300–8310.

Soto, G. E., et al., 1994. Serial section electron tomography: A method for three dimensional reconstruction of large structures. *Neuroimage*, 1, pp.230–243.

Spacek, J. and Harris, K. M., 1998. Three-dimensional organization of cell adhesion junctions at synapses and dendritic spines in area CA1 of the rat hippocampus. *J Comp Neurol*, 393, pp.58–68.

Spirou, G. A., Rowland, K. C. and Berrebi, A. S., 1998. Ultrastructure of neurons and large synaptic terminals in the lateral nucleus. *J Comp Neurol*, 398, pp.257–272.

Sukhorukov, V. M. and Bereiter-Hahn, J., 2009. Anomalous diffusion induced by cristae geometry in the inner mitochondrial membrane. *PLoS One*, 4, p.e4604.

Szabadkai, G., et al., 2006. Mitochondrial dynamics and Ca2+ signaling. *Biochim Biophys Acta*, 1763, pp.442–449.

Takeichi, M., 2007. The cadherin superfamily in neuronal connections and interactions. *Nat Rev Neurosci*, 8, pp.11–20.

Tang, Y. and Zucker, R. S., 1997. Mitochondrial involvement in post-tetanic potentiation of synaptic transmission. *Neuron*, 18, pp.483–491.

Tolbert, L. P. and Morest, D. K., 1982. The neuronal architecture of the anteroventral cochlear nucleus of the cat in the region of the cochlear nerve root: Electron microscopy. *Neuroscience*, 7, pp.3053–3067.

Uchida, N., et al., 1996. The catenin/cadherin adhesion system is localized in synaptic junctions bordering transmitter release zones. *J Cell Biol*, 135, pp.767–779.

Vafai, S. B. and Mootha, V. K., 2012. Mitochondrial disorders as windows into an ancient organelle. *Nature*, 491, pp.374–383.

Vance, J. E., 2014. MAM (mitochondria-associated membranes) in mammalian cells: Lipids and beyond. *Biochim Biophys Acta*, 1841, pp.595–609.

Verstreken P., et al., (2005) Synaptic mitochondria are critical for mobilization of reserve pool vesicles at Drosophila neuromuscular junctions. *Neuron*, 47, pp.365–378.

Vogel, F., et al., 2006. Dynamic subcompartmentalization of the mitochondrial inner membrane. *J Cell Biol*, 175, pp.237–247.

Von der Malsburg, K., et al., 2011. Dual role of mitofilin in mitochondrial membrane organization and protein biogenesis. *Dev Cell*, 21, pp.694–707.

Wurm, C. A. and Jakobs, S., 2006. Differential protein distributions define two sub-compartments of the mitochondrial inner membrane in yeast. *FEBS Lett*, 580, pp.5628–5634.

Yamagata, M., Sanes, J. R. and Weiner, J. A., 2003. Synaptic adhesion molecules. *Curr Opin Cell Biol*, 15, pp.621–632.

Yamamoto, H., et al., 2013. Genetic deletion of afadin causes hydrocephalus by destruction of adherens junctions in radial glial and ependymal cells in the midbrain. *PLoS One*, 8, p.e80356.

Yoshida-Noro, C., Suzuki, N. and Takeichi, M., 1984. Molecular nature of the calcium-dependent cell-cell adhesion system in mouse teratocarcinoma and embryonic cells studied with a monoclonal antibody. *Dev Biol*, 101, pp.19–27.

Zerbes, R. M., et al., 2012a. Role of MINOS in mitochondrial membrane architecture: Cristae morphology and outer membrane interactions differentially depend on mitofilin domains. *J Mol Biol*, 422, pp.183–191.

Zerbes, R. M., et al., 2012b. Mitofilin complexes: Conserved organizers of mitochondrial membrane architecture. *Biol Chem*, 393, pp.1247–1261.

SECTION III

DEFECTIVE MITOCHONDRIAL DYNAMICS AND MITOPHAGY

10

NEURONAL MITOCHONDRIA ARE DIFFERENT: RELEVANCE TO NEURODEGENERATIVE DISEASE

SARAH B. BERMAN

Department of Neurology, Pittsburgh Institute for Neurodegenerative Diseases, University of Pittsburgh, Pittsburgh, Pennsylvania, USA

J. MARIE HARDWICK

W. Harry Feinstone Department of Molecular Microbiology and Immunology, and Department of Neurology, Johns Hopkins University Bloomberg Schools of Public Health and Medicine, Baltimore, Maryland, USA

10.1 INTRODUCTION

Mitochondria are complex double-membrane organelles found in the cytoplasm of nearly all eukaryotic cells, with mammalian red blood cells being a notable exception. Mitochondria are composed of more than 1000 different nuclear-encoded proteins plus 13 proteins encoded by the maternally inherited mitochondrial genome. These 13 proteins are vital to the most widely recognized function of mitochondria—oxidative phosphorylation—carried out by the electron transport (respiratory) chain (complexes I-IV) coupled to the adenosine triphosphate (ATP) synthase (complex V). However, mitochondria also perform many other essential functions. Mitochondria are the sole source of iron-sulfur clusters critical for diverse mitochondrial, cytoplasmic, and nuclear functions. They also perform fatty acid oxidation, as well as cardiolipin, steroid, ketone body, and heme biosynthesis. Mitochondria regulate neuronal activity by

The Functions, Disease-Related Dysfunctions, and Therapeutic Targeting of Neuronal Mitochondria, First Edition. Edited by Valentin K. Gribkoff, Elizabeth A. Jonas and J. Marie Hardwick.

storing and re-releasing calcium and by interconversion of excess glutamine to generate the neurotransmitter glutamate or alpha-ketoglutarate. Mitochondria detoxify ammonia in the urea cycle and produce acetyl-CoA that fuels the tricarboxylic acid (TCA) cycle, also known as the citric acid or Krebs cycle, to provide substrates for energy production via the respiratory chain and for amino acid biosynthesis. Specific mitochondrial components when released into the cytoplasm are important mediators of programmed apoptotic cell death, which is essential for selecting neurons during brain development. There are many gaps in basic knowledge about mitochondria, though new research efforts are beginning to fill these gaps as described in this chapter. Notably, there is little understanding of how mitochondrial organelles differ between neurons and other cell types, or within a single cell type. It is also not clear how these dynamic organelles maintain quality control and how entire organelles or portions thereof are turned over or degraded through mitophagy and other mechanisms. These critical processes of mitochondria are known to underlie specific neurological diseases and are suspected to underlie many more neurological pathologies.

10.2 MITOCHONDRIAL DYNAMICS IN NEURONS AND NEURODEGENERATIVE DISEASE

The dynamic properties of mitochondrial organelles include mitochondrial fission, fusion, transport, biogenesis, and degradation. Mitochondrial dynamics are critical to maintain a functional pool of mitochondria and to maintain neuronal health and function (Detmer and Chan, 2007). Dysregulation of mitochondrial dynamics is linked to compromised quality control and is increasingly implicated in many neurodegenerative diseases. In Parkinson's disease, for example, evidence from genetic and environmental studies suggests that dysregulation of these processes is a critical early feature of the disease (McCoy and Cookson, 2012; Van Laar and Berman, 2009).

Research on neurodegenerative diseases has greatly expanded our understanding of the dynamic mitochondrial processes that maintain neuronal health. Many of the responsible genes and fundamental processes were originally identified through studies using yeast and proliferating mammalian cells. However, neurons have unique properties, including their postmitotic state, the distribution of mitochondria in distal axons and dendrites often far from the nucleus located in cell bodies, and the specialized energetic needs of neurons as a result of synaptic activity. These features of neurons may contribute to their slow rate of mitochondrial turnover under conditions that readily degrade mitochondria in proliferating cells, as discussed in this chapter. Progress toward delineating how mitochondrial dynamics is regulated specifically in neurons and in neurodegenerative disease is hampered in part because the relevant models require more cumbersome research tools.

10.2.1 Quantifying Mitochondrial Dynamics

Despite experimental difficulties, methodologies have been developed to directly quantify rates of mitochondrial fission and fusion in live primary neuron cultures (Berman,

Pineda, and Hardwick, 2008; Berman et al., 2009). Many had assumed that to maintain a constant number of mitochondrial organelles, the rate of mitochondrial fission events would be approximately equal to the rate of fusion events. Live cell microscopy of neuronal mitochondria, combined with mathematical models to fully interpret the data, revealed an imbalance in rates of mitochondrial fission and fusion at steady state. Instead, the rate of mitochondrial fission in neuronal processes was higher than the mitochondrial fusion rate under all situations tested (Berman et al., 2009). Fusion was defined as organelle recombination resulting in the mixing of their mitochondrial contents, a criterion established earlier by yeast biologists. This was achieved by labeling a subset of mitochondria in neuronal processes with a photo-activatable mitochondria-targeted green fluorescent protein that upon organelle fusion would rapidly spread from one labeled mitochondrion to another mitochondrion labeled only with red fluorescent protein. However, few spatial encounters between two mitochondrial organelles in neuronal processes result in a fusion event. Additional experimental strategies were required to determine why fission events were consistently more prevalent than fusion events. The explanation for this imbalance in the fission-to-fusion ratio in normal neurons was explained primarily by two additional parameters not directly observable by live microscopy, organelle growth through elongation (increase in mitochondrial mass) and degradation by mitophagy or by other turnover mechanisms (Berman et al., 2009). These studies illustrated that mitochondrial organelle length is not dependent only on fission and fusion. Instead, multiple dynamic processes including mitochondrial biogenesis, degradation, fission, fusion, and velocity are all integrated to determine organelle morphology. Several different processes likely contribute to mitochondrial turnover rates. Mitochondria can be engulfed by autophagosomes within axons, whereas a large portion of axonal mitochondria in retinal ganglion cells were shown to be transferred to astrocytes for degradation (Ashrafi et al., 2014; Davis et al., 2014). Alternatively, budding of mitochondria-derived vesicles (MDVs), a process involving neurological disease genes that is distinct from mitochondrial fission, appears to tidy the powerhouse without whole organelle turnover (Sugiura et al., 2014). The relative proportion of mitochondrial biomass that is turned over by these different mechanisms is not yet known.

10.2.2 Mutations and Toxins Alter Mitochondrial Dynamics in Neurological Disease

Evidence for critical involvement for dysregulation of mitochondrial dynamics is particularly compelling for Parkinson's disease (Van Laar and Berman, 2009). This chronic, progressive neurodegenerative disease results in degeneration of dopaminergic and other neurons, leading to significant motor, autonomic, and cognitive dysfunction over time. The neurons that are selectively vulnerable to this degeneration throughout the brain are poorly myelinated, with relatively long, thin, energy-requiring axons (Braak et al., 2004). These types of neurons are also likely most dependent on adequate mitochondrial function, distribution, and quality control. In fact, in Parkinson's disease, neurodegeneration of dopaminergic neurons begins at the axon terminals before cell body loss (Braak et al., 2004), suggesting the possibility that the initiating insult begins in these distal regions. In addition, growing evidence indicates that environmental toxins associated with Parkinson's disease disrupt normal mitochondrial fission, fusion, and mitophagy events. Furthermore, many of the causal mutations in familial forms of Parkinson's disease also regulate mitochondrial dynamics. Compelling evidence indicates that the Parkinson's disease-linked gene product Parkin, an E3 ubiquitin ligase, promotes mitophagy after mitochondrial depolarization (Geisler et al., 2010; Kawajiri et al., 2010; Narendra et al., 2008; Vives-Bauza et al., 2010). This process requires the autophagy machinery. Perhaps related to its mitophagy role, Parkin can also influence mitochondrial fission or fusion (Deng et al., 2008; Park, Lee, and Chung, 2009; Poole et al., 2008; Yang et al., 2008).

Other causal mutations in familial forms of Parkinson's disease are found in genes implicated in mitochondrial dynamics, suggesting a possible common pathway (reviewed in Van Laar and Berman, 2013). For example, DJ-1 mutations cause autosomal-recessive Parkinson's disease (Bonifati et al., 2003). DJ-1 undergoes redox-regulated mitochondrial translocation and preserves mitochondrial function (Blackinton et al., 2009; Canet-Aviles et al., 2004). Loss of DJ-1 has also been reported to perturb mitochondrial fusion and mitophagy (Irrcher et al., 2010; Krebiehl et al., 2010; Larsen et al., 2011; Thomas et al., 2011). Likewise, α-synuclein, a gene linked to autosomal dominant Parkinson's disease through both mutation and gene duplication (Chartier-Harlin et al., 2004; Ibanez et al., 2004), appears to decrease mitochondrial fusion, increase mitochondrial fission, and trigger mitophagy (Chinta et al., 2010; Choubey et al., 2011; Kamp et al., 2010; Nakamura et al., 2011). Interestingly, mitochondrial morphology changes induced by α-synuclein or loss of DJ-1 have both been reported to be rescued by the expression of PINK1 and Parkin (Irrcher et al., 2010; Kamp et al., 2010), further suggesting that this pathway is functionally connected to DJ-1.

Although less well-studied, environmental toxins that influence mitochondrial dynamics can also increase the risk for developing Parkinson's disease. Pesticide and herbicide use, including exposure to rotenone, 1,1'-dimethyl-4,4'-bipyridinium dichloride (paraquat), and maneb, has been shown to be risk factors in Parkinson's disease, as has exposures to organic solvents and metals (Brown et al., 2006; Wirdefeldt et al., 2011). Chronic rotenone is a well-established model of Parkinson's disease (described below), recapitulating many of the features of Parkinson's disease neurodegeneration (Betarbet et al., 2000; Cannon et al., 2009; Drolet et al., 2009). The pesticide rotenone, also used in fisheries to eliminate fish populations, inhibits Complex I of the mitochondrial electron transport chain, and induces reactive oxygen species (ROS) production (Sherer et al., 2002). Rotenone also inhibits microtubule function (Choi, Palmiter, and Xia, 2011; Ren et al., 2009; Srivastava and Panda, 2007), thereby contributing to changes in mitochondrial dynamics indirectly. Multiple epidemiological studies have also linked exposure to the herbicide, paraquat, to Parkinson's disease (Berry, La Vecchia, and Nicotera, 2010). Paraquat also is known to have effects on mitochondrial function but has been reported to be only a weak inhibitor of Complex I (Richardson et al., 2005). Paraquat, however, is a potent redox cycler, generates ROS through mitochondrial Complexes I and III (Martinez and Greenamyre, 2012), and may also alter microtubules (Cappelletti, Incani, and

Maci, 1994; Li, Zhao, and Chou, 1987). Chronic administration of paraquat in rats induces dopaminergic neuron loss (Ossowska et al., 2005), and thus, has been used as an experimental model of toxin-induced Parkinson's disease. Paraquat had been shown to reduce mitochondrial branching (Grunewald et al., 2010). Moreover, recent work found that paraquat-induced dopaminergic neuron loss in *Drosophila* was protected by down-regulation of the mitochondrial deubiquitinase USP30, thereby opposing the mitophagy-promoting function of Parkin (Bingol et al., 2014).

Initially, much of the early work examining mitochondrial dynamics in general, and the role of Parkinson's disease-related genes, was performed in cell lines and other proliferating cells. To begin to better evaluate mitochondrial dynamics in chronic neurodegenerative diseases, new methods were developed for integrating the analyses of multiple parameters of mitochondrial dynamics in living neurons. This included assessing mitochondrial transport, density, and degradation through mitophagy, along with dynamic measures of mitochondrial fission and fusion (Arnold et al., 2011). This more Parkinson's disease-relevant model uses environmental toxin exposure, except under more chronic and mild conditions than in previous studies. Treated neurons are then evaluated for changes in mitochondrial dynamics over time, rather than after acute, lethal exposures. Exposure to low concentrations of rotenone, a mitochondrial Complex I inhibitor and pesticide, for several weeks has little effect on cell death of cultured neurons (Arnold et al., 2011). This tractable in vitro model is based on early evidence of dysfunction of Complex I of the mitochondrial electron transport chain in patients with Parkinson's disease (Mizuno et al., 1989; Schapira et al., 1989, 1990). Furthermore, chronic systemic administration of low-doses of rotenone to rats led to pathology typical of human Parkinson's disease (Betarbet et al., 2000). Interestingly, this rat model results in specific neuropathology including the extensive loss of dopaminergic neuron processes before any loss of dopaminergic cells is detected, similar to human Parkinson's disease findings. These rats also exhibit Parkinsonian-like behaviors and nigral neuron inclusions that resemble Lewy bodies. Furthermore, the rotenone-induced degeneration in chronically treated rats appears to begin distally in the axons away from the cell body, similar to the brains of Parkinson's disease patients (Betarbet et al., 2000). Thus, this model more closely mimics the slow, chronic nature of Parkinson's disease neurodegeneration.

Adaptation of this Parkinson's disease model of chronic low-dose rotenone to cultured neurons facilitated the analysis of mitochondrial dynamics by live microscopy (Arnold et al., 2011). Weeks of exposure of cultured cortical neurons to low levels of rotenone (0.1–1 nM) only minimally affected mitochondrial respiration, and ATP levels were unchanged (Arnold et al., 2011). Despite apparently normal mitochondrial morphology, mitochondrial fission, fusion, and transport were measurably altered (Arnold et al., 2011). Therefore, monitoring mitochondrial morphology alone fails to adequately assess mitochondrial dynamics. By integrating additional measures of mitochondrial degradation, morphology, and mitochondrial density over time, a picture begins to emerge. Following an initial wave of rotenone-induced, presumed compensatory mitochondrial fusion in the first week, there was an elevation of mitochondrial fission by the following week (Arnold et al., 2011). Mitochondrial fusion-fission rates are also linked to mitochondrial trafficking.

For example, increased velocity also increases the probability of encountering another mitochondrion, consistent with the observed increase in velocity during early phases of increased mitochondrial fusion (Arnold et al., 2011). Independently of these increased rates, there was an increase in paired fusion-fission events, with fission occurring approximately 2.5 min after a fusion event, similar to mitophagy-associated fission events reported for nonneuronal cells (Twig et al., 2008). However, it is not clear if the fusion-fission events in the setting of rotenone treatment were linked to mitophagy. In addition, chronic rotenone treatment resulted in early increases in distal mitochondrial density independently of corresponding changes in mitochondrial transport (Arnold et al., 2011), suggesting increased mitochondrial growth or biogenesis. Although the contribution of these rotenone-induced alterations in mitochondrial dynamics to neurodegeneration is not yet known, there is evidence that the observed changes in mitochondrial fission and fusion after chronic rotenone may contribute to early neuropathology. Using similar long-term, low-dose exposure to rotenone in differentiated dopaminergic cell lines, it was observed that experimental inhibition of mitochondrial fission protects against rotenone-induced loss of neurites that precedes cell death, suggesting an early and pathologic role for mitochondrial fission (Arnold et al., 2011).

10.3 TRIGGERING MITOPHAGY IN NEURONS VERSUS OTHER CELL TYPES

10.3.1 Parkin Mitophagy Pathway Disease Genes

Insights gained from studies in *Drosophila* on Parkinson's disease genes have brought wide attention to the role of mitochondrial dynamics in neurological disease. Parkin is an E3 ubiquitin ligase which, when mutated, causes autosomal recessive Parkinson's disease (Kitada et al., 1998; Shimura et al., 2000). Parkin mutations disrupt normal mitochondrial morphology, transport, and degradation (Deng et al., 2008; Narendra et al., 2008; Park et al., 2009; Poole et al., 2008; Wang et al., 2011; Yu et al., 2011). Most prominently, Parkin is known to function downstream from another Parkinson's disease-related gene, PINK1, to promote degradation of depolarized mitochondria through mitophagy (Narendra et al., 2008, 2010; Vives-Bauza et al., 2010). PINK1 contains a classical N-terminal mitochondrial import presequence and is apparently rapidly degraded in the mitochondrial matrix (Lazarou et al., 2012), though PINK1 may carry out yet uncharacterized normal functions inside mitochondria as well (Steer, Dail, and Chu, 2015). Because protein import requires a mitochondrial membrane potential, import of PINK1 is compromised in damaged mitochondria. Stabilized PINK1 on the surface of compromised mitochondria serves to recruit Parkin to these damaged organelles (Lazarou et al., 2012). PINK1 recruits and activates the E3 activity of Parkin by phosphorylating ubiquitin on Ser65, and by phosphorylating Parkin on a similar site (also Ser65) in Parkin's ubiquitin-like domain (Kane et al., 2014; Kazlauskaite et al., 2014; Kondapalli et al., 2012; Koyano et al., 2014; Shiba-Fukushima et al., 2012). Different patient mutations impair one or more of these steps.

Detailed mechanisms by which Parkin promotes the removal of mitochondrial organelles are less well understood. Parkin adds ubiquitin to a number of target proteins, including the mitochondrial fusion factors Mfn1 and Mfn2 (Gegg et al., 2010; Ziviani, Tao, and Whitworth, 2010), which are mutated in other neurological disorders (Wallace and Chalkia, 2013). TBC1D15 may participate in steps downstream of Parkin that control autophagosome membrane formation around Parkin-labeled mitochondria (Yamano et al., 2014). This Rab-inhibiting GAP protein and its partner bind to both mitochondria and autophagosome membranes that grow around damaged mitochondria (Yamano et al., 2014). Interestingly, the conserved mitochondrial fission protein Fis1 serves as a mitochondrial receptor for TBC1D15, which is reported to limit the accumulation of ubiquitin-like autophagy factor LC3 on mitochondria during Parkin-mediated mitophagy (Onoue et al., 2013). A role for yeast Fis1 in mitophagy remains controversial and may be condition-dependent (Mao and Klionsky, 2013; Mendl et al., 2011). Genetic studies indicate that yeast Fis1 has additional roles beside mitochondrial fission (Cheng et al., 2008), but its role in mitophagy remains controversial and may be condition-dependent (Mao and Klionsky, 2013; Mendl et al., 2011). Thus, different machinery may delete or preserve mitochondria for different purposes under different basal and stress conditions. Worm and mammalian Fis1 has been implicated in mitophagy based on the analysis of knockout cells (Shen et al., 2014). Though not yet demonstrated for other species, knocking out the gene for Fis1 in yeast potently drives the selection for compensatory secondary mutations in a specific gene, encoding yeast Whi2 (Cheng et al., 2008; Teng et al., 2013). Yeast Whi2 was recently found to be homologous to the human KCTD protein family of unknown function, but mutations in KCTDs are associated with neurological disorders (Golzio et al., 2012; Kousi et al., 2012; Teng et al., 2013; Van Bogaert et al., 2007). Thus, at least in yeast, a Fis1 mutation is often paired with another mutation in KCTD (*WHI2*). This co-occurrence of mutations is not by random chance (although the initial nucleotide mutation occurred by random chance) because one mutation compensates in part for the other, driving the selection for cells with these specific co-occurring mutations (Teng et al., 2013). The concept that one disease mutation could drive the selection for a second disease-related mutation (e.g., during brain development) is largely unstudied. Although genetic mosaicism has been described in the brain associated with disease (Vreeburg and van Steensel, 2012), the prevalence or significance in disease is undetermined. The unexpectedly high prevalence in yeast of gene mutation-driven compensatory mutations (deletion of any one of ~75% of nonessential genes results in a single new mutation in a different gene) raises similar possibilities for humans (taking mutation rates and number of cell divisions into account) (Teng et al., 2013). What limited evidence exists for such events in humans is almost always attributed to accidental coincidences.

10.3.2 Metabolic States of Neurons Modulate Mitophagy Induction

Mitophagy induced by mitochondrial depolarization has been extensively studied in cancer cell lines and other proliferating cells in culture. For example, instantaneous mitochondrial depolarization triggered by protonophores/ionophores, such as CCCP or FCCP in HeLa cells overexpressing Parkin leads to stabilization of PINK1 and rapid disappearance of mitochondrial organelles (Youle and Narendra, 2011). The strong synchronous induction of mitophagy by CCCP/FCCP treatment is an attractive model system because the slow basal turnover rates of mitochondrial organelles impede efforts to study this process in HeLa cells. Intriguingly, applying this model to primary neurons does not fully recapitulate the findings in tumor cell lines (Van Laar et al., 2011). Despite the importance of Parkin in neurons, Parkin translocation to mitochondria and subsequent mitophagy is strongly suppressed in primary neurons when using the CCCP/FCCP model (Van Laar et al., 2011). Although the detailed mechanisms of mitophagy avoidance in neurons are unresolved (see Section 4.1), neuron survival may rely more heavily on one or more of the specialized functions carried out by mitochondrial organelles. The strong dependence of neurons, unlike cancer cells, on mitochondrial OXPHOS for ATP production may be an important contributor to this striking difference. However, the rapid loss of ATP in neurons resulting from collapse of respiratory chain function after mitochondrial depolarization may not be the only factor dictating mitophagy events. When ATP is temporarily preserved in neurons after mitochondrial depolarization (by preventing reversal of the mitochondrial ATP synthase), Parkin translocation becomes detectable in neurons, but mitophagy is still rarely observed (Van Laar et al., 2011). Thus, Parkin translocation and mitophagy are separable events that are regulated differently in neurons. Furthermore, consequences of bioenergetic switching other than global ATP depletion are presumably required to identify and target specific individual mitochondrial organelles destined for degradation.

A bioenergetic influence on mitophagy has been demonstrated in yeast (Kanki and Klionsky, 2008). Starvation readily induces the degradation of yeast mitochondria by mitophagy. However, if yeast is forced to rely on mitochondrial respiration for energy production, levels of mitophagy are barely detectable, even under severe starvation (Kanki and Klionsky, 2008). Similarly, HeLa cells forced to become more OXPHOS-dependent (cultured in galactose/glutamine rather than glucose) do not undergo mitophagy after protonophore-induced mitochondrial depolarization (McCoy et al., 2014; Van Laar et al., 2011). Furthermore, overexpression of the glycolysis enzyme hexokinase 2 promotes translocation of Parkin to mitochondria in neurons, downstream of PINK1 (McCoy et al., 2014). Thus, through unknown mechanisms, yeast and mammalian neurons (and presumably other OXPHOS-dependent cells) preserve mitochondrial organelles to sustain survival, and evade or turn off the mechanisms that would otherwise degrade mitochondria in glycolytic cells.

10.3.3 Neurons Distinguish between Different Types of Mitochondrial Damage

In contrast to mitochondrial depolarization induced by CCCP/FCCP, more physiological models of global depolarization are capable of triggering neuronal mitophagy. Recent studies have further evaluated conditions leading to Parkin translocation after mitochondrial depolarization and subsequent mitophagy in neurons. Interestingly, glutamate excitotoxicity was found to regulate Parkin translocation

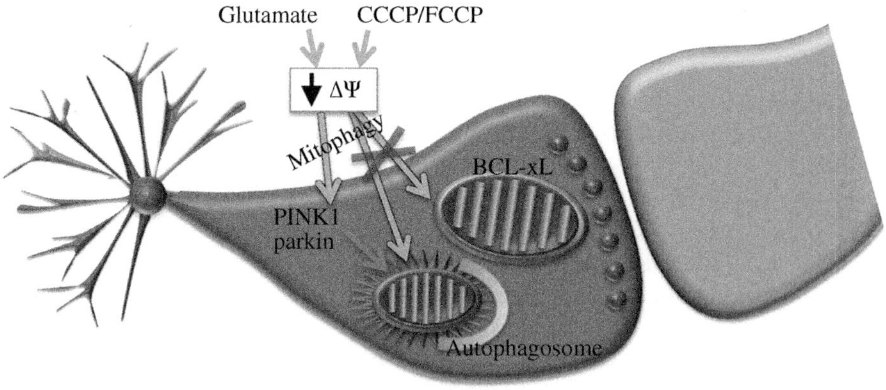

FIGURE 10.1 Mitochondrial organelles in primary neurons differ from those in cancer cell lines by their propensity to avoid Parkin-mediated mitophagy. This mechanism potentially preserves essential mitochondrial functions and delays neurodegeneration as long as possible. Agents that trigger the collapse of mitochondrial inner membrane potential are useful in vitro models and likely mimic in part physiological damage to mitochondria. Depolarization (loss of $\Delta\Psi$) induced by ionophores/protonophores such as CCCP/FCCP potently induce trigger Parkin translocation to mitochondria and mitophagy in HeLa cells but not in neurons. In contrast, transient exposure to the excitatory neurotransmitter glutamate, which potently and rapidly induces mitochondrial depolarization in neurons, induces Parkin translocation to mitochondria but only a fraction of the mitochondria in each neuron. Furthermore, subsequent mitophagy does not occur unless neurons are treated with the antioxidant NAC. NAC-triggered mitophagy is NMDA receptor-dependent and Ca^{2+}-flux dependent in this model. (*See insert for color representation of the figure.*)

to mitochondria and subsequent mitophagy in neurons (Van Laar et al., 2015). The neurotransmitter glutamate activates NMDA receptors, which leads to a local influx of calcium and subsequent uptake of calcium into mitochondria (Stout et al., 1998). This causes mitochondrial depolarization as well as ROS production (Stout et al., 1998; White and Reynolds, 1996). Interestingly, complete mitochondrial depolarization triggered within minutes after a brief exposure to physiological (synaptic) levels of glutamate leads to Parkin accumulation on a subset of neuronal mitochondria in about 30% of neurons within 2 hours, in sharp contrast to CCCP/FCCP-induced depolarization (Fig. 10.1). However, glutamate-induced Parkin accumulation on neuronal mitochondria did not progress to mitophagy. Furthermore, Parkin translocation to mitochondria occurs hours prior to global suppression of ATP levels (Van Laar et al., 2015). Therefore, low ATP is not sufficient to signal Parkin translocation after glutamate treatment. To the contrary, it is possible that sustained ATP is an important facilitator of Parkin translocation in neurons.

The role of ROS appears to be another discordant feature of neuronal mitophagy. Despite a significant increase in Parkin translocation with glutamate-induced mitochondrial depolarization, evidence of subsequent Parkin-associated mitophagy is only readily observed in the presence of the anti-oxidant N-acetylcysteine (NAC), suggesting that ROS under some circumstances can suppress mitophagy (Van Laar et al., 2015). Furthermore, although NAC prevents ROS production, it does not prevent glutamate-induced mitochondrial depolarization, indicating that mitochondrial ROS production is a consequence rather than a contributor to glutamate-induced depolarization (Van Laar et al., 2015). Although ROS production is often used to provoke mitochondrial damage/depolarization and subsequent mitophagy, the downstream actions of glutamate-induced ROS production appear to suppress mitophagy in neurons, although the detailed mechanisms are unknown.

Glutamate excitotoxicity has been implicated in multiple neurodegenerative diseases, including Parkinson's disease (Mehta et al., 2013), and low intrinsic calcium buffering capacity by mitochondria may be one factor linked to the selective vulnerability of specific neuronal populations (Sulzer and Surmeier, 2013). Similarly, the accumulation of Parkin on mitochondria in primary neurons following brief exposure to glutamate requires the NMDA receptor and is calcium-dependent (Van Laar et al., 2015).

In contrast to glutamate toxicity, ROS was previously suggested to contribute to protonophore-induced Parkin translocation to mitochondria in neurons. Removing neuron culture media supplements that contain several anti-oxidants was reported to stimulate Parkin translocation in cortical neurons (Joselin et al., 2012). Parkin translocation and subsequent mitophagy was also reported in neurons after prolonged (24 hour) exposure to CCCP while sustaining cell viability with caspase inhibitors and using lysosomal enzyme inhibitors to prevent degradation during long treatment times (Cai et al., 2012). However, it is difficult to rule out indirect roles of media supplements on other cell functions, and the potential for prolonged exposure to engender action of CCCP on nonmitochondrial membranes is also possible. Furthermore, although valinomycin-induced mitochondrial depolarization causes Parkin-mitochondrial translocation in human induced pluripotent stem cell-derived (iPS) neurons (Seibler et al., 2011), mitophagy was also not detected in that model system (Rakovic et al., 2013), further suggesting variability in neuronal responses to global mitochondrial depolarization.

Most intriguingly, and consistent with the hypothesis that mitochondrial quality control in distal axons might be important in Parkinson's disease-related neurodegeneration, Parkin-associated mitophagy was observed after localized damage to mitochondria specifically in distal axons (Ashrafi et al., 2014). Thus, Parkin can be directed to specific mitochondrial organelles in the absence of a global depolarization signal.

Striking resilience of neuronal mitochondria to induced mitophagy compared to cell lines has been observed in most studies reported to date. This appearance of resilience is unlikely to be explained by rapid mitophagy events that are missed by observers. Furthermore, only a relatively small percentage of depolarized mitochondria accumulate Parkin or undergo mitophagy in neurons compared to proliferating cells. This suggests additional mechanisms in neurons to tightly regulate mitochondrial quality control in these postmitotic, mitochondria-dependent cells. Other factors undoubted are also involved in controlling neuronal mitophagy. For example, differences in PINK1 expression levels can strongly influence baseline Parkin translocation to mitochondria and mitophagy in healthy neurons (Bingol et al., 2014; Van Laar et al., 2015).

Parkin interacts with other organelles as well. Glutamate excitotoxicity triggers Parkin accumulation not only on neuronal mitochondria but also on the endoplasmic reticulum (ER), and specifically at apparent mitochondria-ER junctions (Van Laar et al., 2015). However, this was not observed after protonophore-induced mitochondrial depolarization. The glutamate-induced mitochondria-ER accumulation of Parkin is interesting in light of evidence that mitochondria-associated ER membrane contact sites, also known as MAMs, were found to be the site of mitophagosome formation in yeast and mammals (Bockler and Westermann, 2014). This is also the site of Drp1-dependent mitochondrial fission (Friedman et al., 2011), and during mitophagy, selected mitochondrial outer membrane proteins may be translocated from mitochondria to ER in a Parkin-dependent manner, to avoid degradation during mitophagy (Saita, Shirane, and Nakayama, 2013) (see Section 4.1). Parkin has also been reported to facilitate calcium crosstalk between mitochondria and ER (Cali et al., 2013). Excitotoxicity-induced calcium uptake into mitochondria may trigger complex interactions between Parkin, mitochondria, and ER to coordinate and regulate many aspects of mitochondrial dynamics and quality control.

10.4 BCL-xL: THE GUARDIAN OF MITOCHONDRIA

10.4.1 BCL-xL Regulates Mitochondrial Dynamics and Neuronal Activity

Like Parkin, endogenous BCL-xL is also found on mitochondrial membranes adjacent to ER in brain tissue (Chen et al., 2011; Van Laar et al., 2015). The anti-apoptotic BCL-xL protein has also been shown to antagonize Parkin-mediated mitophagy (Hollville et al., 2014). However, the detailed mechanisms of how this occurs are not understood. The role of BCL-xL in preventing or modulating human neurological disease is also not known, but unlike other BCL-2 family members, BCL-xL is essential for normal brain development and is abundantly expressed in mature neurons throughout life. Early studies analyzing functional point mutants of BCL-xL and BCL-2 provided the first evidence to challenge the general assumption that these proteins function solely by binding and inactivating pro-apoptotic family members (Cheng et al., 1996). In fact, BCL-xL appears to have many other functions particularly in neurons, but also other cell types, and these effects are unlikely to be explained by its anti-apoptotic function (Hardwick, Chen, and Jonas, 2012; Hardwick and Soane 2013). Despite many years of work, there is no clear biochemical function of BCL-2 homologs that explains their shared three-dimensional structure.

Using live microscopy and photo-activatable EGFP targeted to the mitochondrial matrix of cortical neurons (see Section 10.2.1), BCL-xL was found to have many effects on mitochondrial dynamics in healthy neurons. BCL-xL increases the rates and shifts the balance of mitochondrial fission and fusion, and BCL-xL increases mitochondrial biomass in neuronal processes (Berman et al., 2009). Co-expression of BCL-xL with mitochondrial fission factor Drp1 revealed that BCL-xL fulfills these functions by both Drp1-dependent and -independent mechanisms (Berman et al., 2009; Li et al., 2008). In mature hippocampal cultures, expressed and endogenous

BCL-xL increases the number and size of synapses, and increases the number of synapses with an associated mitochondrion, presumably explaining the ability of BCL-xL to significantly increase the synaptic activity of these cells (Jonas et al., 2003, 2004). The overexpression of BCL-xL for several days could potentially alter many downstream cellular processes. However, BCL-xL also has acute effects on neurons. Within a few minutes after microinjection of purified recombinant BCL-xL into the presynaptic cell of fresh squid giant axon preparations, full-length, but not truncated, BCL-xL potently strengthened evoked postsynaptic potentials (Jonas et al., 2003, 2004). These effects were sustained sometimes for hours. Microinjected BCL-xL also rescues declining synaptic activity, for example of oxygen-deprived squid preparations, in contrast to several types of controls (Jonas et al., 2005). Expressed and endogenous BCL-xL also profoundly protects neuronal mitochondria, for example against oligomycin-induced mitochondrial depolarization. How might this work?

10.4.2 BCL-xL Regulates Mitochondrial Energetics

During apoptosis, BCL-2 family proteins are known to act on the surface of mitochondria to control permeability of the outer mitochondrial membrane. However, several biochemical strategies and immuno-electron microscopy studies comparing wild type and BCL-xL knockout neurons in mouse brains yielded compelling evidence that a significant proportion of BCL-xL unexpectedly localized to the inner mitochondrial membrane, in addition to its known localization on the outer mitochondrial membrane and adjacent ER membranes (Chen et al., 2011). Similar results were observed for expressed BCL-xL in cultured hippocampal neurons (Alavian et al., 2011). Although BCL-xL has both N- and C-terminal mitochondrial targeting sequences, the mechanism by which this tail-anchor protein enters mitochondria is not known (McNally et al., 2013). However, other BCL-2 family proteins have been reported to localize inside mitochondria.

The sharing of unpublished immuno-gold EM results showing inner membrane localization of BCL-xL encouraged several labs to retrieve their unexplained findings that BCL-xL could interact with the mitochondrial F_1F_O ATP synthase (Chen et al., 2011; Hardwick et al., 2012). Using several different biochemical purification schemes and additional approaches, it was clear that BCL-xL protein is enriched in highly purified preparations of the F_1F_O ATP synthase, which is embedded in the inner mitochondrial membrane and extends into the matrix compartment (Chen et al., 2011; Hardwick et al., 2012). Unexpected differential effects of oligomycin (inhibits F_O) and aurovertin (inhibits F_1) suggest that the enzymatic F_1 subunit and the F_O proton pump of the ATP synthase are not properly coupled in the absence of BCL-xL (Chen et al., 2011). The exact biochemical mechanism is not known, but mechanisms that are currently under study support a role for BCL-xL in regulating ion flux through a complex containing the F_1F_O ATP synthase on the inner mitochondrial membrane (Alavian et al., 2011, 2014; Chen et al., 2011). This action of BCL-xL closing the wasteful leak of unproductive protons into the mitochondrial matrix could potentially provide considerable energetic savings and reduce expenditures of NADH and ATPduring basal conditions and particularly during stress conditions, consistent with the experimental evidence (Chen et al., 2011). Depolarization of mitochondria in neurons triggered by a brief exposure to glutamate induces ROS production and Parkin translocation to a subset of mitochondria, unlike CCCP/FCCP, which does neither (see Sections 3.2 and 3.3) (Van Laar et al., 2015). It is at least conceivable that the ability of BCL-xL to regulate the flux of ions across the mitochondrial inner membrane in stressed neurons could contribute to its reported ability to suppress Parkin-dependent mitophagy (see Fig. 10.1).

Many other functions have been attributed to BCL-xL, including its ability to suppress autophagy by binding Beclin 1 (Hollville et al., 2014), and its ability to control acetyl-CoA levels (Yi et al., 2011). Acetyl-CoA is a critical molecule in metabolism in the cytosol, in the TCA cycle in mitochondria, in neurotransmitter production and generation of ketone bodies thought to be protective against epilepsy (Hartman and Vining, 2007). Recently it was shown that BCL-xL and its homolog MCL-1 protect depolarized mitochondria from degradation by blocking Parkin translocation from the cytoplasm to mitochondria in a Beclin 1-independent manner (Hollville et al., 2014). Parkin also promotes degradation of MCL-1 (Carroll, Hollville, and Martin, 2014).

BCL-xL also has its dark side. Proteolytic cleavage of BCL-xL generates a toxic fragment, and promotes neuronal injury following ischemic injury based on analysis of uncleavable BCL-xL knockin mice (Cheng et al., 1997; Clem et al., 1998; Ofengeim et al., 2012). How these functions integrate with the ability of BCL-xL to bind and modulate the transcription factor TP53, and to regulate the activity of IP3Rs on ER, and much more, remain unanswered. However, investigators agree that BCL-xL fortifies mitochondria and potently protects neurons and other cell types from death.

ACKNOWLEDGMENTS

This work was supported by National Institutes of Health grants NS37402 and GM077875 (J.M.H.) and NS077964 (S.B.B.).

REFERENCES

Alavian, K. N., et al., 2014. An uncoupling channel within the c-subunit ring of the F1FO ATP synthase is the mitochondrial permeability transition pore. *Proc Natl Acad Sci U S A*, 111, pp.10580–10585.

Alavian, K. N., et al., 2011. Bcl-xL regulates metabolic efficiency of neurons through interaction with the mitochondrial F1FO ATP synthase. *Nat Cell Biol*, 13, pp.1224–1233.

Arnold, B., et al., 2011. Integrating multiple aspects of mitochondrial dynamics in neurons: age-related differences and dynamic changes in a chronic rotenone model. *Neurobiol Dis*, 41, pp.189–200.

Ashrafi, G., et al., 2014. Mitophagy of damaged mitochondria occurs locally in distal neuronal axons and requires PINK1 and Parkin. *J Cell Biol*, 206, pp.655–670.

Berman, S. B., et al., 2009. Bcl-x L increases mitochondrial fission, fusion, and biomass in neurons. *J Cell Biol*, 184, pp.707–719.

Berman, S. B., Pineda, F. J. and Hardwick, J. M., 2008. Mitochondrial fission and fusion dynamics: the long and short of it. *Cell Death Differ*, 15, pp.1147–1152.

Berry, C., La Vecchia, C. and Nicotera, P., 2010. Paraquat and Parkinson's disease. *Cell Death Differ*, 17, pp.1115–1125.

Betarbet, R., et al., 2000. Chronic systemic pesticide exposure reproduces features of Parkinson's disease. *Nat Neurosci*, 3, pp.1301–1306.

Bingol, B., et al., 2014. The mitochondrial deubiquitinase USP30 opposes parkin-mediated mitophagy. *Nature*, 510, pp.370–375.

Blackinton, J., et al., 2009. Formation of a stabilized cysteine sulfinic acid is critical for the mitochondrial function of the parkinsonism protein DJ-1. *J Biol Chem*, 284, pp.6476–6485.

Bockler, S. and Westermann, B., 2014. Mitochondrial ER contacts are crucial for mitophagy in yeast. *Dev Cell*, 28, pp.450–458.

Bonifati, V., et al., 2003. Mutations in the DJ-1 gene associated with autosomal recessive early-onset parkinsonism. *Science*, 299, pp.256–259.

Braak, H., et al., 2004. Stages in the development of Parkinson's disease-related pathology. *Cell Tissue Res*, 318, pp.121–134.

Brown, T. P., et al., 2006. Pesticides and Parkinson's disease—is there a link? *Environ Health Perspect*, 114, pp.156–164.

Cai, Q., et al., 2012. Spatial parkin translocation and degradation of damaged mitochondria via mitophagy in live cortical neurons. *Curr Biol*, 22, pp.545–552.

Cali, T., et al., 2013. Enhanced parkin levels favor ER-mitochondria crosstalk and guarantee Ca(2+) transfer to sustain cell bioenergetics. *Biochim Biophys Acta*, 1832, pp.495–508.

Canet-Aviles, R. M., et al., 2004. The Parkinson's disease protein DJ-1 is neuroprotective due to cysteine-sulfinic acid-driven mitochondrial localization. *Proc Natl Acad Sci U S A*, 101, pp.9103–9108.

Cannon, J. R., et al., 2009. A highly reproducible rotenone model of Parkinson's disease. *Neurobiol Dis*, 34, pp.279–290.

Cappelletti, G., Incani, C. and Maci, R., 1994. Paraquat induces irreversible actin cytoskeleton disruption in cultured human lung cells. *Cell Biol Toxicol*, 10, pp.255–263.

Carroll, R. G., Hollville, E. and Martin, S. J., 2014. Parkin sensitizes toward apoptosis induced by mitochondrial depolarization through promoting degradation of Mcl-1. *Cell Rep*, 9, pp.1538–1553.

Chartier-Harlin, M. C., et al., 2004. Alpha-synuclein locus duplication as a cause of familial Parkinson's disease. *Lancet*, 364, pp.1167–1169.

Chen, Y. B., et al., 2011. Bcl-xL regulates mitochondrial energetics by stabilizing the inner membrane potential. *J Cell Biol*, 195, pp.263–276.

Cheng, E. H., et al., 1997. Conversion of Bcl-2 to a Bax-like death effector by caspases. *Science*, 278, pp.1966–1968.

Cheng, E. H., et al., 1996. Bax-independent inhibition of apoptosis by Bcl-XL. *Nature*, 379, pp.554–556.

Cheng, W. C., et al., 2008. Fis1 deficiency selects for compensatory mutations responsible for cell death and growth control defects. *Cell Death Differ*, 15, pp.1838–1846.

Chinta, S. J., et al., 2010. Mitochondrial alpha-synuclein accumulation impairs complex I function in dopaminergic neurons and results in increased mitophagy in vivo. *Neurosci Lett*, 486, pp.235–239.

Choi, W. S., Palmiter, R. D. and Xia, Z., 2011. Loss of mitochondrial complex I activity potentiates dopamine neuron death induced by microtubule dysfunction in a Parkinson's disease model. *J Cell Biol*, 192, pp.873–882.

Choubey, V., et al., 2011. Mutant A53T alpha-synuclein induces neuronal death by increasing mitochondrial autophagy. *J Biol Chem*, 286, pp.10814–10824.

Clem, R. J., et al., 1998. Modulation of cell death by Bcl-XL through caspase interaction. *Proc Natl Acad Sci U S A*, 95, pp.554–559.

Davis, C. H., et al., 2014. Transcellular degradation of axonal mitochondria. *Proc Natl Acad Sci U S A*, 111, pp.9633–9638.

Deng, H., et al., 2008. The Parkinson's disease genes pink1 and parkin promote mitochondrial fission and/or inhibit fusion in Drosophila. *Proc Natl Acad Sci U S A*, 105, pp.14503–14508.

Detmer, S. A. and Chan, D. C., 2007. Functions and dysfunctions of mitochondrial dynamics. *Nat Rev Mol Cell Biol*, 8, pp.870–879.

Drolet, R. E., et al., 2009. Chronic rotenone exposure reproduces Parkinson's disease gastrointestinal neuropathology. *Neurobiol Dis*, 36, pp.96–102.

Friedman, J. R., et al., 2011. ER tubules mark sites of mitochondrial division. *Science*, 334, pp.358–362.

Gegg, M. E, et al., 2010. Mitofusin 1 and mitofusin 2 are ubiquitinated in a PINK1/parkin-dependent manner upon induction of mitophagy. *Hum Mol Genet*, 19, pp.4861–4870.

Geisler, S., et al., 2010. PINK1/Parkin-mediated mitophagy is dependent on VDAC1 and p62/SQSTM1. *Nat Cell Biol*, 12, pp.119–131.

Golzio, C., et al., 2012. KCTD13 is a major driver of mirrored neuroanatomical phenotypes of the 16p11.2 copy number variant. *Nature*, 485, pp.363–367.

Grunewald, A., et al., 2010. Mutant Parkin impairs mitochondrial function and morphology in human fibroblasts. *PLoS One*, 5, p.e12962.

Hardwick, J. M., Chen, Y. B. and Jonas, E. A., 2012. Multipolar functions of BCL-2 proteins link energetics to apoptosis. *Trends Cell Biol*, 22, pp.318–328.

Hardwick, J. M. and Soane, L., 2013. Multiple functions of BCL-2 family proteins. *Cold Spring Harb Perspect Biol*, 5. doi: 10.1101/cshperspect.a008722.

Hartman, A. L. and Vining, E. P., 2007. Clinical aspects of the ketogenic diet. *Epilepsia*, 48, pp.31–42.

Hollville, E., et al., 2014. Bcl-2 family proteins participate in mitochondrial quality control by regulating Parkin/PINK1-dependent mitophagy. *Mol Cell*, 55, pp.451–466.

Ibanez, P., et al., 2004. Causal relation between alpha-synuclein gene duplication and familial Parkinson's disease. *Lancet*, 364, pp.1169–1171.

Irrcher, I., et al., 2010. Loss of the Parkinson's disease-linked gene DJ-1 perturbs mitochondrial dynamics. *Hum Mol Genet*, 19, pp.3734–3746.

Jonas, E. A., et al., 2004. Proapoptotic N-truncated BCL-xL protein activates endogenous mitochondrial channels in living synaptic terminals. *Proc Natl Acad Sci U S A*, 101, pp.13590–13595.

Jonas, E. A., et al., 2005. Exposure to hypoxia rapidly induces mitochondrial channel activity within a living synapse. *J Biol Chem*, 280, pp.4491–4497.

Jonas, E. A., et al., 2003. Modulation of synaptic transmission by the BCL-2 family protein BCL-xL. *J Neurosci*, 23, pp.8423–8431.

Joselin, A. P., et al., 2012. ROS-dependent regulation of Parkin and DJ-1 localization during oxidative stress in neurons. *Hum Mol Genet*, 21, pp.4888–4903.

Kamp, F., et al., 2010. Inhibition of mitochondrial fusion by alpha-synuclein is rescued by PINK1, Parkin and DJ-1. *EMBO J*, 29, pp.3571–3589.

Kane, L. A., et al., 2014. PINK1 phosphorylates ubiquitin to activate Parkin E3 ubiquitin ligase activity. *J Cell Biol*, 205, pp.143–153.

Kanki, T. and Klionsky, D. J., 2008. Mitophagy in yeast occurs through a selective mechanism. *J Biol Chem*, 283, pp.32386–32393.

Kawajiri, S., et al., 2010. PINK1 is recruited to mitochondria with parkin and associates with LC3 in mitophagy. *FEBS Lett*, 584, pp.1073–1079.

Kazlauskaite, A., et al., 2014. Phosphorylation of Parkin at Serine65 is essential for activation: elaboration of a Miro1 substrate-based assay of Parkin E3 ligase activity. *Open Biol*, 4, p.130213.

Kitada, T., et al., 1998. Mutations in the parkin gene cause autosomal recessive juvenile parkinsonism. *Nature*, 392, pp.605–608.

Kondapalli, C., et al., 2012. PINK1 is activated by mitochondrial membrane potential depolarization and stimulates Parkin E3 ligase activity by phosphorylating Serine 65. *Open Biol*, 2, p.120080.

Kousi, M., et al., 2012. Novel mutations consolidate KCTD7 as a progressive myoclonus epilepsy gene. *J Med Genet*, 49, pp.391–399.

Koyano, F., et al., 2014. Ubiquitin is phosphorylated by PINK1 to activate parkin. *Nature*, 510, pp.162–166.

Krebiehl, G., et al., 2010. Reduced basal autophagy and impaired mitochondrial dynamics due to loss of Parkinson's disease-associated protein DJ-1. *PLoS One*, 5, p.e9367.

Larsen, N. J., et al., 2011. DJ-1 knock-down impairs astrocyte mitochondrial function. *Neuroscience*, 196, pp.251–264.

Lazarou, M., et al., 2012. Role of PINK1 Binding to the TOM complex and alternate intracellular membranes in recruitment and activation of the E3 ligase parkin. *Dev Cell*, 22, pp.320–333.

Li, H., et al., 2008. Bcl-xL induces Drp1-dependent synapse formation in cultured hippocampal neurons. *Proc Natl Acad Sci U S A*, 105, pp.2169–2174.

Li, W. D., Zhao, Y. Z. and Chou, I. N., 1987. Paraquat-induced cytoskeletal injury in cultured cells. *Toxicol Appl Pharmacol*, 91, pp.96–106.

Mao, K. and Klionsky, D. J., 2013. Mitochondrial fission facilitates mitophagy in Saccharomyces cerevisiae. *Autophagy*, 9, pp.1900–1901.

Martinez, T. N. and Greenamyre, J. T., 2012. Toxin models of mitochondrial dysfunction in Parkinson's disease. *Antioxid Redox Signal*, 16, pp.920–934.

McCoy, M. K. and Cookson, M. R., 2012. Mitochondrial quality control and dynamics in Parkinson's disease. *Antioxid Redox Signal*, 16, pp.869–882.

McCoy, M. K., et al., 2014. Hexokinase activity is required for recruitment of parkin to depolarized mitochondria. *Hum Mol Genet*, 23, pp.145–156.

McNally, M. A., et al., 2013. The N-terminal helix of Bcl-xL targets mitochondria. *Mitochondrion*, 13, pp.119–124.

Mehta, A., et al., 2013. Excitotoxicity: bridge to various triggers in neurodegenerative disorders. *Eur J Pharmacol*, 698, pp.6–18.

Mendl, N., et al., 2011. Mitophagy in yeast is independent of mitochondrial fission and requires the stress response gene WHI2. *J Cell Sci*, 124, pp.1339–1350.

Mizuno, Y., et al., 1989. Deficiencies in complex I subunits of the respiratory chain in Parkinson's disease. *Biochem Biophys Res Commun*, 163, pp.1450–1455.

Nakamura, K., et al., 2011. Direct membrane association drives mitochondrial fission by the Parkinson disease-associated protein alpha-synuclein. *J Biol Chem*, 286, pp.20710–20726.

Narendra, D., et al., 2008. Parkin is recruited selectively to impaired mitochondria and promotes their autophagy. *J Cell Biol*, 183, pp.795–803.

Narendra, D. P., et al., 2010. PINK1 is selectively stabilized on impaired mitochondria to activate Parkin. *PLoS Biol*, 8, p.e1000298.

Ofengeim, D., et al., 2012. N-terminally cleaved Bcl-xL mediates ischemia-induced neuronal death. *Nat Neurosci.* 15, 574–580.

Onoue, K., et al., 2013. Fis1 acts as a mitochondrial recruitment factor for TBC1D15 that is involved in regulation of mitochondrial morphology. *J Cell Sci*, 126, pp.176–185.

Ossowska, K., et al., 2005. A slowly developing dysfunction of dopaminergic nigrostriatal neurons induced by long-term paraquat administration in rats: an animal model of preclinical stages of Parkinson's disease? *Eur J Neurosci*, 22, pp.1294–1304.

Park, J., Lee, G. and Chung, J., 2009. The PINK1-Parkin pathway is involved in the regulation of mitochondrial remodeling process. *Biochem Biophys Res Commun*, 378, pp.518–523.

Poole, A. C., et al., 2008. The PINK1/Parkin pathway regulates mitochondrial morphology. *Proc Natl Acad Sci U S A*, 105, pp.1638–1643.

Rakovic, A., et al., 2013. Phosphatase and tensin homolog (PTEN)-induced putative kinase 1 (PINK1)-dependent ubiquitination of endogenous Parkin attenuates mitophagy: study in human primary fibroblasts and induced pluripotent stem cell-derived neurons. *J Biol Chem*, 288, pp.2223–2237.

Ren, Y., et al., 2009. Parkin protects dopaminergic neurons against microtubule-depolymerizing toxins by attenuating microtubule-associated protein kinase activation. *J Biol Chem*, 284, pp.4009–4017.

Richardson, J. R., et al., 2005. Paraquat neurotoxicity is distinct from that of MPTP and rotenone. *Toxicol Sci*, 88, pp.193–201.

Saita, S., Shirane, M. and Nakayama, K. I., 2013. Selective escape of proteins from the mitochondria during mitophagy. *Nat Commun*, 4, p.1410.

Schapira, A. H., et al., 1990. Mitochondrial complex I deficiency in Parkinson's disease. *J Neurochem*, 54, pp.823–827.

Schapira, A. H., et al., 1989. Mitochondrial complex I deficiency in Parkinson's disease. *Lancet*, 1, p.1269.

Seibler, P., et al., 2011. Mitochondrial Parkin recruitment is impaired in neurons derived from mutant PINK1 induced pluripotent stem cells. *J Neurosci*, 31, pp.5970–5976.

Shen, Q., et al., 2014. Mutations in Fis1 disrupt orderly disposal of defective mitochondria. *Mol Biol Cell*, 25, pp.145–159.

Sherer, T. B., et al., 2002. An in vitro model of Parkinson's disease: linking mitochondrial impairment to altered alpha-synuclein metabolism and oxidative damage. *J Neurosci*, 22, pp.7006–7015.

Shiba-Fukushima, K., et al., 2012. PINK1-mediated phosphorylation of the Parkin ubiquitin-like domain primes mitochondrial translocation of Parkin and regulates mitophagy. *Sci Rep*, 2, p. 1002.

Shimura, H., et al., 2000. Familial Parkinson disease gene product, parkin, is a ubiquitin-protein ligase. *Nat Genet*, 25, pp.302–305.

Srivastava, P. and Panda, D., 2007. Rotenone inhibits mammalian cell proliferation by inhibiting microtubule assembly through tubulin binding. *FEBS J*, 274, 4788–4801.

Steer, E. K., Dail, M. K. and Chu, C. T., 2015. Beyond mitophagy: Cytosolic PINK1 as a messenger of mitochondrial health. *Antioxid Redox Signal*, 22, pp.1047–1059.

Stout, A. K., et al., 1998. Glutamate-induced neuron death requires mitochondrial calcium uptake. *Nat Neurosci*, 1, pp.366–373.

Sugiura, A., et al., 2014. A new pathway for mitochondrial quality control: Mitochondrial-derived vesicles. *EMBO J*, 33, pp.2142–2156.

Sulzer, D. and Surmeier, D. J., 2013. Neuronal vulnerability, pathogenesis, and Parkinson's disease. *Mov Disord*, 28, pp.715–724.

Teng, X., et al., 2013. Genome-wide consequences of deleting any single gene. *Mol Cell*, 52, pp.485–494.

Thomas, K. J., et al., 2011. DJ-1 acts in parallel to the PINK1/parkin pathway to control mitochondrial function and autophagy. *Hum Mol Genet*, 20, pp.40–50.

Twig, G., et al., 2008. Fission and selective fusion govern mitochondrial segregation and elimination by autophagy. *EMBO J*, 27, pp.433–446.

Van Bogaert, P., et al., 2007. Mutation of a potassium channel-related gene in progressive myoclonic epilepsy. *Ann Neurol*, 61, pp.579–586.

Van Laar, V. S., et al., 2011. Bioenergetics of neurons inhibit the translocation response of Parkin following rapid mitochondrial depolarization. *Hum Mol Genet*, 20, pp.927–940.

Van Laar, V. S. and Berman, S. B., 2009. Mitochondrial dynamics in Parkinson's disease. *Exp Neurol*, 218, pp.247–256.

Van Laar, V. S. and Berman, S. B., 2013. The interplay of neuronal mitochondrial dynamics and bioenergetics: Implications for Parkinson's disease. *Neurobiol Dis*, 51, pp.43–55.

Van Laar, V. S., et al., 2015. Glutamate excitotoxicity in neurons triggers mitochondrial and endoplasmic reticulum accumulation of Parkin, and, in the presence of N-acetyl cysteine, mitophagy. *Neurobiol Dis*, 74C, pp.180–193.

Vives-Bauza, C., et al., 2010. PINK1-dependent recruitment of Parkin to mitochondria in mitophagy. *Proc Natl Acad Sci U S A*, 107, pp.378–383.

Vreeburg, M. and van Steensel, M. A., 2012. Genodermatoses caused by genetic mosaicism. *Eur J Pediatr*, 171, pp.1725–1735.

Wallace, D. C. and Chalkia, D., 2013. Mitochondrial DNA genetics and the heteroplasmy conundrum in evolution and disease. *Cold Spring Harb Perspect Med*, 3, p.a021220.

Wang, X., et al., 2011. PINK1 and Parkin target Miro for phosphorylation and degradation to arrest mitochondrial motility. *Cell*, 147, pp.893–906.

White, R. J. and Reynolds, I. J., 1996. Mitochondrial depolarization in glutamate-stimulated neurons: an early signal specific to excitotoxin exposure. *J Neurosci*, 16, pp.5688–5697.

Wirdefeldt, K., et al., 2011. Epidemiology and etiology of Parkinson's disease: A review of the evidence. *Eur J Epidemiol*, 26 Suppl 1, pp.S1–S58.

Yamano, K., et al., 2014. Mitochondrial Rab GAPs govern autophagosome biogenesis during mitophagy. *Elife*, 3, p.e01612.

Yang, Y., et al., 2008. Pink1 regulates mitochondrial dynamics through interaction with the fission/fusion machinery. *Proc Natl Acad Sci U S A*, 105, pp.7070–7075.

Yi, C. H., et al., 2011. Metabolic regulation of protein N-alpha-acetylation by Bcl-xL promotes cell survival. *Cell*, 146, pp.607–620.

Youle, R. J. and Narendra, D. P., 2011. Mechanisms of mitophagy. *Nat Rev Mol Cell Biol*, 12, pp.9–14.

Yu, W., et al., 2011. The PINK1/Parkin pathway regulates mitochondrial dynamics and function in mammalian hippocampal and dopaminergic neurons. *Hum Mol Genet*, 20, pp.3227–3240.

Ziviani, E., Tao, R. N. and Whitworth, A. J., 2010. Drosophila parkin requires PINK1 for mitochondrial translocation and ubiquitinates mitofusin. *Proc Natl Acad Sci U S A*, 107, pp.5018–5023.

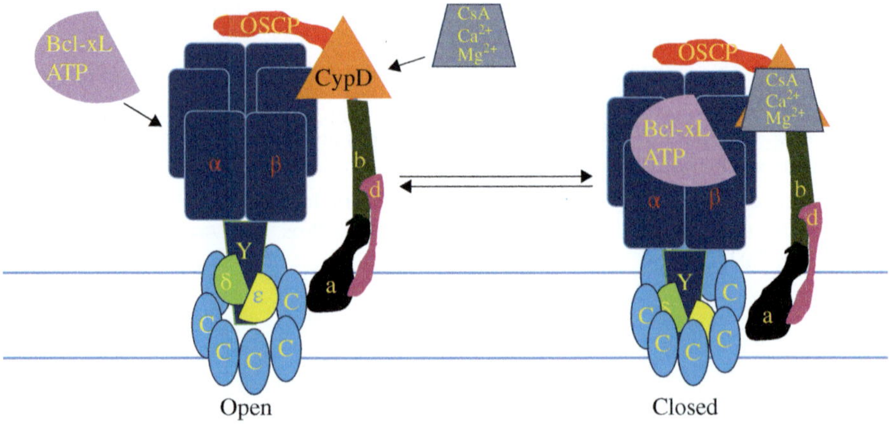

FIGURE 2.2 Model of physiological mPTP activity. β subunit, which binds Bcl-xL and ATP, blocks c-subunit conductance. CsA, inhibitor of CypD, blocks c-subunit conductance. Both Bcl-xL and CsA bind to F1 components during pore closure. The pore diameter also decreases during pore closure. The channel can switch reversibly between these open and closed states.

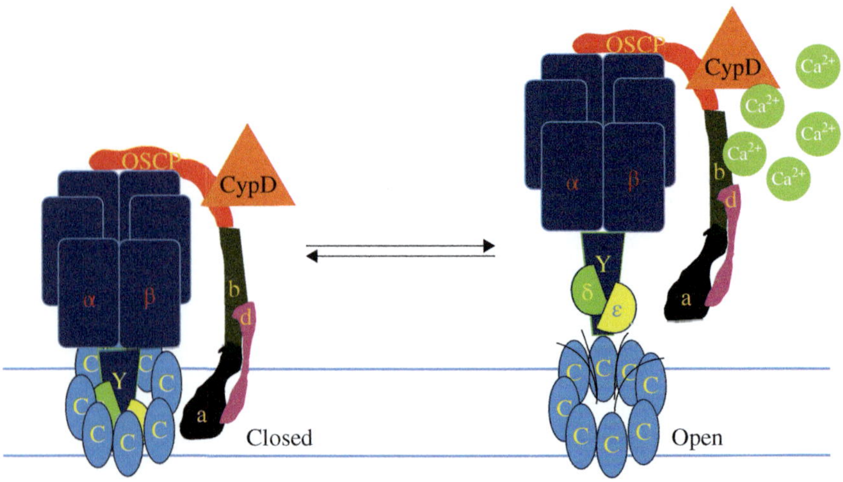

FIGURE 2.3 Model of pathophysiological mPTP activity. Ca^{2+} induces separation of F_1 from F_O during mPTP opening. The pore diameter also expands during Ca^{2+}-induced increases in mPTP conductance.

The Functions, Disease-Related Dysfunctions, and Therapeutic Targeting of Neuronal Mitochondria, First Edition. Edited by Valentin K. Gribkoff, Elizabeth A. Jonas and J. Marie Hardwick.

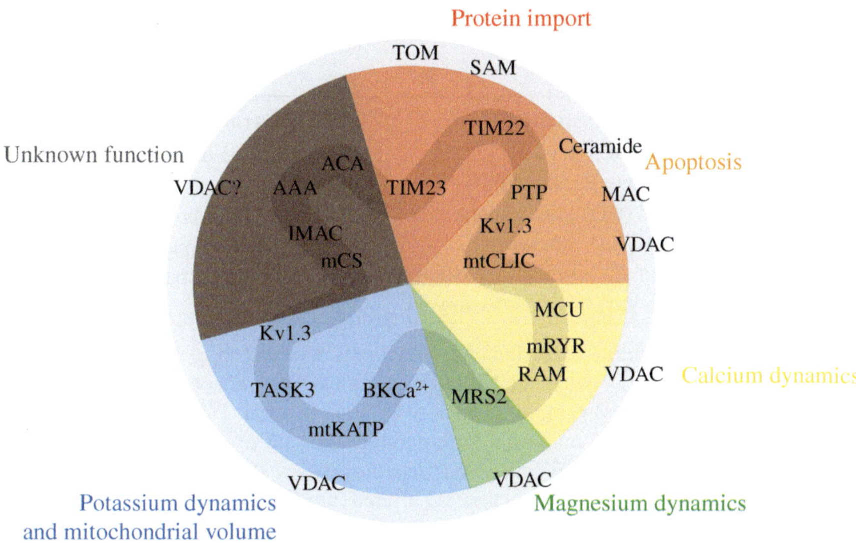

FIGURE 3.1 Mitochondrial ion channels by location and function. The pie chart represents the spatial and functional distribution of ion channels in the outer *(grey circle)* and inner *(shadowed cristae)* mitochondrial membranes. Detailed information and references are compiled in the text.

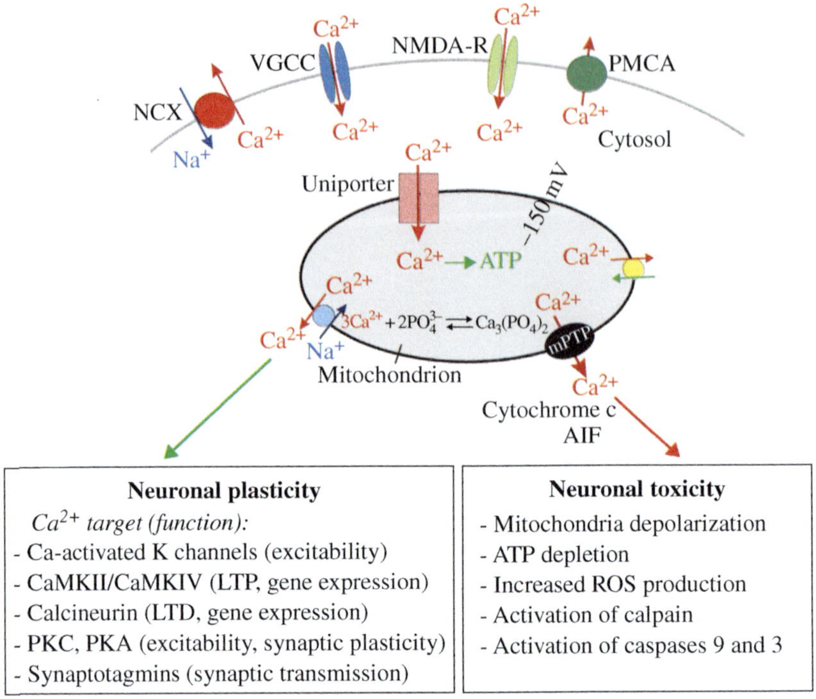

FIGURE 4.1 Mitochondrial Ca^{2+} signaling in neurons. Ca^{2+} entering neurons via voltage-gated Ca^{2+} channels (VGCC), glutamate NMDA receptors (NMDA-R), or other Ca^{2+}-permeable channels (e.g., TRP channels) is rapidly taken up by mitochondria. This Ca^{2+} uptake is mediated by the mitochondrial Ca^{2+} uniporter. Within mitochondria, Ca^{2+} stimulates ATP synthesis through activation of several Ca^{2+}-activated dehydrogenases (Denton, 2009; Glancy and Balaban, 2012). Free Ca^{2+} within the matrix is buffered into Ca-phosphate complexes. Ca^{2+} is released from mitochondria back into the cytosol via at least three mechanisms, Na^+/Ca^{2+} exchanger, H^+/Ca^{2+} exchanger, and mitochondrial permeability transition pore (mPTP). Under physiological conditions, Ca^{2+} mobilization from mitochondria is mediated primarily by the mitochondrial Na^+/Ca^{2+} exchanger. The Ca^{2+} released from mitochondria is ultimately extruded from neurons via the plasma membrane Ca^{2+}-ATPases (PMCA) and Na^+/Ca^{2+} exchangers (NCX). Ca^{2+} uptake and release by mitochondria reduces the amplitude of increase in cytosolic Ca^{2+} concentration ($[Ca^{2+}]_i$), and prolongs the duration of $[Ca^{2+}]_i$ elevation. By shaping $[Ca^{2+}]_i$, mitochondrial Ca^{2+} cycling modulates activity of many neuronal Ca^{2+} targets and influences numerous Ca^{2+}-dependent functions such as excitability, synaptic transmission, and gene regulation. Under pathological conditions, mitochondrial overload with Ca^{2+} triggers opening of the mPTP, a large channel permeable for Ca^{2+} and small proteins including the pro-apoptotic factors cytochrome c and apoptosis-inducing factor (AIF). Opening of the mPTP triggers a cascade of toxic events (e.g., mitochondrial depolarization, collapse of ATP synthesis, increase in reactive oxygen species (ROS) production, Ca^{2+} deregulation, activation of caspases, and the Ca^{2+}-dependent protease calpain), ultimately leading to neuronal demise. The deregulation of mitochondrial Ca^{2+} handling is implicated in neuronal toxicity in stroke, Alzheimer's disease, Parkinson's disease, Huntington's disease, amyotrophic lateral sclerosis (ALS), and other neurodegenerative disorders (Bezprozvanny, 2009; Mattson, Gleichmann, and Cheng, 2008; Pivovarova and Andrews, 2010).

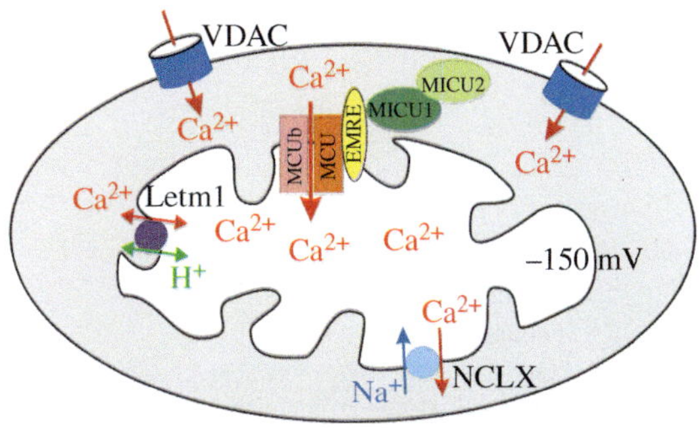

Expression and function of mitochondrial Ca^{2+} transporters in neurons

Protein	Alternative name	Function	Nervous system expression (method)	KD or KO phenotype	References
MCU	CCDC109A	Ca^{2+} uptake	Cortex, cerebellum, hippocacampus, hypothalamus, spinal cord, DRG (mouse gene expression atlas and PCR) Cortex and hippocampus (western blot)	Reduced mitochondrial Ca^{2+} uptake, protection from excitotoxicity	Plovanich et al., 2013 Qiu et al., 2013 Sankak et al., 2013 Shutov et al., 2013
MCUb	CCDC109B	Ca^{2+} uptake	Cortex, cerebellum, hypothalamus, spinal cord (mouse gene expression atlas and qPCR)		Sankak et al., 2013 Raffaello et al., 2013
MICU1	CBARA1	Ca^{2+} uptake regulator	Cortex, cerebellum, hypothalamus, spinal cord, DRG (mouse gene expression atlas and PCR) Cerebellum, ventral tegmental nuclei (immunohistochemistry)	Increased mitochondrial Ca^{2+} uptake, learning impairment, ataxia, neuropathies **loss-of-function mutations	Plovanich et al., 2013 Logan et al., 2014 Sankak et al., 2013 Shutov et al., 2013
MICU2	EFHA1	Ca^{2+} uptake regulator	Hypothalamus, spinal cord (mouse gene expression atlas)		Plovanich et al., 2013 Sankak et al., 2013
MICU3	EFHA2	Ca^{2+} uptake regulator	Cortex, cerebellum, hypothalamus, spinal cord (mouse gene expression atlas)		Plovanich et al., 2013 Sankak et al., 2013
EMRE		Ca^{2+} uptake regulator	Cortex, cerebellum, hypothalamus, spinal cord (mouse gene expression atlas)		Plovanich et al., 2013 Sankak et al., 2013
Letm1		Ca^{2+} uptake and release	Brain (Western blot) DRG (PCR)	Reduced mitochondrial Ca^{2+} uptake, impaired control of ATP synthesis in the brain, increased seizures	Tamai et al., 2008 Jiang et al., 2013 Shutov et al., 2013
NCLX	SLC8B1	Ca^{2+} release	Brain (western blot and immunohistochemistry with EM) DRG (PCR)		Palty et al., 2010 Shutov et al., 2013

FIGURE 4.4 Molecular components of mitochondrial Ca^{2+} transport in neurons. The upper panel shows the main components of the mitochondrial Ca^{2+} transport machinery in the inner mitochondrial membrane. The table summarizes the information about the expression and function of the mitochondrial Ca^{2+} transporters in neurons, including the data from knockdown (KD) and knockout (KO) studies, as well as a recent report on the loss-of-function MICU1 mutations in humans.

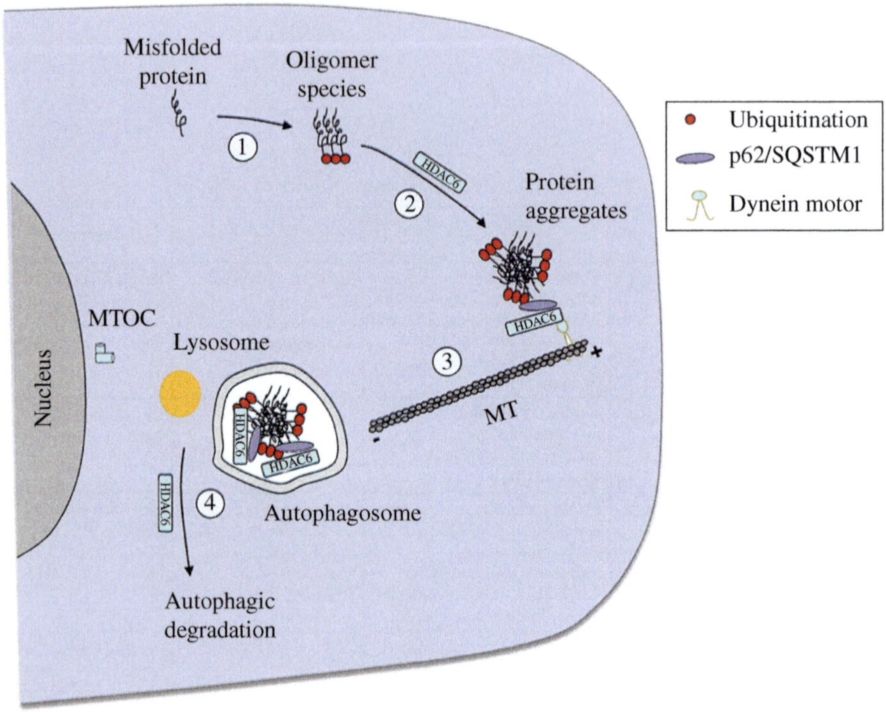

FIGURE 6.3 Schematic processes of HDAC6-mediated autophagic elimination of proteotoxicity. Proteotoxicity elimination requires HDAC6 at multi-levels. Step 1, genetic or environmental causes lead to accumulated misfolded proteins that is prone to aggregate into oligomer species. These oligomer species, if not eliminated, are toxic to the cells leading to eventually cell deaths. Step 2, the toxic oligomers with markers such as specific ubiquitinations can be recognized by the ubiquitin-binding domain of HDAC6. Recruitment of HDAC6 to the oligomers facilitates the formation of high weight protein aggregates. Step 3, the ubiquitin-positive protein aggregates can be transported to the microtubule organizing center (MTOC) by HDAC6-mediated microtubule (MT) network. Step 4, HDAC6 promotes the fusion of protein aggregates-containing autophagosomes and lysosomes, which is essential for the eventual clearance of protein aggregates.

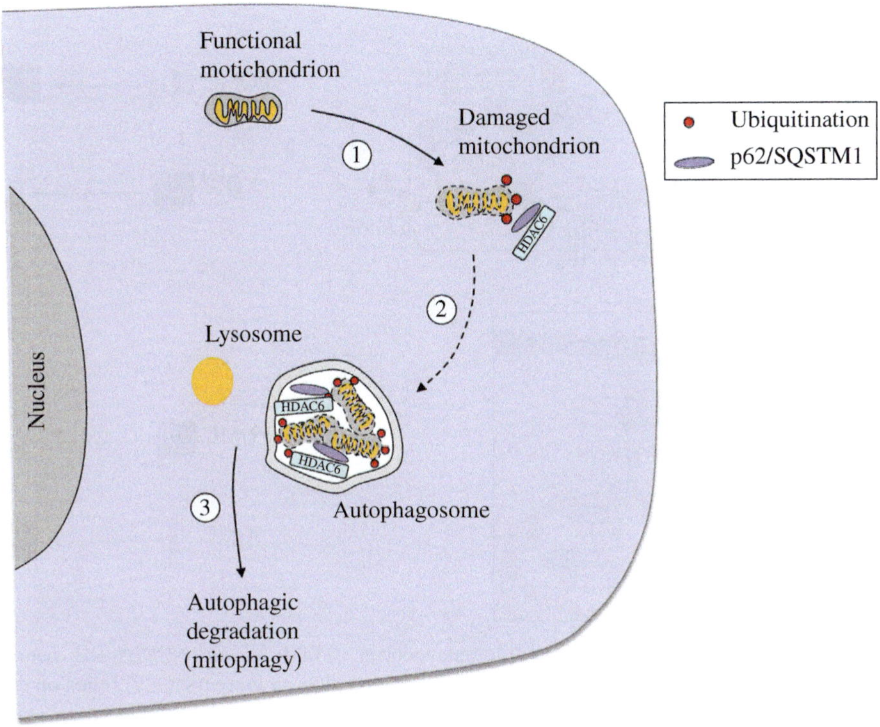

FIGURE 6.4 Schematic processes of HDAC6-mediated clearance of damaged mitochondria. HDAC6 is required for the autophagic clearance of dysfunctional mitochondria. Step 1, the damaged mitochondria that are resulted from causative mutations or environmental factors are accumulated in the cytoplasm, leading to cellular toxicity. Step 2, HDAC6 is recruited to the damaged mitochondria by specific ubiquitin signals and the recruitment of HDAC6 to the dysfunctional mitochondria enhances the autophagic clearance (mitophagy). Step 3, autophagic activities are necessary for the degradation of damaged mitochondria. It is unknown whether HDAC6 is also involved in the fusion of the damaged mitochondria-containing autophagosomes with the lysosomes.

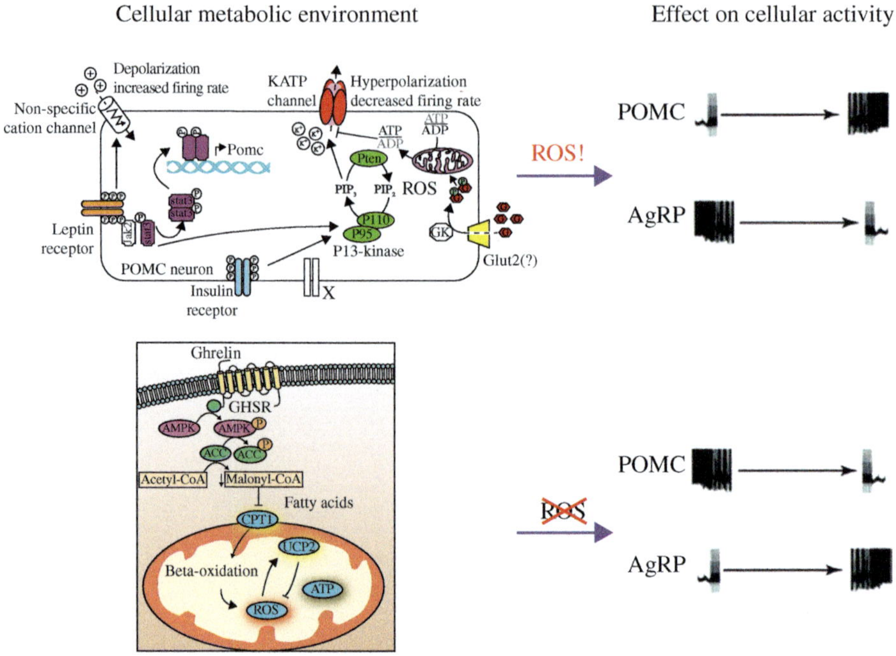

FIGURE 8.2 Regulation of pro-opiomelanocortin (POMC) and NPY/AgRP neuron activation. The firing of POMC neurons during elevated glucose levels (satiety) relies on cellular glucose metabolism and signaling pathways activated by leptin and insulin receptors *(top)*. It appears that a fundamental signaling event that affects action potential generation in these cells is glucose-derived reactive oxygen species (ROS). The same event, however, hyperpolarizes NPY/AgRP neurons and suppresses their firing. During low glucose levels, fatty acid metabolism is increased *(bottom)*. Active phosphorylated AMP-activated protein kinase (AMPK) phosphorylates ACC, leading to a decrease in its enzymatic activity. This latter event decreases levels of malonyl-CoA, releasing CPT1 to uptake fatty acids inside the mitochondria. Fatty acid metabolism, promoted by UCP2 activity, reduces intracellular ROS levels leading to the decreased activity of POMC neurons but elevated firing of NPY/AgRP neurons.

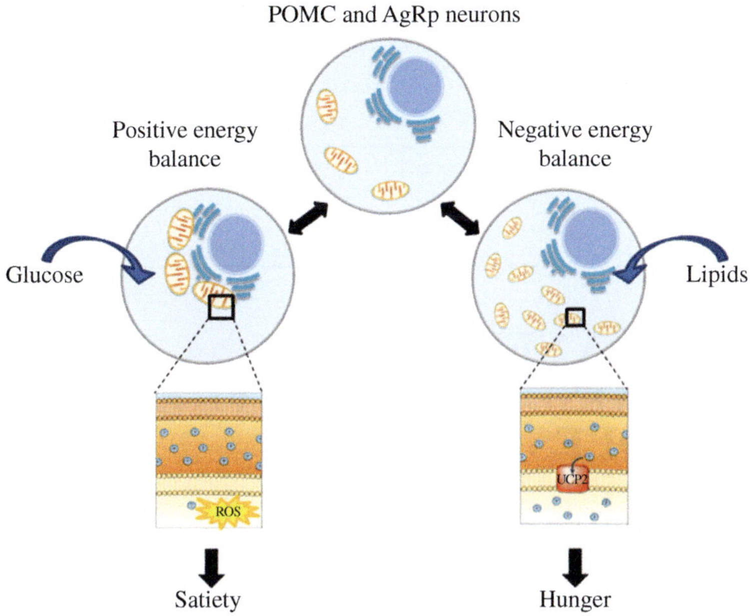

FIGURE 8.3 Feeding states and mitochondrial dynamics in AgRP and pro-opiomelanocortin (POMC) neurons. The activity of POMC neurons promotes satiety associated with glucose utilization, elevated reactive oxygen species (ROS) levels, increase in mitochondrial fusion and contact with the ER. By contrast, the activity of NPY/AgRP neurons during fasting or caloric restriction promotes fat utilization and hunger associated with low ROS levels and increase mitochondrial fission.

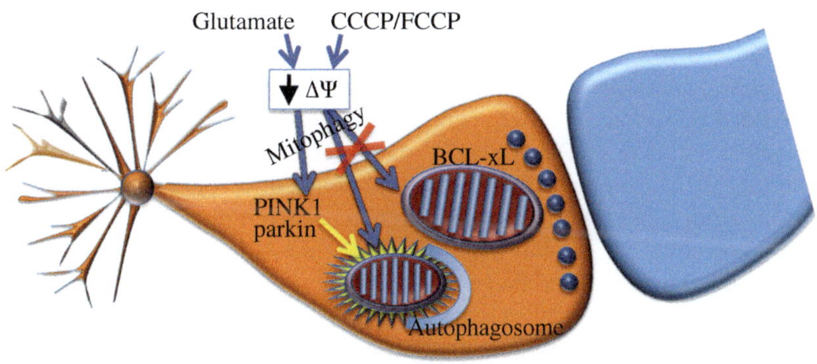

FIGURE 10.1 Mitochondrial organelles in primary neurons differ from those in cancer cell lines by their propensity to avoid Parkin-mediated mitophagy. This mechanism potentially preserves essential mitochondrial functions and delays neurodegeneration as long as possible. Agents that trigger the collapse of mitochondrial inner membrane potential are useful in vitro models and likely mimic in part physiological damage to mitochondria. Depolarization (loss of $\Delta\Psi$) induced by ionophores/protonophores such as CCCP/FCCP potently induce trigger Parkin translocation to mitochondria and mitophagy in HeLa cells but not in neurons. In contrast, transient exposure to the excitatory neurotransmitter glutamate, which potently and rapidly induces mitochondrial depolarization in neurons, induces Parkin translocation to mitochondria but only a fraction of the mitochondria in each neuron. Furthermore, subsequent mitophagy does not occur unless neurons are treated with the antioxidant NAC. NAC-triggered mitophagy is NMDA receptor-dependent and Ca^{2+}-flux dependent in this model.

11

PINK1 AS A SENSOR FOR MITOCHONDRIAL FUNCTION: DUAL ROLES

ERIN STEER, MICHELLE DAIL AND CHARLEEN T. CHU

Department of Pathology, University of Pittsburgh School of Medicine, Pittsburgh, Pennsylvania USA

11.1 INTRODUCTION

Approximately 1% of the population older than the age of 65 is affected by Parkinson's disease (PD), the second-most common neurodegenerative disorder (de Rijk et al., 2000). Loss of dopaminergic neurons from the pars compacta of the substantia nigra, often preceded by formation of intracellular inclusion bodies called Lewy bodies, is a pathological hallmark of PD (Braak et al., 2003). The resulting loss of signaling from the nigrostriatal system leads to the development of motor signs, including: postural instability, bradykinesia, resting tremor, and rigidity. The majority of PD cases are idiopathic, but a small percentage of affected individuals have heritable forms of the disease (Trinh and Farrer, 2013).

Mutations in the gene that encodes for PTEN-induced kinase 1 (PINK1) are responsible for the second-most common cause of autosomal recessive parkinsonism (Ibáñez et al., 2006). The early onset parkinsonism caused by pathogenic PINK1 mutations was first identified in 2004 (Valente et al., 2004) and is characterized clinically by earlier age of onset (range of 9 to 52 years), long disease duration, and sustained response to levodopa treatment (Bonifati et al., 2005; Klein et al., 2005; Leutenegger et al., 2006; Valente et al., 2004). Either reduced kinase activity or

The Functions, Disease-Related Dysfunctions, and Therapeutic Targeting of Neuronal Mitochondria, First Edition. Edited by Valentin K. Gribkoff, Elizabeth A. Jonas and J. Marie Hardwick.

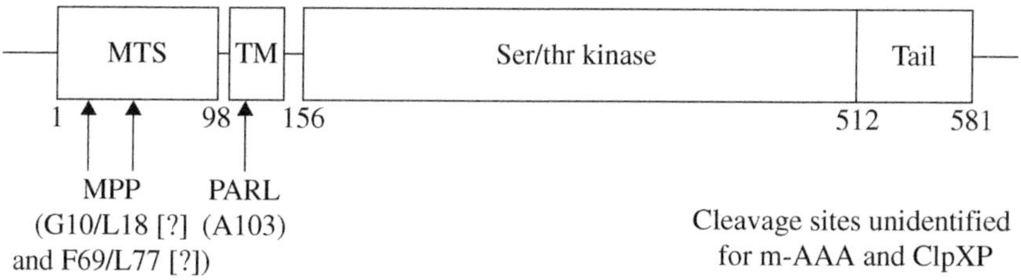

FIGURE 11.1 PINK1 is processed by multiple mitochondrial proteases. PINK1 is cleaved by PARL at A103. Cleavage by m-AAA generates a product that is approximately the same size as the PARL processed fragment (Greene et al., 2012a). Cleavage sites for m-AAA and ClpXP have not been identified. The MTS of PINK1 contains two predicted MPP cleavage motifs (Sim et al., 2012). The first would cleave PINK1 at amino acids G10 and L18, the second at F69 and L77. However, it remains unclear whether MPP is involved in PINK1 processing (Becker et al., 2012; Kato et al., 2013).

decreased protein stability have been demonstrated with the majority of identified mutations, which suggests that loss of PINK1 function is pathogenic.

PINK1 is a 581 amino acid, serine/threonine kinase with homology to the calcium/calmodulin family. It is comprised of a C-terminal regulatory tail, putative transmembrane and kinase domains, and a N-terminal canonical mitochondrial targeting sequence (MTS) (Sim et al., 2012) (Fig. 11.1). In healthy mitochondria, PINK1 is imported and processed, and a significant portion of processed PINK1 appears to be released back into the cytosol (Beilina et al., 2005; Fedorowicz et al., 2014; Haque et al., 2008; Lin and Kang, 2008; Meissner et al., 2011; Takatori, Ito, and Iwatsubo, 2008; Weihofen et al., 2008). However, mitochondrial dysfunction induces the accumulation of full-length, unprocessed PINK1 on the outer mitochondrial membrane (OMM). A growing body of evidence suggests that this switch in subcellular distribution of PINK1 relays information to the cell about the integrity of its mitochondrial networks. Here we identify the shared and unique functions of the mitochondrial and cytosolic pools to describe how PINK1 acts as a sensor of both mitochondrial function and dysfunction (Fig. 11.2).

11.2 PINK1 PROMOTES MITOCHONDRIAL FUNCTION

A growing body of evidence indicates that mitochondrial dysfunction is a key feature of PD pathogenesis. Compromised function of the electron transport chain (ETC) has been observed both in PD patient tissue, and genetic and toxin models of PD (Celardo, Martins, and Gandhi, 2014; Chaturvedi and Flint Beal, 2013). The unique bioenergetic demands of neurons make maintenance of mitochondrial homeostasis crucial for their health and survival. Interestingly, impaired respiration, decreased mitochondrial membrane potential ($\Delta\Psi_m$), and reduced levels of adenosine triphosphate (ATP) have also been reported in PINK1 deficient mammalian cells and animal models. A study employing mitochondria isolated from PINK1 knockout mouse embryonic fibroblasts

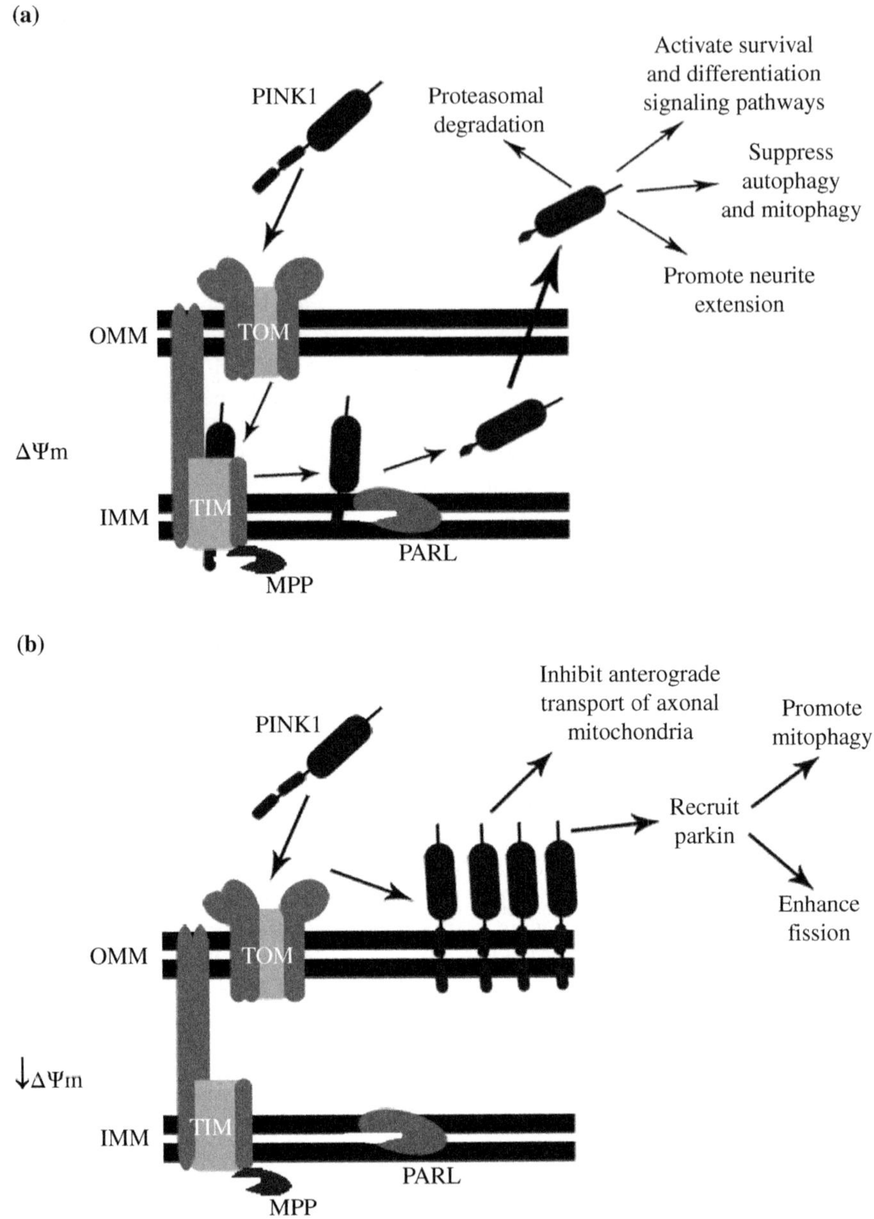

FIGURE 11.2 PINK1 as a sensor of mitochondrial health and damage. **(a)** In healthy mitochondria, PINK1 is imported through the TOM/TIM complexes and processed by several mitochondrial proteases before being released back into the cytosol. This cytosolic pool of PINK1 may serve as a signal to the cell that the integrity of its mitochondrial network remains intact. Processed, cytosolic PINK1 has been implicated in suppressing autophagy/mitophagy, promoting neurite extension, enhancing mitochondrial trafficking into dendrites, and activating pro-survival and differentiation signaling pathways. **(b)** When mitochondria lose their membrane potential, full-length PINK1 accumulates on the OMM. As a result, axonal transport of mitochondria is arrested and the damaged mitochondrion becomes a target for Parkin-mediated degradation.

(MEFs) found that a reduction in substrate oxidation, rather than proton leak, was responsible for the loss of $\Delta\Psi_m$ (Amo et al., 2011). This finding is consistent with reports of reduced ETC complex I (Amo et al., 2011; Gautier, Kitada, and Shen, 2008; Gautier et al., 2012; Morais et al., 2009) and II (Amo et al., 2011; Gautier et al., 2008, 2012) linked respiration in PINK1 deficient neurons and MEFs. However, increased opening of the mitochondrial permeability transition pore (mPTP) (Gautier et al., 2012), and deficient activity of the mitochondrial Na^+/Ca^{2+} exchanger (NCLX) (Gandhi et al., 2009) have also been shown to contribute to loss of $\Delta\Psi_m$ and impaired respiration observed in PINK1 knockout cells. Further, PINK1 deficiency also affects substrate availability (Gandhi et al., 2009) and activity of aconitase (Gautier et al., 2008), an enzymatic component of the tricarboxylic acid (TCA) cycle. Although the mechanism underlying disruption of normal ETC function and ATP production caused by PINK1 deficiency remains unclear, Vitamin K_2 is able to reverse ATP deficiencies in PINK1 null flies by its action as an ETC electron transporter (Vos et al., 2012). This finding has not yet been validated in mammalian systems but may suggest that improving the efficiency of the ETC is sufficient to overcome some bioenergetic consequences of the loss of PINK1 function.

In addition to energy production, the mitochondria's ability to buffer calcium is crucial to neuronal function. This is especially true of neurons that are exposed to frequent and sustained elevations in intracellular calcium concentration, such as the vulnerable populations in PD (reviewed in Chan, Gertler, and Surmeier, 2009). Cytosolic calcium is transported across the IMM into the matrix via the $\Delta\Psi_m$ dependent activity of the mitochondrial calcium uniporter (MCU) (De Stefani et al., 2011), and efflux of calcium from the mitochondria occurs through the NCLX (Palty et al., 2010). Dysregulation of mitochondrial calcium handling has been reported in PINK1 deficient cells (Gandhi et al., 2009; Gautier et al., 2012; Heeman et al., 2011). Elevated levels of calcium in the mitochondrial matrix have been observed in PINK1 deficient neuronal cells and MEFs (Gandhi et al., 2009; Gautier et al., 2012). Gandhi et al. found that mitochondrial calcium extrusion in response to sodium was impaired in PINK1 deficient neurons, and reasoned that accumulation of calcium in the matrix was likely the result of diminished efflux activity of NCLX (Gandhi et al., 2009). It was recently proposed that elevations in mitochondrial matrix calcium concentration through this mechanism leads to increased susceptibility to opening of the mPTP in PINK1 knockout MEFs (Gautier et al., 2012). These findings are inconsistent with a report that PINK1 depletion impairs mitochondrial uptake of calcium upon stimulation (Heeman et al., 2011). The authors of this study suggest that this reduction in mitochondrial uptake of calcium is secondary to the observed loss of $\Delta\Psi_m$, and results in diminished ATP production as result of reduction in enzymatic activity of components of the TCA cycle.

Overexpression of PINK1 protects against a number of apoptotic stimuli with relevance to PD associated injury, including oxidative stress (Pridgeon et al., 2007), proteasomal inhibition (Valente et al., 2004), mitochondrial depolarization (Arena et al., 2013), and ETC complex I inhibition (Haque et al., 2008). Additionally, PINK1 depletion sensitizes cells to apoptosis induced by complex I inhibitors (Deng et al., 2005; Haque et al., 2008) and mitochondrial depolarization (Arena et al., 2013).

PINK1-dependent phosphorylation of the mitochondrial chaperone TNF receptor-associated protein 1 (TRAP1) (Pridgeon et al., 2007), and anti-apoptotic protein Bcl-xL (Arena et al., 2013) have been shown to protect against cell death induced by oxidative stress and mitochondrial depolarization respectively. Further characterization of the upstream and downstream components of these neuroprotective mechanisms will advance the current understanding of how PINK1 protects against apoptotic cell death.

11.3 HEALTHY MITOCHONDRIA IMPORT AND PROCESS PINK1

Virtually all nuclear-encoded mitochondrial proteins are imported across the outer mitochondrial membrane (OMM) via the translocase of the outer membrane (TOM) complex (Dudek, Rehling, and van der Laan, 2013). Two receptor subunits of this complex, TOM20 and TOM22 recognize N-terminal mitochondrial targeting sequences (MTS) and initiate transport through the β-barrel pore of TOM40. Because PINK1 has a predicted N-terminal MTS, and complexes with TOM20, 22, and 40 on depolarized mitochondria (Lazarou et al., 2012), it is likely to be transported across the OMM through this pathway. However, a recent study found that PINK1 import can also occur in the absence of TOM40, through a mechanism that requires TOM70 (Kato et al., 2013). TOM70 is a receptor subunit of the TOM complex that detects internal mitochondrial localization signals (Dudek et al., 2013). Cytosolic chaperones, including known PINK1 interactors, specifically interact with TOM70 to initiate transport of proteins that contain internal mitochondrial localization signals (Zara et al., 2009). This finding provides insight into how trace amounts of PINK1 constructs lacking the predicted N-terminal MTS may be imported into the mitochondria as detected by protease protection assays (Becker et al., 2012; Zhou et al., 2008), and why interaction with CDC37/Hsp90 increases mitochondrial localization of PINK1 (Lin and Kang, 2010; Weihofen et al., 2008). Nevertheless, as only trace levels of overexpressed N-terminal deletions of PINK1 can be detected in mitochondria, it is clear that the major pathway of PINK1 mitochondrial import occurs via the canonical MTS.

The import and insertion of PINK1 into the inner mitochondrial membrane (IMM) is presumed to occur through the translocase of the inner membrane 23 (TIM23) complex in a process driven by $\Delta\Psi_m$. In general, proteins that are imported through the TOM complex are recognized by TIM50, which acts to guide the MTS through TIM23 (Dudek et al., 2013). Proteins destined for the IMM generally have a hydrophobic sequence following the MTS that acts as a stop transfer signal. Upon sequence analysis of PINK1, a putative transmembrane domain that could act as a stop transfer signal has been identified (Sim et al., 2012; Zhou et al., 2008).

11.3.1 Localization and Processing of PINK1 Depends on an Intact $\Delta\Psi_m$

The submitochondrial localization of PINK1, and its concurrent processing by mitochondrial proteases, occurs through a $\Delta\Psi_m$–dependent mechanism. An interference RNA (RNAi) screen suggested that the mitochondrial processing peptidase (MPP) processes PINK1, as it does many other mitochondrial proteins containing an N-terminal MTS (Greene et al., 2012). Sim et al. (2012) have identified two conserved regions of PINK1, corresponding to amino acids 8–18 and 67–77 of human PINK1, that contain potential MPP cleavage motifs. However, it has recently been shown that PINK1 processing was unaffected in isolated mitochondria treated with a divalent ion chelator to inhibit MPP (Becker et al., 2012; Kato et al., 2013), and recombinant MPP did not cleave PINK1 in an in vitro assay (Kato et al., 2013).

The processing of the PINK1 precursor (~60 kDa band) into its ΔN_1 form (~50 kDa) by presenilin-associated rhomboid-like protein (PARL) has been well characterized (Deas et al., 2011; Greene et al., 2012; Jin et al., 2010; Meissner et al., 2011). This integral IMM protease cleaves PINK1 between the alanine-103 and phenylalanine-104 residues, a site within the putative transmembrane domain (Deas et al., 2011). The resultant product is likely targeted for proteasomal degradation (Yamano and Youle, 2013), if not stabilized in a complex with other proteins or further processed by other peptidases. Two ATP-dependent proteases, matrix-AAA (m-AAA) and caseinolytic mitochondrial matrix peptidase (ClpXP) have also been identified as PINK1 processors in a RNAi screen performed by Green et al. (2012). The authors suggest that this cleavage of PINK1 by m-AAA produces a fragment of approximately the same size as the PARL truncated form. The protease responsible for producing the ΔN_2 fragment (~45 kDa band) has not been identified. Following mitochondrial processing, PINK1 is released into the cytosol through an unidentified mechanism.

As discussed, growing evidence indicates that this release of processed PINK1 may serve to communicate mitochondrial status to the rest of the neuron (Chu, 2010a; Gusdon and Chu, 2011). Although failure to process and release PINK1 leads to mitophagy, processed PINK1 signals and maintains a healthy mitochondrial status to suppress mitophagy (Dagda et al., 2009a) and support dendritic outgrowth and neuronal differentiation (Dagda et al., 2013a).

11.4 ACCUMULATION OF FULL LENGTH-PINK1 AS A SENSOR OF MITOCHONDRIAL DYSFUNCTION

In healthy cells, the processes of mitochondrial fission, fusion, and mitochondrial degradation by macroautophagy (mitophagy) are balanced in a dynamic equilibrium necessary for maintaining a healthy mitochondrial network. This is particularly important in neurons, which are dependent on oxidative phosphorylation and require extensive mitochondrial transport among dendrites and axons. Fusion events, mediated by the mitofusins (Mfn1 and 2) and optic atrophy protein (Opa1), allow for sharing of components. Fusion is often followed by fission events, mediated by Drp1, which allow for detachment of mitochondria from the network for transport as well as separation of damaged components for degradation (Chan, 2006). A fraction of fission events may result in daughter mitochondria with different mitochondrial membrane potentials (Twig, Hyde, and Shirihai, 2008b); the daughter with the decreased $\Delta\Psi_m$ is targeted for mitophagy (Twig et al., 2008a). When damaged or dysfunctional mitochondria are unable to repair themselves or undergo fusion, they are also targeted for mitophagy. However, the exact mechanism by which mitochondria are targeted to the autophagosome is an emerging area of research.

In some model systems, PINK1 itself has been identified as a trigger for Parkin-dependent mitochondrial degradation (Matsuda et al., 2010; Narendra et al., 2010a). Treatment of mammalian cells with CCCP causes mitochondrial depolarization, which leads to cessation of import and intramitochondrial processing of PINK1 and results in accumulation of full-length PINK1 on the OMM. Parkin, an E3 ubiquitin ligase, is recruited to the mitochondrion and facilitates its degradation by mitophagy (Greene et al., 2012; Narendra et al., 2008). Parkin recruitment also drives degradation of mitofusin which promotes a pro-fission state (Gegg et al., 2010), although it remains unclear if this occurs through proteasomal, lysosomal or other mechanisms. As Drp1-dependent fission may be necessary for mitophagy (Dagda et al., 2009; Frank et al., 2012), the shift to a pro-fission state would serve to facilitate mitophagy (Twig et al., 2008b). Although the selective autophagy cargo adaptor protein p62 has been implicated in this model of mitophagy (Ding et al., 2010; Geisler et al., 2010), this is still controversial (Narendra et al., 2010b; Okatsu et al., 2010; Youle and Narendra, 2011).

A recent set of *Drosophila* studies also suggests that PINK1 and Parkin are involved in turnover of mitochondrial proteins (Vincow et al., 2013). Interestingly, however, this study shows that Parkin is more tightly linked to Atg7-dependent turnover than PINK1, with PINK1 involved selectively in regulating the apparent stability of respiratory chain proteins. The mismatch between patterns observed with autophagy-deficient Atg7 mutants and Parkin or PINK1 may be explained by additional involvement of the proteasome (Chan et al., 2011; Yoshii et al., 2011) or intrinsic mitochondrial proteases. In any case, the regulation of mitophagy and mitochondrial protein turnover is likely to be complex, and multiple pathways may be engaged.

As previously reviewed by Van Laar and Berman (Van Laar and Berman, 2013b), the mitophagic pathway seems to vary by the bioenergetic state and morphology of the cell, and the type of stressor. The CCCP-PINK Parkin process occurs robustly in non-neuronal, glycolytic cells to result in complete or near-complete mitochondrial clearance. In neuronal and oxidative phosphorylation dependent cells, CCCP treatment does not induce Parkin translocation (Van Laar and Berman, 2013b) until after 24 hours of treatment (Cai et al., 2012). Rotenone and 6-OHDA toxin models did not show accumulation of PINK or Parkin translocation, unlike CCCP (Dagda et al., 2013). The induction of mitophagy in PINK1-deficient animals and cells, dependent on increased superoxide and Drp-mediated fission, suggests additional mechanisms of PINK1-independent mitophagy (Dagda et al., 2009). This may involve externalization of cardiolipin (CL), an inner mitochondrial membrane phospholipid, to the mitochondrial surface, which was shown to occur in rotenone-treated cortical neurons. CL then interacts with microtubule-associated-protein-1-light chain-3 (LC3) on autophagic membranes to mediate mitophagy of the damaged mitochondrion (Dagda et al., 2013).

11.5 CYTOSOLIC PINK1 AS A SENSOR FOR MITOCHONDRIAL FUNCTION

Although a large body of literature indicates that accumulation of full length PINK1 on the mitochondrial surface serves to trigger Parkin-related degradation pathways, less is known about the potential signaling function(s) of correctly processed PINK1. Given that healthy mitochondria process and release PINK1 to the cytosol, we propose that PINK1 could also serve as a sensor for mitochondrial function. In other words, correctly processed PINK1 may be released from healthy mitochondria to regulate other cellular functions.

11.5.1 Cytosolic PINK1 Suppresses Cell Death and Autophagy/Mitophagy

Hacque et al. (2008) were the first to suggest that cytoplasmic activities of PINK1 may be important for its neuroprotective function. They reported that a PINK1 construct lacking the first 111 amino acids (ΔN-PINK1), was able to protect against MPP+ toxicity in vitro and MPTP toxicity in vivo despite a primarily cytosolic localization. It has since been discovered that PINK1 induces activation of the cytosolically localized mammalian target of rapamycin complex 2 (mTORC2)/Akt pathway, and activation of this pathway contributes to PINK1's ability to protect against apoptosis induction by a panel of cellular stressors (Murata et al., 2011).

Moreover, knockdown or knockout studies of PINK1 reveal increased autophagy and mitophagy, suggesting that endogenous PINK1 may serve to suppress autophagy/mitophagy (Cui et al., 2011; Dagda et al., 2009; Liu and Lu, 2010; Qi et al., 2011). As mTOR/Akt represent a major autophagy-suppressive pathway in cells, it is possible that activation of this pathway by PINK1 is not only involved in preventing apoptosis (Murata et al., 2011) but also in suppressing autophagy. We have shown that ΔN-PINK1 is sufficient to suppress macroautophagy induced by PINK1 knockdown (Dagda et al., 2009), or by 6-hydroxydopamine (6-OHDA) treatment (Dagda and Chu, unpublished data). A recent pair of reports suggest that ΔN-PINK1 may also repress mitophagy by directly binding Parkin (Fedorowicz et al., 2014), or by promoting ubiquitination of endogenous parkin (Rakovic et al., 2013), thus preventing its mitochondrial localization. Finally, cytosolic PINK1 promotes the PKA-dependent phosphorylation of LC3 (Cherra and Chu, unpublished data). Phosphorylation of LC3 at this site suppresses injury-induced autophagy (Cherra et al., 2010), and cytosolic PKA is sufficient to inhibit the effects of PINK1 deficiency on autophagy (Dagda et al., 2011). These apparently contradictory effects of PINK1 can be reconciled by the concept of PINK1 processing as a switch. Whereas the mitophagy-promoting effects of PINK1 depend on accumulation of the full-length protein on the surface of dysfunctional mitochondria, processed PINK1 acts as a sensor for functional mitochondria to suppress mitochondrial turnover by autophagy (Fedorowicz et al., 2014; Geisler et al., 2010; Matsuda et al., 2010; Michiorri et al., 2010; Narendra et al., 2010a; Rakovic et al., 2013).

11.5.2 Cytosolic PINK1 Promotes Neurite Extension and Cell Survival

In addition to its cytoprotective activity and autophagy/mitophagy-suppressive effects, recent data indicate that cytosolic PINK1 regulates neurite morphology (Dagda et al., 2013). Knockdown of PINK1 in primary cortical and midbrain neurons reduces neurite length and complexity. Overexpression of either full length or ΔN-PINK1 is able to significantly rescue neurite morphology in these cells. A PINK1 construct targeted to the OMM, despite being able to partially reverse the effects of PINK1 deficiency on mitochondrial homeostasis, had no effect on neurite length. Additionally, ΔN-PINK1 promotes mitochondrial transport into dendrites in a kinase dependent manner (Dagda et al., 2013). Although the mechanism has not been completely elucidated, PKA signaling is required for neurite outgrowth promoted by cytosolic PINK1 (Dagda et al., 2013), suggesting that PINK1 and PKA signaling pathways converge to coregulate neurite extension.

11.6 PINK1 AND MITOCHONDRIAL DYNAMICS

In mammalian cells, loss of PINK1 function has also been associated with the accumulation of fragmented mitochondria caused by deregulation of Dynamin related protein 1 (Drp1) mediated fission (reviewed in Chu, 2010b; Van Laar and Berman, 2013a). It remains unclear whether this aberrant regulation of dynamics induced by PINK1 depletion is a primary or secondary effect. It is conceivable that the reported electron transport chain dysfunction arises from a failure to isolate and degrade dysfunctional mitochondria. However, it is also possible that mitochondrial function is directly impaired by loss of PINK1, and these defects induce fission (Dagda et al., 2009; Exner et al., 2007; Lutz et al., 2009). A secondary effect may also be indirectly supported by observations that although PINK1 promotes mitochondrial fusion or inhibits fission in some model systems (Dagda et al., 2009; Kamp et al., 2010; Sandebring et al., 2009), it has the opposite effect on mitochondrial morphology in other systems (Clark et al., 2006; Yang et al., 2008; Yu et al., 2011). Our proposed model may explain these apparently contradictory effects. Healthy mitochondria would release processed, cytosolic PINK1, which acts to suppress fission and stabilize functional mitochondrial networks. When this signal is absent, as a result of the accumulation of full-length PINK1 on the OMM, the fission/fusion balance is then altered to favor mitochondrial fragmentation, which serves to isolate dysfunctional segments of mitochondria for repair or clearance by mitophagy.

PINK1 has been shown to form a complex with Miro, an outer mitochondrial membrane Rho GTPase, and Milton in *Drosophila*, proteins involved in trafficking of mitochondria (Weihofen et al., 2009). PINK1 together with Parkin down-regulate Miro protein levels in *Drosophila*, leading to a decrease in mitochondrial trafficking in axons (Liu et al., 2012). RNAi knockdown of Miro in CCCP treated HeLa cells leads to increased Parkin-mediated mitophagy (Liu et al., 2012). As postulated by Van Laar and Berman (2013), the ability of PINK1/Parkin to decrease mitochondrial trafficking in axons could isolate the damaged mitochondria from the functional network, prevent the accumulation of damaged mitochondria in high-energy demand areas, such as axon terminals, and allow the mitochondria to be degraded (Van Laar and Berman, 2013).

Alternatively, processed cytoplasmic PINK1 has also been shown to promote mitochondrial transport into dendrites (Dagda et al., 2013). Since the polarity of microtubules differs between axons and dendrites (Franker and Hoogenraad, 2013), kinesin-1/KIF5B-Miro is primarily responsible for axonal anterograde trafficking, whereas the dynein/dynactin-Miro complex is responsible for anterograde dendritic trafficking (van Spronsen et al., 2013). The difference in effects in axons and dendrites may be related to different interactions of mammalian Miro-1 and 2 and trafficking kinesin adaptor proteins (TRAK), the mammalian orthologues of Milton, with transport machinery (van Spronsen et al., 2013). More fundamentally, the differences in effects of full length PINK1/Parkin versus processed cytosolic PINK1 may reflect the ability of PINK1 to not only sense mitochondrial dysfunction but also to signal mitochondrial health to the rest of the cell.

11.7 DUAL ROLES FOR PINK1 AS A SENSOR OF MITOCHONDRIAL FUNCTION AND DYSFUNCTION

In summary, the potential biological roles of PINK1, as with any other kinase, vary depending upon subcellular localization and accessibility to downstream signaling partners. A better understanding of the processing and regulation of PINK1 in the presence of functional mitochondria provide important information for a spectrum of PINK1-dependent functions that may be disrupted by recessive PD-linked mutations. Although much has been learned about the catabolic function of PINK1 in response to mitochondrial depolarization, another emerging body of data suggests that the export of N-terminally cleaved PINK1 from healthy mitochondria may function as a signal of energetic competence to promote neuronal trophic responses (see Fig. 11.2). The cytosolic pool of PINK1 relays this message by suppressing autophagy, activating pro-survival pathways, regulating mitochondrial transport and engaging pro-differentiation signaling pathways. Further investigation is needed to determine whether manipulation of these pathways may have a therapeutic benefit.

REFERENCES

Amo, T., et al., 2011. Mitochondrial membrane potential decrease caused by loss of PINK1 is not due to proton leak, but to respiratory chain defects. *Neurobiol Dis*, 4, pp.111–118.

Arena, G., et al., 2013. PINK1 protects against cell death induced by mitochondrial depolarization, by phosphorylating Bcl-xL and impairing its pro-apoptotic cleavage. *Cell Death Differ*, 20, pp.920–930.

Becker, D., et al., 2012. Pink1 kinase and its membrane potential (Deltaψ)-dependent cleavage product both localize to outer mitochondrial membrane by unique targeting mode. *J Biol Chem*, 287, pp.22969–22987.

Beilina, A., et al., 2005. Mutations in PTEN-induced putative kinase 1 associated with recessive parkinsonism have differential effects on protein stability. *Proc Natl Acad Sci U S A*, 102, pp.5703–5708.

Bonifati, V., et al., 2005. Early-onset parkinsonism associated with PINK1 mutations: frequency, genotypes, and phenotypes. *Neurology*, 65, pp.87–95.

Braak, H., et al., 2003. Staging of brain pathology related to sporadic Parkinson's disease. *Neurobiol Aging*, 24, pp.197–211.

Cai, Q., et al., 2012. Spatial parkin translocation and degradation of damaged mitochondria via mitophagy in live cortical neurons. *Curr Biol*, 22, pp.545–552.

Celardo, I., Martins, L. and Gandhi, S., 2014. Unravelling mitochondrial pathways to Parkinson's disease. *Br J Pharmacol*, 171, pp.1943–1957.

Chan, C., Gertler, T. and Surmeier, D., 2009. Calcium homeostasis, selective vulnerability and Parkinson's disease. *Trends Neurosci*, 32, pp.249–256.

Chan, D. C., 2006. Mitochondrial fusion and fission in mammals. *Annu Rev Cell Dev Biol*, 22, pp.79–99.

Chan, N., et al., 2011. Broad activation of the ubiquitin-proteasome system by Parkin is critical for mitophagy. *Hum Mol Gen*, 20, pp.1726–1737.

Chaturvedi, R. and Flint Beal, M., 2013. Mitochondrial diseases of the brain. *Free Radic Biol Med*, 63, pp.1–29.

Cherra, S., et al., 2010. Regulation of the autophagy protein LC3 by phosphorylation. *J Cell Biol*, 190, pp.533–539.

Chu, C., 2010a. A pivotal role for PINK1 and autophagy in mitochondrial quality control: implications for Parkinson disease. *Hum Mole Gen*, 19, p.37.

Chu, C., 2010b. Tickled PINK1: mitochondrial homeostasis and autophagy in recessive Parkinsonism. *Biochimi Biophys Acta*, 1802, pp.20–28.

Clark, I., et al., 2006. Drosophila pink1 is required for mitochondrial function and interacts genetically with parkin. *Nature*, 441, pp.1162–1166.

Cui, T., et al., 2011. Silencing of PINK1 induces mitophagy via mitochondrial permeability transition in dopaminergic MN9D cells. *Brain Res*, 1394, pp.1–13.

Dagda, R., et al., 2009. Loss of PINK1 function promotes mitophagy through effects on oxidative stress and mitochondrial fission. *J Biol Chem*, 284, pp.13843–13855.

Dagda, R., et al., 2011. Mitochondrially localized PKA reverses mitochondrial pathology and dysfunction in a cellular model of Parkinson's disease. *Cell Death Differ*, 18, pp.1914–1923.

Dagda, R., et al., 2013. Beyond the mitochondrion: Cytosolic PINK1 remodels dendrites through protein kinase A. *J Neurochem*, 128, pp.867–877.

de Rijk, M., et al., 2000. Prevalence of Parkinson's disease in Europe: A collaborative study of population-based cohorts. Neurol Diseases in the Elderly Research Group. *Neurology*, 54, p.S21–S23.

De Stefani, D., et al., 2011. A forty-kilodalton protein of the inner membrane is the mitochondrial calcium uniporter. *Nature*, 476, pp.336–340.

Deas, E., et al., 2011. PINK1 cleavage at position A103 by the mitochondrial protease PARL. *Hum Mole Gen*, 20, pp.867–879.

Deng, H., et al., 2005. Small interfering RNA targeting the PINK1 induces apoptosis in dopaminergic cells SH-SY5Y. *Biochem Biophys Res Commun*, 337, pp.1133–1138.

Ding, W. X., et al., 2010. Nix is critical to two distinct phases of mitophagy, reactive oxygen species-mediated autophagy induction and Parkin-ubiquitin-p62-mediated mitochondrial priming. *J Biol Chem*, 285, pp.27879–27890.

Dudek, J., Rehling, P. and van der Laan, M., 2013. Mitochondrial protein import: common principles and physiological networks. *Biochimica et biophysica acta*. 1833, pp.274–285.

Exner, N., et al., 2007. Loss-of-function of human PINK1 results in mitochondrial pathology and can be rescued by parkin. *J Neurosci*, 27, pp.12413–12418.

Fedorowicz, M., et al., 2014. Cytosolic cleaved PINK1 represses Parkin translocation to mitochondria and mitophagy. *EMBO Reports*, 15, pp.86–93.

Frank, M., et al., 2012. Mitophagy is triggered by mild oxidative stress in a mitochondrial fission dependent manner. *Biochim Biophysica Acta*, 1823, pp.2297–2310.

Franker, M. A. and Hoogenraad, C. C., 2013. Microtubule-based transport—basic mechanisms, traffic rules and role in neurological pathogenesis. *J Cell Sci*, 126, pp.2319–2329.

Gandhi, S., et al., 2009. PINK1-associated Parkinson's disease is caused by neuronal vulnerability to calcium-induced cell death. *Molecular Cell*, 33, pp.627–638.

Gautier, C., Kitada, T. and Shen, J., 2008. Loss of PINK1 causes mitochondrial functional defects and increased sensitivity to oxidative stress. *Proc Natl Acad Sci U S A*, 105, pp.11364–11369.

Gautier, C., et al., 2012. Regulation of mitochondrial permeability transition pore by PINK1. *Molecular Neurodegeneration*, 7, p.22.

Gegg, M. E., et al., 2010. Mitofusin 1 and mitofusin 2 are ubiquitinated in a PINK1/parkin-dependent manner upon induction of mitophagy. *Hum Mol Genet*, 19, pp.4861–4870.

Geisler, S., et al., 2010. PINK1/Parkin-mediated mitophagy is dependent on VDAC1 and p62/SQSTM1. *Nat Cell Biol*, 12, pp.119–131.

Greene, A., et al., 2012a. Mitochondrial processing peptidase regulates PINK1 processing, import and Parkin recruitment. *EMBO Rep*, 13, pp.378–385.

Greene, A. W., et al., 2012b. Mitochondrial processing peptidase regulates PINK1 processing, import and Parkin recruitment. *EMBO Rep*, 13, pp.378–385.

Gusdon, A. and Chu, C., 2011. To eat or not to eat: neuronal metabolism, mitophagy, and Parkinson's disease. *Antioxidants Redox Sign*, 14, pp.1979–1987.

Haque, M., et al., 2008. Cytoplasmic Pink1 activity protects neurons from dopaminergic neurotoxin MPTP. *Proc Natl Acad Sci U S A*, 105, pp.1716–1721.

Heeman, B., et al., 2011. Depletion of PINK1 affects mitochondrial metabolism, calcium homeostasis and energy maintenance. *J Cell Sci*, 124, pp.1115–1125.

Ibáñez, P., et al., 2006. Mutational analysis of the PINK1 gene in early-onset parkinsonism in Europe and North Africa. *Brain J Neurol*, 129, pp.686–694.

Jin, S., et al., 2010. Mitochondrial membrane potential regulates PINK1 import and proteolytic destabilization by PARL. *J Cell Biol*, 191, pp.933–942.

Kamp, F., et al., 2010, Inhibition of mitochondrial fusion by α-synuclein is rescued by PINK1, Parkin and DJ-1. *EMBO J*, 29, pp.3571–3589.

Kato, H., et al., 2013. Tom70 is essential for PINK1 import into mitochondria. *PloS one*, 8, p. 58435.

Klein, C., et al., 2005. PINK1, Parkin, and DJ-1 mutations in Italian patients with early-onset parkinsonism. *Eur J Hum Genet*, 13, pp.1086–1093.

Lazarou, M., et al., 2012. Role of PINK1 binding to the TOM complex and alternate intracellular membranes in recruitment and activation of the E3 ligase Parkin. *Developmental Cell*, 22, pp.320–333.

Leutenegger, A.-L., et al., 2006. Juvenile-onset Parkinsonism as a result of the first mutation in the adenosine triphosphate orientation domain of PINK1. *Arch Neurol*, 63, pp.1257–1261.

Lin, W. and Kang, U., 2008. Characterization of PINK1 processing, stability, and subcellular localization. *J Neurochem*, 106, pp.464–474.

Lin, W. and Kang, U., 2010. Structural determinants of PINK1 topology and dual subcellular distribution. *BMC Cell Biol*, 11, p.90.

Liu, S. and Lu, B., 2010. Reduction of protein translation and activation of autophagy protect against PINK1 pathogenesis in Drosophila melanogaster. *PLoS Genet*, 6, pp.e1001237.

Liu, S., et al., 2012. Parkinson's disease-associated kinase PINK1 regulates Miro protein level and axonal transport of mitochondria. *PLoS Genet*, 8, p.e1002537.

Lutz, A., et al., 2009. Loss of parkin or PINK1 function increases Drp1-dependent mitochondrial fragmentation. *J Biologic Chem*, 284, pp.22938–22951.

Matsuda, N., et al., 2010. PINK1 stabilized by mitochondrial depolarization recruits Parkin to damaged mitochondria and activates latent Parkin for mitophagy. *J Cell Biol*, 189, pp.211–221.

Meissner, C., et al., 2011. The mitochondrial intramembrane protease PARL cleaves human Pink1 to regulate Pink1 trafficking. *J Neurochem*, 117, pp.856–867.

Michiorri, S., et al., 2010. The Parkinson-associated protein PINK1 interacts with Beclin1 and promotes autophagy. *Cell Death Differ*, 17, pp.962–974.

Morais, V., et al., 2009. Parkinson's disease mutations in PINK1 result in decreased Complex I activity and deficient synaptic function. *EMBO Mole Med.* 1, pp.99–111.

Murata, H., et al., 2011. A new cytosolic pathway from a Parkinson disease-associated kinase, BRPK/PINK1: Activation of AKT via mTORC2. *J Bio Chem*, 286, pp.7182–7189.

Narendra, D., et al., 2008. Parkin is recruited selectively to impaired mitochondria and promotes their autophagy. *J Cell Biol*, 183, pp.795–803.

Narendra, D., et al., 2010a. PINK1 is selectively stabilized on impaired mitochondria to activate Parkin. *PLoS biology*, 8, p. e1000298.

Narendra, D., et al., 2010b. p62/SQSTM1 is required for Parkin-induced mitochondrial clustering but not mitophagy; VDAC1 is dispensable for both. *Autophagy*, 6, pp.1090–1106.

Okatsu, K., et al., 2010. p62/SQSTM1 cooperates with Parkin for perinuclear clustering of depolarized mitochondria. *Genes Cells*, 15, pp.887–900.

Palty, R., et al., 2010. NCLX is an essential component of mitochondrial Na+/Ca2+ exchange. *Proc Natl Acad Sci U S A*, 107, pp.436–441.

Pridgeon, J., et al., 2007. PINK1 protects against oxidative stress by phosphorylating mitochondrial chaperone TRAP1. *PLoS Bio*, 5.

Qi, Z., et al., 2011. Loss of PINK1 function decreases PP2A activity and promotes autophagy in dopaminergic cells and a murine model. *Neurochem Int*, 59, pp.572–581.

Rakovic, A., et al., 2013. Phosphatase and tensin homolog (PTEN)-induced putative kinase 1 (PINK1)-dependent ubiquitination of endogenous parkin attenuates mitophagy: Study in human primary fibroblasts and induced pluripotent stem dell-derived neurons. *J Biol Chem*, 288, pp.2223–2237.

Sandebring, A., et al., 2009. Mitochondrial alterations in PINK1 deficient cells are influenced by calcineurin-dependent dephosphorylation of dynamin-related protein 1. *PloS One*, 4.

Sim, C., et al., 2012. Analysis of the regulatory and catalytic domains of PTEN-induced kinase-1 (PINK1). *Human Mutation*, 33, pp.1408–1422.

Takatori, S., Ito, G. and Iwatsubo, T., 2008. Cytoplasmic localization and proteasomal degradation of N-terminally cleaved form of PINK1. *Neurosci Lett*, 430, pp.13–17.

Trinh, J. and Farrer, M., 2013. Advances in the genetics of Parkinson disease. Nature reviews. *Neurology*, 9, pp.445–454.

Twig, G., et al., 2008a. Fission and selective fusion govern mitochondrial segregation and elimination by autophagy. *EMBO J*, 27, pp.433–446.

Twig, G., Hyde, B. and Shirihai, O.S., 2008b. Mitochondrial fusion, fission and autophagy as a quality control axis: the bioenergetic view. *Biochim Biophys Acta*, 1777, pp.1092–1097.

Valente, E., et al., 2004. Hereditary early-onset Parkinson's disease caused by mutations in PINK1. *Science*, 304, pp.1158–1160.

Van Laar, V. S. and Berman, S. B, 2013a. The interplay of neuronal mitochondrial dynamics and bioenergetics: implications for Parkinson's disease. *Neurobiol Dis*, 51, pp.43–55.

Van Laar, V. S. and Berman, S. B., 2013b. The interplay of neuronal mitochondrial dynamics and bioenergetics: implications for Parkinson's disease. *Neurobiol Dis*, 51, pp.43–55.

van Spronsen, M., et al., 2013. TRAK/Milton motor-adaptor proteins steer mitochondrial trafficking to axons and dendrites. *Neuron*, 77, pp.485–502.

Vincow, E. S., et al., 2013. The PINK1-Parkin pathway promotes both mitophagy and selective respiratory chain turnover in vivo. *Proc Natl Acad Sci U S A*, 110, pp.6400–6405.

Vos, M., et al., 2012. Vitamin K2 is a mitochondrial electron carrier that rescues pink1 deficiency. *Science*, 336, pp.1306–1310.

Weihofen, A., et al., 2008. Pink1 Parkinson mutations, the Cdc37/Hsp90 chaperones and Parkin all influence the maturation or subcellular distribution of Pink1. *Hum Mole Genet*, 17, pp.602–616.

Weihofen, A., et al., 2009. Pink1 forms a multiprotein complex with Miro and Milton, linking Pink1 function to mitochondrial trafficking. *Biochemistry*, 48, pp.2045–2052.

Yamano, K. and Youle, R., 2013. PINK1 is degraded through the N-end rule pathway. *Autophagy*, 9, pp.1758–1769.

Yang, Y., et al., 2008. Pink1 regulates mitochondrial dynamics through interaction with the fission/fusion machinery. *Proc Natl Acad Sci U S A*, 105, pp.7070–7075.

Yoshii, S., et al., 2011. Parkin mediates proteasome-dependent protein degradation and rupture of the outer mitochondrial membrane. *J Bio Chem* 286, pp.19630–19640.

Youle, R. J. and Narendra, D. P., 2011. Mechanisms of mitophagy. *Nat Rev Mol Cell Biol*, 12, pp.9–14.

Yu, W., et al., 2011. The PINK1/Parkin pathway regulates mitochondrial dynamics and function in mammalian hippocampal and dopaminergic neurons. *Hum Mole Gen*, 20, pp.3227–3240.

Zara, V., et al., 2009. Mitochondrial carrier protein biogenesis: role of the chaperones Hsc70 and Hsp90. *Biochem J*, 419, pp.369–375.

Zhou, C., et al., 2008. The kinase domain of mitochondrial PINK1 faces the cytoplasm. *Proc Natl Acad Sci U S A*, 105, pp.12022–12027.

12

A GET-TOGETHER TO TEAR IT APART: THE MITOCHONDRION MEETS THE CELLULAR TURNOVER MACHINERY

GIAN-LUCA MCLELLAND AND EDWARD A. FON
Montreal Neurological Institute, McGill University, Montreal, Quebec, Canada

12.1 MITOCHONDRIAL QUALITY CONTROL IN NEURODEGENERATION

Given their high metabolic demand, neurons depend on an efficient and effective mitochondrial network capable of producing the vast amounts of ATP required by these cells, while simultaneously sequestering excess cellular calcium and acting as the site of catalysis for a plethora of biosynthetic reactions. To this effect, improperly functioning mitochondria are thought to underlie the pathogenesis of several neurodegenerative diseases, most notably Parkinson's disease (PD), in which deficits in mitochondrial complex I (CI) activity were first reported a quarter-century ago in samples from patient brains and skeletal muscle (Bindoff et al., 1989; Schapira et al., 1989). Alzheimer's disease (AD), the most prevalent neurodegenerative disease worldwide, has also been associated with dysfunctions in mitochondrial metabolism for more than two decades (Kish et al., 1992; Parker, Filley, and Parks, 1990; Parker et al., 1994), with links to CIV and F_1F_0-ATPase dysfunction (Bosetti et al., 2002; Cardoso et al., 2004). Common cellular pathologies in AD and PD include an accumulation of both abnormalities in the mitochondrial genome (mtDNA) and

The Functions, Disease-Related Dysfunctions, and Therapeutic Targeting of Neuronal Mitochondria, First Edition. Edited by Valentin K. Gribkoff, Elizabeth A. Jonas and J. Marie Hardwick.

elemental iron in diseased brains and at-risk neuronal populations (Barnham and Bush, 2008; Bender et al., 2006; Dexter et al., 1992; Mecocci, MacGarvey, and Beal, 1994; Trimmer et al., 2004). Indeed, mutations in *POLG*, encoding the mtDNA polymerase, have been associated with parkinsonism (Anvret et al., 2010; Davidzon et al., 2006; Eerola et al., 2010; Gui et al., 2012; Luoma et al., 2007).

Recently, studies delving into mechanism have proposed that genes linked to familial forms of PD function in "quality control" pathways designed to maintain the integrity of the mitochondrial population via the degradation of damaged proteins, lipids, and whole organelles. Although a buildup of damaged mitochondria—concomitant with a decline in mitochondrial function—may be observed in the disease state, this dysfunction may be nucleated early on by loss of mitochondrial quality control pathways. Moreover, evidence for a loss of mitochondrial quality control has been suggested to at least contribute to many degenerative disorders of the nervous system (Rugarli and Langer, 2012), demonstrating the importance of these pathways as potential therapeutic targets.

Although the cell employs several distinct but overlapping mitochondrial quality control pathways, this chapter will provide an overview on mechanisms that involve interplay between cytosolic and mitochondrial factors to process substrates for degradation. Additionally, mitochondria employ intrinsic "housekeeping" mechanisms—namely, mitochondrial proteases. Although mutations in the protease-encoding genes *AFG3L2* and *SPG7*, for example, lead to spinocerebellar ataxia and hereditary spastic paraplegia, respectively, this chapter will focus on crosstalk between the organelle and the extra-mitochondrial cellular turnover machinery, and how miscommunication can lead to disease (Casari et al., 1998; Di Bella et al., 2010). Indeed, dysfunctions in disease-linked proteins, such as Mfn2 and PINK1 at the mitochondrion, as well as extra-mitochondrial factors that include parkin, Vps35, and VCP, have been shown to impinge on quality control pathways. This chapter will first discuss the ubiquitination of mitochondrial proteins and their subsequent degradation, then focus on turnover of the entire organelle by the process of macroautophagy, and conclude with more recent data concerning alternative quality control pathways.

12.2 AN OVERVIEW OF THE UBIQUITIN-PROTEASOME SYSTEM

Since its characterization was begun toward the end of the 1970s, the contemporary view of the ubiquitin-proteasome system (UPS) is that it is the major pathway for turnover of short-lived proteins in the cell (Arrigo et al., 1988; Hershko et al., 1980). Canonical UPS-mediated protein degradation involves the conjugation of a substrate protein with ubiquitin, a small 76-residue polypeptide, via the formation of a covalent bond between the C-terminal glycine residue of mature ubiquitin and the ε-amino group of a target lysine (K) residue of the substrate protein. Ubiquitin can also be conjugated to other ubiquitin molecules, forming chains originating from the initial substrate. The seven lysine residues in ubiquitin allow for the formation of an assortment of linkages: K6, K11, K27, K29, K33, K48, and K63, as well as linear conjugation of the C-terminal glycine to the first methionine of another ubiquitin molecule. Branching, as well as the incorporation of several linkages into a "mixed" chain, adds to the complexity of the so-called ubiquitin code (Trempe, 2011).

Conjugation of a single ubiquitin directly onto substrates (monoubiquitination) or the incorporation of many ubiquitin moieties into chains (polyubiquitination) is mediated by ubiquitin ligases. This process first involves the "activation" of ubiquitin through the formation of a thioester bond with an E1-activating enzyme in an ATP-dependent manner. This ubiquitin is then transferred to an E2-conjugating enzyme, which acts as a ubiquitin source for the ubiquitination reaction carried out by the E3 ligase. HECT-type ligases covalently bind the activated ubiquitin before transferring it to a substrate protein, while RING-type ligases mediate ubiquitin transfer by bringing the E2-ubiqutin and substrate protein into close proximity (Metzger, Hristova, and Weissman, 2012). Several E3 ubiquitin ligases, including the PD-linked ligase parkin, function as HECT/RING hybrids (Wenzel et al., 2011).

While polyubiquitination has been implicated in a plethora of cellular processes, K48-linked chain formation is most closely associated with substrate degradation by the 26S proteasome—an enormous multiprotein complex consisting of a 20S core particle and a 19S regulatory particle. Although the core particle, shaped like a barrel composed of four heptameric rings, is responsible for the proteolytic activity of the complex, the regulatory particle binds ubiquitin via the receptors Rpn10 and Rpn13 and is also comprised of AAA-ATPases that unfold and remove ubiquitin from substrate proteins as they are fed into the core particle for proteolysis (Besche, Peth, and Goldberg, 2009; Goldberg, 2007). While proteasomal degradation of K48-linked polyubiquitinated proteins has been studied in various cellular activities, there are exceptions to the rule; a classic example being the proteasomal degradation of ornithine decarboxylase in the absence of its ubiquitination (Murakami et al., 1992).

12.3 ACTIVITIES OF THE CYTOSOLIC PROTEASOME AT THE OUTER MITOCHONDRIAL MEMBRANE

Traditionally thought of as a diffusely cytosolic (and nuclear) complex, proteasomal activity at various subcellular organelles occurs regularly. The most well-known example of this localized activity is in the process of endoplasmic reticulum-associated degradation (ERAD), an endoplasmic reticulum (ER) quality control pathway. Indeed, a significant amount of proteasomal activity occurs at the ER (Enenkel, Lehmann, and Kloetzel, 1998; Stafford and Bonifacino, 1991). ERAD of lumenal proteins involves the recognition of misfolded polypeptides in the ER lumen by chaperones, which is followed by unfolding and dislocation back across the ER membrane (Smith Ploegh, and Weissman, 2011). This dislocation is an ATP-dependent process driven by the AAA-ATPase valosin-containing protein (VCP/p97/Cdc48), a multimeric complex involved in many aspects of ubiquitin-related activity and mutated in inclusion body myopathy associated with Paget's disease of the bone and frontotemporal dementia (IBMPFD), a degenerative condition affecting muscle, bone and the nervous system (Custer et al., 2010; Guinto et al., 2007; Meyer, Bug, and Bremer, 2012; Watts et al., 2004). As the polypeptide emerges from the dislocation pore, it is rapidly polyubiquitinated and ultimately degraded by the proteasome, thus preventing an accumulation of misfolded proteins in the ER (Smith et al., 2011).

A number of more recent studies have described a similar ubiquitin-dependent process occurring across the outer mitochondrial membrane (OMM), dubbed mitochondria-associated degradation (MAD) or OMM-associated degradation (OMMAD). MITOL/March5, an E3 ubiquitin ligase localized to the OMM, was shown to regulate mitochondrial dynamics by interacting with and ubiquitinating Fis1 and Drp1, leading to their turnover (Nakamura et al., 2006; Yonashiro et al., 2006). Although MITOL was also shown to ubiquitinate mitofusin 2 (Mfn2), this was later demonstrated to serve a nondegradative role in maintaining homotypic Mfn2 contacts with the ER (Sugiura et al., 2013). Mfn2 is mutated in Charcot-Marie-Tooth type 2A (CMT2A), a peripheral neuropathy, and Fzo1, the ancestral yeast mitofusin, can be modified with K48-linked chains by Mdm30 and turned over by the proteasome (Cohen et al., 2008; Neutzner, Youle, and Karbowski, 2007). It is important to note, however, that the proteasome-independent degradation of ubiquitinated Fzo1 has also been described, supporting the existence of multiple, overlapping turnover pathways for OMM proteins such as Fzo1 (Escobar-Henriques, Westermann, and Langer, 2006). Interestingly, a mutant form of Fzo1, analogous to the CMT2A mutant Mfn2^{I213T}, showed decreased Mdm30-dependent ubiquitination, suggesting a role for aberrant Mfn2 ubiquitination and turnover in CMT2A (Amiott et al., 2009).

Although those observations support proteasomal degradation of OMM proteins, is there any evidence for the dislocation of inner mitochondrial proteins to the OMM? Initial studies reported the steady-state ubiquitination and proteasomal degradation of proteins localized to the inner mitochondrial membrane (IMM)—uncoupler proteins (UCPs) 2 and 3, as well as the oligomycin sensitivity-conferring protein (OSCP) of the F_1F_0-ATPase—suggesting a retrotranslocation mechanism analogous to ERAD (Azzu and Brand, 2010; Azzu, Mookerjee, and Brand, 2010; Margineantu et al., 2007). Each of these studies detected a buildup of the full-length protein on proteasomal inhibition, implying dislocation and degradation of the complete polypeptide; essentially, in the absence of processing by mitochondrial proteases. Moreover, although off-target inhibition of mitochondrial proteases by proteasomal inhibitors such as MG132 has been reported (Granot et al., 2007), each study was able to establish a role for ubiquitination in substrate degradation—K48-linked polyubiquitination for UCP2, in the only study that tested linkages—and reconstitute turnover in vitro with added proteasome. As ubiquitin is localized to the cytosol, its substrate conjugation would be presumed to occur on the cytosolic side of the OMM. Accordingly, both OSCP and the CIV subunit COX2 (also normally localized to the IMM) were detected at the OMM upon treatment of cells with MG132 (Margineantu et al., 2007).

Thus, inner mitochondrial proteins can be dislocated to the OMM, presumably in a chaperone- and likely ATP-dependent process, where they are polyubiquitinated and degraded by the proteasome. Proteomic studies in yeast, humans and *Arabidopsis thaliana* have revealed that mitochondrial proteins within the matrix, IMM and intermembrane space (IMS) are ubiquitinated, indicating that much more work is needed to elucidate the precise mechanism of protein/chaperone recognition, unfolding and dislocation (Maor et al., 2007; Matsumoto et al., 2005; Peng et al., 2003; Tagwerker et al., 2006; Weekes et al., 2003). Indeed, biochemical studies in yeast have suggested that proteasomal turnover of immature IMS proteins may act as an import quality control mechanism, turning over proteins that fail to fold properly (Bragoszewski et al., 2013; Pearce and Sherman, 1997; Radke et al., 2008). Disruption of these steady-state MAD/OMMAD pathways may be prevalent in disease states. Accordingly, several UCPs are robustly expressed in dopaminergic neurons of the *substantia nigra pars compacta* (SNpc), the major neuronal population at risk in PD, and are responsible for transient mitochondrial depolarizations in these cells, phenomenona that are critical for their survival (Guzman et al., 2010). Thus, as an example, dysregulation of proper UCP turnover pathways may contribute to mitochondrial dysfunction in diseases such as PD.

How might these ubiquitinated proteins be extracted from the OMM? Reminiscent of ERAD, steady-state degradation of the OMM proteins Mcl-1 and Mfn1 were shown to involve VCP (Xu et al., 2011). Mcl-1 is first recognized and ubiquitinated by Mule/ARF-BP1, a HECT-type ligase residing on the OMM, to facilitate cell death (Zhong et al., 2005). Further supporting a role for Mule-/ARF-BP1-dependent ubiquitination at the OMM during apoptosis, another study identified Mfn2 as a Mule/ARF-BP1 substrate, demonstrating that the proteasomal degradation of Mfn2 promoted mitochondrial fragmentation and cell death (Leboucher et al., 2012). Using dominant-negative VCP mutants and siRNA-mediated loss-of-function techniques, it was shown that VCP plays an ATPase-dependent role in the removal (and subsequent degradation) of mitochondrial proteins from the OMM (Xu et al., 2011). Like Mcl-1, many other pro- and anti-apoptotic factors are regulated via ubiquitination and proteasomal turnover at the OMM, including activated Bax, Bcl-2, and its related family members, among others, in both heterologous cells and cultured neurons (Benard et al., 2010; Beverly et al., 2012; Kutuk and Letai, 2008; Magiera et al., 2013; Meller et al., 2006; Pang et al., 2014; Tait et al., 2007; van de Kooij et al., 2013). Thus, efficient MAD/OMMAD pathways must be in place to ensure control over the balance between cell death and survival.

Although the precise regulatory mechanisms governing basal VCP translocation to mitochondria remain to be elucidated, recent work has shed light on possible pathways of recruitment. At the ER retrotranslocation complex, consisting of the Sec61 pore and Derlin-family proteins, resident proteins such as Derlin-1 and VIMP (among others) act to tether VCP and cluster E3 ubiquitin ligases necessary for ERAD at these sites (Carvalho, Goder, and Rapoport, 2006; Denic, Quan, and Weissman, 2006; Greenblatt, Olzmann, and Kopito, 2011; Lilley and Ploegh, 2004; Oda et al., 2006; Vembar and Brodsky, 2008; Ye et al., 2005). This model implies that VCP binding partners present on specific organelles can lead to its localized activity. Along these lines, Vms1 was shown to translocate to mitochondria during periods of mitochondrial stress in both yeast and *Caenorhabditis elegans*, and deletion of Vms1 caused a reduction in mitochondrial function and a loss of organismal viability (Heo et al., 2010).

Indeed, VCP preferentially associates with S-nitrosylated mitochondrial proteins in mammalian cells, indicative of a role in quality control (Benischke et al., 2014). Vms1 contains a VCP-interacting motif, composed of an α-helix of predominantly basic and nonpolar residues and acts to recruit Cdc48p (the yeast VCP homologue) and its cofactor Npl4 to mitochondria in times of stress (Ballar et al., 2006; Heo et al., 2010). Further work elucidating the mechanism of Vms1 translocation to mitochondria revealed that oxidative stress resulted in the exposure of a C-terminal mitochondrial-targeting domain in Vms1, which is likely repressed at the steady-state by its N-terminus (Heo et al., 2013). Thus, the complex regulation of specific factors harboring motifs capable of interacting with VCP may in turn modulate its translocation to mitochondria, governing its activity at the OMM.

Although the control of substrate polyubiquitination, leading to its extraction from the OMM, represents a crucial locus of regulation in MAD/OMMAD, the reverse reaction, involving the removal of ubiquitin moieties from a chain or substrate, termed *deubiquitination*, is of equal importance. Deubiquitination is catalyzed by deubiquitinases (DUBs), which are capable of recognizing specific ubiquitin linkages, chains, or substrates (Komander Clague, and Urbe, 2009). For example, Usp9X, a predominantly cytoplasmic DUB implicated in Lewy body pathology, antagonizes Mcl-1 K48-linked polyubiquitination, protecting it from proteasomal degradation via MAD/OMMAD and promoting cell survival (Rott et al., 2011; Schwickart et al., 2010). In yeast, two DUBs, Ubp2 and Ubp12, were shown to deubiquitinate Fzo1, with opposing effects on mitochondrial morphology. Although Ubp2 removed degradative ubiquitin chains from Fzo1—thus protecting it from proteasomal turnover via MAD/OMMAD and promoting mitochondrial fusion—Ubp12 antagonized the formation of oligomer-stabilizing ubiquitin chains, which prevented fusion (Anton et al., 2013). Thus, the ubiquitination of OMM substrates, such as Fzo1, is subject to intense regulation by many factors, with MAD/OMMAD representing but one endpoint of this highly dynamic process.

In mammals, only one DUB, Usp30, is currently known to be a resident OMM protein. Usp30 acts on nondegradative ubiquitin chains on Mfn1 and Mfn2, preventing their pro-fusion capabilities (Nakamura and Hirose, 2008; Yue et al., 2014). Indeed, small-molecule inhibition or siRNA-mediated depletion of Usp30 results in an elongated mitochondrial reticulum (Nakamura and Hirose, 2008; Yue et al., 2014). Although it is currently unclear if Usp30 plays a role in the steady-state MAD/OMMAD mechanisms mentioned previously, it is evident that, like VCP, cytosolic regulatory and core components of both the ubiquitination and deubiquitination system can translocate to mitochondria. Although mitochondrial recruitment and activity of VCP seems to be involved in most MAD-/OMMAD-related pathways, translocation of other cytosolic machinery can occur at both the steady-state (as was the case for UCP2, etc.) and under distinct stimuli, the latter of which range from mitochondrial fusion (Mdm30), to activation of survival-promoting signaling processes (Usp9X), to oxidative stress (Vms1). However, with a strong enough induction, signaling mechanisms are in place to engage a completely different, but not totally independent, cellular turnover pathway.

12.4 THE TURNOVER OF WHOLE MITOCHONDRIA BY MITOPHAGY

Macroautophagy (commonly referred to as *autophagy*) is a bulk degradation process, ultimately ending with the lysosomal turnover of substrate proteins or organelles (Yang and Klionsky, 2010a). This process, conserved throughout *Eukaryota*, is initially triggered by the activation of several upstream kinase complexes that regulate nutrient-sensing, as well as lipid phosphorylation and signaling (Yang and Klionsky, 2010a). Cargo is sequestered by an expanding, double membrane, termed the *phagophore*. The resulting autophagosome undergoes a maturation process (Filimonenko et al., 2007), ultimately fusing with lysosomes for degradation. As it was first used as a term to describe the presence of subcellular structures in double membrane vesicles, as had been observed in electron micrographs, autophagy as a term has been around for more than 50 years (Yang and Klionsky, 2010b). However, insight into the precise mechanism of cargo sequestration and specificity, autophagosome formation and lysosomal targeting has only arisen in the past two decades, stemming from an initial screen in yeast that identified mutant strains defective for autophagy, followed by characterization of the first autophagy gene, *Atg1* (Matsuura et al., 1997; Tsukada and Ohsumi, 1993).

Mitophagy, the selective turnover of mitochondria by autophagy, represents a cellular mechanism by which entire mitochondria can be lysosomally degraded. In yeast, Atg32 localizes to mitochondria, where, upon phosphorylation, it recruits the autophagic machinery to organelles targeted for degradation (Aoki et al., 2011; Kanki et al., 2009; Okamoto, Kondo-Okamoto, and Ohsumi, 2009). Atg32 interacts directly with Atg11, an adaptor protein involved in selective autophagy, and Atg8, a component localized to the phagophore, thereby acting as a scaffold to initiate mitophagy (Kanki et al., 2009; Kondo-Okamoto et al., 2012; Okamoto et al., 2009). Atg8, the yeast orthologue of human microtubule-associated protein 1, light chain 3 (MAP1LC3, or simply LC3) family proteins, is inserted into the elongating phagophore membrane by its stepwise C-terminal cleavage and conjugation to the phospholipid phosphatidylethanolamine, and remains associated with mature autophagosomes (Yang and Klionsky, 2010a). Although it acts to recruit core autophagy machinery, Atg11 also associates with Dnm1, the yeast orthologue of the fission GTPase Drp1 (Mao et al., 2013). Mitophagy is repressed in yeast strains lacking Dnm1 (or one of its binding partners, Fis1), indicating that mitochondrial fission is required in order to turn over mitochondria by autophagy (Mao et al., 2013). Presumably, once a target mitochondrion is identified, the fission complex is locally assembled so that it can be separated from the reticulum before degradation (Mao et al., 2013). Indeed, mitochondrial fusion and fission are believed to aid in the steady-state sequestration of mitochondrial components for mitophagy (Abeliovich et al., 2013; Twig et al., 2008).

Mammalian cell culture systems offer much complexity in the variety of upstream signals that regulate mitochondrial autophagy, as well as the array of factors that recruit and modulate the autophagic machinery. However, the same principles that govern mitophagy in yeast—identification, isolation and degradation of target mitochondria—still hold true in mammals. For example, hypoxia-driven mitophagy is dependent on the OMM phosphoprotein FUNDC1. On exposure to hypoxic conditions, FUNDC1 is rapidly dephosphorylated, allowing its direct association with LC3 and initiation of autophagosome formation (Chen et al., 2014; Liu et al., 2012). Thus, FUNDC1 represents a crucial locus of regulation in hypoxia-induced mitophagy, much like Atg32 regulates general mitophagy in yeast. Although numerous intrinsic mechanisms that govern the mitochondrial recruitment of core autophagic machinery exist—such as the binding of LC3 to externalized cardiolipin at the OMM, or the autophagy of respiring mitochondria via the small GTPase Rheb and the mitophagy receptor Nix—they are not discussed here (Chu et al., 2013; Melser et al., 2013; Novak et al., 2010). The importance of functional mitophagic pathways in higher organisms is highlighted by the accumulation of defective and deformed mitochondria in mice lacking key autophagy genes, as has been observed in hepatocytes isolated from both Ulk1$^{-/-}$ (a mammalian orthologue of Atg1) and Atg7$^{-/-}$ mice (Egan et al., 2011; Komatsu et al., 2005).

Although a role for mitophagy has been proposed in many physiological processes—autophagic clearance of paternal mitochondria during oocyte fertilization has been described in both *C. elegans* and *Drosophila melanogaster*, for example—the lysosomal degradation of mitochondria via autophagy has been implicated in many neurodegenerative diseases (Al Rawi et al., 2011; Politi et al., 2014; Sato and Sato, 2011). Autophagy itself is upregulated in AD—noted by the increased amount of autophagic vacuoles observed in diseased brains—and plays a role in Aβ peptide generation (Nixon et al., 2005; Yu et al., 2004, 2005). Moreover, disruption of the autophagy-lysosome system in isolated neuronal culture systems can recapitulate certain disease phenotypes (Boland et al., 2008; Lee et al., 2010a; Lee, Sato, and Nixon, 2011). Although increased mitophagy has been reported in brains of patients with AD, as well as skeletal muscle from transgenic mouse models of amyotropic lateral sclerosis, this is thought to be a cellular defense mechanism, with the goal of ridding the cell of defective mitochondria (Moreira et al., 2007; Wong et al., 2013; Youle and van der Bliek, 2012). Indeed, while defective mitophagy has been implicated in neurodegenerative disorders, such as xeroderma pigmentosum group A (Fang et al., 2014), most of our understanding concerning the molecular mechanisms of mitophagy, and how dysfunction of these pathways may be a causative factor in neurodegenerative disease, comes from studies involving genes linked to PD.

12.5 PROTEASOMES AND PHAGOPHORES CONVERGE IN THE PINK1/PARKIN PATHWAY

Although mitochondrial dysfunction and PD have been linked for quite some time, precise connections between disease pathogenesis and the mitochondrion itself have only emerged in the past decade, stemming from the study of genes linked to autosomal recessive juvenile parkinsonism (AR-JP), an early-onset, recessive form of PD. Genes linked to AR-JP include *PARK2/PARKIN*, which encodes a cytosolic E3 ubiquitin ligase, and *PARK6/PINK1*, encoding a mitochondrially targeted serine/threonine kinase (*p*hosphatase and tensin homologue-*in*duced *k*inase 1) (Kitada et al., 1998;

Valente et al., 2004). Mutations in *PARKIN* are the most common cause of AR-JP, and patients with PD carrying parkin mutations have a mitochondrial functional deficit comparable to patients suffering from sporadic forms of PD (Mann et al., 1992; Muftuoglu et al., 2004; West and Maidment, 2004).

Despite mutations in parkin and PINK1 leading to AR-JP in humans, parkin$^{-/-}$ and PINK1$^{-/-}$ mice exhibit no obvious neurodegenerative phenotype. These mice do, however, display mild perturbations in mitochondrial metabolism and susceptibility to oxidative stress (Gautier, Kitada, and Shen, 2008; Gispert et al., 2009; Goldberg et al., 2003; Kitada et al., 2007, 2009; Palacino et al., 2004; Perez and Palmiter, 2005). Most of our initial insight into the in vivo functions of parkin and PINK1 has come from studying their homologues in *Drosophila*. Early reports on PINK1 and parkin single-mutant flies revealed that genetic ablation of either of these genes resulted in male sterility, and degeneration of the indirect flight muscle and a subset of dopaminergic neurons—all of which were tied, at early stages, to morphological, bioenergetic and stress defects of mitochondria (Clark et al., 2006; Greene et al., 2003; Park et al., 2006; Yang et al., 2006). The most revelatory finding from these studies, however, indicated that loss of PINK1 could be rescued genetically by over-expression of parkin, implicating both in a common genetic pathway—with parkin acting downstream of PINK1—and that loss of this pathway leads to mitochondrial defects in mutant flies, and potentially in patients with AR-JP as well (Clark et al., 2006; Park et al., 2006; Yang et al., 2006).

What is the nature of the PINK1/parkin pathway that acts on mitochondria? Fly studies first suggested that it might impinge on mitochondrial morphology, as genetically gearing the mitochondrial network towards fission—by either overex-pressing Drp1, or ablating the expression of pro-fusion genes such as Opa1 or the *Drosophila* mitofusin homologue—suppressed PINK1/parkin mutant phenotypes (Deng et al., 2008; Poole et al., 2008; Yang et al., 2008). However, replication of a pro-fission role for parkin and PINK1 yielded contradictory results in heterologous mammalian cultured cells, isolated neurons and patient fibroblasts, suggesting that aberrant mitochondrial morphology may be secondary to an underlying effect resulting from the loss of these genes in these systems (Exner et al., 2007; Lutz et al., 2009; Mortiboys et al., 2008; Yang et al., 2008; Yu et al., 2011).

Studies seeking suppressors of the PINK1/parkin phenotype found that the mito-chondrial degeneration and cell loss in *Drosophila* could be rescued by increasing the antioxidant capacity of the organism, either by overexpressing a ROS scavenger or chaperone, or by feeding flies antioxidants directly (Greene et al., 2005; Trinh et al., 2008; Wang et al., 2006; Whitworth et al., 2005). Moreover, parkin expression reduced oxidative and cellular stress in numerous paradigms, and S-nitrosylation of parkin was found to modulate its ubiquitin ligase activity (Chung et al., 2004; Hyun et al., 2002; Muller-Rischart et al., 2013; Yao et al., 2004). Thus, it was conceivable that parkin and PINK1 function in a pathway related to mitochondrial stress.

Accordingly, mammalian cell culture studies demonstrated a role for parkin and PINK1 in mitochondrial quality control, involving the recognition and autophagic turnover of damaged mitochondria. At healthy mitochondria, newly synthesized PINK1 is imported, sequentially processed by mitochondrial proteases—first by the mitochondrial processing peptidase (MPP), and then by PARL and/or AFG3L2 in a possibly-redundant second step—and then rapidly exported and degraded in the cytosol by the proteasome (Greene et al., 2012; Jin et al., 2010). Further study revealed that PARL cleavage exposes a destabilizing amino acid at the N-terminus of PINK1 (arginine-98, a so-called N-degron), resulting in ubiquitination of the truncated protein by the cytosolic E3 ligases UBR1, UBR2 and UBR4 and proteasomal turnover via the N-end rule pathway (Yamano and Youle, 2013).

Although PINK1 may act on substrates within the mitochondrion during its life cycle—possibly regulating mitochondrial processes such as respiration and calcium flux—its association with parkin occurs on the surface of damaged mitochondria (Gandhi et al., 2009; Heeman et al., 2011; Morais et al., 2014). Here, PINK1 accumulates on the OMM at or near import channels, as its $\Delta\psi$-dependent import is abrogated (Jin et al., 2010; Lazarou et al., 2012; Narendra et al., 2010b; Vives-Bauza et al., 2010). Release of full-length PINK1 into the OMM from the TOM complex occurs by an unknown mechanism but has been shown to require the TOM7 subunit of the TOM complex (Hasson et al., 2013). Although initial studies concerning PINK1 import used direct ablation of $\Delta\psi$ by an uncoupling agent to promote PINK1 accumulation on the OMM, more recent studies have studied PINK1 buildup with less extreme stimuli; for example, accumulation of unfolded proteins in the mitochondrial matrix was enough to stabilize PINK1 on the OMM without depolarization (Jin and Youle, 2013). Moreover, modest reduction of MPP was shown to be sufficient to selectively stabilize PINK1, demonstrating that the import and degradation kinetics for PINK1 are much more rapid in comparison to other mitochondrial proteins (Greene et al., 2012).

Once stabilized on the OMM of damaged mitochondria, PINK1 acts to locally recruit parkin from the cytosol in an activity-dependent manner, thus acting as a sensor for damaged mitochondria (Narendra et al., 2010b). To achieve this, PINK1 directly phosphorylates both ubiquitin and the ubiquitin-like (Ubl) domain of parkin at serine-65 (S65)—a residue conserved in both proteins—leading to the activation of the E3 ubiquitin ligase activity of parkin at the OMM (Kane et al., 2014; Kazlauskaite et al., 2014a, 2014b; Kondapalli et al., 2012; Koyano et al., 2014; Shiba-Fukushima et al., 2012). At the steady-state, parkin is kept in an inactive conformation, and phosphorylation of the Ubl combined with binding to phosphoubiquitin present on the OMM may induce a conformational change in parkin, resulting in its activation, and possibly its oligomerization, at damaged mitochondria (Koyano et al., 2014; Lazarou et al., 2013; Trempe et al., 2013). Indeed, parkin translocation to damaged mitochondria is coupled to its ability to ligate ubiquitin, and its ability to form an ubiquitin-ester (a ubiquitination reaction intermediate) is dependent on PINK1-mediated Ubl phosphorylation (Iguchi et al., 2013; Zheng and Hunter, 2013). The observation that phosphorylated ubiquitin lacking C-terminal residues critical for its incorporation into polyubiquitin chains was still able to activate parkin ligase activity supports the notion that phosphoubiquitin may act as an allosteric activator of parkin (Koyano et al., 2014). PINK1 may phosphorylate free ubiquitin in the proximity of mitochondria, or ubiquitin already conjugated to proteins on the OMM; this latter scenario would explain the observation that PINK1 ectopically targeted to other organelles is able to recruit parkin (Kane et al., 2014; Lazarou et al., 2012).

Recruited to and activated at damaged mitochondria, parkin ubiquitinates various OMM substrates associated with many functions, including mitochondrial dynamics, transport, and metabolism, ultimately promoting the isolation and subsequent autophagic degradation of damaged mitochondria (Geisler et al., 2010a; Narendra et al., 2008; Sarraf et al., 2013). Although parkin-catalyzed ubiquitination of the mitofusins leads to their turnover in both *Drosophila* and mammalian cells, which may act to sequester a target mitochondrion from the reticulum by preventing fusion, parkin-mediated turnover of Miro proteins, which are involved in microtubule-based mitochondrial transport, may also function to isolate damaged mitochondria, thus facilitating engulfment by the autophagosome (Gegg et al., 2010; Poole et al., 2010; Tanaka et al., 2010; Wang et al., 2011; Ziviani, Tao, and Whitworth, 2010). Moreover, autophagy of damaged mitochondria was inhibited in cells deficient in mitochondrial fission; that is, Drp1$^{-/-}$ MEFs or HeLa cells in which Drp1 was silenced (Tanaka et al., 2010). Interestingly, this had no effect on parkin recruitment to depolarized mitochondria or mitofusin turnover, suggesting that the convergence of Drp1-dependent fission and parkin-dependent ubiquitination of OMM substrates is required for proper turnover of entire mitochondria (Tanaka et al., 2010). Given the dependency on Dnm1 in Atg32-mediated mitophagy in yeast, there is, accordingly, an apparent conservation across *Eukaryota* of the requirement for mitochondrial fission during mitophagy (Mao et al., 2013).

How are parkin substrates on the OMM turned over once they are ubiquitinated? Using Mfn1 as a model substrate, it was shown that, on mitochondrial depolarization, turnover occurs by the proteasome in a parkin- and, perhaps unsurprisingly, VCP-dependent manner, suggesting the involvement of a MAD-/OMMAD-type pathway (Chan et al., 2011; Tanaka et al., 2010). Expression of a dominant-negative VCP mutant, defective in ATP hydrolysis, was enough to suppress Mfn1 turnover in depolarized cells (Tanaka et al., 2010; Ye, Meyer, and Rapoport, 2003). The buildup of polyubiquitinated Mfn1 in mitochondrial fractions from these cells was consistent with a MAD/OMMAD pathway in which Mfn1 (as well as other OMM substrates) is first ubiquitinated by parkin, then extracted from the OMM by VCP, and finally degraded by the proteasome (Tanaka et al., 2010). Accordingly, VCP, as well as its cofactors Npl4 and p47, were shown to translocate to mitochondria on depolarization in a parkin-dependent manner, and this translocation was abrogated in cells expressing IBMPFD-linked pathogenic VCP mutations (Kim et al., 2013; Kimura et al., 2013; Tanaka et al., 2010). Moreover, silencing of VCP cofactors in parkin-expressing cells blocked both Mfn1 turnover and mitophagy (Kim et al., 2013; Kimura et al., 2013). These data indicated that VCP-dependent clearance of ubiquitinated OMM proteins is a crucial step in PINK1-/parkin-mediated mitophagy. This is further supported by the observation that VCP interacts with this pathway genetically in *Drosophila* (Kim et al., 2013).

Although parkin-mediated ubiquitination on the OMM serves to isolate target organelles from the rest of the reticulum, it also functions to recruit machinery necessary for autophagic turnover. p62/sequestosome 1 contains a polyubiquitin-binding UBA domain, mutations in which can lead to Paget's disease of bone, as well as an LC3-interacting region (LIR), a stretch of more than 20 amino acids that binds directly to LC3 (Cavey et al., 2005; Kirkin et al., 2009; Pankiv et al., 2007). Thus, p62 acts as an autophagic receptor, linking polyubiquitination to the autophagic machinery. p62 was first implicated in parkin-mediated mitophagy when it was observed that p62 translocated to mitochondria in parkin-expressing cells treated with an uncoupling agent (Geisler et al., 2010a). Although the requirement of p62 in this pathway remains unclear, further work has shown that other potential ubiquitin-binding autophagic receptors, such as CALCOCO2 and TAX1BP1/NDP52, associate with parkin during mitophagy, suggesting possible redundancy at this level of the pathway (Narendra et al., 2010a; Sarraf et al., 2013).

In addition to the parkin-dependent recruitment of ubiquitin-binding autophagic receptors, other upstream autophagy components, such as Atg9A-containing structures and the Ulk1 complex, translocate independently of LC3 to depolarized mitochondria, suggesting a convergence of the autophagic machinery at these organelles, possibly at sites of ER-mitochondrial contact, an alleged location of autophagosome biogenesis (Hamasaki et al., 2013; Itakura et al., 2012; Yang and Yang, 2013). Moreover, recent work has demonstrated that TBC1D15, a Rab-GAP bound to Fis1 on the surface of depolarized mitochondria, binds LC3-family proteins to act as a regulator of autophagosome biogenesis (Yamano et al., 2014). Thus, it evident that, at damaged mitochondria, parkin coordinates both the proteasomal turnover of certain OMM proteins by MAD/OMMAD, as well as whole organellar turnover through the assembly of the autophagic machinery. Although it is clear that these two parkin-mediated processes are interdependent—blocking MAD/OMMAD abrogates mitophagy—the precise, stepwise mechanism of the autophagy of damaged mitochondria remains to be elucidated.

12.6 IMPLICATIONS OF PINK1-/PARKIN-DEPENDENT MITOPHAGY IN THE BRAIN AND IN PD

Although depolarization robustly triggers PINK1- and parkin-dependent mitophagy in heterologous cell culture, the study of this pathway in neurons has remained complicated (Grenier, McLelland, and Fon, 2013). This may owe, at least in part, to neuronal bioenergetics; unlike astrocytes, cortical neurons are incapable of switching their metabolism toward glycolysis during periods of mitochondrial stress (Almeida et al., 2001; Almeida, Moncada, and Bolanos, 2004). In these cells, glucose is typically shunted toward the pentose-phosphate pathway in an effort to maintain a high fraction of reduced glutathione—thereby heavily relying on oxidative phosphorylation for ATP production (Bolanos, Almeida, and Moncada, 2010; Herrero-Mendez et al., 2009). Recapitulating this type of bioenergetic situation—by forcing non-neuronal cells in culture to respire—prevents mitophagy in cultured cells (MacVicar and Lane, 2014; Van Laar et al., 2011). Thus, cellular bioenergetics play a pivotal role in allowing the collapse of $\Delta\psi$ and mitophagy to proceed.

Despite an apparent discrepancy in assay robustness between neurons and glycolytic cells in culture, several studies have still been able to recapitulate elements of the PINK1/parkin pathway in either isolated or iPSC-derived neurons. Initial studies in isolated neuronal cultures demonstrated that the removal the antioxidant component from the neuronal medium, or co-incubation of cells with anti-apoptotic agents along with an uncoupler, promoted parkin translocation (Cai et al., 2012; Joselin et al., 2012). In these studies, however, parkin was found to be recruited to mitochondria in only a fraction (25 to 50%) of neurons, despite prolonged periods of uncoupling (Cai et al., 2012; Joselin et al., 2012; Koyano et al., 2013). Similar to assays in cell culture, this modest effect was found to be dependent on the presence of functional PINK1, which was also shown to accumulate in rat cortical neurons when $\Delta\psi$ was ablated (Joselin et al., 2012; Narendra et al., 2010b; Seibler et al., 2011). Accordingly, ubiquitination of OMM substrates as well as LC3 recruitment to parkin-positive mitochondria was also observed (Cai et al., 2012; Koyano et al., 2013). These data suggest that both parkin-dependent MAD/OMMAD and mitophagy can be triggered by depolarization in neurons, though the extent of turnover by this pathway remains to be elucidated. Moreover, although endogenous parkin can mediate depolarization-dependent ubiquitination of OMM substrates, such as Mfn2, in neurons, it remains to be resolved whether or not this leads to mitophagy (Koyano et al., 2013; Rakovic et al., 2013). Intriguingly, a recent study has demonstrated PINK1- and parkin-dependent delivery of mitochondria to the lysosome in neurons at the steady-state, suggesting that there is basal mitophagic turnover in this cell type (Bingol et al., 2014).

Clearly, the exact function of the PINK1/parkin pathway in neurons, and its relevance to PD, has yet to be decoded. Many PD-linked mutations in parkin and PINK1 disrupt mitochondrial turnover at distinct levels, indicating that this pathway is potentially dysfunctional in the diseased state (Geisler et al., 2010b; Lee et al., 2010b). Equally, it remains to be answered why neurons of the SNpc are selectively vulnerable to dysregulation of PINK1-/parkin-dependent MAD/OMMAD and mitophagy. The intrinsic, vigorous pacemaking activity of dopaminergic SNpc neurons may make them more vulnerable to oxidative stress (Chan et al., 2007; Guzman et al., 2009, 2010). Accordingly, an analysis of neuronal subtypes in *Drosophila* demonstrated that dopaminergic neurons were selectively susceptible to loss of parkin (Burman et al., 2012). PINK1- and parkin-mediated autophagy, however, has been linked to an array of other disease states and dysfunctions, including lysosomal storage disease, sepsis, infectious disease, mutations in mtDNA, and dysfunction in cardiac and muscle cells (de Pablo-Latorre et al., 2012; Huang et al., 2011; Kubli, Quinsay, and Gustafsson, 2013; Manzanillo et al., 2013; Piquereau et al., 2013; Suen et al., 2010). Disruption of mitochondrial quality control mediated by the parkin/PINK1 pathway may therefore severely affect other systems, beyond the degeneration of dopaminergic neurons seen in patients with AR-JP carrying PINK1 or parkin mutations. Indeed, the observations that parkin confers resistance to *Mycobacterium tuberculosis* in mice, and that polymorphisms in the *PARKIN* gene are associated with increased vulnerability to infectious disease in humans, suggest that the parkin/PINK1 pathway may play a role in infection or immunity

(Ali et al., 2006; Manzanillo et al., 2013; Mira et al., 2004). Thus, more work is required to elucidate the breadth of human disease affected by parkin/PINK1 mitochondrial quality control.

12.7 EMERGING MITOCHONDRIAL QUALITY CONTROL MECHANISMS

Although autophagy serves to envelop damaged mitochondria and transport them to lysosomes for degradation, several recent studies have demonstrated that direct contact and trafficking between mitochondria and lysosomes may mediated another form of mitochondrial quality control that is independent of mitophagy and MAD/OMMAD. Functional connectivity between these two organelles is highlighted by the observation that yeast vacuolar pH is a critical regulator of mitochondrial function, and lysosomal-mitochondrial crosstalk has been implicated in inflammasome activation, iron metabolism, and apoptotic signaling in mammalian cells (Heid et al., 2013; Hughes and Gottschling, 2012; Repnik, Cesen, and Turk., 2013; Richardson et al., 2010; Weber and Schilling, 2014; Zhou et al., 2011).

Although the vacuole-mitochondrion contact site has been recently elucidated in yeast, the precise molecular nature of mitochondrion-lysosome contacts in mammalian cells remains to be determined (Elbaz-Alon et al., 2014; Honscher et al., 2014). Analogous to its role in mitochondrion-ER contact sites, Mfn2 has been implicated in the tethering of melanosomes—lysosome-like organelles present in specialized cells such as melanocytes and retinal pigment epithelial cells—to mitochondria (Daniele et al., 2014; de Brito and Scorrano, 2008). Similarly, time-lapse microscopy of the lysosome-associated Rab26 demonstrated its association and overlap with mitochondria (Jin and Mills, 2014). Lysosome-mitochondrion contact sites may serve metabolic functions such as lipid transfer; however, they may also function, under certain stimuli, in the degradation of whole or portions of mitochondria. For example, in a cellular model of cardiac ischemia and reoxygenation, large lysosomal-like structures, containing components of the endocytic pathway, were seen to associate with mitochondria and promote mitochondrial protein turnover independently of the autophagy pathway (Yogalingam et al., 2013).

In addition to the potential for protein turnover at lysosome-mitochondrial contact sites, mitochondrial-derived vesicles (MDVs) constitute an alternative route to the lysosome for mitochondrial proteins and lipids (Sugiura et al., 2014). Recent work has shown that these vesicles bud off mitochondria independently of Drp1-mediated fission, contain selective mitochondrial cargo—a hallmark of vesicular trafficking—and target to lysosomes independently of autophagy machinery such as Atg5 and beclin-1 (McLelland et al., 2014; Soubannier et al., 2012a). MDV biogenesis is triggered by reactive oxygen species formation, and in vitro reconstitution experiments have shown that MDVs incorporate oxidized cargo (Soubannier et al., 2012b). Thus, MDVs represent a quality control mechanism, which selectively extracts damaged mitochondrial components and targets them for turnover, thus preserving the functionality of the organelle. Additionally, MDV transport to lysosomes was shown to require both parkin and PINK1, suggesting some redundancy between machinery required for MDV formation and that required for the autophagy of damaged mitochondria (McLelland et al., 2014). Consequently, kinetic experiments indicated that parkin-dependent mitochondrial vesicle formation represents a "first response" to mitochondrial damage, compared to the kinetically slower process of mitophagy (McLelland et al., 2014).

The involvement of parkin and PINK1 in MDV transport to lysosomes enlarges the parkin/PINK1 pathway beyond mitophagy. Supporting this latter point, a recent proteomic study in *Drosophila* demonstrated that, in vivo, parkin and PINK1 mediate selective turnover of transmembrane electron transport chain components, in addition to autophagic degradation of whole mitochondria (Vincow et al., 2013). Additionally, MDV-based transport to peroxisomes—comprising its own repertoire of cargo that is distinct from MDVs marked for lysosomal turnover—requires the retromer subunit Vps35 (Braschi et al., 2010; Neuspiel et al., 2008). Although the function of peroxisomally targeted MDVs remains unclear—one possibility includes the β-oxidation of fatty acids—mutations in *VPS35* have been linked to a form of late-onset PD (Vilarino-Guell et al., 2011; Zimprich et al., 2011). Thus, in line with the observation that disease-linked parkin mutations impede vesicle formation, defects in MDV-based transport from mitochondria to other organelles may contribute to PD pathogenesis (McLelland et al., 2014; Sugiura et al., 2014).

Over the past two decades, the field of mitochondrial cell biology has challenged the traditional notion that mitochondria are completely autonomous organelles, inert within the cellular milieu surrounding them. Although mitochondria form functional contacts with both the ER and lysosomes, they also serve as a signaling platform, playing crucial roles in apoptosis and innate immunity. Along these lines, the cellular degradation toolbox—the proteasome, autophagosomes, and lysosomes—has an intimate relationship with mitochondria. Functioning at the steady-state or under distinct stimuli, these machinery are crucial effectors of mitochondrial quality control pathways that ensure organellar health. Consequently, defects in these quality control mechanisms may underlie the pathogeneses of many neurodegenerative diseases. Future work will aim to identify the precise molecular players, as well as their respective roles, in these mechanisms, and elucidate the function of these mitochondrial quality control pathways in neuronal systems, with the ultimate goal of establishing druggable, therapeutic targets to treat neurodegenerative disease.

REFERENCES

Abeliovich, H., et al., 2013. Involvement of mitochondrial dynamics in the segregation of mitochondrial matrix proteins during stationary phase mitophagy. *Nat Commun*, 4, p.2789.

Al Rawi, S., et al., 2011. Postfertilization autophagy of sperm organelles prevents paternal mitochondrial DNA transmission. *Science*, 334, pp.1144–1147.

Ali, S., et al., 2006. PARK2/PACRG polymorphisms and susceptibility to typhoid and paratyphoid fever. *Clin Exp Immunol*, 144, pp.425–431.

Almeida, A., et al., 2001. Different responses of astrocytes and neurons to nitric oxide: the role of glycolytically generated ATP in astrocyte protection. *Proc Natl Acad Sci U S A*, 98, pp.15294–15299.

Almeida, A., Moncada, S. and Bolanos, J. P., 2004. Nitric oxide switches on glycolysis through the AMP protein kinase and 6-phosphofructo-2-kinase pathway. *Nat Cell Biol*, 6, pp.45–51.

Amiott, E. A., et al., 2009. A mutation associated with CMT2A neuropathy causes defects in Fzo1 GTP hydrolysis, ubiquitylation, and protein turnover. *Mol Biol Cell*, 20, pp.5026–5035.

Anton, F., et al., 2013. Two deubiquitylases act on mitofusin and regulate mitochondrial fusion along independent pathways. *Mol Cell*, 49, pp.487–498.

Anvret, A., et al., 2010. Variations of the CAG trinucleotide repeat in DNA polymerase gamma (POLG1) is associated with Parkinson's disease in Sweden. *Neurosci Lett*, 485, pp.117–120.

Aoki, Y., et al., 2011. Phosphorylation of Serine 114 on Atg32 mediates mitophagy. *Mol Biol Cell*, 22, pp.3206–3217.

Arrigo, A. P., et al., 1988. Identity of the 19S 'prosome' particle with the large multifunctional protease complex of mammalian cells (the proteasome). *Nature*, 331, pp.192–194.

Azzu, V. and Brand, M. D., 2010. Degradation of an intramitochondrial protein by the cytosolic proteasome. *J Cell Sci*, 123, pp.578–585.

Azzu, V., Mookerjee, S. A. and Brand, M. D., 2010. Rapid turnover of mitochondrial uncoupling protein 3. *Biochem J*, 426, pp.13–17.

Ballar, P., et al., 2006. The role of a novel p97/valosin-containing protein-interacting motif of gp78 in endoplasmic reticulum-associated degradation. *J Biol Chem*, 281, pp.35359–35368.

Barnham, K. J. and Bush, A. I., 2008. Metals in Alzheimer's and Parkinson's diseases. *Curr Opin Chem Biol*, 12, pp.222–228.

Benard, G., et al., 2010. IBRDC2, an IBR-type E3 ubiquitin ligase, is a regulatory factor for Bax and apoptosis activation. *EMBO J*, 29, pp.1458–1471.

Bender, A., et al., 2006. High levels of mitochondrial DNA deletions in substantia nigra neurons in aging and Parkinson disease. *Nat Genet*, 38, pp.515–517.

Benischke, A. S., et al., 2014. Proteasome-mediated quality control of S-nitrosylated mitochondrial proteins. *Mitochondrion*, 17, pp.182–186.

Besche, H. C., Peth, A. and Goldberg, A. L., 2009. Getting to first base in proteasome assembly. *Cell*, 138, pp.25–28.

Beverly, L. J., et al., 2012. Ubiquitination, localization, and stability of an anti-apoptotic BCL2-like protein, BCL2L10/BCLb, are regulated by Ubiquilin1. *Proc Natl Acad Sci U S A*, 109, pp.E119–E126.

Bindoff, L. A., et al., 1989. Mitochondrial function in Parkinson's disease. *Lancet*, 2, p.49.

Bingol, B., et al., 2014. The mitochondrial deubiquitinase USP30 opposes parkin-mediated mitophagy. *Nature*, 509, pp.370–375.

Boland, B., et al., 2008. Autophagy induction and autophagosome clearance in neurons: relationship to autophagic pathology in Alzheimer's disease. *J Neurosci*, 28, pp.6926–6937.

Bolanos, J. P., Almeida, A. and Moncada, S., 2010. Glycolysis: a bioenergetic or a survival pathway? *Trends Biochem Sci*, 35, pp.145–149.

Bosetti, F., et al., 2002. Cytochrome c oxidase and mitochondrial F1F0-ATPase (ATP synthase) activities in platelets and brain from patients with Alzheimer's disease. *Neurobiol Aging*, 23, pp.371–376.

Bragoszewski, P., et al., 2013. The ubiquitin-proteasome system regulates mitochondrial intermembrane space proteins. *Mol Cell Biol*, 33, pp.2136–2148.

Braschi, E., et al., 2010. Vps35 mediates vesicle transport between the mitochondria and peroxisomes. *Curr Biol*, 20, pp.1310–1315.

Burman, J. L., et al., 2012. Analysis of neural subtypes reveals selective mitochondrial dysfunction in dopaminergic neurons from parkin mutants. *Proc Natl Acad Sci U S A*, 109, pp.10438–10443.

Cai, Q., et al., 2012. Spatial parkin translocation and degradation of damaged mitochondria via mitophagy in live cortical neurons. *Curr Biol*, 22, pp.545–552.

Cardoso, S. M., et al., 2004. Mitochondria dysfunction of Alzheimer's disease cybrids enhances Abeta toxicity. *J Neurochem*, 89, pp.1417–1426.

Carvalho, P., Goder, V. and Rapoport, T. A., 2006. Distinct ubiquitin-ligase complexes define convergent pathways for the degradation of ER proteins. *Cell*, 126, pp.361–373.

Casari, G., et al., 1998. Spastic paraplegia and OXPHOS impairment caused by mutations in paraplegin, a nuclear-encoded mitochondrial metalloprotease. *Cell*, 93, pp.973–983.

Cavey, J. R., et al., 2005. Loss of ubiquitin-binding associated with Paget's disease of bone p62 (SQSTM1) mutations. *J Bone Miner Res*, 20, pp.619–624.

Chan, C. S., et al., 2007. "Rejuvenation" protects neurons in mouse models of Parkinson's disease. *Nature*, 447, pp.1081–1086.

Chan, N. C., et al., 2011. Broad activation of the ubiquitin-proteasome system by Parkin is critical for mitophagy. *Hum Mol Genet*, 20, pp.1726–1737.

Chen, G., et al., 2014. A regulatory signaling loop comprising the PGAM5 phosphatase and CK2 controls receptor-mediated mitophagy. *Mol Cell*, 54, pp.362–377.

Chu, C. T., et al., 2013. Cardiolipin externalization to the outer mitochondrial membrane acts as an elimination signal for mitophagy in neuronal cells. *Nat Cell Biol*, 15, pp.1197–1205.

Chung, K. K., et al., 2004. S-nitrosylation of parkin regulates ubiquitination and compromises parkin's protective function. *Science*, 304, pp.1328–1331.

Clark, I. E., et al., 2006. Drosophila pink1 is required for mitochondrial function and interacts genetically with parkin. *Nature*, 441, pp.1162–1166.

Cohen, M. M., et al., 2008. Ubiquitin-proteasome-dependent degradation of a mitofusin, a critical regulator of mitochondrial fusion. *Mol Biol Cell*, 19, pp.2457–2464.

Custer, S. K., et al., 2010. Transgenic mice expressing mutant forms VCP/p97 recapitulate the full spectrum of IBMPFD including degeneration in muscle, brain and bone. *Hum Mol Genet*, 19, pp.1741–1755.

Daniele, T., et al., 2014. Mitochondria and melanosomes establish physical contacts modulated by Mfn2 and involved in organelle biogenesis. *Curr Biol*, 24, pp.393–403.

Davidzon, G., et al., 2006. Early-onset familial parkinsonism due to POLG mutations. *Ann Neurol*, 59, pp.859–862.

de Brito, O. M. and Scorrano, L., 2008. Mitofusin 2 tethers endoplasmic reticulum to mitochondria. *Nature*, 456, pp.605–610.

de Pablo-Latorre, R., et al., 2012. Impaired parkin-mediated mitochondrial targeting to autophagosomes differentially contributes to tissue pathology in lysosomal storage diseases. *Hum Mol Genet*, 21, pp.1770–1781.

Deng, H., et al., 2008. The Parkinson's disease genes pink1 and parkin promote mitochondrial fission and/or inhibit fusion in Drosophila. *Proc Natl Acad Sci U S A*, 105, pp.14503–14508.

Denic, V., Quan, E. M. and Weissman, J. S., 2006. A luminal surveillance complex that selects misfolded glycoproteins for ER-associated degradation. *Cell*, 126, pp.349–359.

Dexter, D. T., et al., 1992. Alterations in levels of iron, ferritin, and other trace metals in neurodegenerative diseases affecting the basal ganglia. The Royal Kings and Queens Parkinson's Disease Research Group. *Ann Neurol*, 32 Suppl, pp.S94–S100.

Di Bella, D., et al., 2010. Mutations in the mitochondrial protease gene AFG3L2 cause dominant hereditary ataxia SCA28. *Nat Genet*, 42, pp.313–321.

Eerola, J., et al., 2010. POLG1 polyglutamine tract variants associated with Parkinson's disease. *Neurosci Lett*, 477, pp.1–5.

Egan, D. F., et al., 2011. Phosphorylation of ULK1 (hATG1) by AMP-activated protein kinase connects energy sensing to mitophagy. *Science*, 331, pp.456–461.

Elbaz-Alon, Y., et al., 2014. A dynamic interface between vacuoles and mitochondria in yeast. *Dev Cell*, 30, pp.95–102.

Enenkel, C., Lehmann, A. and Kloetzel, P. M., 1998. Subcellular distribution of proteasomes implicates a major location of protein degradation in the nuclear envelope-ER network in yeast. *EMBO J*, 17, pp.6144–6154.

Escobar-Henriques, M., Westermann, B. and Langer, T., 2006. Regulation of mitochondrial fusion by the F-box protein Mdm30 involves proteasome-independent turnover of Fzo1. *J Cell Biol*, 173, pp.645–650.

Exner, N., et al., 2007. Loss-of-function of human PINK1 results in mitochondrial pathology and can be rescued by parkin. *J Neurosci*, 27, pp.12413–12418.

Fang, E. F., et al., 2014. Defective mitophagy in XPA via PARP-1 hyperactivation and NAD(+)/SIRT1 reduction. *Cell*, 157, pp.882–896.

Filimonenko, M., et al., 2007. Functional multivesicular bodies are required for autophagic clearance of protein aggregates associated with neurodegenerative disease. *J Cell Biol*, 179, pp.485–500.

Gandhi, S., et al., 2009. PINK1-associated Parkinson's disease is caused by neuronal vulnerability to calcium-induced cell death. *Mol Cell*, 33, pp.627–638.

Gautier, C. A., Kitada, T. and Shen, J., 2008. Loss of PINK1 causes mitochondrial functional defects and increased sensitivity to oxidative stress. *Proc Natl Acad Sci U S A*, 105, pp.11364–11369.

Gegg, M. E., et al., 2010. Mitofusin 1 and mitofusin 2 are ubiquitinated in a PINK1/parkin-dependent manner upon induction of mitophagy. *Hum Mol Genet*, 19, pp.4861–4870.

Geisler, S., et al., 2010a. PINK1/Parkin-mediated mitophagy is dependent on VDAC1 and p62/SQSTM1. *Nat Cell Biol*, 12, pp.119–131.

Geisler, S., et al., 2010b. The PINK1/Parkin-mediated mitophagy is compromised by PD-associated mutations. *Autophagy*, 6, pp.871–878.

Gispert, S., et al., 2009. Parkinson phenotype in aged PINK1-deficient mice is accompanied by progressive mitochondrial dysfunction in absence of neurodegeneration. *PLoS One*, 4, p.e5777.

Goldberg, A. L., 2007. Functions of the proteasome: from protein degradation and immune surveillance to cancer therapy. *Biochem Soc Trans*, 35, pp.12–17.

Goldberg, M. S., et al., 2003. Parkin-deficient mice exhibit nigrostriatal deficits but not loss of dopaminergic neurons. *J Biol Chem*, 278, pp.43628–43635.

Granot, Z., et al., 2007. Turnover of mitochondrial steroidogenic acute regulatory (StAR) protein by Lon protease: the unexpected effect of proteasome inhibitors. *Mol Endocrinol*, 21, pp.2164–2177.

Greenblatt, E. J., Olzmann, J. A. and Kopito, R. R., 2011. Derlin-1 is a rhomboid pseudoprotease required for the dislocation of mutant alpha-1 antitrypsin from the endoplasmic reticulum. *Nat Struct Mol Biol*, 18, pp.1147–1152.

Greene, A. W., et al., 2012. Mitochondrial processing peptidase regulates PINK1 processing, import and Parkin recruitment. *EMBO Rep*, 13, pp.378–385.

Greene, J. C., et al., 2003. Mitochondrial pathology and apoptotic muscle degeneration in Drosophila parkin mutants. *Proc Natl Acad Sci U S A*, 100, pp.4078–4083.

Greene, J. C., et al., 2005. Genetic and genomic studies of Drosophila parkin mutants implicate oxidative stress and innate immune responses in pathogenesis. *Hum Mol Genet*, 14, pp.799–811.

Grenier, K., McLelland, G. L. and Fon, E. A., 2013. Parkin- and PINK1-dependent mitophagy in neurons: Will the real pathway please stand up? *Front Neurol*, 4, p.100.

Gui, Y. X., et al., 2012. Association of mitochondrial DNA polymerase gamma gene POLG1 polymorphisms with parkinsonism in Chinese populations. *PLoS One*, 7, p.e50086.

Guinto, J. B., et al., 2007. Valosin-containing protein and the pathogenesis of frontotemporal dementia associated with inclusion body myopathy. *Acta Neuropathol*, 114, pp.55–61.

Guzman, J. N., et al., 2009. Robust pacemaking in substantia nigra dopaminergic neurons. *J Neurosci*, 29, pp.11011–11019.

Guzman, J. N., et al., 2010. Oxidant stress evoked by pacemaking in dopaminergic neurons is attenuated by DJ-1. *Nature*, 468, pp.696–700.

Hamasaki, M., et al., 2013. Autophagosomes form at ER-mitochondria contact sites. *Nature*, 495, pp.389–393.

Hasson, S. A., et al., 2013. High-content genome-wide RNAi screens identify regulators of parkin upstream of mitophagy. *Nature*, 504, pp.291–295.

Heeman, B., et al., 2011. Depletion of PINK1 affects mitochondrial metabolism, calcium homeostasis and energy maintenance. *J Cell Sci*, 124, pp.1115–1125.

Heid, M. E., et al., 2013. Mitochondrial reactive oxygen species induces NLRP3-dependent lysosomal damage and inflammasome activation. *J Immunol*, 191, pp.5230–5238.

Heo, J. M., et al., 2010. A stress-responsive system for mitochondrial protein degradation. *Mol Cell*, 40, pp.465–480.

Heo, J. M., et al., 2013. Intramolecular interactions control Vms1 translocation to damaged mitochondria. *Mol Biol Cell*, 24, pp.1263–1273.

Herrero-Mendez, A., et al., 2009. The bioenergetic and antioxidant status of neurons is controlled by continuous degradation of a key glycolytic enzyme by APC/C-Cdh1. *Nat Cell Biol*, 11, pp.747–752.

Hershko, A., et al., 1980. Proposed role of ATP in protein breakdown: conjugation of protein with multiple chains of the polypeptide of ATP-dependent proteolysis. *Proc Natl Acad Sci U S A*, 77, pp.1783–1786.

Honscher, C., et al., 2014. Cellular metabolism regulates contact sites between vacuoles and mitochondria. *Dev Cell*, 30, pp.86–94.

Huang, C., et al., 2011. Preconditioning involves selective mitophagy mediated by Parkin and p62/SQSTM1. *PLoS One*, 6, p.e20975.

Hughes, A. L. and Gottschling, D. E., 2012. An early age increase in vacuolar pH limits mitochondrial function and lifespan in yeast. *Nature*, 492, pp.261–265.

Hyun, D. H., et al., 2002. Effect of wild-type or mutant Parkin on oxidative damage, nitric oxide, antioxidant defenses, and the proteasome. *J Biol Chem*, 277, pp.28572–28577.

Iguchi, M., et al., 2013. Parkin-catalyzed ubiquitin-ester transfer is triggered by PINK1-dependent phosphorylation. *J Biol Chem*, 288, pp.22019–22032.

Itakura, E., et al., 2012. Structures containing Atg9A and the ULK1 complex independently target depolarized mitochondria at initial stages of Parkin-mediated mitophagy. *J Cell Sci*, 125, pp.1488–1499.

Jin, R. U. and Mills, J. C., 2014. RAB26 coordinates lysosome traffic and mitochondrial localization. *J Cell Sci*, 127, pp.1018–1032.

Jin, S. M. and Youle, R. J., 2013. The accumulation of misfolded proteins in the mitochondrial matrix is sensed by PINK1 to induce PARK2/Parkin-mediated mitophagy of polarized mitochondria. *Autophagy*, 9, pp.1750–1757.

Jin, S. M., et al., 2010. Mitochondrial membrane potential regulates PINK1 import and proteolytic destabilization by PARL. *J Cell Biol*, 191, pp.933–942.

Joselin, A. P., et al., 2012. ROS-dependent regulation of Parkin and DJ-1 localization during oxidative stress in neurons. *Hum Mol Genet*, 21, pp.4888–4903.

Kane, L. A., et al., 2014. PINK1 phosphorylates ubiquitin to activate Parkin E3 ubiquitin ligase activity. *J Cell Biol*, 205, pp.143–153.

Kanki, T., et al., 2009. Atg32 is a mitochondrial protein that confers selectivity during mitophagy. *Dev Cell*, 17, pp.98–109.

Kazlauskaite, A., et al., 2014a. Phosphorylation of Parkin at Serine65 is essential for activation: elaboration of a Miro1 substrate-based assay of Parkin E3 ligase activity. *Open Biol*, 4, p/130213.

Kazlauskaite, A., et al., 2014b. Parkin is activated by PINK1-dependent phosphorylation of ubiquitin at Ser65. *Biochem J*, 460, pp.127–139.

Kim, N. C., et al., 2013. VCP is essential for mitochondrial quality control by PINK1/Parkin and this function is impaired by VCP mutations. *Neuron*, 78, pp.65–80.

Kimura, Y., et al., 2013. Different dynamic movements of wild-type and pathogenic VCPs and their cofactors to damaged mitochondria in a Parkin-mediated mitochondrial quality control system. *Genes Cells*, 18, pp.1131–1143.

Kirkin, V., et al., 2009. A role for ubiquitin in selective autophagy. *Mol Cell*, 34, pp.259–269.

Kish, S. J., et al., 1992. Brain cytochrome oxidase in Alzheimer's disease. *J Neurochem*, 59, pp.776–779.

Kitada, T., et al., 1998. Mutations in the parkin gene cause autosomal recessive juvenile parkinsonism. *Nature*, 392, pp.605–608.

Kitada, T., et al., 2007. Impaired dopamine release and synaptic plasticity in the striatum of PINK1-deficient mice. *Proc Natl Acad Sci U S A*, 104, pp.11441–11446.

Kitada, T., et al., 2009. Absence of nigral degeneration in aged parkin/DJ-1/PINK1 triple knockout mice. *J Neurochem*, 111, pp.696–702.

Komander, D., Clague, M. J. and Urbe, S., 2009. Breaking the chains: structure and function of the deubiquitinases. *Nat Rev Mol Cell Biol*, 10, pp.550–563.

Komatsu, M., et al., 2005. Impairment of starvation-induced and constitutive autophagy in Atg7-deficient mice. *J Cell Biol*, 169, pp.425–434.

Kondapalli, C., et al., 2012. PINK1 is activated by mitochondrial membrane potential depolarization and stimulates Parkin E3 ligase activity by phosphorylating Serine 65. *Open Biol*, 2, p.120080.

Kondo-Okamoto, N., et al., 2012. Autophagy-related protein 32 acts as autophagic degron and directly initiates mitophagy. *J Biol Chem*, 287, pp.10631–10638.

Koyano, F., et al., 2013. The principal PINK1 and Parkin cellular events triggered in response to dissipation of mitochondrial membrane potential occur in primary neurons. *Genes Cells*, 18, pp.672–681.

Koyano, F., et al., 2014. Ubiquitin is phosphorylated by PINK1 to activate parkin. *Nature*, 510, pp.162–166.

Kubli, D. A., Quinsay, M. N. and Gustafsson, A. B., 2013. Parkin deficiency results in accumulation of abnormal mitochondria in aging myocytes. *Commun Integr Biol*, 6, p.e24511.

Kutuk, O. and Letai, A., 2008. Regulation of Bcl-2 family proteins by posttranslational modifications. *Curr Mol Med*, 8, pp.102–118.

Lazarou, M., et al., 2012. Role of PINK1 binding to the TOM complex and alternate intracellular membranes in recruitment and activation of the E3 ligase Parkin. *Dev Cell*, 22, pp.320–333.

Lazarou, M., et al., 2013. PINK1 drives Parkin self-association and HECT-like E3 activity upstream of mitochondrial binding. *J Cell Biol*, 200, pp.163–172.

Leboucher, G. P., et al., 2012. Stress-induced phosphorylation and proteasomal degradation of mitofusin 2 facilitates mitochondrial fragmentation and apoptosis. *Mol Cell*, 47, pp.547–557.

Lee, J. H., et al., 2010a. Lysosomal proteolysis and autophagy require presenilin 1 and are disrupted by Alzheimer-related PS1 mutations. *Cell*, 141, pp.1146–1158.

Lee, J. Y., et al., 2010b. Disease-causing mutations in parkin impair mitochondrial ubiquitination, aggregation, and HDAC6-dependent mitophagy. *J Cell Biol*, 189, pp.671–679.

Lee, S., Sato, Y. and Nixon, R. A., 2011. Lysosomal proteolysis inhibition selectively disrupts axonal transport of degradative organelles and causes an Alzheimer's-like axonal dystrophy. *J Neurosci*, 31, pp.7817–7830.

Lilley, B. N. and Ploegh, H. L., 2004. A membrane protein required for dislocation of misfolded proteins from the ER. *Nature*, 429, pp.834–840.

Liu, L., et al., 2012. Mitochondrial outer-membrane protein FUNDC1 mediates hypoxia-induced mitophagy in mammalian cells. *Nat Cell Biol*, 14, pp.177–185.

Luoma, P. T., et al., 2007. Mitochondrial DNA polymerase gamma variants in idiopathic sporadic Parkinson disease. *Neurology*, 69, pp.1152–1159.

Lutz, A. K., et al., 2009. Loss of parkin or PINK1 function increases Drp1-dependent mitochondrial fragmentation. *J Biol Chem*, 284, pp.22938–22951.

MacVicar, T. D. and Lane, J. D., 2014. Impaired OMA1-dependent cleavage of OPA1 and reduced DRP1 fission activity combine to prevent mitophagy in cells that are dependent on oxidative phosphorylation. *J Cell Sci*, 127, pp.2313–2325.

Magiera, M. M., et al., 2013. Trim17-mediated ubiquitination and degradation of Mcl-1 initiate apoptosis in neurons. *Cell Death Differ*, 20, pp.281–292.

Mann, V. M., et al., 1992. Brain, skeletal muscle and platelet homogenate mitochondrial function in Parkinson's disease. *Brain*, 115 (Pt 2), pp.333–342.

Manzanillo, P. S., et al., 2013. The ubiquitin ligase parkin mediates resistance to intracellular pathogens. *Nature*, 501, pp.512–516.

Mao, K., et al., 2013. The scaffold protein Atg11 recruits fission machinery to drive selective mitochondria degradation by autophagy. *Dev Cell*, 26, pp.9–18.

Maor, R., et al., 2007. Multidimensional protein identification technology (MudPIT) analysis of ubiquitinated proteins in plants. *Mol Cell Proteomics*, 6, pp.601–610.

Margineantu, D. H., et al., 2007. Hsp90 inhibition decreases mitochondrial protein turnover. *PLoS One*, 2, p.e1066.

Matsumoto, M., et al., 2005. Large-scale analysis of the human ubiquitin-related proteome. *Proteomics*, 5, pp.4145–4151.

Matsuura, A., et al., 1997. Apg1p, a novel protein kinase required for the autophagic process in Saccharomyces cerevisiae. *Gene*, 192, pp.245–250.

McLelland, G. L., et al., 2014. Parkin and PINK1 function in a vesicular trafficking pathway regulating mitochondrial quality control. *EMBO J*, 33, pp.282–295.

Mecocci, P., MacGarvey, U. and Beal, M. F., 1994. Oxidative damage to mitochondrial DNA is increased in Alzheimer's disease. *Ann Neurol*, 36, pp.747–751.

Meller, R., et al., 2006. Rapid degradation of Bim by the ubiquitin-proteasome pathway mediates short-term ischemic tolerance in cultured neurons. *J Biol Chem*, 281, pp.7429–7436.

Melser, S., et al., 2013. Rheb regulates mitophagy induced by mitochondrial energetic status. *Cell Metab*, 17, pp.719–730.

Metzger, M. B., Hristova, V. A. and Weissman, A. M., 2012. HECT and RING finger families of E3 ubiquitin ligases at a glance. *J Cell Sci*, 125, pp.531–537.

Meyer, H., Bug, M. and Bremer, S., 2012. Emerging functions of the VCP/p97 AAA-ATPase in the ubiquitin system. *Nat Cell Biol*, 14, pp.117–123.

Mira, M. T., et al., 2004. Susceptibility to leprosy is associated with PARK2 and PACRG. *Nature*, 427, pp.636–640.

Morais, V. A., et al., 2014. PINK1 loss-of-function mutations affect mitochondrial complex I activity via NdufA10 ubiquinone uncoupling. *Science*, 344, pp.203–207.

Moreira, P. I., et al., 2007. Autophagocytosis of mitochondria is prominent in Alzheimer disease. *J Neuropathol Exp Neurol*, 66, pp.525–532.

Mortiboys, H., et al., 2008. Mitochondrial function and morphology are impaired in parkin-mutant fibroblasts. *Ann Neurol*, 64, pp.555–565.

Muftuoglu, M., et al., 2004. Mitochondrial complex I and IV activities in leukocytes from patients with parkin mutations. *Mov Disord*, 19, pp.544–548.

Muller-Rischart, A. K., et al., 2013. The E3 ligase parkin maintains mitochondrial integrity by increasing linear ubiquitination of NEMO. *Mol Cell*, 49, pp.908–921.

Murakami, Y., et al., 1992. Ornithine decarboxylase is degraded by the 26S proteasome without ubiquitination. *Nature*, 360, pp.597–599.

Nakamura, N., et al., 2006. MARCH-V is a novel mitofusin 2- and Drp1-binding protein able to change mitochondrial morphology. *EMBO Rep*, 7, pp.1019–1022.

Nakamura, N. and Hirose, S., 2008. Regulation of mitochondrial morphology by USP30, a deubiquitinating enzyme present in the mitochondrial outer membrane. *Mol Biol Cell*, 19, pp.1903–1911.

Narendra, D., et al., 2008. Parkin is recruited selectively to impaired mitochondria and promotes their autophagy. *J Cell Biol*, 183, pp.795–803.

Narendra, D., et al., 2010a. p62/SQSTM1 is required for Parkin-induced mitochondrial clustering but not mitophagy; VDAC1 is dispensable for both. *Autophagy*, 6, pp.1090–1106.

Narendra, D. P., et al., 2010b. PINK1 is selectively stabilized on impaired mitochondria to activate Parkin. *PLoS Biol*, 8, p.e1000298.

Neuspiel, M., et al., 2008. Cargo-selected transport from the mitochondria to peroxisomes is mediated by vesicular carriers. *Curr Biol*, 18, pp.102–108.

Neutzner, A., Youle, R. J. and Karbowski, M., 2007. Outer mitochondrial membrane protein degradation by the proteasome. *Novartis Found Symp*, 287, pp.4–14; 14–20.

Nixon, R. A., et al., 2005. Extensive involvement of autophagy in Alzheimer disease: an immuno-electron microscopy study. *J Neuropathol Exp Neurol*, 64, pp.113–122.

Novak, I., et al., 2010. Nix is a selective autophagy receptor for mitochondrial clearance. *EMBO Rep*, 11, pp.45–51.

Oda, Y., et al., 2006. Derlin-2 and Derlin-3 are regulated by the mammalian unfolded protein response and are required for ER-associated degradation. *J Cell Biol*, 172, pp.383–393.

Okamoto, K., Kondo-Okamoto, N. and Ohsumi, Y., 2009. Mitochondria-anchored receptor Atg32 mediates degradation of mitochondria via selective autophagy. *Dev Cell*, 17, pp.87–97.

Palacino, J. J., et al., 2004. Mitochondrial dysfunction and oxidative damage in parkin-deficient mice. *J Biol Chem*, 279, pp.18614–18622.

Pang, X., et al., 2014. The carboxyl-terminal tail of noxa regulates the stability of noxa and Mcl-1. *J Biol Chem*, 289, pp.17802–17811.

Pankiv, S., et al., 2007. p62/SQSTM1 binds directly to Atg8/LC3 to facilitate degradation of ubiquitinated protein aggregates by autophagy. *J Biol Chem*, 282, pp.24131–24145.

Park, J., et al., 2006. Mitochondrial dysfunction in Drosophila PINK1 mutants is complemented by parkin. *Nature*, 441, pp.1157–1161.

Parker, W. D., Jr., Filley, C. M. and Parks, J. K., 1990. Cytochrome oxidase deficiency in Alzheimer's disease. *Neurology*, 40, pp.1302–1303.

Parker, W. D., Jr., et al., 1994. Electron transport chain defects in Alzheimer's disease brain. *Neurology*, 44, pp.1090–1096.

Pearce, D. A. and Sherman, F., 1997. Differential ubiquitin-dependent degradation of the yeast apo-cytochrome c isozymes. *J Biol Chem*, 272, pp.31829–31836.

Peng, J., et al., 2003. A proteomics approach to understanding protein ubiquitination. *Nat Biotechnol*. 21, pp.921–926.

Perez, F. A. and Palmiter, R. D., 2005. Parkin-deficient mice are not a robust model of parkinsonism. *Proc Natl Acad Sci U S A*, 102, pp.2174–2179.

Piquereau, J., et al., 2013. Protective role of PARK2/Parkin in sepsis-induced cardiac contractile and mitochondrial dysfunction. *Autophagy*, 9, pp.1837–1851.

Politi, Y., et al., 2014. Paternal mitochondrial destruction after fertilization is mediated by a common endocytic and autophagic pathway in Drosophila. *Dev Cell*, 29, pp.305–320.

Poole, A. C., et al., 2008. The PINK1/Parkin pathway regulates mitochondrial morphology. *Proc Natl Acad Sci U S A*, 105, pp.1638–1643.

Poole, A. C., et al., 2010. The mitochondrial fusion-promoting factor mitofusin is a substrate of the PINK1/parkin pathway. *PLoS One*, 5, p.e10054.

Radke, S., et al., 2008. Mitochondrial protein quality control by the proteasome involves ubiquitination and the protease Omi. *J Biol Chem*, 283, pp.12681–12685.

Rakovic, A., et al., 2013. Phosphatase and tensin homolog (PTEN)-induced putative kinase 1 (PINK1)-dependent ubiquitination of endogenous Parkin attenuates mitophagy: study in human primary fibroblasts and induced pluripotent stem cell-derived neurons. *J Biol Chem*, 288, pp.2223–2237.

Repnik, U., Cesen, M. H. and Turk, B., 2013. The endolysosomal system in cell death and survival. *Cold Spring Harb Perspect Biol*, 5, p.a008755.

Richardson, D. R., et al., 2010. Mitochondrial iron trafficking and the integration of iron metabolism between the mitochondrion and cytosol. *Proc Natl Acad Sci U S A*, 107, pp.10775–10782.

Rott, R., et al., 2011. alpha-Synuclein fate is determined by USP9X-regulated monoubiquitination. *Proc Natl Acad Sci U S A*, 108, pp.18666–18671.

Rugarli, E. I. and Langer, T., 2012. Mitochondrial quality control: A matter of life and death for neurons. *EMBO J*, 31, pp.1336–1349.

Sarraf, S. A., et al., 2013. Landscape of the PARKIN-dependent ubiquitylome in response to mitochondrial depolarization. *Nature*, 496, pp.372–376.

Sato, M. and Sato, K., 2011. Degradation of paternal mitochondria by fertilization-triggered autophagy in *C. elegans* embryos. *Science*, 334, pp.1141–1144.

Schapira, A. H., et al., 1989. Mitochondrial complex I deficiency in Parkinson's disease. *Lancet*, 1, p.1269.

Schwickart, M., et al., 2010. Deubiquitinase USP9X stabilizes MCL1 and promotes tumour cell survival. *Nature*, 463, pp.103–107.

Seibler, P., et al., 2011. Mitochondrial Parkin recruitment is impaired in neurons derived from mutant PINK1 induced pluripotent stem cells. *J Neurosci*, 31, pp.5970–5976.

Shiba-Fukushima, K., et al., 2012. PINK1-mediated phosphorylation of the Parkin ubiquitin-like domain primes mitochondrial translocation of Parkin and regulates mitophagy. *Sci Rep*, 2, p.1002.

Smith, M. H., Ploegh, H. L. and Weissman, J. S., 2011. Road to ruin: targeting proteins for degradation in the endoplasmic reticulum. *Science*, 334, pp.1086–1090.

Soubannier, V., et al., 2012a. A vesicular transport pathway shuttles cargo from mitochondria to lysosomes. *Curr Biol*, 22, pp.135–141.

Soubannier, V., et al., 2012b. Reconstitution of mitochondria derived vesicle formation demonstrates selective enrichment of oxidized cargo. *PLoS One*, 7, p.e52830.

Stafford, F. J. and Bonifacino, J. S., 1991. A permeabilized cell system identifies the endoplasmic reticulum as a site of protein degradation. *J Cell Biol*, 115, pp.1225–1236.

Suen, D. F., et al., 2010. Parkin overexpression selects against a deleterious mtDNA mutation in heteroplasmic cybrid cells. *Proc Natl Acad Sci U S A*, 107, pp.11835–11840.

Sugiura, A., et al., 2013. MITOL regulates endoplasmic reticulum-mitochondria contacts via Mitofusin2. *Mol Cell*, 51, pp.20–34.

Sugiura, A., et al., 2014. A new pathway for mitochondrial quality control: Mitochondrial-derived vesicles. *EMBO J*, 33, pp.2142–2156.

Tagwerker, C., et al., 2006. A tandem affinity tag for two-step purification under fully denaturing conditions: application in ubiquitin profiling and protein complex identification combined with in vivocross-linking. *Mol Cell Proteomics*, 5, pp.737–748.

Tait, S. W., et al., 2007. Apoptosis induction by Bid requires unconventional ubiquitination and degradation of its N-terminal fragment. *J Cell Biol*, 179, pp.1453–1466.

Tanaka, A., et al., 2010. Proteasome and p97 mediate mitophagy and degradation of mitofusins induced by Parkin. *J Cell Biol*, 191, pp.1367–1380.

Trempe, J. F., 2011. Reading the ubiquitin postal code. *Curr Opin Struct Biol*, 21, pp.792–801.

Trempe, J. F., et al., 2013. Structure of parkin reveals mechanisms for ubiquitin ligase activation. *Science*, 340, pp.1451–1455.

Trimmer, P. A., et al., 2004. Mitochondrial abnormalities in cybrid cell models of sporadic Alzheimer's disease worsen with passage in culture. *Neurobiol Dis*, 15, pp.29–39.

Trinh, K., et al., 2008. Induction of the phase II detoxification pathway suppresses neuron loss in Drosophila models of Parkinson's disease. *J Neurosci*, 28, pp.465–472.

Tsukada, M. and Ohsumi, Y., 1993. Isolation and characterization of autophagy-defective mutants of Saccharomyces cerevisiae. *FEBS Lett*, 333, pp.169–174.

Twig, G., et al., 2008. Fission and selective fusion govern mitochondrial segregation and elimination by autophagy. *EMBO J*, 27, pp.433–446.

Valente, E. M., et al., 2004. Hereditary early-onset Parkinson's disease caused by mutations in PINK1. *Science*, 304, pp.1158–1160.

van de Kooij, B., et al., 2013. Polyubiquitination and proteasomal turnover controls the anti-apoptotic activity of Bcl-B. *Oncogene*, 32, pp.5439–5448.

Van Laar, V. S., et al., 2011. Bioenergetics of neurons inhibit the translocation response of Parkin following rapid mitochondrial depolarization. *Hum Mol Genet*, 20, pp.927–940.

Vembar, S. S. and Brodsky, J. L., 2008. One step at a time: endoplasmic reticulum-associated degradation. *Nat Rev Mol Cell Biol*, 9, pp.944–957.

Vilarino-Guell, C., et al., 2011. VPS35 mutations in Parkinson disease. *Am J Hum Genet*, 89, pp.162–167.

Vincow, E. S., et al., 2013. The PINK1-Parkin pathway promotes both mitophagy and selective respiratory chain turnover in vivo. *Proc Natl Acad Sci U S A*, 110, pp.6400–6405.

Vives-Bauza, C., et al., 2010. PINK1-dependent recruitment of Parkin to mitochondria in mitophagy. *Proc Natl Acad Sci U S A*, 107, pp.378–383.

Wang, D., et al., 2006. Antioxidants protect PINK1-dependent dopaminergic neurons in Drosophila. *Proc Natl Acad Sci U S A*, 103, pp.13520–13525.

Wang, X., et al., 2011. PINK1 and Parkin target Miro for phosphorylation and degradation to arrest mitochondrial motility. *Cell*, 147, pp.893–906.

Watts, G. D., et al., 2004. Inclusion body myopathy associated with Paget disease of bone and frontotemporal dementia is caused by mutant valosin-containing protein. *Nat Genet*, 36, pp.377–381.

Weber, K. and Schilling, J. D., 2014. Lysosomes integrate metabolic-inflammatory cross-talk in primary macrophage inflammasome activation. *J Biol Chem*, 289, pp.9158–9171.

Weekes, J., et al., 2003. Hyperubiquitination of proteins in dilated cardiomyopathy. *Proteomics*, 3, pp.208–216.

Wenzel, D. M., et al., 2011. UBCH7 reactivity profile reveals parkin and HHARI to be RING/HECT hybrids. *Nature*, 474, pp.105–108.

West, A. B. and Maidment, N. T., 2004. Genetics of parkin-linked disease. *Hum Genet*, 114, pp.327–336.

Whitworth, A. J., et al., 2005. Increased glutathione S-transferase activity rescues dopaminergic neuron loss in a Drosophila model of Parkinson's disease. *Proc Natl Acad Sci U S A*, 102, pp.8024–8029.

Wong, M., et al., 2013. Mitochondrial DNMT3A and DNA methylation in skeletal muscle and CNS of transgenic mouse models of ALS. *Front Cell Neurosci*, 7, p.279.

Xu, S., et al., 2011. The AAA-ATPase p97 is essential for outer mitochondrial membrane protein turnover. *Mol Biol Cell*, 22, pp.291–300.

Yamano, K. and Youle, R. J., 2013. PINK1 is degraded through the N-end rule pathway. *Autophagy*, 9, pp.1758–1769.

Yamano, K., et al., 2014. Mitochondrial Rab GAPs govern autophagosome biogenesis during mitophagy. *Elife*, 3, p.e01612.

Yang, J. Y. and Yang, W. Y., 2013. Bit-by-bit autophagic removal of parkin-labelled mitochondria. *Nat Commun*, 4, p.2428.

Yang, Y., et al., 2006. Mitochondrial pathology and muscle and dopaminergic neuron degeneration caused by inactivation of Drosophila Pink1 is rescued by Parkin. *Proc Natl Acad Sci U S A*, 103, pp.10793–10798.

Yang, Y., et al., 2008. Pink1 regulates mitochondrial dynamics through interaction with the fission/fusion machinery. *Proc Natl Acad Sci U S A*, 105, pp.7070–7075.

Yang, Z. and Klionsky, D. J., 2010a. Mammalian autophagy: core molecular machinery and signaling regulation. *Curr Opin Cell Biol*, 22, pp.124–131.

Yang, Z. and Klionsky, D. J., 2010b. Eaten alive: a history of macroautophagy. *Nat Cell Biol*, 12, pp.814–822.

Yao, D., et al., 2004. Nitrosative stress linked to sporadic Parkinson's disease: S-nitrosylation of parkin regulates its E3 ubiquitin ligase activity. *Proc Natl Acad Sci U S A*, 101, pp.10810–10814.

Ye, Y., Meyer, H. H. and Rapoport, T. A., 2003. Function of the p97-Ufd1-Npl4 complex in retrotranslocation from the ER to the cytosol: dual recognition of nonubiquitinated polypeptide segments and polyubiquitin chains. *J Cell Biol*, 162, pp.71–84.

Ye, Y., et al., 2005. Recruitment of the p97 ATPase and ubiquitin ligases to the site of retrotranslocation at the endoplasmic reticulum membrane. *Proc Natl Acad Sci U S A*, 102, pp.14132–14138.

Yogalingam, G., et al., 2013. Glyceraldehyde-3-phosphate dehydrogenase (GAPDH) phosphorylation by protein kinase Cdelta (PKCdelta) inhibits mitochondria elimination by lysosomal-like structures following ischemia and reoxygenation-induced injury. *J Biol Chem*, 288, pp.18947–18960.

Yonashiro, R., et al., 2006. A novel mitochondrial ubiquitin ligase plays a critical role in mitochondrial dynamics. *EMBO J*, 25, pp.3618–3626.

Youle, R. J. and van der Bliek, A. M., 2012. Mitochondrial fission, fusion, and stress. *Science*, 337, pp.1062–1065.

Yu, W., et al., 2011. The PINK1/Parkin pathway regulates mitochondrial dynamics and function in mammalian hippocampal and dopaminergic neurons. *Hum Mol Genet*, 20, pp.3227–3240.

Yu, W. H., et al., 2004. Autophagic vacuoles are enriched in amyloid precursor protein-secretase activities: implications for beta-amyloid peptide over-production and localization in Alzheimer's disease. *Int J Biochem Cell Biol*, 36, pp.2531–2540.

Yu, W. H., et al., 2005. Macroautophagy—a novel Beta-amyloid peptide-generating pathway activated in Alzheimer's disease. *J Cell Biol*, 171, pp.87–98.

Yue, W., et al., 2014. A small natural molecule promotes mitochondrial fusion through inhibition of the deubiquitinase USP30. *Cell Res*, 24, pp.482–496.

Zheng, X. and Hunter, T., 2013. Parkin mitochondrial translocation is achieved through a novel catalytic activity coupled mechanism. *Cell Res*, 23, pp.886–897.

Zhong, Q., et al., 2005. Mule/ARF-BP1, a BH3-only E3 ubiquitin ligase, catalyzes the polyubiquitination of Mcl-1 and regulates apoptosis. *Cell*, 121, pp.1085–1095.

Zhou, R., et al., 2011. A role for mitochondria in NLRP3 inflammasome activation. *Nature*, 469, pp.221–225.

Zimprich, A., et al., 2011. A mutation in VPS35, encoding a subunit of the retromer complex, causes late-onset Parkinson disease. *Am J Hum Genet*, 89, pp.168–175.

Ziviani, E., Tao, R. N. and Whitworth, A. J., 2010. Drosophila parkin requires PINK1 for mitochondrial translocation and ubiquitinates mitofusin. *Proc Natl Acad Sci U S A*, 107, pp.5018–5023.

13

MITOCHONDRIAL INVOLVEMENT IN NEURODEGENERATIVE DEMENTIA

LAURA BONANNI, VALERIO FRAZZINI, ASTRID THOMAS AND MARCO ONOFRJ

Department of Neuroscience, Imaging and Clinical Sciences, University G.d'Annunzio of Chieti-Pescara, Chieti, Italy

13.1 INTRODUCTION

Neurodegenerative disorders include progressive, age-dependent brain disorders, clinically characterized by cognitive decline, variable patterns of motor impairment and irreversible changes in behavior and personality. Pathological hallmarks of these disorders including Dementia with Lewy Bodies (DLB), Alzheimer's disease (AD), Huntington's disease, and Amyotrophic Lateral Sclerosis (ALS) are accumulations in specific brain areas of mutant proteins such as α-synuclein, extracellular β-amyloid, and intracellular hyperphosporylated tau, mutant huntingtin, TAR DNA binding protein and superoxide dismutase respectively. Oxidative stress, inflammation, mitochondrial dysfunction and excitotoxicity have been identified as contributing to the pathogenesis of neurodegenerative disorders. However, it is unclear whether mitochondrial impairment and oxidative stress are actually involved in the onset and progression of neurodegenerative disorders.

The role of mitochondrial dysfunction in neurodegenerative disease is considered so important as to lead to the tentative definition of "neurodegenerative mitochondriopathy" to identify the ensemble of different neurodegenerative diseases that recognize

The Functions, Disease-Related Dysfunctions, and Therapeutic Targeting of Neuronal Mitochondria, First Edition. Edited by Valentin K. Gribkoff, Elizabeth A. Jonas and J. Marie Hardwick.

mitochondrial involvement in their pathogenesis (Swerdlow, 2009). Indeed, there is growing evidence that mitochondrial dysfunction and impairment of the respiratory complexes are associated with neuronal loss in different neurodegenerative diseases. Decreased complex IV activity has been reported in Alzheimer's disease (Chagnon et al., 1995) and decreased complex I activity is reported in the *substantia nigra* of postmortem samples in PD (Mann et al., 1992; Mizuno et al., 1989).

In this chapter we will report the evidence for mitochondrial involvement inthe most common form of neurodegenerative dementia, represented by AD and DLB.

13.2 MITOCHONDRIAL DYSFUNCTION IN ALZHEIMER DISEASE

AD is the most common cause of dementia, affecting more than 15 million individuals world-wide (Querfurth and LaFerla, 2010; Reiman, 2014). The pathological hallmarks of AD are brain deposition of β-amyloid (Aβ) in senile plaques, the appearance of neurofibrillary tangles (NFTs) made of hyperphosphorylated tau protein, synaptic dystrophy and neuronal degeneration. In accordance with the amyloid cascade hypothesis, it has been proposed that Aβ accumulation, as a result of an imbalance between its production and clearance, may represent the initiating factor in Alzheimer's disease pathogenesis (Hardy and Selkoe, 2002). Aβ peptides result from sequential proteolysis of the amyloid precursor protein (APP) by the enzymatic activity of beta-site amyloid precursor protein–cleaving enzyme 1 (BACE-1), or β-secretase, and γ-secretase, a protein complex with presenilin 1 at its catalytic core. It has been supposed that overproduction and accumulation of soluble Aβ is followed by the appearance of low molecular weight oligomers that rapidly associate in higher order aggregates, finally precipitating to form senile plaques (Hardy and Selkoe, 2002).

However, mitochondrial roles in cellular physiology actually extend far beyond the traditional role in ATP synthesis. Mitochondria are recognized as critical modulators of a great variety of neuronal functions including intraneuronal Ca^{2+} homeostasis, synaptic transmission and plasticity, neurotrophic signaling, regulation of gene transcription and formation of contact sites with the endoplasmic reticulum (ER) (Rizzuto et al., 1999; Rizzuto, Bernardi, and Pozzan, 2000; Sabatini, Maravall, and Svoboda, 2001; Vos, Lauwers, and Verstreken, 2010; Zucker, 1999). As will be discussed, altered mitochondrial dynamics may comprise a key event in AD pathogenesis. Recent molecular, cellular, animal model, and postmortem brain studies support the notion that mitochondrial dysfunction represents a key factor in the progressive neuronal damage and consequent cognitive decline observed in AD. Mitochondrial dysfunction in AD pathogenesis was described several years ago and continues to be the focus of intense investigations. Mitochondrial deficits were described in autoptic samples of postmortem brains from patients with AD, in preclinical animal models of AD, as well as cellular models prone to overexpress Aβ peptides.

13.3 MITOCHONDRIAL DYSFUNCTION, BIOENERGETIC DEFICITS, AND OXIDATIVE STRESS IN AD

Human mitochondrial DNA (mtDNA) encodes 13 proteins, which are essential for the proper functioning of mitochondrial respiratory complexes. Given the small size of the mtDNA, it is highly susceptible to mutations arising from its close proximity to the site of production of reactive oxygen species (ROS), ant its repair mechanisms are critical for maintaining normal mitochondrial function (Kazak, Reyes, and Holt, 2012). Moderate levels of ROS production have been shown to perform physiological neuronal signaling. However, ROS overproduction leads to damage of intracellular molecules and organelles including mitochondria themselves (Droge, 2002; Massaad and Klann, 2011; Starkov, 2008; Valko et al., 2007). mtDNA damage may result in decreased mitochondrial respiratory activity and eventually neuronal death. Previous studies demonstrated an increase in somatic mtDNA point mutations and functional deficits in mitochondrial transcription and replication in postmortem brain tissue from patients with AD (Coskun, Beal, and Wallace, 2004; Lin et al., 2002). These results are in line with previous data from other laboratories showing differential expression of mitochondrial oxidative phosphorylation genes, such as cytochrome oxidase enzyme, in AD postmortem brains and in those from AD transgenic mice (Chandrasekaran et al., 1994, 1996; Manczak et al., 2004). Another study conducted on transgenic animal models of AD observed an up-regulation of gene products related to mitochondrial energy metabolism together with decreased cytochrome oxidase activity, increased free radicals and apoptosis in young animals compared to the age-matched wild-type mice, suggesting that mitochondrial dysfunction is an early event in AD progression (Reddy et al., 2004). Decreased density of mitochondrial respiratory complex II in the CA1 hippocampal region of transgenic mice has also been described (Fattoretti et al., 2010). Accordingly, precocious mitochondrial dysfunction and consequent bioenergetic dysfunction and ROS overproduction has been demonstrated in the triple transgenic mouse model of AD in an early disease phase (i.e., before the appearance of Aβ plaques and neurofibrillary tangles) corroborating the notion that mitochondrial dysfunction is an early event in AD pathology (Resende et al., 2008; Yao et al., 2009). A recent paper, using intravital multiphoton imaging, explored mitochondrial structural and functional changes in the vicinity of Aβ plaques in an AD mice model. These modifications included reduction of mitochondrial numbers and membrane potential as well as the emergence of morphological modifications such as fragmented mitochondria (Xie et al., 2013). Analyses of mRNA expression of several mitochondrial-encoded genes from the frontal cortex of patients with AD and healthy subjects revealed a down-regulation of mitochondrial genes in complex I of the electron transport chain genes and increased mRNA expression of complexes III and IV in the brain specimens from patients with AD with respect to controls (Manczak et al., 2004). Importantly, Aβ accumulates at mitochondria and inhibits respiratory complex IV, increasing ROS production and thereby increasing the frequency of mtDNA mutations. Aβ has been also reported to accumulate in neuronal subcellular compartments, including mitochondria, to impair the normal function of neurons. Aβ can target mitochondria, blocking the mitochondrial import channels (TOM40 and TIM23) and preventing the import of nuclear-encoded complex IV subunits (Devi et al., 2006). It also alters mitochondrial dynamics by negatively interfering with proteins essential for mitochondrial fission and fusion, promoting mitochondrial fragmentation (Wang et al., 2008a, 2008b, 2009).

13.4 MITOCHONDRIAL FRAGMENTATION IN AD

Mitochondria are extremely dynamic organelles that undergo cycles of fission and fusion to ensure their biogenesis. Mitochondrial fission and fusion allow the mixing of metabolites and mtDNA, the proliferation and distribution of mitochondria and cellular adaptation to changing energy demands (Knott et al., 2008). Briefly, mitochondrial fission consists of the physical separation of one mitochondrion into two new, independent, mitochondria. This process is dependent on the coordinated activity of the mitochondrial-localized Mff protein that recruits the cytosolic dynamin-related Drp1 to the mitochondrial outer membrane. Drp1 cycles in and out of contact with mitochondrial outer membranes, interacting indirectly with the outer-membrane-associated protein hFis1. Once it is bound to the mitochondrial outer membrane, Drp1 forms large clusters that mediate membrane fission (Chang and Blackstone, 2010; Knott et al., 2008; Otera et al., 2010). The opposite process of mitochondrial fission is fusion, where two physically separate mitochondria combine to become a single mitochondrion. This represents a complex mechanism that in mammalian cells is regulated by three large GTPases. Mitofusins1 and 2 (MFN1 and MFN2) on the outer membranes of adjacent mitochondria dimerize and induce outer membrane fusion, followed by fusion of the inner membranes mediated by Optic Atrophy 1 (OPA1) (Knott et al., 2008; Song et al., 2009). Impaired balance of fission and fusion has been reported in AD. Decreased levels of mitochondrial fission transcripts/proteins, together with a strong increase in Fis1 expression were reported in AD brains (Manczak, Calkins, and Reddy, 2011; Wang et al., 2009). Physical interaction between Aβ and DRP1 has also been reported (Manczak et al., 2011). Exposure of primary neuronal cultures to oligomeric forms of Aβ has been shown to promote a significant reduction in mitochondrial length in neurites, suggesting enhanced mitochondrial fragmentation, decreased mitochondrial density and altered synaptic localization (Du et al., 2008, 2010; Wang et al., 2008b, 2009).

13.5 SYNAPTIC MITOCHONDRIA IN AD

Synaptic damage is considered the earliest cellular event in AD pathogenesis, and synaptic loss is considered the best marker of cognitive impairment in AD (DeKosky and Scheff, 1990; Terry et al., 1991). Moreover, synapses and neurites are mostly damaged in the vicinity of Aβ plaques (Meyer-Luehmann et al., 2009). Synaptic mitochondria differ from nonsynaptic mitochondria in size, motility, life span, and other properties. Synaptic mitochondria are synthesized in neuronal soma and are transported to nerve terminals where they have an important role in maintaining normal synaptic function through their ability to meet the high energy demands of synapses while maintaining synaptic calcium levels. However, synaptic mitochondria are subjects of intense oxidation during aging and contain higher levels of the matrix protein cyclophilin D (CypD), which makes them more susceptible to calcium insults (Naga, Sullivan, and Geddes, 2007). CypD is a key component of the mitochondrial permeability transition pore (mPTP) complex, and the level of CypD influences mPTP formation. Low expression of CypD in brain mitochondria increases resistance to mPTP (Eliseev et al., 2007), whereas high levels of CypD in neuronal mitochondria have been shown to result in greater vulnerability to mPTP. Du et al. (2010) observed that the progressive accumulation of intraneuronal Aβ is associated with several abnormalities in synaptic mitochondria isolated from AD mice. AD mitochondria overexpressed significantly higher levels of CypD than synaptic mitochondria isolated from age-matched controls. These results suggest that synaptic mitochondria from AD neurons are more susceptible to mPTP, lowered calcium handling ability and enhanced mitochondrial oxidative stress. Interestingly, CypD-deficient cortical mitochondria are resistant to Aβ-induced mitochondrial deficits, permeability transition, Ca^{2+} dyshomeostasis and ROS production. Furthermore, CypD deficiency substantially improves learning and memory and synaptic function in an Alzheimer's disease mouse model (Du et al., 2008, 2010).

13.6 MITOCHONDRIAL DYSFUNCTION AND CATIONIC DYSHOMEOSTASIS IN AD

Ca^{2+} ions act as ubiquitous intracellular messengers to regulate a large array of physiological functions. In excitable cells, and particularly in neurons, Ca^{2+} ions are subjected to complex homeostatic machinery. The discovery that several intramitochondrial dehydrogenases are regulated by intramitochondrial Ca^{2+} suggested that Ca^{2+} influx could regulate NADH production and an increase in the overall rate of oxidative ATP synthesis. Dysfunction in mitochondrial Ca^{2+} signaling is considered a prominent feature in AD. Previous studies performed on mitochondria isolated from fibroblasts of patients with AD exhibit reduced mitochondrial Ca^{2+} uptake when compared to age-matched controls, however, when exposed to oxidative stress mitochondria from AD fibroblasts have been shown to undergo increases in Ca^{2+} uptake suggesting a higher sensitivity toward oxidative stress (Kumar, Dunlop, and Richardson, 1994). Mitochondria with overexpression of human APP also show a lower Ca^{2+} capacity compared to nontransgenic mitochondria (Du et al., 2008). Aβ1–42 oligomer treatment induces Ca^{2+} overload in mitochondria in both cortical and cerebellar granular neurons (Sanz-Blasco et al., 2008). The increase is limited to a pool of mitochondria close to the sites of Ca^{2+} entry and release (Sanz-Blasco et al., 2008). Ca^{2+} overload in mitochondria has been shown to promote increased ROS production and impairment of bioenergetic metabolism. Mutations in presenilins may promote mitochondrial dysfunction by perturbing ER Ca^{2+} handling, which promotes synaptic mitochondrial Ca^{2+} overload and in turn triggers apoptosis. A novel Ca^{2+} channel calcium homeostasis modulator 1 (CALHM1) generates cytosolic Ca^{2+} transients that have been shown to regulate Aβ production. An interesting study reports that P86L-CALHM1, a mutation associated with an increased risk of AD development, may slow kinetics of mitochondrial Ca^{2+} uptake and release, increasing the risk of mitochondrial Ca^{2+} overload and causing the cells to become more vulnerable to apoptotic stimuli (Moreno-Ortega et al., 2010). ER-mitochondrial contact sites represent critical structures for optimal mitochondrial functioning. Several reports have suggested that these sites are also enriched in PS1 and PS2 (Area-Gomez et al., 2009) and that γ-secretase activity at ER-mitochondrial interfaces is the main producer of Aβ (Area-Gomez et al., 2009). This supports the hypothesis that in AD PS1 or PS2 activity may promote Aβ overproduction at ER-mitochondrial contact sites promoting mitochondria Aβ accumulation, consequent organelle dysfunction, mtDNA mutations and decreased respiratory complex activity (Manczak et al., 2006). Therefore, mutations that lead to increased Aβ production would first impair mitochondrial function, thereby affecting cellular calcium and lipid homeostasis, which may have downstream effects leading to increased Aβ and tau aggregation, and ultimately neuronal death (Schon and Area-Gomez, 2013). In this scenario, AD-associated mutations directly impair the function of mitochondria and ER-mitochondrial contact sites leading to downstream effects characteristic of AD, thus establishing mitochondria as executors of AD pathogenesis. Other divalent ions have been involved in AD related mitochondrial dysfunction. Zn^{2+} is the second-most abundant metal in the body after iron and the brain possesses a large pool of endogenous Zn^{2+}. Like Ca^{2+}, Zn^{2+} is involved in a wide array of physiological functions and, as with Ca^{2+} its deregulation is actually considered to play an important role in AD pathogenesis. Zn^{2+} is often localized, together with glutamate, within synaptic vesicles of a subset of glutamatergic axon terminals, where it is transported into presynaptic vesicles by the neuronal-specific Zn^{2+} transporter, ZnT^3, and coreleased with glutamate in the synaptic cleft during neuronal activity. Both excitatory and inhibitory synaptic transmission are affected by Zn^{2+} (for review on Zn^{2+} neurobiology see Sensi et al., 2009). Zn^{2+} can be sequestered into mitochondria, inducing organelle dysfunction and promoting mitochondrial depolarization and further generation of ROS (Bonanni et al., 2006; Dineley et al., 2005; Sensi et al., 1999; Sensi, Yin, and Weiss, 2000). Such Zn^{2+}-dependent mitochondrial ROS production can promote further intraneuronal Zn^{2+} mobilization. This phenomenon has been found to be particularly high in AD neurons expressing mutant APP, PS1, and h-tau (Sensi et al., 2008). ROS-driven Zn^{2+} mobilization has also been shown to promote intraneuronal Aβ aggregation (Cherny et al., 1999). Furthermore, Zn^{2+} can induce a multiconductance cation channel activity in mitochondrial membranes that is consistent with the activation of the mPTP (Bonanni et al., 2006; Gazaryan et al., 2007; Jiang et al., 2001), thereby leading to release of pro-apoptotic factors. Zn^{2+} can also profoundly affect mitochondrial trafficking and morphology (Malaiyandi et al., 2005).

13.7 MITOCHONDRIAL DYSFUNCTION IN DLB

DLB is the second-most common form of dementia in older people, accounting for up to 36% of all demented cases at autopsy (McKeith et al., 2000). Neuropathologically, DLB is characterized by the presence of intraneuronal inclusions, Lewy Bodies (LBs), found in the brain stem and in various cortical areas, and Lewy neurites (LN), predominantly present in the CA2 hippocampal region (Mukaetova-Ladinska and McKeith, 2006).

LBs are composed of ubiquitin, neurofilaments, tubulin, parkin, and α-Synuclein. α-Synuclein, a synaptic protein, has been identified as the principal component of the intraneuronal LBs and LN. Aggregation of α-Synuclein is facilitated by mutations within the α-synuclein gene (Mukaetova-Ladinska and McKeith, 2006). α-synuclein is predominantly a presynaptic protein; it is mainly a cytosolic protein but binds rapidly and specifically with mitochondrial membranes, especially under conditions of cellular stress or cytosolic acidification (Cole et al., 2008; Liu et al., 2009; Nakamura et al., 2008, 2011). Mitochondrial α-synuclein has been localized predominantly to the outer membrane, although under favorable energy and pH conditions it appears to migrate to the inner membrane (Cole et al., 2008; Liu et al., 2009; Shavali et al., 2008; Zhang et al., 2008). The interaction between synuclein and mitochondrial membranes appears to be specific and independent of changes in membrane potential. Migration into the mitochondria is energy dependent and appears to be mediated through the translocase of the outer membrane (TOM) 40. α-synuclein can be found in the mitochondria of dopaminergic neurons in postmortem human PD brains (Devi et al., 2008). It localizes to neuronal and glial cells of the substantia nigra, thalamus, neocortex and hippocampus.

The function of α-synuclein is unclear. It appears to be particularly prone to rapid conformational changes and has been implicated in a range of functions including the ubiquitin–proteasome system, maintenance of synaptic functionality and regulation of oxidative stress (Clayton and George, 1998; Greten-Harrison et al., 2010; Nemani et al., 2010; Tompa, 2005). In cell models, overexpression of α-synuclein results in inhibition of complex I, mitochondrial membrane depolarization (Hsu et al., 2000), and formation of giant mitochondria containing laminated bodies, autophagosomes, and decreased MTT levels.

Transgenic mice overexpressing A53T α-synuclein mutant show neurodegeneration characterized by intraneuronal inclusions, neuronal death and mitochondrial pathology, including altered morphology and mitochondrial DNA damage (Martin, 2007), with impairment of complex I activity in the Substantia nigra (Stichel et al., 2007) and with an increase in mitochondrial autophagy. In vitro culture of *Caenorhabditis elegans* overexpressing synuclein leads to fragmentation of mitochondria with inhibition of mitochondrial membrane fusion. Conversely, siRNA knock down of synuclein results in elongated mitochondria (Kamp et al., 2010). Interestingly, α-synuclein knockout mice are resistant to the effects of MPTP, a complex I inhibitor, suggesting that α-synuclein is required downstream of complex I (Dauer et al., 2002). Thus, α-synuclein directly damages mitochondria, and appears to do so also through interference with mitophagy and mitochondrial fusion. The mitochondrial dysfunction and oxidative stress found in LBD promote α-synuclein aggregation from its monomeric state to oligomers (Conway et al., 2000). Therefore, α-synuclein causes mitochondrial dysfunction, and this in turn exacerbates the misfolding of α-synuclein, leading to the formation of LBs.

Recently, impairments of tissue respiration, state 3 mitochondrial respiration with malate–glutamate as substrates and decreased complex I activity associated with oxidative damage were found in the frontal cortex (area 8) of patients with DLB (Navarro et al., 2009). The observed mitochondrial impairment was described as a reduced tissue respiration of frontal cortex with reduced mitochondrial respiration with NAD-dependent substrates and with a marked decrease in complex I activity. The decrease in respiratory rates and in enzymatic activities was associated with oxidative damage, as determined by an increased content of phospholipid and protein oxidation products. Cortical mitochondrial dysfunction is now added to the classically recognized mitochondrial dysfunction in the dopaminergic neurons of *substantia nigra* in Parkinson's disease (Navarro and Boveris, 2010).

LBD, including Parkinson's disease, is now considered to be not only characterized by *substantia nigra* dysfunction but also by involvement of the frontal cortex with cognitive decline at the early stages of parkinsonism (McNamara, Durso, and Harris, 2007). Oxidative and nitrosative damage and mitochondrial dysfunction in the human frontal cortex are to be considered factors leading to impaired cognition in LBD.

13.8 LRRK2 MUTATIONS, MITOCHONDRIA AND DLB

Leucine-rich repeat kinase 2 (LRRK2) is a large multidomain protein kinase, which localizes to the cytosol and associates with the mitochondrial outer membrane (Mata et al., 2006). LRRK2 is suggested to play a crucial role in the pathology of both familial and sporadic α-synucleinopathies, including DLB. Both α-synuclein and LRRK2 are involved in various pathologies such as impairment of cytoskeleton dynamics and dysregulation of the protein degradation system (Cookson, 2010). Mitochondria may also be a common target of α-synuclein and LRRK2, since immunoreactivities of LRRK2 are associated with mitochondria and with other intracellular structures, including lysosomes, endosomes, and transport vesicles (Biskup et al., 2006). Furthermore, it has been shown that LRRK2 regulates mitochondrial dynamics by increasing mitochondrial dynamin-like kinase 1 (DLP1) through a direct interaction with DLP1 (Wang et al., 2012).

13.9 AKINETIC CRISIS IN SYNUCLEINOPATHIES IS LINKED TO GENETIC MUTATIONS INVOLVING MITOCHONDRIAL PROTEINS

From a clinical point of view the involvement of mitochondrial pathology in synucleinopathy is highlighted by the recent finding of an unexpectedly high incidence of akinetic crises in patients with PD/DLB who are carriers of genetic mutations involving mitochondrial proteins.

Akinetic crisis (AC), also termed *malignant syndrome* or *Parkinson hyperpyrexia* syndrome, neuroleptic like malignant syndrome, or acute akinesia, is a complication that appears in parkinsonism because of treatment manipulations or withdrawal, infectious diseases, trauma, or gastrointestinal tract diseases (Onofrj and Thomas, 2005). AC is the most severe complication of Lewy Body disease occurring with an annual incidence of 3 cases per 1000 parkinsonian patients and consists of acute motor symptom worsening characterized by akinetic state with dysphagia, hyperthermia, increment of serum CPK and myoglobin, dysautonomia, and transient unresponsiveness to current antiparkinsonian treatment or to an increase of dopaminergic drug doses (Onofrj and Thomas, 2005; Thomas and Onofrj, 2005). Its symptoms are identical to symptoms of neuroleptic malignant syndrome (Onofrj and Thomas, 2005; Thomas and Onofrj, 2005; Ueda et al., 1999), and it was hypothesized that both represent idiosyncratic severe complications induced by heterogeneous causes (Margetic and Aukst-Margetic, 2010).

Cohort or multicenter studies highlighted the relevance of this complication of LBD (Takubo et al., 2003). In a recent work by our group (Bonanni et al., 2014), in a cohort of 756 patients with parkinsonism, AC occurred in 30 patients for an incidence of 2.8‰ person/year (PY), and was lethal in 7 (23%), not dissimilarly from the known incidence of this complication. Yet 6 of 30 patients were carriers of genetic mutations, one GBA, one LRRK2, one POLG1, and three PINK1. In POLG1 and PINK1 carriers, the syndrome was recurrent and was fatal in 3. The incidence of AC was 3.0‰ in parkinsonism, 21.2‰ in genetic parkinsonisms. Our findings suggest that the incidence of AC is remarkably increased in carriers of genetic mutations linked to mitochondrial protein expression. The striking coincidence of AC with the genetic diagnoses obtained in the 6 patients is evidence of a disproportionate prevalence of this crisis in comparison with the population of patients with PD. The incidence of AC in the described genetic mutation carriers (POLG1, PINK1, LRRK2, GBA) was a multiple (i.e., 21.2 versus 3.0‰) of the known incidence of AC in parkinsonian populations. Moreover, AC episodes were recurrent in 4 out of the 6 patients, all four were carriers of mutations involving mitochondrial function: POLG1 (found in one of the four) is the gene that codes for the catalytic subunit of mitochondrial DNA polymerase, PINK1 (three patients) is the gene encoding a mitochondrial serine/threonine-protein kinase. Three of the four patients died in the last crisis. This finding suggests that mitochondrial dysfunction may represent a predisposition to AC and to its possible lethal outcome.

Therefore, our findings point to mitochondrial-genetically driven dysfunction as one of the probable underlying causes of AC. We suggest that in a cell environment where mitochondrial buffering power is defective, stressful conditions, such as the precipitating events described in our case cohort (infections, trauma, etc.) could activate a stress cascade, altering the cellular energy state by inhibiting the electron transport chain. We suggest therefore that the determinant factor in AC might be akin to inhibition of complex I activity observed in brain mitochondria of rats exposed to neuroleptics (Burkhardt et al., 1993) or observed in post mortem evaluations of patients affected by neuroleptic malignant syndrome (Maurer and Moller, 1997).

13.10 CONCLUSIONS

There is increasing evidence that mitochondrial dysfunction plays a critical role in the pathogenesis of neurodegenerative disorders. Studies in cybrids suggest direct involvement of mitochondria in the progression of neurodegenerative disorders. In some neurodegenerative diseases such as Friedreich's ataxia or Huntington disease, the mitochondrial involvement is direct. In other neurodegenerative disorders such as LBD and AD the involvement of mitochondria is more indirect. In AD, the pathogenic protein Aβ may induce mitochondrial dysfunction by directly binding to the mitochondria and to mitochondrial proteins, leading to reduced enzymatic activity of complexes III and IV and attenuated mitochondrial respiration. In PD, α-synuclein and LRRK2 cause mitochondrial dysfunction by association with the mitochondria. Therefore, a remarkable number of proteins including β-amyloid and α-synuclein that cause neurodegeneration in their aggregated forms interact with mitochondria or affect mitochondrial function. It is important that studies of the mitochondria-neurodegeneration link continue for many reasons. Such studies could provide the basis for new pharmacological treatments.

REFERENCES

Area-Gomez, E., et al., 2009. Presenilins are enriched in endoplasmic reticulum membranes associated with mitochondria. *Am J Pathol*, 175, pp.1810–1816.

Biskup, S., et al., 2006. Localization of LRRK2 to membranous and vesicular structures in mammalian brain. *Ann Neurol*, 60, pp.557–569.

Bonanni, L., et al., 2006. Zinc-dependent multi-conductance channel activity in mitochondria isolated from ischemic brain. *J Neurosci*, 26, pp.6851–6862.

Bonanni, L., et al., 2014. Recurrent and fatal akinetic crisis in genetic-mitochondrial parkinsonisms. *Eur J Neurol*, 21, pp.1242–1246.

Burkhardt, C., et al., 1993. Neuroleptic medications inhibit complex I of the electron transport chain. *Ann Neurol*, 33, pp.512–517.

Chagnon, P., et al., 1995. Distribution of brain cytochrome oxidase activity in various neurodegenerative diseases. *Neuroreport*, 6, pp.711–715.

Chandrasekaran, K., et al., 1994. Impairment in mitochondrial cytochrome oxidase gene expression in Alzheimer disease. *Brain Res Mol Brain Res*, 24, pp.336–340.

Chandrasekaran, K., et al., 1996. Evidence for physiological down-regulation of brain oxidative phosphorylation in Alzheimer's disease. *Exp Neurol*, 142, pp.80–88.

Chang, C. R. and Blackstone, C., 2010. Dynamic regulation of mitochondrial fission through modification of the dynamin-related protein Drp1. *Ann N Y Acad Sci*, 1201, pp.34–39.

Cherny, R. A., et al., 1999. Aqueous dissolution of Alzheimer's disease Abeta amyloid deposits by biometal depletion. *J Biol Chem*, 274, pp.23223–23228.

Clayton, D. F. and George, J. M., 1998. The synucleins: a family of proteins involved in synaptic function, plasticity, neurodegeneration and disease. *Trends Neurosci*, 21, pp.249–254.

Cole, N. B., et al., 2008. Mitochondrial translocation of alpha-synuclein is promoted by intracellular acidification. *Exp Cell Res*, 314, pp.2076–2089.

Conway, K. A., et al., 2000. Acceleration of oligomerization, not fibrillization, is a shared property of both alpha-synuclein mutations linked to early-onset Parkinson's disease: implications for pathogenesis and therapy. *Proc Natl Acad Sci U S A*, 97, pp.571–576.

Cookson, M. R., 2010. The role of leucine-rich repeat kinase 2 (LRRK2) in Parkinson's disease. *Nat Rev Neurosci*, 11, pp.791–797.

Coskun, P. E., Beal, M. F. and Wallace, D. C., 2004. Alzheimer's brains harbor somatic mtDNA control-region mutations that suppress mitochondrial transcription and replication. *Proc Natl Acad Sci U S A*, 101, pp.10726–10731.

Dauer, W., et al., 2002. Resistance of alpha-synuclein null mice to the parkinsonian neurotoxin MPTP. *Proc Natl Acad Sci U S A*, 99, pp.14524–14529.

DeKosky, S. T. and Scheff, S. W., 1990. Synapse loss in frontal cortex biopsies in Alzheimer's disease: Correlation with cognitive severity. *Ann Neurol*, 27, pp.457–464.

Devi, L., et al., 2006. Accumulation of amyloid precursor protein in the mitochondrial import channels of human Alzheimer's disease brain is associated with mitochondrial dysfunction. *J Neurosci*, 26, pp.9057–9068.

Devi, L., et al., 2008. Mitochondrial import and accumulation of alpha-synuclein impair complex I in human dopaminergic neuronal cultures and Parkinson disease brain. *J Biol Chem*, 283, pp.9089–9100.

Dineley, K. E., et al., 2005. Zinc causes loss of membrane potential and elevates reactive oxygen species in rat brain mitochondria. *Mitochondrion*, 5, pp.55–65.

Droge, W., 2002. Free radicals in the physiological control of cell function. *Physiol Rev*, 82, pp.47–95.

Du, H., et al., 2008. Cyclophilin D deficiency attenuates mitochondrial and neuronal perturbation and ameliorates learning and memory in Alzheimer's disease. *Nat Med*, 14, pp.1097–1105.

Du, H., et al., 2010. Early deficits in synaptic mitochondria in an Alzheimer's disease mouse model. *Proc Natl Acad Sci U S A*, 107, pp.18670–18675.

Eliseev, R. A., et al., 2007. Role of cyclophilin D in the resistance of brain mitochondria to the permeability transition. *Neurobiol Aging*, 28, pp.1532–1542.

Fattoretti, P., et al., 2010. Decreased numeric density of succinic dehydrogenase-positive mitochondria in CA1 pyramidal neurons of 3xTg-AD mice. *Rejuvenation Res*, 13, pp.144–147.

Gazaryan, I. G., et al., 2007. Zinc irreversibly damages major enzymes of energy production and antioxidant defense prior to mitochondrial permeability transition. *J Biol Chem*, 282, pp.24373–24380.

Greten-Harrison, B., et al., 2010. alphabetagamma-Synuclein triple knockout mice reveal age-dependent neuronal dysfunction. *Proc Natl Acad Sci U S A*, 107, pp.19573–19578.

Hardy, J. and Selkoe, D. J., 2002. The amyloid hypothesis of Alzheimer's disease: Progress and problems on the road to therapeutics. *Science*, 297, pp.353–356.

Hsu, L. J., et al., 2000. Alpha-synuclein promotes mitochondrial deficit and oxidative stress. *Am J Pathol*, 157, pp.401–410.

Jiang, D., et al., 2001. Zn(2+) induces permeability transition pore opening and release of pro-apoptotic peptides from neuronal mitochondria. *J Biol Chem*, 276, pp.47524–47529.

Kamp, F., et al., 2010. Inhibition of mitochondrial fusion by alpha-synuclein is rescued by PINK1, Parkin and DJ-1. *EMBO J*, 29, pp.3571–3589.

Kazak, L., Reyes, A. and Holt, I. J., 2012. Minimizing the damage: repair pathways keep mitochondrial DNA intact. *Nat Rev Mol Cell Biol*, 13, pp.659–671.

Knott, A. B., et al., 2008. Mitochondrial fragmentation in neurodegeneration. *Nat Rev Neurosci*, 9, pp.505–518.

Kumar, U., Dunlop, D. M. and Richardson, J. S., 1994. Mitochondria from Alzheimer's fibroblasts show decreased uptake of calcium and increased sensitivity to free radicals. *Life Sci*, 54, pp.1855–1860.

Lin, M. T., et al., 2002. High aggregate burden of somatic mtDNA point mutations in aging and Alzheimer's disease brain. *Hum Mol Genet*, 11, pp.133–145.

Liu, G., et al., 2009. Alpha-synuclein is differentially expressed in mitochondria from different rat brain regions and dose-dependently down-regulates complex I activity. *Neurosci Lett*, 454, pp.187–192.

Malaiyandi, L. M., et al., 2005. Zn2+ inhibits mitochondrial movement in neurons by phosphatidylinositol 3-kinase activation. *J Neurosci*, 25, pp.9507–9514.

Manczak, M., et al., 2004. Differential expression of oxidative phosphorylation genes in patients with Alzheimer's disease: implications for early mitochondrial dysfunction and oxidative damage. *Neuromolecular Med*, 5, pp.147–162.

Manczak, M., et al., 2006. Mitochondria are a direct site of a beta accumulation in Alzheimer's disease neurons: Implications for free radical generation and oxidative damage in disease progression. *Hum Mol Genet*, 15, pp.1437–1449.

Manczak, M., Calkins, M. J. and Reddy, P. H., 2011. Impaired mitochondrial dynamics and abnormal interaction of amyloid beta with mitochondrial protein Drp1 in neurons from patients with Alzheimer's disease: Implications for neuronal damage. *Hum Mol Genet*, 20, pp.2495–2509.

Mann, V. M., et al., 1992. Brain, skeletal muscle and platelet homogenate mitochondrial function in Parkinson's disease. *Brain*, 115 (Pt 2), pp.333–342.

Margetic, B. and Aukst-Margetic, B., 2010. Neuroleptic malignant syndrome and its controversies. *Pharmacoepidemiol Drug Saf*, 19, pp.429–435.

Martin, L. J., 2007. Transgenic mice with human mutant genes causing Parkinson's disease and amyotrophic lateral sclerosis provide common insight into mechanisms of motor neuron selective vulnerability to degeneration. *Rev Neurosci*, 18, pp.115–136.

Massaad, C. A. and Klann, E., 2011. Reactive oxygen species in the regulation of synaptic plasticity and memory. *Antioxid Redox Signal*, 14, pp.2013–2054.

Mata, I. F., et al., 2006. LRRK2 in Parkinson's disease: Protein domains and functional insights. *Trends Neurosci*, 29, pp.286–293.

Maurer, I. and Moller, H. J., 1997. Inhibition of complex I by neuroleptics in normal human brain cortex parallels the extrapyramidal toxicity of neuroleptics. *Mol Cell Biochem*, 174, pp.255–259.

McKeith, I. G., et al., 2000. Prospective validation of consensus criteria for the diagnosis of dementia with Lewy bodies. *Neurology*, 54, pp.1050–1058.

McNamara, P., Durso, R. and Harris, E., 2007. "Machiavellianism" and frontal dysfunction: evidence from Parkinson's disease. *Cogn Neuropsychiatry*, 12, pp.285–300.

Meyer-Luehmann, M., et al., 2009. A reporter of local dendritic translocation shows plaque-related loss of neural system function in APP-transgenic mice. *J Neurosci*, 29, pp.12636–12640.

Mizuno, Y., et al., 1989. Deficiencies in complex I subunits of the respiratory chain in Parkinson's disease. *Biochem Biophys Res Commun*, 163, pp.1450–1455.

Moreno-Ortega, A. J., et al., 2010. Mitochondria sense with different kinetics the calcium entering into HeLa cells through calcium channels CALHM1 and mutated P86L-CALHM1. *Biochem Biophys Res Commun*, 391, pp.722–726.

Mukaetova-Ladinska, E. B. and McKeith, I. G., 2006. Pathophysiology of synuclein aggregation in Lewy body disease. *Mech Ageing Dev*, 127, pp.188–202.

Naga, K. K., Sullivan, P. G. and Geddes, J. W., 2007. High cyclophilin D content of synaptic mitochondria results in increased vulnerability to permeability transition. *J Neurosci*, 27, pp.7469–7475.

Nakamura, K., et al., 2008. Optical reporters for the conformation of alpha-synuclein reveal a specific interaction with mitochondria. *J Neurosci*, 28, pp.12305–12317.

Nakamura, K., et al., 2011. Direct membrane association drives mitochondrial fission by the Parkinson disease-associated protein alpha-synuclein. *J Biol Chem*, 286, pp.20710–20726.

Navarro, A., et al., 2009. Human brain cortex: mitochondrial oxidative damage and adaptive response in Parkinson disease and in dementia with Lewy bodies. *Free Radic Biol Med*, 46, pp.1574–1580.

Navarro, A. and Boveris, A., 2010. Brain mitochondrial dysfunction in aging, neurodegeneration, and Parkinson's disease. *Front Aging Neurosci.* 2. doi: 10.3389/fnagi.2010.00034.

Nemani, V. M., et al., 2010. Increased expression of alpha-synuclein reduces neurotransmitter release by inhibiting synaptic vesicle reclustering after endocytosis. *Neuron*, 65, pp.66–79.

Onofrj, M. and Thomas, A., 2005. Acute akinesia in Parkinson disease. *Neurology*, 64, pp.1162–1169.

Otera, H., et al., 2010. Mff is an essential factor for mitochondrial recruitment of Drp1 during mitochondrial fission in mammalian cells. *J Cell Biol*, 191, pp.1141–1158.

Querfurth, H. W. and LaFerla, F. M., 2010. Alzheimer's disease. *N Engl J Med*, 362, pp.329–344.

Reddy, P. H., et al., 2004. Gene expression profiles of transcripts in amyloid precursor protein transgenic mice: Up-regulation of mitochondrial metabolism and apoptotic genes is an early cellular change in Alzheimer's disease. *Hum Mol Genet*, 13, pp.1225–1240.

Reiman, E. M., 2014. Alzheimer's disease and other dementias: Advances in 2013. *Lancet Neurol*, 13, pp.3–5.

Resende, R., et al., 2008. Brain oxidative stress in a triple-transgenic mouse model of Alzheimer disease. *Free Radic Biol Med*, 44, pp.2051–2057.

Rizzuto, R., et al., 1999. Mitochondria as biosensors of calcium microdomains. *Cell Calcium*, 26, pp.193–199.

Rizzuto, R., Bernardi, P. and Pozzan, T., 2000. Mitochondria as all-round players of the calcium game. *J Physiol*, 529 (Pt 1), pp.37–47.

Sabatini, B. L., Maravall, M. and Svoboda, K., 2001. Ca(2+) signaling in dendritic spines. *Curr Opin Neurobiol*, 11, pp.349–356.

Sanz-Blasco, S., et al., 2008. Mitochondrial Ca2+ overload underlies Abeta oligomers neurotoxicity providing an unexpected mechanism of neuroprotection by NSAIDs. *PLoS One*, 3, p.e2718.

Schon, E. A. and Area-Gomez, E., 2013. Mitochondria-associated ER membranes in Alzheimer disease. *Mol Cell Neurosci*, 55, pp.26–36.

Sensi, S. L., et al., 1999. Preferential Zn2+ influx through Ca2+-permeable AMPA/kainate channels triggers prolonged mitochondrial superoxide production. *Proc Natl Acad Sci U S A*, 96, pp.2414–2419.

Sensi, S. L., Yin, H. Z. and Weiss, J. H., 2000. AMPA/kainate receptor-triggered Zn2+ entry into cortical neurons induces mitochondrial Zn2+ uptake and persistent mitochondrial dysfunction. *Eur J Neurosci*, 12, pp.3813–3818.

Sensi, S. L., et al., 2008. Altered oxidant-mediated intraneuronal zinc mobilization in a triple transgenic mouse model of Alzheimer's disease. *Exp Gerontol*, 43, pp.488–492.

Sensi, S. L., et al., 2009. Zinc in the physiology and pathology of the CNS. *Nat Rev Neurosci*, 10, pp.780–791.

Shavali, S., et al., 2008. Mitochondrial localization of alpha-synuclein protein in alpha-synuclein overexpressing cells. *Neurosci Lett*, 439, pp.125–128.

Song, Z., et al., 2009. Mitofusins and OPA1 mediate sequential steps in mitochondrial membrane fusion. *Mol Biol Cell*, 20, pp.3525–3532.

Starkov, A. A., 2008. The role of mitochondria in reactive oxygen species metabolism and signaling. *Ann N Y Acad Sci*, 1147, pp.37–52.

Stichel, C. C., et al., 2007. Mono- and double-mutant mouse models of Parkinson's disease display severe mitochondrial damage. *Hum Mol Genet*, 16, pp.2377–2393.

Swerdlow, R. H., 2009. The neurodegenerative mitochondriopathies. *J Alzheimers Dis*, 17, pp.737–751.

Takubo, H., et al., 2003. A collaborative study on the malignant syndrome in Parkinson's disease and related disorders. *Parkinsonism Relat Disord*, 9 Suppl 1, pp.S31–S41.

Terry, R. D., et al., 1991. Physical basis of cognitive alterations in Alzheimer's disease: synapse loss is the major correlate of cognitive impairment. *Ann Neurol*, 30, pp.572–580.

Thomas, A. and Onofrj, M., 2005. Akinetic crisis, acute akinesia, neuroleptic malignant-like syndrome, Parkinsonism-hyperpyrexia syndrome, and malignant syndrome are the same entity and are often independent of treatment withdrawal. *Mov Disord*, 20, pp.1671–1672.

Tompa, P., 2005. The interplay between structure and function in intrinsically unstructured proteins. *FEBS Lett*, 579, pp.3346–3354.

Ueda, M., et al., 1999. Susceptibility to neuroleptic malignant syndrome in Parkinson's disease. *Neurology*, 52, pp.777–781.

Valko, M., et al., 2007. Free radicals and antioxidants in normal physiological functions and human disease. *Int J Biochem Cell Biol*, 39, pp.44–84.

Vos, M., Lauwers, E. and Verstreken, P., 2010. Synaptic mitochondria in synaptic transmission and organization of vesicle pools in health and disease. *Front Synaptic Neurosci*, 2, p.139.

Wang, X., et al., 2008a. Dynamin-like protein 1 reduction underlies mitochondrial morphology and distribution abnormalities in fibroblasts from sporadic Alzheimer's disease patients. *Am J Pathol*, 173, pp.470–482.

Wang, X., et al., 2008b. Amyloid-beta overproduction causes abnormal mitochondrial dynamics via differential modulation of mitochondrial fission/fusion proteins. *Proc Natl Acad Sci U S A*, 105, pp.19318–19323.

Wang, X., et al., 2009. Impaired balance of mitochondrial fission and fusion in Alzheimer's disease. *J Neurosci*, 29, pp.9090–9103.

Wang, X., et al., 2012. LRRK2 regulates mitochondrial dynamics and function through direct interaction with DLP1. *Hum Mol Genet*, 21, pp.1931–1944.

Xie, H., et al., 2013. Mitochondrial alterations near amyloid plaques in an Alzheimer's disease mouse model. *J Neurosci*, 33, pp.17042–17051.

Yao, J., et al., 2009. Mitochondrial bioenergetic deficit precedes Alzheimer's pathology in female mouse model of Alzheimer's disease. *Proc Natl Acad Sci U S A*, 106, pp.14670–14675.

Zhang, L., et al., 2008. Semi-quantitative analysis of alpha-synuclein in subcellular pools of rat brain neurons: an immunogold electron microscopic study using a C-terminal specific monoclonal antibody. *Brain Res*, 1244, pp.40–52.

Zucker, R. S., 1999. Calcium- and activity-dependent synaptic plasticity. *Curr Opin Neurobiol*, 9, pp.305–313.

SECTION IV

MITOCHONDRIA-TARGETED THERAPEUTICS AND MODEL SYSTEMS

14

NEURONAL MITOCHONDRIA AS A TARGET FOR THE DISCOVERY AND DEVELOPMENT OF NEW THERAPEUTICS

Valentin K. Gribkoff

Department of Medicine, Section of Endocrinology, Yale University School of Medicine, New Haven, Connecticut, USA

14.1 NEURODEGENERATIVE DISORDERS AND THE STATUS OF DRUG DISCOVERY

Diseases of the central nervous system (CNS) have proven to be difficult targets for the successful development of new and effective medications. The lack of animal models of CNS conditions and diseases, or at least models having predictive validity for the development of treatments, as well as the relative difficulty encountered in getting treatments into the target organ and the problems of both on- and off-target side effects, although not unique to diseases of the CNS, are more pronounced when dealing with the complexity and range of functions of neurons. This has been particularly true for neurodegenerative disorders (NDDs), including strokes and neural trauma and their sequelae (which are acute events with short- and long-term consequences), and chronic disorders such as Alzheimer's disease (AD), Parkinson's disease (PD), and amyotrophic lateral sclerosis (ALS; Lou Gehrig's disease). In this discussion we will not include consideration of multiple sclerosis (MS) or other forms of demyelination disease, and for which a number of relatively effective

The Functions, Disease-Related Dysfunctions, and Therapeutic Targeting of Neuronal Mitochondria, First Edition. Edited by Valentin K. Gribkoff, Elizabeth A. Jonas and J. Marie Hardwick.

therapeutic approaches have been developed (Broadley et al., 2014; Carrithers, 2014; Cross and Naismith, 2014). Collectively, the non-MS NDDs and a constellation of similar disorders affect millions of people, with only ALS in this high-profile group affecting relatively small numbers of patients. In all cases the effects of an NDD are devastating to both the patients and their caregivers. Long-term care of patients is expensive, even in the case of ALS, where the typically rapid progress of the disease results in patient death within several years of definitive diagnosis. The unmet medical need for these NDDs and others is significant, yet there has to date been little success in developing effective treatments.

Unfortunately, the well-publicized failures of clinical candidates in these disorders have resulted in a significant change in current efforts to discover and develop new medicines. Although palliative treatments are available for most long-term degenerative disorders, they are variably effective (almost ineffective in the case of the only approved treatment for ALS), and with the passage of time they all lose their efficacy as the underlying disease processes proceed unchecked. Only in the acute treatment of thrombotic stroke has a disease modifying treatment, consisting of thrombolytic agents with a limited temporal window of opportunity (Dorado, Millan, and Davalos, 2014; Wardlaw et al., 2014), been shown to be effective and approved. Even with thrombolytic therapy in stroke, the majority of patients either do not reach treatment in time to prevent neuronal damage, are ineligible for treatment because of the presence or suspicion of cerebral hemorrhage, or cerebrovascular integrity is incompletely restored (Alderazi and Grotta, 2014; Dorado et al., 2014; Wardlaw et al., 2014; Yaghi, Eisenberger, and Willey, 2014).

In chronic NDDs the discovery of disease groups that have a clear genetic component (familial forms) led to an early burst of optimism that treatment approaches for each of the major NDDs would soon lead to effective disease-modifying treatments; unfortunately, this has thus far proven elusive. In AD a small percentage of cases were linked to mutations leading to altered production or accumulation of toxic forms of β-amyloid protein, and for more than two decades, drug discovery programs have targeted various aspects of toxic β-amyloid protein production, distribution, or accumulation. Despite these efforts, all of these programs have failed to date, although in some cases the effects of treatments were in the direction predicted (a reduction in soluble toxic β-amyloid protein, for example) (Berk and Sabbagh, 2013; Mullane and Williams, 2013; Rosenblum, 2014; Schneider et al., 2014). Although there is still hope for this approach, and clinical candidates aimed at this and related targets are still being actively pursued, these failures have dashed hopes for any simple and straightforward approach to this and probably any chronic NDD. In general, the targeting of the underlying causes of the familial forms of NDDs, which generally account directly for 10% or less of the total patient populations, probably underestimates the complexity of the "sporadic" disease population that is responsible for the vast majority of cases. Although the same familial causal factors may be involved in sporadic disease via some epigenetic or other mechanism, they may not be, or they may be only one of several contributors.

Repeated failure in NDD treatment has resulted in some real diminution in the level of interest in future drug discovery efforts and has even resulted in the failure of biotech companies focused on their development. As this chapter will show, failure is a continuing and vexing problem but must be considered part of the drug discovery and development process, as it always has been. The more we know about the underlying biology the better we will be at creating drugs and other treatments, and better technology will provide more rapid access to clinical candidates in any disease. Anyone that claims to have a magic solution to the likelihood of failure in treating acute or chronic NDDs is either delusional or a fraud. Each clinical failure, however, although incredibly disappointing to both patients and researchers, provides at least a small amount of data informing us about how to proceed with a better probability of success in the future, and in that sense is never a complete loss. Setbacks have been well publicized, and in most cases some types of resources targeting these diseases have been reduced (particularly at the preclinical level), but there are still many ongoing efforts to develop therapeutics for these diseases, and demonstrations of a greater probability of success would lead quickly to a larger-scale resumption of therapeutic efforts. Current efforts are led by AD and PD because of the significant increase in the number of current and future patients resulting from an aging population (social pressure), but testable new approaches, or a better understanding of why treatment failures occurred in the past, could lead to breakthroughs in all of the NDDs.

In particular, combination approaches may prove fruitful, using agents that target multiple pathways, including novel targets, as has been proposed for AD (Fan and Chiu, 2014; Sobow, 2010). The key to combination therapy is to admit that a single agent and approach may not be sufficient to adequately address the severity and complexity of any NDD. Promising agents, which have not been shown to be independently effective but which are supported by strong preclinical proof of concept and have demonstrated adequate clinical safety, may be combined to attack the NDD from multiple pathways. This is a variant of what has proven so successful over the last couple of decades in antiviral therapy and oncologic medicine; the exception is that in those therapeutic areas the individual components have shown some degree of independent (and usually independently approvable) efficacy, and the combination is additive, or perhaps even synergistic, but not wholly emergent. This would take a change of thinking on the part of the Food and Drug Administration (FDA), which has shown some willingness to consider such an approach only recently, and on the part of pharmaceutical/biotech companies, who would have to collaborate in the CNS area to a greater degree than is customary. This has been happening with increasing frequency in oncology and anti-infective drug development, which should allow for better collaboration on NDDs.

One such approach that may prove useful for the development of stand-alone therapeutics or components of combination therapies is to target dysfunctional mitochondria. This chapter will focus on a recent effort, a small-molecule mitochondrial modulator for ALS. Although the clinical experience with the compound ultimately resulted in treatment failure in Phase 3, despite initial positive clinical results, much has been learned from the studies, and the drug could conceivably still contribute as part of a future combination therapy. One really must ask oneself while looking at this program, and any similar approach, whether it is (or was) reasonable to expect significant clinical efficacy, given that the probable mechanism of action is the ability to modestly reduce mitochondrial inefficiency by potently and effectively reducing a source of metabolic leak at the level of the inner mitochondrial membrane. Could a modest increase in bioenergetic efficiency overcome the cascade of contributors to motor neuron death leading to the characteristic near complete loss of voluntary muscle control characteristic of ALS? Could this ever work as a disease-modifying agency in an NDD where the proximal "cause(s)" of the disease may not be directly affected by the treatment, regardless of the eventual distal contribution of mitochondrial dysfunction to neuronal death?

14.2 MITOCHONDRIA AS TARGETS FOR THE DEVELOPMENT OF NEW NDD THERAPIES

As suggested by many of the chapters in this book, as well as by other recent reviews (Andersen, 2004; Chaturvedi and Beal, 2013; Faes and Callewaert, 2011; Gonzalez-Lima, Barksdale, and Rojas, 2014; Muller et al., 2010; Muyderman and Chen, 2014; Peixoto, Dejean, and Kinnally, 2012; Schapira et al., 2014; Swerdlow, 2007), mitochondrial dysfunction is a probable contributing factor, or a final common pathway, in many and possibly all NDDs. This holistic appraisal, although accurate, does not necessarily simplify the search for effective therapies. Mitochondria can become dysfunctional and reduce the probability of neuronal survival for many reasons, and cell death can often be the result of multiple insults. Mitochondrial production of reactive oxygen species can contribute to cellular damage during oxidative stress, effects that are exacerbated if one or more of the natural antioxidant mechanisms are affected by disease. Mitochondria can be structurally damaged by a multitude of insults, including accumulation of misfolded proteins, ROS-induced damage, disruptions of the production and processing of mitochondrial proteins, or the import of extra-mitochondrial proteins. Damage can lead not only to mitophagy, but to cell death through the activation of pro-apoptotic mechanisms or other death pathways. In some disease states, several avenues of amelioration of mitochondrial damage can be compromised, such as mitochondrial fission and fusion, mitochondrial motility and mitophagy, leading to improper distribution or accumulation of dysfunctional mitochondria, decreased cellular energy production, promotion of pro-apoptotic mechanisms, and eventual cell compromise and death (Andersen, 2004; Breuer et al., 2013; Celardo, Martins, and Gandhi, 2014; Dias, Junn, and Mouradian, 2013; Faes and Callewaert, 2011; Gonzalez-Lima, et al., 2014; Muller et al., 2010; Wang et al., 2014; Yin, Boveris, and Cadenas, 2014; Zsurka and Kunz, 2013). Neurons have high energy demands. They have complex morphologies and are largely postmitotic, and therefore are not usually replaced upon cell death at the relevant late stage of life where most NDDs occur. This makes any compromise in energy production and cell homeostasis a potentially catastrophic event with long-term effects. Although both phylogeny and ontogeny provide for considerable redundancy in most nuclei and the pathways they create, disease states eventually lead to functional compromise and ultimately, in many diseases, to severe dysfunction or death of the individual.

The unique morphologies and functions of neurons require the differential subcellular distribution or creation of fully functional mitochondria in response to activity-dependent energy demands within regions of each cell. Given this, it is no surprise that mitochondrial dysfunction has been implicated in many NDDs, and that targeting mitochondria represents a rich, relatively recent hunting ground for the development of new therapies. The example in the following section will illustrate a mitochondrial approach. The development of dexpramipexole (DEX) for the treatment of ALS illustrates the logic of the approach, and helped define a target that may be useful with other, more effective small molecules. It demonstrated a promising level of effect in early clinical trials, but, as mentioned, ultimately failed when tested in a larger population of patients with ALS.

14.3 THE EFFECTS OF DEXPRAMIPEXOLE ON MITOCHONDRIAL CONDUCTANCES: AN EXAMPLE OF AN APPROACH FOR ALS AND OTHER NDDS

14.3.1 ALS as a Therapeutic Target

ALS is a devastating NDD characterized by rapid and progressive loss of almost all voluntary muscle control; it is generally fatal within several years, with death usually resulting from respiratory failure or related complications. ALS is an orphan disease, but nevertheless is the most common motor neuron disease. There are approximately 30,000 patients in the United States and similar rates of occurrence in the European Union (~1-2/1000,000 population). Patients suffer a progressive loss of voluntary muscle control resulting from the death of motor neurons, but in most cases cognitive function and sensation remains largely intact (Chio et al., 2013; Gordon, 2013a). Despite similarities between all definitive cases of ALS, there are subtypes of the disease, usually categorized on the basis of the order in which different muscle groups are affected; these different populations have different profiles of disease progression, a fact that may have significant implications for drug discovery and development efforts. In addition, there are many conditions that mimic some of the symptoms of ALS, particularly early in the disease, making for heterogeneous patient populations in those clinical trials that enroll patients before definitive diagnosis.

The need for effective disease modifying therapies has been a focus for several decades, with an initial boost following the discovery of the efficacy, however modest, of the benzothiazole riluzole in the 1990s. This remains the single approved treatment for ALS. Rilutek® (riluzole; 2-amino-6-(trifluoromethoxy)benzothiazole) is apparently non-specific, is not disease-modifying according to most operational definitions, and is at best only modestly effective, with a continuing consensus that it may prolong patient life by 2 to 3 months (Bensimon, Lacomblez, and Meininger, 1994; Lacomblez et al., 1996a, 1996b; Miller, Mitchell, and Moore, 2012; Miller et al., 1996, 2003, 2007). The mechanism of action of this drug is unknown, although there have been many hypotheses advanced over the last two decades. Significantly, it has been known for several decades that the compound reduced sodium currents, specifically persistent sodium current (Benoit and Escande, 1991; Doble, 1996; Kuo et al., 2006; Lamanauskas and Nistri, 2008; MacIver et al., 1996; Urbani and Belluzzi, 2000; Wu et al., 2005; Yvon, Czarnecki, and Streit, 2007) and is often used as a tool for studying these currents in neural networks. It also affects other ion channels, such as calcium-activated potassium channels and has many other effects on currents that may or may not result in neuroprotection via a reduction in glutamate-induced hyperexcitability and excitotoxicity (Bellingham, 2011; Daniel et al., 2013; Doble, 1996; Frizzo et al., 2004; He et al., 2002; Hubert et al., 1994; Lamanauskas and Nistri, 2008; Richter, Schaefer, and Hill, 2014; Wu and Li, 1999; Xu, Enyeart, and Enyeart, 2001).

A second major support for ALS drug discovery efforts, similar to discoveries that propelled AD efforts, was the discovery of a number of mutations (Leblond et al., 2014; Mackenzie et al., 2007; Renton, Chio, and Traynor, 2014; Tan et al., 2007), most notably in the gene encoding the largely cytosolic superoxide dismutase 1 (SOD1) (Bowling et al., 1993; Nakano et al., 1994; Orrell, Marklund, and deBelleroche, 1997; Rainero et al., 1994; Robberecht et al., 1994; Rosen et al., 1993, 1994; Sapp et al., 1995; Siddique and Hentati, 1995; Siddique, Nijhawan, and Hentati., 1996; Tsuda et al., 1994). Mutations of this gene do not result in disease because of a reduction in the enzymatic activity of mutant SOD1, which is modest, but by a gain of toxic function possibly resulting from the disposition of the mis-folded SOD1 (Bruijn et al., 1998; Williamson et al., 2000). The known mutations account for about 20% of known familial cases of ALS (mutations in other genes have more recently been associated with other types of familial ALS), or about 2% of the total patient population. This means that some 90% of ALS cases are currently classified as sporadic, and 98% have no known connection with a mutation in the gene encoding SOD1. Some susceptibility mutations have been associated with other genes in sporadic ALS, but this has not been well established at this point (Dunckley et al., 2007; Greenway et al., 2006).

Despite these percentages, animal models have been created which express a mutated SOD1 homolog, and these animals present with an animal model of the disease characterized by the hallmark loss of motor neurons and voluntary muscle control, leading to death, observed in ALS (Bruijn and Cleveland, 1996; Doble and Kennel, 2000; Gonzalez de Aguilar et al., 1999; Julien and Kriz, 2006). This represents a high degree of face validity and is therefore a potentially good thing because an animal model expressing both functional decline and death as measurable variables for testing drug candidates is an important component of ALS drug discovery efforts. However, as in many cases, there are complications. Notable among these are the number of highly effective candidates (effective in the animal models) that had no significant effect when examined in large Phase 3 clinical trials (Beghi et al., 2007; Benatar, 2007; Berk and Sabbagh, 2013; Gordon and Meininger, 2011), and the observation that the single approved compound riluzole has only modest or no significant effects in these models (Scott et al., 2008). The meaning of this is unclear, and could mean that normal model variability is too great to use small animal numbers to test compounds, as suggested by Scott et al. (2008), or that the animal models do not adequately represent the human disease, despite their apparent face validity. More likely, these models represent a small subgroup of human cases, and that effects of potential drugs in animals expressing mutant SOD1 may be self-selecting for efficacy in a small percentage of subjects in large registrational clinical trials. This depends on whether the mechanism of the drug is specific for interfering with SOD1's toxicity, or interrupts processes leading to cell death that apply more broadly to the death of motor neurons at risk. Clearly, there are impediments to drug discovery in ALS that have slowed the successful search for effective disease-modifying treatments, and new approaches, even if initially unsuccessful, could prove useful. One such general approach is to target dysfunctional mitochondria in ALS, which could represent a final pathway for motor neuron death in this disease, as has been suggested for other cell types in other NDDs.

14.3.2 Mitochondrial Dysfunction in ALS

Mitochondrial dysfunction has long been implicated in ALS (Cozzolino and Carri, 2012; Faes and Callewaert, 2011; Swerdlow et al., 1998). Mitochondrial profiles are altered in postmortem motor neurons from patients with ALS and in motor neurons from presymptomatic mice expressing SOD1 mutations (Bendotti et al., 2001; Sasaki and Iwata, 2007; Xu et al., 2004). In these mice, increasing proportions of motor neuron mitochondria appear swollen with vacuoles that can be quite large (and can contain mutant SOD1 protein, not normally associated with mitochondria), with larger numbers of points of contact between the IMM and the OMM, potential loci for complexes mediating the release of substrates that can initiate cell death. In cells expressing mutant SOD1, ATP production is reduced, and ROS generation is increased, both of which suggest mitochondrial dysfunction. Axonal transport of mitochondria is impaired in ALS motor neurons (Bilsland et al., 2010; Collard, Cote, and Julien, 1995); as a result of the very long axons in these cells, any diminution of the ability to transport mitochondria and other substrates to nerve terminals will put motor neurons at risk. Mutant protein aggregates have been found in neuronal mitochondria in human ALS and in neurons from genetic models. Most directly, mutant SOD1 has been shown to interact with a number of mitochondrial components, and may be associated with specific mitochondrial dysfunctions in model systems (Carri and Cozzolino, 2011; Kawamata and Manfredi, 2010). Although it is unlikely that ALS is a mitochondrial disease in the usual sense, investigators believe that increasing proportions of dysfunctional mitochondria in ALS doom motor neurons. However, we do not yet understand where or how sensitive the tipping point is, or whether preserving intact mitochondria, or increasing mitochondrial function in a proportion of dysfunctional mitochondria, can significantly affect disease progression. Previous attempts to stimulate mitochondrial function, or ameliorate the effects of ROS overproduction, have been unsuccessful (Beal, 2002; Zhu et al., 2002), as have attempts to block cell death pathways initiated by mitochondrial dysfunction (Martin, 2010). This was the background available when preparations for the development of dexpramipexole for ALS were initiated in 2006.

14.3.3 Dexpramipexole and Bioenergetic Efficiency: Preclinical Studies

Dexpramipexole (DEX; KNS-760704; (6R)-4,5,6,7-tetrahydro-N6-propyl-2,6-benzothiazole-diamine) is a benzothiazole structural analog of riluzole, although it was not synthesized to mimic riluzole's efficacy in ALS by exploiting this structural similarity. Instead, it is the R(+) enantiomer of the effective dopamine D2 receptor agonist Pramipexole (Mirapex®), a widely used anti-Parkinsonian drug (Antonini et al., 2010; Biglan and Holloway, 2002; Perez-Lloret and Rascol, 2010; Yamamoto and Schapira, 2008) also approved for restless legs syndrome (Ferini-Strambi and Marelli, 2014; Kushida, 2006; McCormack and Siddiqui, 2007). Dexpramipexole was first recognized as a potential therapy for ALS by a number of groups who found that DEX's S(-) enantiomer pramipexole, the potent dopamine agonist, and DEX, which is not a potent dopamine agonist, had equivalent ability to protect cells from a variety of insults, presumably by a relatively low-potency but non-dopaminergic unknown mechanism (Ferrari-Toninelli et al., 2010; Nakayama et al., 2009), as reviewed previously (Gribkoff and Bozik, 2008). This was an important observation, replicated in a number of systems. Although pramipexole is not a good candidate neuroprotectant because of the prevalence of dopamine receptor-mediated side-effects (its neuroprotective potency is much lower than its dopamine receptor agonist affinity), DEX potentially could be. DEX can be dosed at much higher doses without significant dopamine receptor interaction, doses that would be intolerable with pramipexole; pramipexole requires dose titration even at low doses. In addition, initial studies suggested that the compound readily entered the brain and also entered mitochondria and could reduce ROS production (Abramova et al., 2002; Danzeisen et al., 2006).

Bennett et al. (Wang et al., 2008), who first recognized the potential for the use of DEX in ALS, decided to do a clinical test in patients with ALS to determine if they could observe any treatment benefit of the drug, as well as further establish its safety. The results of this open label trial were promising; although there are many caveats in the interpretation of open label clinical trial results, it appeared that the rate of decline in neuromuscular function was reduced in these patients, and the drug appeared safe even at high doses.

14.3.3a Effects of DEX on Mitochondrial Conductance

Initial insights into a possible mechanism of action for DEX came from a study by Sayeed et al. (2006) that examined the mitochondrial effects of pramipexole. They found that in liver mitoplasts, pramipexole produced a rapid, reversible and concentration-dependent inhibition of large-conductance, Ca^{2+}-stimulated membrane ion channels. They also presented evidence of suppression of mitochondrial permeability transition by pramipexole. The implication was that the membrane channels inhibited by pramipexole were mitochondrial permeability transition pores (mPTP; PTP). The mPTP is an incompletely characterized large-conductance membrane complex involved in initiating cell death. Components of this pore may have other, less suicidal roles in regulating mitochondrial function and could represent a viable target for drugs ameliorating cancer, cardiovascular dysfunction and NDDs (Alavian et al., 2014; Bernardi and Di Lisa, 2014; Morciano et al., 2015; Rao, Carlson, and Yan, 2014). Current evidence strongly suggests that the large-conductance pathway may be part of the complex of proteins that form the ATP synthase (Alavian et al., 2014; Bernardi and Di Lisa, 2014; Morciano et al., 2015). Given that DEX and pramipexole appear to have identical effects except where their relative affinities to dopamine receptor subtypes are concerned, and the effects described in liver mitoplasts were unlikely to involve dopamine receptor agonism, it seemed a logical stepping-off point for further studies on DEX. However, although suppression of the mPTP and subsequent mPTP-initiated cell death would provide a ready mechanism of action for DEX in reducing cell death, the role of mitochondrial permeability transition in NDDs at this point is still an area of intense study (see Chapter 3), and, specifically, apoptotic profiles have not been demonstrated in neurons in ALS (Bendotti et al., 2001; He and Strong, 2000). Nevertheless, the suppression of large-conductance ion channel activity in mitochondria provided an anchor point for further work in our group. Specifically, we sought to determine if the effects of DEX and other drugs, as manifested by the putative inhibition of mitochondrial large-conductance channel activity, may reflect an interaction with the ATP synthase complex. If true, we wanted to determine if the interaction could increase bioenergetic efficiency in a manner and to a degree that would provide significant neuroprotection.

We recently found that DEX, like pramipexole, produced a reduction in large-conductance ion channel activity that was initiated by high Ca^{2+} in neuronal mitochondria (Alavian et al., 2012) (Fig. 14.1). In addition, in mitochondria isolated from the brains of rats that were pretreated with the ubiquitin proteasome inhibitor PSI (Z-Ile-Glu(OtBu)-Ala-Leu-al), large-conductance channels were encountered during patch clamp recording (without added Ca^{2+}) at a significantly higher frequency than in mitochondria isolated from control rats. In PSI-treated mitochondria, DEX potently and reversibly inhibited the large-conductance activity. Cyclosporine A (CSA), an inhibitor of mitochondrial transition and the mPTP, was also effective in these models. However, we also examined Ca^{2+}-induced mPTP transition in liver mitochondria; whereas CSA was effective in blocking mPTP transition, neither DEX nor pramipexole under our experimental conditions were able to inhibit this phenomenon, even at high concentrations. These data suggested that although the large conductance channels in neuronal mitochondria sensitive to DEX and pramipexole shared some characteristics with the mitochondrial transition pore, including sensitivity to CSA, their inhibition did not block transition in liver mitochondria, in contrast with what Sayeed and colleagues (2006) had previously reported in a similar preparation.

14.3.3b Effects on Cellular Metabolism

In subsequent experiments we examined the metabolic consequences of DEX applied to neurons and other cells in culture. In individual hippocampal neurons, using oxygen-sensitive electrodes and a luciferase assay to assess oxygen use and ATP levels, DEX produced small but significant increases in ATP levels, and oxygen uptake was significantly reduced. We also measured the simultaneous average oxygen consumption rate (OCR) and extracellular acidification rate (ECAR) in culture plates using a Seahorse® extracellular flux analyzer. Although the drug had no effect on basal OCR in cortical neurons, stimulation with either the metabolic substrate succinate or the ATP precursor ADP resulted in DEX-dependent significant reductions in OCR. ATP levels were not reduced under these conditions. In both of these models, neither of which were models of *dysfunctional* cells, dexpramipexole

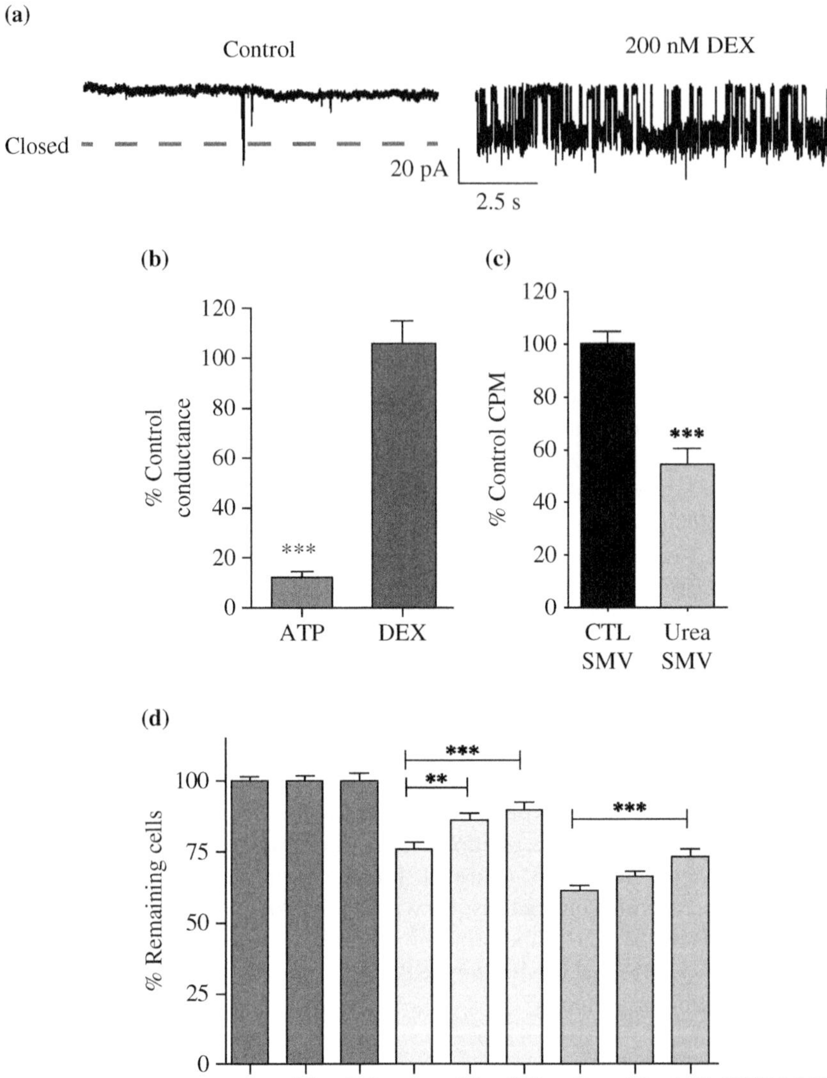

FIGURE 14.1 Effects of Dexpramipexole (DEX) in isolated mitochondrial preparations and stressed cells. **a,** Following treatment of rats in vivo with a proteasome inhibitor (Z-Ile-Glu(OtBu)-Ala-Leu-al; PSI), isolated brain-derived mitochondria exhibited large-conductance channel activity (organelle-attached patch clamp). Application of DEX produced a concentration-dependent and reversible inhibition of this activity; in the example shown the holding potential was +80 mV, and the channel(s) was nearly constantly in the open state. DEX at 200 nM produced a greater than 50% reduction in the open probability. **b,** In brain-derived submitochondrial vesicles (SMVs), application of DEX produced a concentration dependent inhibition of membrane conductances (*not shown*). Similar to brain-derived mitochondria

increased the efficiency of ATP production because the amount of oxygen consumed in the presence of the drug, relative to vehicle controls, was reduced while ATP levels were increased or maintained at previous levels.

To produce a model of metabolism in stressed cells, we treated C2C12 myoblasts in vitro with a sublethal concentration of the proteasome inhibitor PSI. In control cells DEX reduced OCR; the changes were small, <10%, but significant. PSI treatment did not produce a significant effect on OCR, but when PSI-treated cells were exposed to DEX, the OCR decrease was nearly twice as great as the decrease in control cells. Interestingly, in PSI-treated cells ECAR, an indirect index of glycolysis levels, was *greatly* increased. This suggested that these stressed cells resorted to glycolysis to maintain energy levels. When DEX was applied to PSI-treated cells, as noted, OCR was significantly decreased, whereas ECAR was largely normalized. Although ATP levels could not be observed in these cells for technical reasons, we did observe the effects of DEX on ATP levels in SH-SY5Y neuroblastoma cells in both normal medium and in medium containing only galactose as a substrate (which cannot be used by glycolysis). Under both incubation conditions DEX produced a concentration-dependent and significant increase in cellular ATP levels (maximum increases ~18%). What are the consequences of this for cells under toxic conditions? As had been shown by others for pramipexole, we found that in SH-SY5Y neuroblastoma cells treated with toxic levels of PSI, dexpramipexole produced a concentration-dependent and significant reduction in cell death (see Fig. 14.1).

Our conclusion was that dexpramipexole acted by an unknown mechanism to increase the bioenergetic efficiency of stressed, and to a lesser extent, "normal" cells. Given similar profiles of concentration-response in different experimental systems (current inhibition versus effects on metabolism), we believed the metabolic effects were linked to the inhibition of membrane channels in stressed mitochondria. It is important to note that although many of the effects of DEX in highly isolated preparations were quite potent, the drug's level of potency and maximal efficacy in cells (as opposed to direct inhibition of large-conductance channels) was modest. A rightward shift in potency when moving from isolated preparations to whole cells, and then another rightward shift when moving to even more complicated systems

---

isolated from rats that were pretreated with PSI, the effects were potent (EC_{50} ~100 nM for mitochondria, ~110 nM in SMVs). Treatment of these SMVs with urea (shown), resulted in complete elimination of current inhibition by DEX (200 nM, which was >EC_{50}), but no effect on the inhibition produced by ATP (1.0 mM). This indicated that some non-membrane bound component expressed on the inner membrane of the (inside-out) complex-V enriched SMVs, which was denatured by urea treatment, was required for current inhibition by DEX but not ATP. This conclusion was supported by the experiments shown in **c,** wherein [^{14}C]DEX binding in SMVs was significantly reduced by urea treatment. **d,** Treatment of SH-SY5Y cells in culture with the proteasome inhibitor PSI stresses cells, resulting in cell compromise and death. Pretreatment of these cells for 24 hours with DEX at the indicated concentrations, before 24 hour treatment with PSI (at 150 or 650 nM), resulted in significant reductions in the levels of cell death. ** P<0.05; *** P<0.001; in **b** and **c**, t-test, in **d**, Bonferroni-corrected t-tests following significant ANOVA. Adapted from Alavian et al., 2012 and 2014a, with permission.

such as whole tissues or, even more likely and pronounced, whole animals is a common and expected observation. Although consistent with previous observations in cytoprotection assays using either pramipexole enantiomer, the modest cellular efficacy in our experiments was more concerning. Nevertheless, beyond imposing some elements of mitochondrial dysfunction, we do not really know the degree to which any of our experimental manipulations were relevant to stressors that place neurons at risk in ALS, and we were not studying motor neurons. Given this, the primary conclusions that should be drawn from these data was that DEX was neuroprotective, and likely acted through an improvement in metabolic efficiency mediated by inhibition or regulation of mitochondrial large-conductance currents.

14.3.3c Molecular Targets of DEX

To further examine and define the molecular target(s) of DEX, we used other models, including submitochondrial vesicles (SMVs) enriched in F_1F_O ATP synthase, and preparations of isolated subunits of complex V, the ATP synthase complex (Sacchetti et al., 2013). In SMVs, which are brain-derived isolated and inverted inner membrane vesicles that contain detergent-resistant mitochondrial components enriched in F_1F_O ATP synthase, DEX and CSA inhibited large-conductance currents as observed in previous mitochondrial membrane preparations, with an EC_{50} that was remarkably close to the potency seen in whole mitochondrial current inhibition (~100 nm) (Alavian et al., 2015). ATP also inhibited these currents, with much lower potency, and calcium modestly increased them. Because SMVs contain high levels of F_1F_O ATP synthase we could measure ATP hydrolysis by DEX and CSA, and both compounds enhanced hydrolysis of ATP. Exploiting the inside-out predominate orientation of SMVs (Chan, Greenawalt, and Pedersen, 1970; Ko et al., 2003), which exposes the F1 head of the complex to chemical manipulation, we treated these preparations with urea. This removed the F1 head, as well as other exposed protein, but preserved membrane bound components. This was determined by western blot analysis, which confirmed that the F1 β-subunit was removed, while the membrane-inserted adenine nucleotide translocator (ANT) was unaffected. Urea-treated SMVs no longer hydrolyzed ATP, and although they still displayed large-conductance membrane currents, these currents could no longer be inhibited by DEX. They were still inhibited by ATP (see Fig. 14.1). ^{14}C-labeled DEX bound to untreated SMVs, but binding was significantly reduced (by an average of 42%) following urea treatment. DEX, CSA, and ATP therefore appeared to interact with complex V to inhibit large-conductance membrane currents in brain-derived mitochondria and SMVs, while removal of extramembrane components, including F1, eliminated DEX's (but not ATP's) ability to inhibit currents and greatly reduced the binding of radiolabeled DEX. These data suggest that DEX interacts with a component of F1 to block a large-conductance current, which may represent a metabolic leak current. Subsequent experiments, using individual subunits of complex V expressed in 293T cells, demonstrated that incubation in ^{14}C-DEX resulted in significantly higher levels of binding to affinity gels bound to two subunits, b and the oligomycin-sensitivity conferring protein (OSCP), when compared with those incubated with the untransfected control cell lysate or other complex V subunits (Alavian et al., 2015). Co-incubation of these gels with a high concentration of unlabeled ("cold") DEX reduced binding to levels seen in untransfected controls. These results indicate that DEX bound in a specific and competitive manner to subunits that comprise the stator stalk of complex V. It appears probable that the channel being inhibited by this interaction is similar or identical to the channel previously shown to be formed by the c-subunit of the F_1F_O ATP synthase (Alavian et al., 2014); this c-subunit channel is inhibited by ATP, and as a component of the entire complex, this channel can be inhibited by compounds that alter the function of other subunits. In the case of DEX this appears to be the b and OSCP that form part of the stator stalk. These data serve to provide the initial structure of a model for understanding how drug-mediated modulation of complex V can alter bioenergetic efficiency in a manner that could ameliorate disease-induced mitochondrial dysfunction. The most important question, of course, is whether an interesting but relatively modest re-invigoration of mitochondrial function can alter the time course or outcome of an intractable NDD such as ALS. As mentioned previously, none of these data were obtained in specific models of ALS-affected cells or mitochondria, and in fact the compound, like riluzole, had little effect in mouse or cellular models of ALS (Vieira et al., 2014; Yang et al., 2013). Other than the early promising open-label results with DEX, the interesting effects of the compound on mitochondria and its modest level of neuroprotection, there was no specific reason to suspect that DEX would have efficacy in ALS.

14.3.4 Dexpramipexole in the Clinic

As discussed previously, the early open-label study of DEX in subjects with ALS was promising, indicating possible drug-related improvement, and long-term safety at the doses tested (Wang et al., 2008). The compound was licensed by the small biotech company Knopp Neurosciences (now Knopp Biosciences LLC) with the specific goal of developing DEX as a treatment for ALS. Further development followed a more rigorous path that could eventually lead to the submission of a New Drug Application (NDA) for the US Food and Drug Administration (FDA) and its equivalent for the European Medicines Agency (EMA). Following completion of a series of mandated animal safety and pharmacokinetic studies and the submission of an Investigation New Drug application (IND), a clinical program was started that continued to a large Phase 3 trial designed to determine the degree of safety and efficacy required for eventual approval.

14.3.4a Phase 1 Studies

In two Phase 1 studies, the safety, tolerability and pharmacokinetics of single doses of 50 mg, 150 mg, or 300 mg, or twice daily doses of 50 mg, 100 mg, or 150 mg, were tested in healthy human volunteers. DEX was shown to be safe and well tolerated, was rapidly absorbed, with an estimated half-life of 6.4 to 8.1 hours under fasted conditions (Bozik et al., 2011). No food effect was observed on the single-dose pharmacokinetics. These data were considered strongly supportive of advancement of DEX into studies in subjects with ALS. A Phase 2 study in subjects with ALS was designed that could further study the safety of the drug at divided daily doses to a maximum of 300 mg/day, as well as examine the efficacy in several measures of disease progression. As will be discussed, the daily maximal dose of 300 mg/day was not determined because this study or any of the preceding animal studies determined a maximum tolerated or maximally effective dose but rather was the result of an abundance of caution. There were several unexplained animal deaths at high doses, and although there was no obvious cause, and no certainty that they were drug-related, a maximum dose of 300 mg/day was determined. It was felt at the time that this should adequately test the hypothesis that DEX can reduce the rate of functional decline in subjects with ALS. Such an assumption is always a risk in clinical development because the cost of failure in an expensive clinical program, even if the underlying assumption may not have been valid, is so great as to almost always be fatal to further development.

14.3.4b Phase 2 Trial

The results of the Phase 2 study were largely positive (Cudkowicz et al., 2011; Rudnicki et al., 2013), in fact at the time they represented the most positive data set for any drug tested in ALS; given its safety and apparent efficacy, the compound showed greater promise following Phase 2 than any previous compound at an equivalent point in clinical development. The trial was a two-part, multisite double-blind safety and tolerability study with a preliminary assessment of the efficacy of DEX on the decline of neuromuscular function and mortality. In part 1, 102 subjects with ALS were randomized to receive either placebo or one of three doses of DEX (50, 150, or 300 total mg/day as divided doses), for 12 weeks; there were no deaths in part 1, but several of the subjects did not progress into part 2 because of disease progression or withdrawal of consent. In part 2, 97 of the subjects participating in part 1 entered a 4-week single-blind washout period, and 92 of these subjects (3 subjects died during the washout and 2 additional subjects discontinued) were re-randomized into two active treatment, double-blind groups (50 mg or 300 mg total mg/day as divided doses) for a 24-week treatment period. Data were based on 90 subjects from whom at least one postbaseline clinical evaluation was obtained. Analysis of the data demonstrated a dose-dependent trend in the reduction of the revised ALS functional rating scale (ALSFRS-R) in part 1, with some apparent drug-dependent differences seen in different measures of loss of function (Fig. 14.2**a**). In part 2 the highest dose (300 mg/day) was compared to the lowest dose (50 mg/day); 50 mg/day had not shown any differentiation from placebo in part 1, and the ability of all subjects to continue on what theoretically could be an active dose, despite its previous lack of efficacy in Part 1, was an inducement for patients to continue on the longer component of the study in Part 2. Fewer subjects died in the 300 mg/day group, relative to the low-dose group (Fig. 14.2**b**). Most important, a statistically significant difference between the dose groups ($p = 0.046$) was observed in a joint rank scale, which is a method of ranking outcomes in each dose group by rating death as the worst outcome, and then ranking all survivors based on their degree of change from baseline in the ALSFRS-R (Fig. 14.2**c**). At this point a decision was made to advance the compound into a larger, longer duration Phase 3 trial, and development proceeded in partnership with a larger pharmaceutical/biotech company (Biogen Idec).

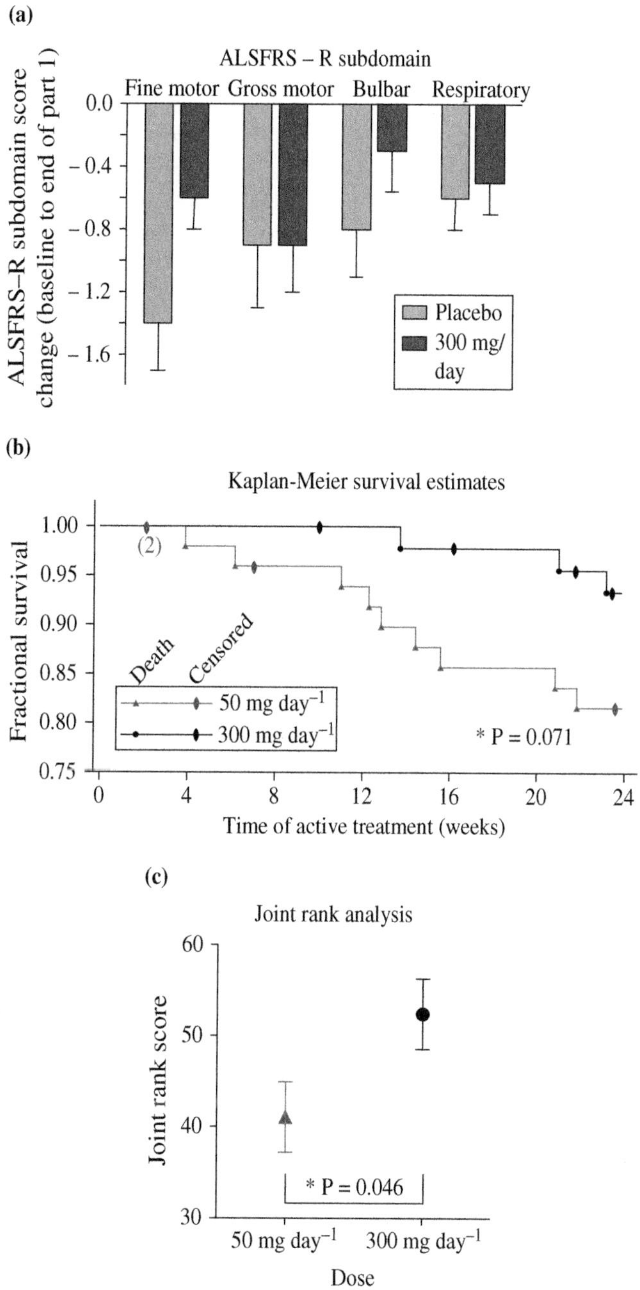

FIGURE 14.2 Effects of dexpramipexole in a phase 2 study in human subjects with amyotrophic lateral sclerosis. **a,** The bar chart demonstrates the differential effect of the highest dose of DEX (300 mg/day) versus placebo on measures of the loss of function, as indicated by ALSFRS-R decline, in the fine motor, gross motor, bulbar and respiratory subdomains measured during the course of Part 1 in the phase 2 study of safety and efficacy in subjects with ALS. **b,** Kaplan-Meier estimates of the probability of survival in part 2 of the study. Censored patients were those that discontinued for any reason other than death, $P = 0.07$, log rank test. **c,** Mean (± SEM) values from a joint rank analysis (see text) of death and change from baseline in ALSFRS-R scores in survivors comparing treatment groups in part 2 of the study; there was no placebo group in part 2. Adapted with permission from Cudkowicz et al., 2011.

14.3.4c Phase 3 Trial

The Phase 3 study of DEX (EMPOWER) enrolled 943 subjects with possible, probable, or definite ALS in a double-blind, placebo-controlled trial, with random assignment to one of two groups. The groups were DEX at a total daily dose of 300 mg (as 150 mg divided doses every 12 hours; 474 subjects), or placebo (468 subjects; one subject withdrew consent). Subjects were followed for 12 months; just as in the Phase 2 trial, a combined assessment of function and survival (CAFS; called joint rank analysis in Phase 2) was used as the primary endpoint. Effects on the rate of functional decline and mortality were also independently assessed using the standard metrics, along with a number of other measures. Blood levels of drug in samples of subjects from both groups confirmed the presence of expected drug levels of DEX in the treated group, and the absence of DEX in the placebo group. Although there were some small differences from the Phase 2 study (greater concomitant riluzole use, for example), the study was designed and carried out as an attempt to replicate the Phase 2 study design, and overall analyses of the study indicate that it adequately tested the efficacy and safety of DEX in a large sample of subjects with possible, probable or confirmed ALS. Unfortunately *there was no evident effect of DEX* in this study on any measure, when all of the subjects' data was used for the comparisons between groups (Cudkowicz et al., 2013); the compound was again found to be generally safe and well-tolerated, but the promising Phase 2 data were not replicated. This was a troubling result, given that the previous data had been unprecedented in its apparent effect, many of the same clinical investigators were involved in both studies, and the trial was designed to be as similar to the Phase 2 trial as was practical.

14.3.4d Subsequent Analyses

As always happens when there is an initial signal that is lost when a drug is tested in a larger, somewhat more diverse population of subjects with a disease (and particularly ALS and other NDDs, where this has happened repeatedly), there has been much discussion about whether Phase 2 trial data should ever be used as a basis for advancing compounds into Phase 3. At the very least we are told to be more careful in the use of efficacy data obtained as secondary endpoints (secondary to safety) as any part of the decision making process for moving into Phase 3 (Berk and Sabbagh, 2013; Berry, Cudkowicz, and Shefner, 2014; Gordon, 2013b). Given the great expense of Phase 3 trials, and the number of failures of compounds regardless of proposed mechanism or Phase 2 data strength, it is currently unclear how much value Phase 2 efficacy data has (I will discuss this again). On the other hand it is just as likely that if Phase 2 efficacy data were not obtained, and after determining that Phase 2 safety data and all other preclinical and clinical data pointed to a reasonable hypothesis to test in the clinic in patients with ALS, a subsequent Phase 3 trial resulted in failure, the opposite argument would be vigorously prosecuted. Phase 2 safety studies nearly as complicated and of similar duration still have to be undertaken; perhaps the number of subjects with ALS would not be as great, but the difference would not be a game-changer, and there always exists the possibility that the different results in the two trials actually means something. The possibility certainly exists that some aspect of the study design, enrollment or other factor in Phase 3 was a major contributor to masking or otherwise obscuring what would have been a more positive result. In support of this idea subsequent post hoc analyses, with all their attendant caveats, have found that potential clinical benefits of DEX were present in a subgroup of riluzole-treated, short-symptom duration study participants who had definite ALS at recruitment (as opposed to possible or probable ALS) (Bozik et al., 2014). It also needs to be reiterated that neither a maximum tolerated dose, nor a maximally effective dose, was ever tested in the clinic because of several serious but possibly spurious cases of toxicity seen in one of the toxicology species (the minipig), and the highest dose in humans was restricted to 300 mg/day. Possibly more important, although there may be no reason to assume that (at least at the doses tested) DEX as a stand-alone therapy would ever be a rational future direction for the drug, it could be an ideal candidate for a combination therapy with one or more other drugs that have different mechanistic rationales for having benefit in ALS.

14.4 WHAT IS THE FUTURE OF A MITOCHONDRIAL APPROACH FOR NDD THERAPY?

Whether DEX or any other individual drug is successful is not the primary focus of this chapter, and further analyses may suggest a path for further development of DEX, although as discussed, the activation energy barrier for such a program after a failed trial likely precludes any further investigation. To researchers in the area of mitochondria-targeted therapeutics, the question is whether its failure says anything significant about targeting mitochondria in ALS or other NDDs. First, DEX is an interesting compound, and the study of its properties suggest a unique site of interaction in the ATP synthase that could be related to the non-dopaminergic neuroprotective properties of both pramipexole and its enantiomer DEX (Alavian et al., 2012, 2015). However, the efficacy of the compound as a cytoprotectant in our experiments was modest, and required concentrations between 30 and 100 μM in cellular assays, concentrations of drug that were not achieved in the clinical trials. Recently, one independent group failed to find benefit of DEX in properly controlled preclinical animal studies (maintenance of body weight and measures of survival as endpoints in high-copy number B6-SJL-SOD1^{G93A}/Gur1 mice), although there were nonsignificant drug effects suggestive of some degree of drug-related amelioration of disease symptoms and progression. Nor did they find a meaningful benefit in primary cortical neurons transfected with wild-type or mutant Tar DNA binding protein (TDP43). TDP43 (or TDP-43) is mutated in some cases of familial ALS, and is most closely associated with frontotemporal dementia (FTD) (Arai et al., 2006; Neumann et al., 2006; Tan et al., 2007). With several other recently identified mutations associated with familial ALS, it represents another genetically linked target for the potential development of ALS therapies. In this cellular assay, in which overexpression of wild-type TDP43 or either of two disease-linked mutations (TDP43A^{A315T} and TDP43^{M337V}) leads to death of the cortical neurons, there was a modest but significant effect of DEX on cells overexpressing wild-type TDP43 and TDP43^{M337V} variants, but only at 10 μM (1-50 μM concentrations were tested, the result suggestive of an inverted U-function, which we observed in other cytoprotection assays). Another recent study by Yang and colleagues (2013) failed to find any neuroprotective effect of DEX in a stem cell-derived cultured motor neuron screening assay. However, it appears that this study employed concentrations between 0.1 and 10 μM in a screening mode, concentrations that are probably too low to detect an effect in a low-replicate experiment. Nevertheless, these data contribute to a general conclusion, increasingly validated by both preclinical and clinical data, that DEX itself is unlikely to find success in ALS as a stand-alone drug, and may have no therapeutic role in any human NDD as a monotherapy. This does not mean, however, that the decision tree used to advance the compound into the clinic was ill-conceived, or that nothing was learned, or can be further learned, from the program.

It is worthwhile to remember that during the conduct of the Phase 3 clinical trials, independent review concluded that the totality of information available to that point was the most supportive of any compound yet offered to treat this devastating disease (Corcia and Gordon, 2012). In the final analysis this particular molecule failed to significantly decrease the rate of disease progression, and increase patient survival, in ALS, which is one of the most rapid and devastating of all NDDs. It definitely *does not* mean that its effects on stressed and dysfunctional mitochondria, a likely result of its interaction with specific components of a tightly-regulated complex-V-associated mitochondrial inner membrane conductance, are to be dismissed lightly as a valuable approach to protecting neurons at risk. DEX is possibly not effective enough, but its potential clinical utility was discovered serendipitously, and its mitochondrial effects are still being investigated. It was not the product of a targeted mechanism-based search and pharmacophore optimization program, yet it produced more positive clinical data in this most difficult disease than any other molecule to date. The data obtained really suggest, and strongly so, that DEX was a pathfinder molecule, validating novel mitochondria-based therapeutic targeting, and that we should ultimately use the data obtained to find more effective molecules that can better exploit what was learned. Also remember, and this can not be stressed enough, that *no one* currently knows whether a single molecule, or a single approach, will be sufficient to produce significant benefit in ALS or any of the major NDDs. As discussed, DEX itself could be useful in combination with other therapies targeting other components of the disease pathway. My thinking is certainly tempered with a thorough knowledge of the costs of developing drugs to test hypotheses, but nevertheless one can hope that the economics of drug discovery and development do not completely outweigh the promise and necessity of pursuing new targets for protecting neurons. The growing body of evidence linking all NDDs to significant and varied forms of mitochondrial dysfunction should not be derailed by the failure of DEX, but should focus our efforts on the target that yielded the greatest degree of clinical success yet observed in ALS.

ACKNOWLEDGMENTS

I would like to Thank Professor Elizabeth A. Jonas, and Drs. Kambiz N. Alavian and Steven I. Dworetzky for their contributions to the original work that was reviewed in this chapter.

REFERENCES

Abramova, N. A., et al., 2002. Inhibition by R(+) or S(−) pramipexole of caspase activation and cell death induced by methylpyridinium ion or beta amyloid peptide in SH-SY5Y neuroblastoma. *J Neurosci Res*, 67, pp.494–500.

Alavian, K. N., et al., 2012. Effects of dexpramipexole on brain mitochondrial conductances and cellular bioenergetic efficiency. *Brain Res*, 1446, pp.1–11.

Alavian, K. N., et al., 2014. An uncoupling channel within the c-subunit ring of the F1FO ATP synthase is the mitochondrial permeability transition pore. *Proc Natl Acad Sci U S A*, 111, pp.10580–1085.

Alavian, K. N., et al., 2015. The mitochondrial complex V-associated large-conductance inner membrane current is regulated by cyclosporine and dexpramipexole. *Mol Pharmacol*, 87, pp.1–8.

Alderazi, Y. J. and Grotta, J. C., 2014. Acute antithrombotic treatment of ischemic stroke. *Curr Vasc Pharmacol*, 12, pp.353–364.

Andersen, J. K., 2004. Oxidative stress in neurodegeneration: cause or consequence? *Nat Med*, 10 Suppl, pp.S18–S25.

Antonini, A., et al., 2010. Role of pramipexole in the management of Parkinson's disease. *CNS Drugs*, 24, pp.829–841.

Arai, T., et al., 2006. TDP-43 is a component of ubiquitin-positive tau-negative inclusions in frontotemporal lobar degeneration and amyotrophic lateral sclerosis. *Biochem Biophys Res Commun*, 351, pp.602–611.

Beal, M. F., 2002. Coenzyme Q10 as a possible treatment for neurodegenerative diseases. *Free Radic Res*, 36, pp.455–460.

Beghi, E., et al., 2007. The heterogeneity of amyotrophic lateral sclerosis: a possible explanation of treatment failure. *Curr Med Chem*, 14, pp.3185–3200.

Bellingham, M. C., 2011. A review of the neural mechanisms of action and clinical efficiency of riluzole in treating amyotrophic lateral sclerosis: What have we learned in the last decade? *CNS Neurosci Ther*, 17, pp.4–31.

Benatar, M., 2007. Lost in translation: treatment trials in the SOD1 mouse and in human ALS. *Neurobiol Dis*, 26, pp.1–13.

Bendotti, C., et al., 2001. Early vacuolization and mitochondrial damage in motor neurons of FALS mice are not associated with apoptosis or with changes in cytochrome oxidase histochemical reactivity. *J Neurol Sci*, 191, pp.25–33.

Benoit, E. and Escande, D., 1991. Riluzole specifically blocks inactivated Na channels in myelinated nerve fibre. *Pflugers Arch*, 419, pp.603–609.

Bensimon, G., Lacomblez, L. and Meininger, V., 1994. A controlled trial of riluzole in amyotrophic lateral sclerosis. ALS/Riluzole Study Group. *N Engl J Med*, 330, pp.585–591.

Berk, C. and Sabbagh, M. N., 2013. Successes and failures for drugs in late-stage development for Alzheimer's disease. *Drugs Aging*, 30, pp.783–792.

Bernardi, P. and Di Lisa, F., 2014. The mitochondrial permeability transition pore: Molecular nature and role as a target in cardioprotection. *J Mol Cell Cardiol*, 78, pp.100–106.

Berry, J. D., Cudkowicz, M. E. and Shefner, J M., 2014. Predicting success: optimizing phase II ALS trials for the transition to phase III. *Amyotroph Lateral Scler Frontotemporal Degener*, 15, pp.1–8.

Biglan, K. M. and Holloway, R. G., 2002. A review of pramipexole and its clinical utility in Parkinson's disease. *Expert Opin Pharmacother* 3, pp.197–210.

Bilsland, L. G., et al., 2010. Deficits in axonal transport precede ALS symptoms in vivo. *Proc Natl Acad Sci U S A*, 107, pp.20523–20528.

Bowling, A. C., et al., 1993. Superoxide dismutase activity, oxidative damage, and mitochondrial energy metabolism in familial and sporadic amyotrophic lateral sclerosis. *J Neurochem*, 61, pp.2322–2325.

Bozik, M. E., et al., 2011. Safety, tolerability, and pharmacokinetics of KNS-760704 (dexpramipexole) in healthy adult subjects. *J Clin Pharmacol*, 51, pp.1177–1185.

Bozik, M. E., et al., 2014. A post hoc analysis of subgroup outcomes and creatinine in the phase III clinical trial (EMPOWER) of dexpramipexole in ALS. *Amyotroph Lateral Scler Frontotemporal Degener*, 15, pp.406–413.

Breuer, M. E., et al., 2013. The role of mitochondrial OXPHOS dysfunction in the development of neurologic diseases. *Neurobiol Dis*, 51, pp.27–34.

Broadley, S. A., et al., 2014. Therapeutic approaches to disease modifying therapy for multiple sclerosis in adults: An Australian and New Zealand perspective Part 2 New and emerging therapies and their efficacy. *J Clin Neurosci*, 21, pp.1847–1856.

Bruijn, L. I. and Cleveland, D. W., 1996. Mechanisms of selective motor neuron death in ALS: insights from transgenic mouse models of motor neuron disease. *Neuropathol Appl Neurobiol*, 22, pp.373–387.

Bruijn, L. I., et al., 1998. Aggregation and motor neuron toxicity of an ALS-linked SOD1 mutant independent from wild-type SOD1. *Science*, 281, pp.1851–1854.

Carri, M. T. and Cozzolino, M., 2011. SOD1 and mitochondria in ALS: a dangerous liaison. *J Bioenerg Biomembr*, 43, pp.593–599.

Carrithers, M. D., 2014. Update on disease-modifying treatments for multiple sclerosis. *Clin Ther*, 36, pp.1938–1945.

Celardo, I., Martins, L. M. and Gandhi, S., 2014. Unravelling mitochondrial pathways to Parkinson's disease. *Br J Pharmacol*, 171, pp.1943–1957.

Chan, T. L., Greenawalt, J. W. and Pedersen, P. L., 1970. Biochemical and ultrastructural properties of a mitochondrial inner membrane fraction deficient in outer membrane and matrix activities. *J Cell Biol*, 45, pp.291–305.

Chaturvedi, R. K. and Beal, M. F., 2013. Mitochondria targeted therapeutic approaches in Parkinson's and Huntington's diseases. *Mol Cell Neurosci*, 55, pp.101–114.

Chio, A., et al., 2013. Global epidemiology of amyotrophic lateral sclerosis: a systematic review of the published literature. *Neuroepidemiology*, 41, pp.118–130.

Collard, J. F., Cote, F. and Julien, J. P., 1995. Defective axonal transport in a transgenic mouse model of amyotrophic lateral sclerosis. *Nature*, 375, pp.61–64.

Corcia, P. and Gordon, P. H., 2012. Amyotrophic lateral sclerosis and the clinical potential of dexpramipexole. *Ther Clin Risk Manag*, 8, pp.359–366.

Cozzolino, M. and Carri, M. T., 2012. Mitochondrial dysfunction in ALS. *Prog Neurobiol*, 97, pp.54–66.

Cross, A. H. and Naismith, R. T., 2014. Established and novel disease-modifying treatments in multiple sclerosis. *J Intern Med*, 275, pp.350–363.

Cudkowicz, M., et al., 2011. The effects of dexpramipexole (KNS-760704) in individuals with amyotrophic lateral sclerosis. *Nat Med*, 17, pp.1652–1656.

Cudkowicz, M. E., et al., 2013. Dexpramipexole versus placebo for patients with amyotrophic lateral sclerosis (EMPOWER): A randomised, double-blind, phase 3 trial. *Lancet Neurol*, 12, pp.1059–1067.

Daniel, B., et al., 2013. Riluzole increases the rate of glucose transport in L6 myotubes and NSC-34 motor neuron-like cells via AMPK pathway activation. *Amyotroph Lateral Scler Frontotemporal Degener*, 14, pp.434–443.

Danzeisen, R., et al., 2006. Targeted antioxidative and neuroprotective properties of the dopamine agonist pramipexole and its nondopaminergic enantiomer SND919CL2x [(+)2-amino-4,5,6,7-tetrahydro-6-Lpropylamino-benzathiazole dihydrochloride]. *J Pharmacol Exp Ther*, 316, pp.189–199.

Dias, V., Junn, E. and Mouradian, M. M., 2013. The role of oxidative stress in Parkinson's disease. *J Parkinsons Dis*, 3, pp.461–491.

Doble, A., 1996. The pharmacology and mechanism of action of riluzole. *Neurology*, 47, pp.S233–S241.

Doble, A. and Kennel, P., 2000. Animal models of amyotrophic lateral sclerosis. *Amyotroph Lateral Scler Other Motor Neuron Disord*, 1, pp.301–312.

Dorado, L., Millan, M. and Davalos, A., 2014. Reperfusion therapies for acute ischemic stroke: An update. *Curr Cardiol Rev*, 10, pp.327–335.

Dunckley, T., et al., 2007. Whole-genome analysis of sporadic amyotrophic lateral sclerosis. *N Engl J Med*, 357, pp.775–788.

Faes, L. and Callewaert, G., 2011. Mitochondrial dysfunction in familial amyotrophic lateral sclerosis. *J Bioenerg Biomembr*, 43, pp.587–592.

Fan, L. Y. and Chiu, M. J., 2014. Combotherapy and current concepts as well as future strategies for the treatment of Alzheimer's disease. *Neuropsychiatr Dis Treat*, 10, pp.439–451.

Ferini-Strambi, L. and Marelli, S., 2014. Pharmacotherapy for restless legs syndrome. *Expert Opin Pharmacother*, 15, pp.1127–1138.

Ferrari-Toninelli, G., et al., 2010. Mitochondria-targeted antioxidant effects of S(−) and R(+) pramipexole. *BMC Pharmacol*, 10, p.2.

Frizzo, M. E., et al., 2004. Riluzole enhances glutamate uptake in rat astrocyte cultures. *Cell Mol Neurobiol*, 24, pp.123–128.

Gonzalez de Aguilar, J. L., et al., 1999. A mouse model of familial amyotrophic lateral sclerosis expressing a mutant superoxide dismutase 1 shows evidence of disordered transport in the vasopressin hypothalamo-neurohypophysial axis. *Eur J Neurosci*, 11, pp.4179–4187.

Gonzalez-Lima, F., Barksdale, B. R. and Rojas, J. C., 2014. Mitochondrial respiration as a target for neuroprotection and cognitive enhancement. *Biochem Pharmacol*, 88, pp.584–593.

Gordon, P. H. and Meininger, V., 2011. How can we improve clinical trials in amyotrophic lateral sclerosis? *Nat Rev Neurol*, 7, pp.650–654.

Gordon, P. H., 2013a. Amyotrophic lateral sclerosis: An update for 2013 clinical features, pathophysiology, management and therapeutic trials. *Aging Dis*, 4, pp.295–310.

Gordon, P. H., 2013b. The murky path to drug discovery in ALS becomes clearer. *Lancet Neurol*, 12, pp.1037–1038.

Greenway, M. J., et al., 2006. ANG mutations segregate with familial and 'sporadic' amyotrophic lateral sclerosis. *Nat Genet*, 38, pp.411–413.

Gribkoff, V. K. and Bozik, M. E., 2008. KNS-760704 [(6R)-4,5,6,7-tetrahydro-N6-propyl-2, 6-benzothiazole-diamine dihydrochloride monohydrate] for the treatment of amyotrophic lateral sclerosis. *CNS Neurosci Ther*, 14, pp.215–226.

He, B. P. and Strong, M. J., 2000. Motor neuronal death in sporadic amyotrophic lateral sclerosis (ALS) is not apoptotic. A comparative study of ALS and chronic aluminium chloride neurotoxicity in New Zealand white rabbits. *Neuropathol Appl Neurobiol*, 26, pp/.150–160.

He, Y., et al., 2002. Neuroprotective agent riluzole potentiates postsynaptic GABA(A) receptor function. *Neuropharmacology*, 42, pp.199–209.

Hubert, J. P., et al., 1994. Antagonism by riluzole of entry of calcium evoked by NMDA and veratridine in rat cultured granule cells: evidence for a dual mechanism of action. *Br J Pharmacol*, 113, pp.261–267.

Julien, J. P. and Kriz, J., 2006. Transgenic mouse models of amyotrophic lateral sclerosis. *Biochim Biophys Acta*, 1762, pp.1013–1024.

Kawamata, H. and Manfredi, G., 2010. Import, maturation, and function of SOD1 and its copper chaperone CCS in the mitochondrial intermembrane space. *Antioxid Redox Signal*, 13, pp.1375–1384.

Ko Y. H., et al., 2003. Mitochondrial ATP synthasome. Cristae-enriched membranes and a multiwell detergent screening assay yield dispersed single complexes containing the ATP synthase and carriers for Pi and ADP/ATP. *J Biol Chem*, 278, pp.12305–12309.

Kuo, J. J., et al., 2006. Essential role of the persistent sodium current in spike initiation during slowly rising inputs in mouse spinal neurones. *J Physiol*, 574, pp.819–834.

Kushida, C. A., 2006. Pramipexole for the treatment of restless legs syndrome. *Expert Opin Pharmacother*, 7, pp.441–451.

Lacomblez, L., et al., 1996a. Dose-ranging study of riluzole in amyotrophic lateral sclerosis. Amyotrophic Lateral Sclerosis/Riluzole Study Group II. *Lancet*, 347, pp.1425–1431.

Lacomblez, L., et al., 1996b. A confirmatory dose-ranging study of riluzole in ALS. ALS/ Riluzole Study Group-II. *Neurology*, 47, pp.S242–S250.

Lamanauskas, N. and Nistri, A., 2008. Riluzole blocks persistent Na+ and Ca2+ currents and modulates release of glutamate via presynaptic NMDA receptors on neonatal rat hypoglossal motoneurons in vitro. *Eur J Neurosci*, 27, pp.2501–2514.

Leblond, C. S., et al., 2014. Dissection of genetic factors associated with amyotrophic lateral sclerosis. *Exp Neurol*, 262PB, pp.91–101.

MacIver, M. B., et al., 1996. Riluzole anesthesia: Use-dependent block of presynaptic glutamate fibers. *Anesthesiology*, 85, pp.626–634.

Mackenzie, I. R., et al., 2007. Pathological TDP-43 distinguishes sporadic amyotrophic lateral sclerosis from amyotrophic lateral sclerosis with SOD1 mutations. *Ann Neurol*, 61, pp.427–434.

Martin, L. J., 2010. Olesoxime, a cholesterol-like neuroprotectant for the potential treatment of amyotrophic lateral sclerosis. *IDrugs*, 13, pp.568–580.

McCormack, P. L. and Siddiqui, M. A., 2007. Pramipexole: In restless legs syndrome. *CNS Drugs*, 21, pp.429–437; 438–440.

Miller, R. G., et al., 1996. Clinical trials of riluzole in patients with ALS. ALS/Riluzole Study Group-II. *Neurology*, 47, S86–S90; S90–S92.

Miller, R. G., et al., 2003. Riluzole for amyotrophic lateral sclerosis (ALS)/motor neuron disease (MND). *Amyotroph Lateral Scler Other Motor Neuron Disord*, 4, pp.191–206.

Miller, R. G., et al., 2007. Riluzole for amyotrophic lateral sclerosis (ALS)/motor neuron disease (MND). *Cochrane Database Syst Rev*, CD001447.

Miller, R. G., Mitchell, J. D. and Moore, D. H., 2012. Riluzole for amyotrophic lateral sclerosis (ALS)/motor neuron disease (MND). *Cochrane Database Syst Rev*, 3, p.CD001447.

Morciano, G., et al., 2015. Molecular identity of the mitochondrial permeability transition pore and its role in ischemia-reperfusion injury. *J Mol Cell Cardiol*, 78, pp.142–153.

Mullane, K. and Williams, M., 2013. Alzheimer's therapeutics: continued clinical failures question the validity of the amyloid hypothesis-but what lies beyond? *Biochem Pharmacol*, 85, pp.289–305.

Muller, W. E., et al., 2010. Mitochondrial dysfunction: common final pathway in brain aging and Alzheimer's disease—therapeutic aspects. *Mol Neurobiol*, 41, pp.159–171.

Muyderman, H. and Chen, T., 2014. Mitochondrial dysfunction in amyotrophic lateral sclerosis—a valid pharmacological target? *Br J Pharmacol*, 171, pp.2191–2205.

Nakano, R., et al., 1994. A novel mutation in Cu/Zn superoxide dismutase gene in Japanese familial amyotrophic lateral sclerosis. *Biochem Biophys Res Commun*, 200, pp.695–703.

Nakayama, H., et al., 2009. Neuroprotective effects of pramipexole against tunicamycin-induced cell death in PC12 cells. *Clin Exp Pharmacol Physiol*, 36, pp.1183–1185.

Neumann, M., et al., 2006. Ubiquitinated TDP-43 in frontotemporal lobar degeneration and amyotrophic lateral sclerosis. *Science*, 314, pp.130–133.

Orrell, R. W., Marklund, S. L. and deBelleroche, J. S., 1997. Familial ALS is associated with mutations in all exons of SOD1: a novel mutation in exon 3 (Gly72Ser). *J Neurol Sci*, 153, pp.46–49.

Peixoto, P. M., Dejean, L. M. and Kinnally, K. W., 2012. The therapeutic potential of mitochondrial channels in cancer, ischemia-reperfusion injury, and neurodegeneration. *Mitochondrion*, 12, pp.14–23.

Perez-Lloret, S. and Rascol, O., 2010. Dopamine receptor agonists for the treatment of early or advanced Parkinson's disease. *CNS Drugs*, 24, pp.941–968.

Rainero, I., et al., 1994. SOD1 missense mutation in an Italian family with ALS. *Neurology*, 44, pp.347–349.

Rao, V. K., Carlson, E. A. and Yan, S. S., 2014. Mitochondrial permeability transition pore is a potential drug target for neurodegeneration. *Biochim Biophys Acta*, 1842, pp.1267–1272.

Renton, A. E., Chio, A. and Traynor, B. J., 2014. State of play in amyotrophic lateral sclerosis genetics. *Nat Neurosci*, 17, pp.17–23.

Richter, J. M., Schaefer, M. and Hill, K., 2014. Riluzole activates TRPC5 channels independently of PLC activity. *Br J Pharmacol*, 171, pp.158–170.

Robberecht, W., et al., 1994. Cu/Zn superoxide dismutase activity in familial and sporadic amyotrophic lateral sclerosis. *J Neurochem*, 62, pp.384–387.

Rosen, D. R., et al., 1993. Mutations in Cu/Zn superoxide dismutase gene are associated with familial amyotrophic lateral sclerosis. *Nature*, 362, pp.59–62.

Rosen, D. R., et al., 1994. A frequent ala 4 to val superoxide dismutase-1 mutation is associated with a rapidly progressive familial amyotrophic lateral sclerosis. *Hum Mol Genet*, 3, pp.981–987.

Rosenblum, W. I., 2014. Why Alzheimer trials fail: removing soluble oligomeric beta amyloid is essential, inconsistent, and difficult. *Neurobiol Aging*, 35, pp.969–974.

Rudnicki, S. A., et al., 2013. Dexpramipexole effects on functional decline and survival in subjects with amyotrophic lateral sclerosis in a Phase II study: Subgroup analysis of demographic and clinical characteristics. *Amyotroph Lateral Scler Frontotemporal Degener*, 14, pp.44–51.

Sacchetti, S., et al., 2013. F1FO ATPase vesicle preparation and technique for performing patch clamp recordings of submitochondrial vesicle membranes. *J Vis Exp*, e4394.

Sapp, P. C., et al., 1995. Identification of three novel mutations in the gene for Cu/Zn superoxide dismutase in patients with familial amyotrophic lateral sclerosis. *Neuromuscul Disord*, 5, pp.353–357.

Sasaki, S. and Iwata, M., 2007. Mitochondrial alterations in the spinal cord of patients with sporadic amyotrophic lateral sclerosis. *J Neuropathol Exp Neurol*, 66, pp.10–16.

Sayeed, I., et al., 2006. Patch clamp reveals powerful blockade of the mitochondrial permeability transition pore by the D2-receptor agonist pramipexole. *FASEB J*, 20, pp.556–558.

Schapira, A. H., et al., 2014. Slowing of neurodegeneration in Parkinson's disease and Huntington's disease: future therapeutic perspectives. *Lancet*, 384, pp.545–555.

Schneider, L. S., et al., 2014. Clinical trials and late-stage drug development for Alzheimer's disease: An appraisal from 1984 to 2014. *J Intern Med*, 275, pp.251–283.

Scott, S., et al., 2008. Design, power, and interpretation of studies in the standard murine model of ALS. *Amyotroph Lateral Scler*, 9, pp.4–15.

Siddique, T. and Hentati, A., 1995. Familial amyotrophic lateral sclerosis. *Clin Neurosci*, 3, pp.338–347.

Siddique, T., Nijhawan, D. and Hentati, A., 1996. Molecular genetic basis of familial ALS. *Neurology*, 47, S27–S34; S34–S5.

Sobow, T., 2010. Combination treatments in Alzheimer's disease: Risks and benefits. *Expert Rev Neurother*, 10, pp.693–702.

Swerdlow, R. H., et al., 1998. Mitochondria in sporadic amyotrophic lateral sclerosis. *Exp Neurol*, 153, pp.135–142.

Swerdlow, R. H., 2007. Treating neurodegeneration by modifying mitochondria: Potential solutions to a "complex" problem. *Antioxid Redox Signal*, 9, pp.1591–1603.

Tan, C. F., et al., 2007. TDP-43 immunoreactivity in neuronal inclusions in familial amyotrophic lateral sclerosis with or without SOD1 gene mutation. *Acta Neuropathol*, 113, pp.535–542.

Tsuda, T., et al., 1994. Analysis of the functional effects of a mutation in SOD1 associated with familial amyotrophic lateral sclerosis. *Neuron*, 13, pp.727–736.

Urbani, A. and Belluzzi, O., 2000. Riluzole inhibits the persistent sodium current in mammalian CNS neurons. *Eur J Neurosci*, 12, pp.3567–3574.

Vieira, F. G., et al., 2014. Dexpramipexole is ineffective in two models of ALS related neurodegeneration. *PLoS One*, 9, p.e91608.

Wang, D. B., et al., 2014. p53 and mitochondrial function in neurons. *Biochim Biophys Acta*, 1842, pp.1186–1197.

Wang, H., et al., 2008. R+ pramipexole as a mitochondrially focused neuroprotectant: initial early phase studies in ALS. *Amyotroph Lateral Scler*, 9, pp.50–58.

Wardlaw, J. M., et al., 2014. Thrombolysis for acute ischaemic stroke. *Cochrane Database Syst Rev*, 7, p.CD000213.

Williamson, T. L., et al., 2000. Toxicity of ALS-linked SOD1 mutants. *Science*, 288, p.399.

Wu, N., et al., 2005. Persistent sodium currents in mesencephalic v neurons participate in burst generation and control of membrane excitability. *J Neurophysiol*, 93, pp.2710–2722.

Wu, S. N. and Li, H. F., 1999. Characterization of riluzole-induced stimulation of large-conductance calcium-activated potassium channels in rat pituitary GH3 cells. *J Investig Med*, 47, pp.484–495.

Xu, L., Enyeart, J. A. and Enyeart, J. J., 2001. Neuroprotective agent riluzole dramatically slows inactivation of Kv1.4 potassium channels by a voltage-dependent oxidative mechanism. *J Pharmacol Exp Ther*, 299, pp.227–237.

Xu, Z., et al., 2004. Mitochondrial degeneration in amyotrophic lateral sclerosis. *J Bioenerg Biomembr*, 36, pp.395–399.

Yaghi, S., Eisenberger, A. and Willey, J. Z., 2014. Symptomatic intracerebral hemorrhage in acute ischemic stroke after thrombolysis with intravenous recombinant tissue plasminogen activator: a review of natural history and treatment. *JAMA Neurol*, 71, pp.1181–1185.

Yamamoto, M. and Schapira, A. H., 2008. Dopamine agonists in Parkinson's disease. *Expert Rev Neurother*, 8, pp.671–677.

Yang, Y. M., et al., 2013. A small molecule screen in stem-cell-derived motor neurons identifies a kinase inhibitor as a candidate therapeutic for ALS. *Cell Stem Cell*, 12, pp.713–726.

Yin, F., Boveris, A. and Cadenas, E., 2014. Mitochondrial energy metabolism and redox signaling in brain aging and neurodegeneration. *Antioxid Redox Signal*, 20, pp.353–371.

Yvon, C., Czarnecki, A. and Streit, J., 2007. Riluzole-induced oscillations in spinal networks. *J Neurophysiol*, 97, pp.3607–3620.

Zhu, S., et al., 2002. Minocycline inhibits cytochrome c release and delays progression of amyotrophic lateral sclerosis in mice. *Nature*, 417, pp.74–78.

Zsurka, G. and Kunz, W. S., 2013. Mitochondrial involvement in neurodegenerative diseases. *IUBMB Life*, 65, pp.263–272.

15

MITOCHONDRIA AS A THERAPEUTIC TARGET FOR ALZHEIMER'S DISEASE

Clara Hiu-Ling Hung

Laboratory of Neurodegenerative Diseases, Department of Anatomy, Li Ka Shing Faculty of Medicine, Hong Kong, China

Sally Shuk-Yee Cheng

Laboratory of Neurodegenerative Diseases, Department of Anatomy, Li Ka Shing Faculty of Medicine, Hong Kong, China

Simon Ming-Yuen Lee

Institute of Chinese Medical Sciences, University of Macau, Macau SAR, China

Raymond Chuen-Chung Chang

Laboratory of Neurodegenerative Diseases, Department of Anatomy, Li Ka Shing Faculty of Medicine and Research Centre of Heart, Brain, Hormone and Healthy Aging, Medicine, Li Ka Shing Faculty of Medicine and State Key Laboratory of Brain and Cognitive Sciences, University of Hong Kong, Pokfulam, Hong Kong SAR, China

15.1 INTRODUCTION

Alzheimer's disease (AD) is a devastating age-related neurodegenerative disorder currently affecting more than 44 million people worldwide (Bright Focus Foundation, 2014) and 5.2 million people in the United States (Alzheimer's Association, 2013). AD is characterized by progressive memory loss and cognitive impairment as a result of neurodegeneration in various brain regions, especially in the hippocampus

The Functions, Disease-Related Dysfunctions, and Therapeutic Targeting of Neuronal Mitochondria, First Edition. Edited by Valentin K. Gribkoff, Elizabeth A. Jonas and J. Marie Hardwick.

(Ball, 1977; Hyman et al., 1984). The etiology of AD remains largely unclear despite extensive research efforts on the disorder. Mitochondrial abnormality is a predominant pathological feature in aging and neurodegenerative diseases including AD. In this chapter, we will review AD-related pathological changes in mitochondrial structure and function and discuss the potential of targeting mitochondria as an effective treatment for the disease.

15.2 MITOCHONDRIAL ABNORMALITIES AND DYSFUNCTION IN ALZHEIMER'S DISEASE

Mitochondria are cytoplasmic organelles that are vital for regulating cell survival and death pathways including energy production, calcium ion (Ca^{2+}) homeostasis, and regulation of apoptosis. Mitochondria are specialized membrane-bound organelles comprised of the outer mitochondrial membrane (OMM), the intermembrane space, the inner mitochondrial membrane (IMM), which is folded into numerous cristae, and the mitochondrial matrix (Fig. 15.1). The electron transport chain (ETC) located at the IMM is essential for the generation of ATP via oxidative phosphorylation. The abundance and distribution of mitochondria varies among tissues and cells depending on their energy requirement. With the high energy demand of synaptic transmission and excitability, neurons are particularly susceptible to mitochondrial impairments. A broad range of AD-related mitochondrial abnormalities have been documented, including defective mitochondrial structure and dynamics, impaired activity of the ETC and excessive production of reactive oxygen species (ROS) (Reddy and Reddy, 2011; Silva et al., 2012).

15.2.1 Mitochondrial Morphology and Ultrastructure

Alterations in mitochondrial morphology and structure from filamentous to round swelling mitochondria with abnormal cristae have been observed in postmortem brain from patients with AD using electron microscopy (Baloyannis, Costa, and Michmizos, 2004; Baloyannis, 2006; Hirai et al., 2001). Recently, Xie et al. (2013) found a massive mitochondrial loss with fragmented mitochondria and mitochondrial swelling near amyloid plaques in vivo in a transgenic AD mouse model using intravital imaging.

15.2.2 Beta Amyloid, Tau, and Mitochondria

Accumulation of beta amyloid (Aβ) senile plaques is one of the most characterized histopathological hallmarks in AD. Direct deposition of Aβ in mitochondria has been observed in brains from transgenic AD mice and patients with AD (Caspersen et al., 2005; Lustbader et al., 2004; Manczak et al., 2006), with Aβ binding to alcohol dehydrogenase to inhibit mitochondrial activity such as the generation of ATP (Lustbader et al., 2004). In addition, $A\beta_{25-35}$ inhibits the activity of complex IV within the mitochondrial respiratory complex (Canevari, Clark, and Bates, 1999).

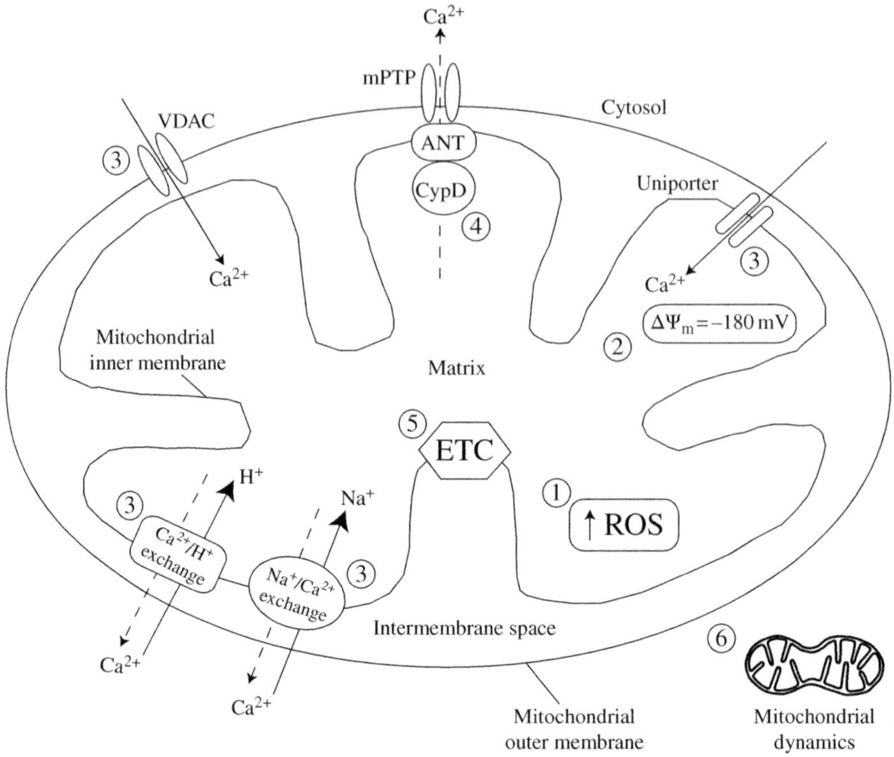

FIGURE 15.1 Current and potential pharmacological targets in mitochondria. Current and potential therapeutic target are labeled with numbers. (1) Mitochondria-targeted antioxidants aim to abolish oxidative stress; (2) manipulation of mitochondrial membrane potential and (3) modulation of Ca^{2+} signaling pathways to attenuate Ca^{2+} overload in mitochondria; (4) inhibition of opening of the mitochondrial permeability transition pore; (5) enhancement of mitochondrial metabolism, and (6) manipulation of mitochondrial dynamics (fission, fusion, and transport). $\Delta\Psi_m$, mitochondrial membrane potential; ANT, adenine nucleotide translocase; CypD, cyclophilin D; ETC, electron transport chain; mPTP, mitochondrial permeability transition pore; ROS, reactive oxygen species; VDAC, voltage-dependent anion channel.

Interestingly, mitochondria may be an amyloid precursor protein (APP) processing site, as the γ-secretase complex involved in the cleavage of APP into Aβ has been found in rat brain mitochondria (Hansson et al., 2004).

Tau is a microtubule-associated protein that stabilizes the cytoskeleton. In AD, hyperphosphorylated tau protein is the major component of the neurofibrillary tangles observed in brains of patients with AD. A cytotoxic N-terminal NH2-truncated tau fragment has been found to be enriched in mitochondria from synaptosomes of AD brains, with its level of accumulation correlating with synaptic impairment (Amadoro et al., 2010). In addition, overexpression of the Asp421-cleaved truncated tau has been shown to induce mitochondrial fragmentation in primary cortical neurons (Quintanilla et al., 2012). Asp421-cleaved tau enhanced mitochondrial dysfunction induced by Aβ_{1-40}, resulting in a reduction in mitochondrial membrane potential, increased levels of mitochondrial oxidative stress and impaired mitochondrial transport (Quintanilla et al., 2012). Furthermore, injection of tau oligomers has been shown to reduce the level of complex I of the ETC in mice brain in vivo (Lasagna-Reeves et al., 2011).

In sum, Aβ and tau are both closely linked to mitochondrial deficits found in AD, making mitochondria an important therapeutic target.

15.2.3 Defective Mitochondria at Synapses

Synaptic degeneration is an early pathological feature in AD. Indeed, recent studies have suggested that synaptic degeneration is closely associated with the level of cognitive impairment in AD. Mitochondria play a prominent role in maintaining proper synaptic activity. ATP is needed at the synapses for different steps of synaptic transmission such as tethering and refilling of synaptic vesicles. In addition, Ca^{2+} homeostasis at the synapses is important for docking of synaptic vesicles as well as reuptake of neurotransmitters. Synaptic mitochondria have been shown to be more vulnerable to Aβ accumulation that nonsynaptic mitochondria (Du et al., 2010). Synaptic mitochondrial malfunction can lead to severe consequences in synaptic transmission, leading to neurodegeneration.

15.2.4 Impaired Mitochondrial Dynamics

Mitochondrial fission, fusion, and transport, collectively known as *mitochondrial dynamics*, are important in the regulation of mitochondrial morphology, turnover, quality control, and distribution within the cell.

Recently, various studies have implicated abnormalities in the fission/fusion balance in different AD models. Mitochondrial fragmentation resulting from extensive fission has been observed in neurons treated with Aβ_{25-35} (Barsoum et al., 2006) and in primary neuronal cultures derived from Tg2576 transgenic AD mice (Calkins et al., 2011). In addition, overexpression of APP has been shown to impair the fission/fusion balance, which results in mitochondrial fragmentation and abnormal distribution of mitochondria in neurons (Wang et al., 2008). Aβ has been shown to cause S-nitrosylation of the mitochondrial fission protein dynamin related protein 1 (Drp1), leading to extensive fission and synaptic loss (Cho et al., 2009). In summary, there appears to be a shift toward fission, resulting in abnormal mitochondrial morphology and distribution.

Because the synthesis of mitochondria mainly takes place in the soma, mitochondria need to be transported to synapses. The transport of mitochondria within neurons is regulated by the Ca^{2+}-dependent kinesin motors Miro and Milton (Liu and Hajnoczky, 2009). Aβ_{1-42} has been shown to induce acute impairment of mitochondrial transport in neurons (Rui et al., 2006). Anterograde transport of mitochondria has also been found to be impaired in primary neurons from APP Tg2576 transgenic AD mice (Calkins et al., 2011).

15.2.5 Oxidative Stress

Mitochondria are regulated by both the mitochondrial and nuclear genome. The majority of mitochondrial proteins required for oxidative phosphorylation are encoded by the nuclear genome. The number of mitochondrial DNA copies and number of mitochondria per cell are highly related to the energy demands of the cell. In AD, abnormal expression of oxidative phosphorylation genes has been observed in postmortem AD brains and APP transgenic mice (Manczak et al., 2004; Reddy et al., 2004).

Oxidative stress has been suggested as one of the earliest pathological features in AD (Nunomura et al., 2001). Decreased ATP production, increased ROS production, and lipid peroxidation have been consistently observed in postmortem brains of patients with AD and in transgenic AD mice. ROS are generated as by-products within mitochondria during oxidative phosphorylation. However, as ROS begins to accumulate within the mitochondria, a number of stress-signaling pathways are activated, resulting in impairment of mitochondrial function and overall cell viability. Therefore, mitochondria are both the primary source and target of oxidative stress.

Specifically in AD, Aβ has been shown to enter mitochondria and induce ROS production. A number of studies have found mitochondrial impairment to promote the formation of Aβ by altering APP processing (Gabuzda et al., 1994; Webster et al., 1998). Oxidative stress has also been considered to alter tau processing by activating glycogen synthase kinase 3, which is an important kinase for tau phosphorylation (Lovell et al., 2004). With the presence of oxidative stress preceding that of severe Aβ and tau pathology in AD, targeting mitochondrial production of ROS may be an effective therapeutic target in AD and other neurodegenerative disorders.

15.2.6 Ca^{2+} Dysregulation in Mitochondria

Ca^{2+} homeostasis in mitochondria is important for various cellular processes. The uptake of Ca^{2+} by mitochondria is essential for the metabolic enzymes involved in the Krebs cycle. Also, mitochondrial Ca^{2+} is important for regulating axonal transport of mitochondria (Liu and Hajnoczky, 2009). However, Ca^{2+} overload in mitochondria can induce the opening of the mitochondrial permeability transition pore, leading to apoptosis.

In AD, perturbed Ca^{2+} homeostasis in mitochondria has been implicated as an important pathological feature. Firstly, mitochondria isolated from the fibroblasts of patients with AD showed a reduced Ca^{2+} buffering capacity when compared to age-matched controls (Kumar, Dunlop, and Richardson, 1994). In addition, mitochondria derived from mice overexpressing human APP have been shown to exhibit a lower Ca^{2+} capacity when compared to mitochondria from controls (Du et al., 2008). $A\beta_{1-42}$ oligomers have also been shown to induce Ca^{2+} overload within mitochondria (Sanz-Blasco et al., 2008). Lastly, many AD-related genetic mutations are linked with Ca^{2+} regulation. For example, an AD-related gene mutation in CALMH1 has also been shown to cause slower kinetics of mitochondrial Ca^{2+} uptake and release (Moreno-Ortega et al., 2010).

15.2.7 Mitochondrial Permeability Transition Pore

The mitochondrial permeability transition pore (mPTP) is a highly conductive multimeric complex consisting of the voltage dependent anion channel (VDAC) in the OMM, the integral protein adenine nucleotide translocase (ANT) in the IMM and cyclophilin D (CypD) in the matrix and recently the c-subunit of complex V (Alavian et al., 2014; Azarashvili et al., 2014; Bonora et al., 2013). Excessive oxidative stress and/or mitochondrial Ca^{2+} overload can trigger the opening of mPTP, leading to mitochondrial permeability transition. Opening of the mPTP leads to depolarization of the mitochondrial membrane potential and allows uncontrolled entry of solutes and water into the mitochondrial matrix.

Recently, interaction between VDAC1 and Aβ and between VDAC1 and phosphorylated tau was observed (Manczak and Reddy, 2012). The formation of these complexes might block the mitochondrial pore on the OMM and impair proper VDAC conductance as well as transport of mitochondrial proteins and other metabolites, resulting in impaired mitochondrial function.

15.3 MITOCHONDRIA AS A DRUG TARGET

Mitochondrial dysfunction has been implicated in various human pathological conditions such as neurodegenerative diseases, metabolic diseases (type 2 diabetes), cardiovascular diseases, as well as cancer. Modulation of mitochondrial function is emerging as an attractive therapeutic approach for a number of diseases, including AD. This review will focus on how mitochondrial-targeted therapeutic approaches may be beneficial specifically for AD.

15.3.1 Targeting Drugs to Mitochondria

Mitochondria are membrane-bound compartmentalized organelles. Although the porous OMM is permeable to molecules below 5 kDa, entry to the mitochondrial matrix is tightly controlled. Currently, there are two main approaches to target compounds specifically to mitochondria: conjugation to (i) lipophilic cations and (ii) mitochondria-targeted peptides (Smith, Hartley, and Murphy, 2011) (Table 15.1).

Small molecules conjugated with lipophilic cations such as triphenylphosphonium (TPP^+) can accumulate in the mitochondrial matrix as driven by the negative mitochondrial membrane potential (–180 mV). A wide range of mitochondria-targeted antioxidants currently available are conjugated to TPP^+, including MitoTEMPO, MitoVitE, and MitoQ (Dikalova et al., 2010; Kelso et al., 2001; Smith et al., 1999; Smith et al., 2011).

Bioactive molecules can also be targeted to mitochondria by binding with mitochondria-targeted peptides such as the mitochondrial-penetrating peptides and Szeto-Schiller (SS) peptides (Horton et al., 2008; Szeto, 2006; Zhao et al., 2004). The charge and hydrophilic feature of mitochondrial-penetrating peptides facilitates their accumulation into mitochondria, whereas the SS peptides are the first class of compounds that specifically target and accumulate in the IMM independent of

TABLE 15.1 List of Current Mitochondria-targeted Neuroprotective Compounds for Alzheimer's Disease.

Compound	Mode of action/ Target	AD model	Reference
MitoQ	Antioxidant	Primary neurons (Tg 2576 line AD mice) with $A\beta_{25-35}$ treatment	Manczak et al., 2010
		N2a neuroblastoma cells with $A\beta_{25-35}$ treatment	Manczak et al., 2010
		Primary cortical neurons (3xTg-AD mice)	McManus et al., 2011
		3xTg-AD mice	McManus et al., 2011
		Transgenic AD *C. elegans*	Ng et al., 2004
SS31	Antioxidant	Primary neurons (Tg 2576 line AD mice) with $A\beta_{25-35}$ treatment	Manczak et al., 2010
		N2a neuroblastoma cells with $A\beta_{25-35}$ treatment	Manczak et al., 2010
		Primary neurons (Tg 2576 line AD mice)	Calkins et al., 2011
R(+) Pramipexole	Antioxidant	SH-SY5Y neuroblastoma cells with $A\beta_{25-35}$ treatment	Abramova et al., 2002
			Phase 2 Clinical Trial (NCT01388478)
FCCP	$\Delta\Psi_m$	Rat cerebellar granule and cortical neurons with $A\beta_{1-42}$ oligomers treatment	Sanz-Blasco et al., 2008
NSAIDs	$\Delta\Psi_m$	Rat cerebellar granule with $A\beta_{1-42}$ oligomers treatment	Sanz-Blasco et al., 2008
Dimebon	$\Delta\Psi_m$	Rat liver mitochondria with $A\beta_{25-35}$ treatment	Bachurin et al., 2003
Cyclosporine A	mPTP	Mouse cortical mitochondria (APP transgenic mice)	Du et al., 2008
Mdivi-1	Mitochondrial Fission	AD cybrid cells	Gan et al., 2014
Methylene Blue	Mitochondrial metabolism	APP/PS1 AD transgenic mice	Paban et al., 2014
AC-1204	Mitochondrial metabolism		Phase 2/3 Clinical Trial (NCT01741194)

$\Delta\Psi_m$, mitochondrial membrane potential; AD, Alzheimer's disease; APP, amyloid precursor protein; FCCP, Carbonyl cyanide p-trifluoro-methoxy-phenyl-hydrazine; Mdivi-1, mitochondrial division inhibitor 1; mPTP, mitochondria permeability transition pore; NSAIDs, Nonsteroidal anti-inflammatory drugs; PS1, presenilin 1.

TABLE 15.2 List of Potential Mitochondria-targeted Neuroprotective Compounds.

Compound/ Intervention	Mode of action/ Target	Model	Reference
MitoTEMPO	Mitochondrial superoxide scavenger	Angiotensin II-induced hypertension	Dikalova et al., 2010
Mdivi-1	Mitochondrial Fission	HT-22 cells/Primary cortical neurons (Glutamate/Oxygen-glucose deprivation)	Grohm et al., 2010
Mitochondrial Fusion Promoter M1 hydrazone	Mitochondrial fusion	SH-SY5Y neuroblastoma cells (MPP^+)	Wang et al., 2014
Minocycline	VDAC	Rat cerebellar granule neurons (NMDA)	Garcia-Martinez et al., 2010
KB-R7943	Mitochondrial Na^+/Ca^{2+} exchanger	Rat cerebellar granule neurons (Glutamate)	Storozhevykh et al., 2009
Photomodulation	Mitochondrial metabolism	C2C12 myotubes (Electrical stimulation)	Xu et al., 2008
Transcranial Low Light/Laser Therapy	Mitochodnrial metabolism	Patients with traumatic brain injury	Naeser et al., 2011
		Healthy human subject	Barrett & Gonzalez-Lima, 2013

Mdivi-1, mitochondrial division inhibitor 1; MPP^+, 1-methyl-4-phenylpyridinium.

mitochondrial membrane potential (Szeto, 2006; Zhao et al., 2004) (Table 15.2). This nature of SS peptides is of particular advantage as diseased mitochondria tend to be depolarized and may not be able to take up compounds conjugated to lipophilic cations efficiently. However, the exact mechanism of how SS peptides can be selectively taken up by mitochondria is still unclear. In addition, SS peptides can readily cross the blood–brain barrier, making them a promising neuroprotective agent. An example of compounds that bind with SS peptides is the mitochondria-targeted antioxidant SS31 (Zhao et al., 2005).

Recently, novel delivery systems such as liposome-based MITO-Porters have been developed, which enable delivery of macromolecules into mitochondria via membrane fusion (Yamada et al., 2008).

15.3.2 Mitochondria-Targeted Antioxidants

Because oxidative stress plays a major role in the pathogenesis of AD, antioxidants such as Coenzyme Q10, vitamin E, and lipoic acid have been tested in different clinical trials for AD (Galasko et al., 2012). However, the results have been largely discouraging despite their beneficial effects in in vitro studies.

As mitochondria are the primary source of ROS, several mitochondria-targeted antioxidants such as MitoQ and SS31 have been developed in an attempt to provide potent antioxidant effects. Currently, MitoQ and SS31 are the most studied mitochondria-targeted antioxidants in AD. MitoQ and SS31 have been shown to protect against Aβ toxicity in neurons from an AD mouse model (McManus, Murphy, and Franklin, 2011). In a transgenic *Caenorhabditis elegans* model of AD, MitoQ was also shown to extend the life span and improve overall health (Ng et al., 2014). Furthermore, MitoQ has been shown to prevent cognitive decline and synaptic loss in the 3xTg-AD mouse model (McManus et al., 2011). MitoQ has been tested in various animal and human studies and is in clinical trials as a therapy for Parkinson's disease (NCT00329056).

Pramipexole is a dopamine agonist for treatment of Parkinson's disease. Pramipexole has been shown to be neuroprotective against Aβ-induced cell death in SH-SY5Y neuroblastoma cells (Abramova et al., 2002). Recently, pramipexole was reported to exert its neuroprotective effects by acting as a mitochondria-targeted antioxidant (Ferrari-Toninelli et al., 2010). The R (+) enantiomer of pramipexole is neuroprotective but is devoid of the dopamine agonist effects and was used to target ALS in recent clinical trials (see Chapter 14).

New generations of mitochondria-targeted antioxidants such as MitoPBN, MitoTEMPO, and MitoPeroxidase exert their antioxidant effects by acting as specific ROS scavengers, which may provide more effective antioxidative effects (Dikalova et al., 2010; Filipovska et al., 2005; Murphy et al., 2003). MitoPBN has been found to be protective against neurodegeneration and ischemia as well as to provide improvement of overall life span in rodents through its antioxidant effects and possible inhibition of inflammation (Sheu, Nauduri, and Anders, 2006). Furthermore, MitoTEMPO is a specific superoxide scavenger which has been reported to attenuate angiotensin II-induced hypertension in mice in vivo (Dikalova et al., 2010).

Based on these findings, mitochondria-targeted antioxidants may be a promising neuroprotective agent for AD.

15.3.3 Mitochondrial Ca^{2+} Pathways

Currently, there are several FDA-approved drugs, such as Memantine, as well as drugs in clinical trial for AD that target Ca^{2+} signaling and homeostasis. Mitochondrial Ca^{2+} homeostasis is crucial for proper neuronal function; hence Ca^{2+} influx and efflux pathways in mitochondria may act as potential therapeutic targets for AD (Hung, Ho, and Chang, 2010).

Ca^{2+} overload in mitochondria can induce the opening of mPTP and subsequently trigger apoptotic signaling. Therefore, it is proposed that the risk of mitochondrial Ca^{2+} overload could be decreased by inhibiting Ca^{2+} entry into mitochondria. The uptake of Ca^{2+} into mitochondria is highly dependent on the mitochondrial membrane potential, which is established by the uniporter on the IMM. The uniporter is readily inhibited by Ruthenium Red and is also regulated by adenine nucleotides, plant flavonoids, and protein kinases such as the p38 mitogen-activated protein (MAP) kinase and protein kinase C (Bernardi, 1999; Litsky and Pfeiffer, 1997; Montero et al., 2002, 2004). FCCP, an uncoupler and protonphore, has been shown to depolarize the mitochondrial membrane potential, thereby inhibiting mitochondrial Ca^{2+} elevation elicited by $A\beta_{1\text{-}42}$ oligomers (Sanz-Blasco et al., 2008). A number of potential neuroprotective agents including minocycline and nonsteroidal anti-inflammatory drugs (NSAIDs) such as salicylate, R-flurbiprofen, and indomethacin have been shown to inhibit mitochondrial Ca^{2+} uptake by inducing mild mitochondrial depolarization (Garcia-Martinez et al., 2010; Sanz-Blasco et al., 2008). Furthermore, inhibition of the mitochondrial Na^{+}-Ca^{2+}-exchanger on the IMM by the selective inhibitor KB-R7943 has also been reported to show neuroprotective effects against glutamate excitotoxicity, induce depolarization of brain mitochondria, and reduce Ca^{2+} uptake into mitochondria (Storozhevykh et al., 2010).

Besides reducing Ca^{2+} uptake into mitochondria to prevent Ca^{2+} overload, enhancing the Ca^{2+} buffering capacity of mitochondria can also prevent disruption of organelle functions. Agents such as antioxidants and taurine have been shown to increase the capacity of mitochondria to sequester Ca^{2+} by enhancing mitochondrial function (El Idrissi, 2008). However, the actual mechanisms by which antioxidants and taurine exert these effects are still unclear.

15.3.4 Mitochondrial Permeability Transition Pore

The Russian antihistamine drug Dimebon, also known as latrepirdine, has been shown to enhance cognition and is now reconsidered as a treatment for AD (Bachurin et al., 2001; Doody et al., 2008). Dimebon has been reported to inhibit the opening of mPTP induced by $A\beta_{25\text{–}35}$ in an early study (Bachurin et al., 2003).

The immunosuppressant cyclosporine A (CsA) inhibits the translocation of CypD to the IMM and prevents the opening of mPTP. Pretreatment with CsA has also been shown to elevate mitochondrial Ca^{2+} buffering capacity in both wild-type and transgenic mice expressing mutant APP (Du et al., 2008). More importantly, the inhibition of CypD in AD mice was found to improve learning and memory (Du et al., 2008). With better understanding of the structure of the mPTP, more specific inhibitors and modulators of the mPTP can be developed.

15.3.5 Mitochondrial Dynamics

Mitochondrial dynamics (fission, fusion, and transport) have been proposed to provide novel mechanisms for study in the field of neurodegeneration. Fission/fusion imbalance may lead to mitochondrial dysfunction, which affects synaptic transmission and consequently neurodegeneration. Recently, several compounds that are able to modulate mitochondrial fission and fusion have been identified. It has been proposed that these compounds could modulate mitochondrial dynamics to exert potential neuroprotective effects. In fact, inhibition of the fission protein Drp1 has been shown to be neuroprotective both in vivo and in vitro (Grohm et al., 2012). Mitochondrial division inhibitor mdivi-1 was the first small molecule identified to inhibit Drp1 assembly at mitochondria and inhibit Bax-mediated mitochondrial outer membrane permeabilization during apoptosis (Cassidy-Stone et al., 2008). Also, mdivi-1 was found to abolish mitochondrial dysfunction in AD cybrid cells (Gan et al., 2014). Also, the mitochondrial fusion promoter M1 hydrazone has been shown to be neuroprotective against MPP+-induced neurotoxicity in SH-SY5Y neuroblastoma cells (Wang et al., 2012). With better understanding of the molecular mechanism of fission and fusion, more pharmacological targets can be developed to modulate mitochondrial dynamics as a therapeutic approach.

15.3.6 Mitochondrial Metabolism

Manipulation of mitochondrial respiration has also been proposed as a therapeutic target for cognitive enhancement as well as neuroprotection (Gonzalez-Lima, Barksdale, and Rojas, 2014). Pharmacologically, agents such as methylene blue have been shown to improve mitochondrial respiration. Methylene blue is an FDA-approved drug that readily crosses the blood–brain barrier. It is a synthetic auto-oxidizable compound that can act as a metabolic enhancer and has a high affinity to neuronal tissue, making it a promising neuroprotective agent for AD. Several studies have implicated the memory-enhancing and neuroprotective effects of low-dose methylene blue in AD (Atamna and Kumar, 2010; Paban et al., 2014). A purified form of methylene blue (methylthioninium chloride; TRx0237) is now undergoing Phase 3 clinical trials for AD (NCT01689246, NCT01689233, NCT01626378)

Recently, the use of low power light in the red to near-infrared wavelengths (600 to 1000 nm) from lasers or light-emitting diodes (LEDs) has emerged as a direct and noninvasive approach to improve cognitive function and promote neuroprotection. Transcranial low level laser therapy (LLLT) has been reported to improve cognitive function in healthy human subjects as well as patients with chronic traumatic brain injury (Barrett and Gonzalez-Lima, 2013; Naeser et al., 2011). The principle of LLLT is based on photomodulation of the mitochondrial enzyme cytochrome oxidase. Absorption of light by cytochrome oxidase facilitates electron transfer in the IMM and enhances ATP production. LLLT has been shown to improve mitochondrial function and attenuate mitochondrial oxidative stress induced by electrical stimulation in C2C12 myotubes (Xu et al., 2008). Transcranial LLLT provides a direct approach to modulate mitochondrial function in humans.

15.3.7 Mitochondrial Biogenesis

Mitochondrial biogenesis is the process in which new mitochondria are formed in the cell. The peroxisome proliferator-activated receptor γ coactivator 1α (PGC-1α) is a key regulator of mitochondrial biogenesis. PGC-1α activates the nuclear respiratory factors NRF1 and NRF2 to regulate the expression of mitochondrial transcription factor A (Tfam) to promote transcription of mitochondrial DNA. Reduction of PGC-1α expression has been observed in postmortem AD brains (Knott et al., 2008).

Overexpression of PGC-1α has been shown to protect neurons from mutant α-synuclein or mutant huntington-induced mitochondrial loss (Wareski et al., 2009). Modulation of PGC-1α activity may promote mitochondrial biogenesis, which may be beneficial for neuronal survival.

Rosiglitazone, an antidiabetic drug, is a potent PPARg agonist and insulin sensitizer. Rosiglitazone has been shown to promote mitochondrial biogenesis in mouse brain (Strum et al., 2007). Although rosiglitazone has been shown to improve the cognitive function of patients with AD, the clinical trial has been discontinued.

15.3.8 Limitations of Mitochondrial-Targeted Drugs

The majority of the current approaches targeting drugs to mitochondria have shown that the drugs only accumulate in the mitochondrial matrix and inner mitochondrial membrane facing the matrix. However, there is also important machinery regulating mitochondrial function in other parts of mitochondria. Furthermore, excessive uptake of lipophilic cations to the matrix can depolarize mitochondria, leading to adverse effects. Many approaches are not organ specific and the drugs or compounds might easily accumulate in other tissues with high abundance of mitochondria (e.g., liver and heart), which might lead to significant adverse side effects. With better understanding of mitochondrial structure, more advanced drug delivery approaches can be developed. As mitochondria-targeted drugs are likely to affect other functions of mitochondria, a combination of drugs targeting specific pathways (antioxidants, Ca^{2+}) with drugs enhancing mitochondrial metabolism and biogenesis might be a possible approach for future development of mitochondria-targeted therapies.

15.4 CONCLUSIONS

Mitochondria play a central role in the pathogenesis of AD. Drugs targeted to mitochondria serve as an attractive and novel therapeutic intervention as discussed in this chapter. Currently, most mitochondria-targeted compounds are antioxidants, which aim to alleviate the production of ROS as an approach to protect neurons against neurodegeneration. However, other mitochondrial functions such as Ca^{2+} signaling and mitochondrial respiration also serve as attractive drug targets. With better understanding of other unique feature of mitochondria such as mitochondrial dynamics and mitophagy, novel drug targets are waiting to be discovered.

ACKNOWLEDGMENTS

The study of mitochondrial function in this laboratory is partly supported by the University of Hong Kong Alzheimer's Disease Research Network under Strategic Research Theme on Aging, and a generous donation by Ms. Kit-Wan Chow.

REFERENCES

Abramova, N. A., et al., 2002. Inhibition by R(+) or S(−) pramipexole of caspase activation and cell death induced by methylpyridinium ion or beta amyloid peptide in SH-SY5Y neuroblastoma. *J Neurosci Res*, 67, pp.494–500.

Alavian, K. N., et al., 2014. An uncoupling channel within the c-subunit ring of the F1FO ATP synthase is the mitochondrial permeability transition pore. *Proc Natl Acad Sci U S A*, 111, pp.10580–10585.

Alzheimer's Association, 2013. Alzheimer's disease facts and fgures. *Alzheimer's & Dementia*, 9.

Amadoro, G., et al., 2010. A NH2 tau fragment targets neuronal mitochondria at AD synapses: possible implications for neurodegeneration. *J Alzheimers Dis*, 21, pp.445–470.

Atamna, H. and Kumar, R., 2010. Protective role of methylene blue in Alzheimer's disease via mitochondria and cytochrome c oxidase. *J Alzheimers Dis*, 20 Suppl , pp.S439–Sf52.

Azarashvili, T., et al., 2014. Potential role of subunit c of F0F1-ATPase and subunit c of storage body in the mitochondrial permeability transition. Effect of the phosphorylation status of subunit c on pore opening. *Cell Calcium*, 55, pp.69–77.

Bachurin, S., et al., 2001. Antihistamine agent Dimebon as a novel neuroprotector and a cognition enhancer. *Ann N Y Acad Sci*, 939, pp.425–435.

Bachurin, S. O., et al., 2003. Mitochondria as a target for neurotoxins and neuroprotective agents. *Ann N Y Acad Sci.* 993, pp.334–344; 345–349.

Ball, M. J., 1977. Neuronal loss, neurofibrillary tangles and granulovacuolar degeneration in the hippocampus with ageing and dementia. A quantitative study. *Acta Neuropathol*, 37, pp.111–118.

Baloyannis, S. J., Costa, V. and Michmizos, D., 2004. Mitochondrial alterations in Alzheimer's disease. *Am J Alzheimers Dis Other Demen*, 19, pp.89–93.

Baloyannis, S. J., 2006. Mitochondrial alterations in Alzheimer's disease. *J Alzheimers Dis*, 9, pp.119–126.

Barrett, D. W. and Gonzalez-Lima, F., 2013. Transcranial infrared laser stimulation produces beneficial cognitive and emotional effects in humans. *Neuroscience*, 230, pp.13–23.

Barsoum, M. J., et al., 2006. Nitric oxide-induced mitochondrial fission is regulated by dynamin-related GTPases in neurons. *EMBO J*, 25, pp.3900–3911.

Bernardi, P., 1999. Mitochondrial transport of cations: channels, exchangers, and permeability transition. *Physiol Rev*, 79, pp.1127–1155.

Bonora, M., et al., 2013. Role of the c subunit of the FO ATP synthase in mitochondrial permeability transition. *Cell Cycle*, 12, pp.674–683.

Bright Focus Foundation, 2014. *The facts on Alzheimer's disease*.

Calkins, M. J., et al., 2011. Impaired mitochondrial biogenesis, defective axonal transport of mitochondria, abnormal mitochondrial dynamics and synaptic degeneration in a mouse model of Alzheimer's disease. *Hum Mol Genet*, 20, pp.4515–4529.

Canevari, L., Clark, J. B. and Bates, T. E., 1999. beta-Amyloid fragment 25-35 selectively decreases complex IV activity in isolated mitochondria. *FEBS Lett*, 457, pp.131–134.

Caspersen, C., et al., 2005. Mitochondrial Abeta: A potential focal point for neuronal metabolic dysfunction in Alzheimer's disease. *FASEB J*, 19, pp.2040–2041.

Cassidy-Stone, A., et al., 2008. Chemical inhibition of the mitochondrial division dynamin reveals its role in Bax/Bak-dependent mitochondrial outer membrane permeabilization. *Dev Cell*, 14, pp.193–204.

Cho, D. H., et al., 2009. S-nitrosylation of Drp1 mediates beta-amyloid-related mitochondrial fission and neuronal injury. *Science*, 324, pp.102–105.

Dikalova, A. E., et al., 2010. Therapeutic targeting of mitochondrial superoxide in hypertension. *Circ Res*, 107, pp.106–116.

Doody, R. S., et al., 2008. Effect of dimebon on cognition, activities of daily living, behaviour, and global function in patients with mild-to-moderate Alzheimer's disease: a randomised, double-blind, placebo-controlled study. *Lancet*, 372, pp.207–215.

Du, H., et al., 2008. Cyclophilin D deficiency attenuates mitochondrial and neuronal perturbation and ameliorates learning and memory in Alzheimer's disease. *Nat Med*, 14, pp.1097–1105.

Du, H., et al., 2010. Early deficits in synaptic mitochondria in an Alzheimer's disease mouse model. *Proc Natl Acad Sci U S A*, 107, pp.18670–18675.

El Idrissi, A., 2008. Taurine increases mitochondrial buffering of calcium: role in neuroprotection. *Amino Acids*, 34, pp.321–328.

Ferrari-Toninelli, G., et al., 2010. Mitochondria-targeted antioxidant effects of S(-) and R(+) pramipexole. *BMC Pharmacol*, 10, p.2.

Filipovska, A., et al., 2005. Synthesis and characterization of a triphenylphosphonium-conjugated peroxidase mimetic. Insights into the interaction of ebselen with mitochondria. *J Biol Chem*, 280, pp.24113–24126.

Gabuzda, D., et al., 1994. Inhibition of energy metabolism alters the processing of amyloid precursor protein and induces a potentially amyloidogenic derivative. *J Biol Chem*, 269, pp.13623–13628.

Galasko, D. R., et al., 2012. Antioxidants for Alzheimer disease: A randomized clinical trial with cerebrospinal fluid biomarker measures. *Arch Neurol*, 69, pp.836–841.

Gan, X., et al., 2014. Inhibition of ERK-DLP1 signaling and mitochondrial division alleviates mitochondrial dysfunction in Alzheimer's disease cybrid cell. *Biochim Biophys Acta*, 1842, pp.220–231.

Garcia-Martinez, E. M., et al., 2010. Mitochondria and calcium flux as targets of neuroprotection caused by minocycline in cerebellar granule cells. *Biochem Pharmacol*, 79, pp.239–250.

Gonzalez-Lima, F., Barksdale, B. R. and Rojas, J. C., 2014. Mitochondrial respiration as a target for neuroprotection and cognitive enhancement. *Biochem Pharmacol*, 88, pp.584–593.

Grohm, J., et al., 2012. Inhibition of Drp1 provides neuroprotection in vitro and in vivo. *Cell Death Differ*, 19, pp.1446–1458.

Hansson, C. A., et al., 2004. Nicastrin, presenilin, APH-1, and PEN-2 form active gamma-secretase complexes in mitochondria. *J Biol Chem*, 279, pp.51654–51560.

Hirai, K., et al., 2001. Mitochondrial abnormalities in Alzheimer's disease. *J Neurosci*, 21, pp.3017–30123.

Horton, K. L., et al., 2008. Mitochondria-penetrating peptides. *Chem Biol*, 15, pp.375–382.

Hung, C. H., Ho, Y. S. and Chang, R. C., 2010. Modulation of mitochondrial calcium as a pharmacological target for Alzheimer's disease. *Ageing Res Rev*, 9, pp.447–456.

Hyman, B. T., et al., 1984. Alzheimer's disease: Cell-specific pathology isolates the hippocampal formation. *Science*, 225, pp.1168–1170.

Kelso, G. F., et al., 2001. Selective targeting of a redox-active ubiquinone to mitochondria within cells: antioxidant and antiapoptotic properties. *J Biol Chem*, 276, pp.4588–4596.

Knott, A. B., et al., 2008. Mitochondrial fragmentation in neurodegeneration. *Nat Rev Neurosci*, 9, pp.505–518.

Kumar, U., Dunlop, D. M. and Richardson, J. S., 1994. Mitochondria from Alzheimer's fibroblasts show decreased uptake of calcium and increased sensitivity to free radicals. *Life Sci*, 54, pp.1855–1860.

Lasagna-Reeves, C. A., et al., 2011. Tau oligomers impair memory and induce synaptic and mitochondrial dysfunction in wild-type mice. *Mol Neurodegener*, 6, p.39.

Litsky, M. L. and Pfeiffer, D. R., 1997. Regulation of the mitochondrial Ca2+ uniporter by external adenine nucleotides: The uniporter behaves like a gated channel which is regulated by nucleotides and divalent cations. *Biochemistry*, 36, pp.7071–7080.

Liu, X. and Hajnoczky, G., 2009. Ca2+-dependent regulation of mitochondrial dynamics by the Miro-Milton complex. *Int J Biochem Cell Biol*, 41, pp.1972–1976.

Lovell, M. A., et al., 2004. Induction of hyperphosphorylated tau in primary rat cortical neuron cultures mediated by oxidative stress and glycogen synthase kinase-3. *J Alzheimers Dis*, 6, pp.659–671; 673–681.

Lustbader, J. W., et al., 2004. ABAD directly links Abeta to mitochondrial toxicity in Alzheimer's disease. *Science*, 304, pp.448–452.

Manczak, M., et al., 2004. Differential expression of oxidative phosphorylation genes in patients with Alzheimer's disease: implications for early mitochondrial dysfunction and oxidative damage. *Neuromolecular Med*, 5, pp.147–162.

Manczak, M., et al., 2006. Mitochondria are a direct site of A beta accumulation in Alzheimer's disease neurons: implications for free radical generation and oxidative damage in disease progression. *Hum Mol Genet*, 15, pp.1437–1449.

Manczak, M. and Reddy, P. H., 2012. Abnormal interaction of VDAC1 with amyloid beta and phosphorylated tau causes mitochondrial dysfunction in Alzheimer's disease. *Hum Mol Genet*, 21, pp.5131–5146.

McManus, M. J., Murphy, M. P. and Franklin, J. L., 2011. The mitochondria-targeted antioxidant MitoQ prevents loss of spatial memory retention and early neuropathology in a transgenic mouse model of Alzheimer's disease. *J Neurosci*, 31, pp.15703–15715.

Montero, M., et al., 2002. A novel regulatory mechanism of the mitochondrial Ca2+ uniporter revealed by the p38 mitogen-activated protein kinase inhibitor SB202190. *FASEB J*, 16, pp.1955–1957.

Montero, M., et al., 2004. Direct activation of the mitochondrial calcium uniporter by natural plant flavonoids. *Biochem J*, 384, pp.19–24.

Moreno-Ortega, A. J., et al., 2010. Mitochondria sense with different kinetics the calcium entering into HeLa cells through calcium channels CALHM1 and mutated P86L-CALHM1. *Biochem Biophys Res Commun*, 391, pp.722–726.

Murphy, M. P., et al., 2003. Superoxide activates uncoupling proteins by generating carbon-centered radicals and initiating lipid peroxidation: studies using a mitochondria-targeted spin trap derived from alpha-phenyl-N-tert-butylnitrone. *J Biol Chem*, 278, pp.48534–48545.

Naeser, M. A., et al., 2011. Improved cognitive function after transcranial, light-emitting diode treatments in chronic, traumatic brain injury: Two case reports. *Photomed Laser Surg*, 29, pp.351–358.

Ng, L. F., et al., 2014. The mitochondria-targeted antioxidant MitoQ extends lifespan and improves healthspan of a transgenic Caenorhabditis elegans model of Alzheimer disease. *Free Radic Biol Med*, 71, pp.390–401.

Nunomura, A., et al., 2001. Oxidative damage is the earliest event in Alzheimer disease. *J Neuropathol Exp Neurol*, 60, pp.759–767.

Paban, V., et al., 2014. Therapeutic and preventive effects of methylene blue on Alzheimer's disease pathology in a transgenic mouse model. *Neuropharmacology*, 76 Pt A, pp.68–79.

Quintanilla, R. A., et al., 2012. Truncated tau and Abeta cooperatively impair mitochondria in primary neurons. *Neurobiol Aging*, 33, pp.619.e25–619.e35.

Reddy, P. H., et al., 2004. Gene expression profiles of transcripts in amyloid precursor protein transgenic mice: Up-regulation of mitochondrial metabolism and apoptotic genes is an early cellular change in Alzheimer's disease. *Hum Mol Genet*, 13, pp.1225–1240.

Reddy, P. H. and Reddy, T. P., 2011. Mitochondria as a therapeutic target for aging and neurodegenerative diseases. *Curr Alzheimer Res*, 8, pp.393–409.

Rui, Y., et al., 2006. Acute impairment of mitochondrial trafficking by beta-amyloid peptides in hippocampal neurons. *J Neurosci*, 26, pp.10480–10487.

Sanz-Blasco, S., et al., 2008. Mitochondrial Ca2+ overload underlies Abeta oligomers neurotoxicity providing an unexpected mechanism of neuroprotection by NSAIDs. *PLoS One*, 3, p.e2718.

Sheu, S. S., Nauduri, D. and Anders, M. W., 2006. Targeting antioxidants to mitochondria: A new therapeutic direction. *Biochim Biophys Acta*, 1762, pp.256–265.

Silva, D. F., et al., 2012. Mitochondrial abnormalities in Alzheimer's disease: Possible targets for therapeutic intervention. *Adv Pharmacol*, 64, pp.83–126.

Smith, R. A., et al., 1999. Selective targeting of an antioxidant to mitochondria. *Eur J Biochem*, 263, pp.709–716.

Smith, R. A., Hartley, R. C. and Murphy, M. P., 2011. Mitochondria-targeted small molecule therapeutics and probes. *Antioxid Redox Signal*, 15, pp.3021–3038.

Storozhevykh, T. P., et al., 2010. Neuroprotective effect of KB-R7943 against glutamate excitotoxicity is related to mild mitochondrial depolarization. *Neurochem Res*, 35, pp.323–335.

Strum, J. C., et al., 2007. Rosiglitazone induces mitochondrial biogenesis in mouse brain. *J Alzheimers Dis*, 11, pp.45–51.

Szeto, H. H., 2006. Mitochondria-targeted peptide antioxidants: novel neuroprotective agents. *AAPS J*, 8, pp.E521–E531.

Wang, D., et al., 2012. A small molecule promotes mitochondrial fusion in mammalian cells. *Angew Chem Int Ed Engl*, 51, pp.9302–9305.

Wang, X., et al., 2008. Amyloid-beta overproduction causes abnormal mitochondrial dynamics via differential modulation of mitochondrial fission/fusion proteins. *Proc Natl Acad Sci U S A*, 105, pp.19318–19323.

Wareski, P., et al., 2009. PGC-1{alpha} and PGC-1{beta} regulate mitochondrial density in neurons. *J Biol Chem*, 284, pp.21379–21385.

Webster, M. T., et al., 1998. The effects of perturbed energy metabolism on the processing of amyloid precursor protein in PC12 cells. *J Neural Transm*, 105, pp.839–853.

Xie, H., et al., 2013. Mitochondrial alterations near amyloid plaques in an Alzheimer's disease mouse model. *J Neurosci*, 33, pp.17042–17051.

Xu, X., et al., 2008. Low-intensity laser irradiation improves the mitochondrial dysfunction of C2C12 induced by electrical stimulation. *Photomed Laser Surg*, 26, pp.197–202.

Yamada, Y., et al., 2008. MITO-Porter: A liposome-based carrier system for delivery of macromolecules into mitochondria via membrane fusion. *Biochim Biophys Acta*, 1778, pp.423–432.

Zhao, K., et al., 2004. Cell-permeable peptide antioxidants targeted to inner mitochondrial membrane inhibit mitochondrial swelling, oxidative cell death, and reperfusion injury. *J Biol Chem*, 279, pp.34682–34690.

16

MITOCHONDRIA IN PARKINSON'S DISEASE

GIUSEPPE ARENA

Institut de Recherche en Cancérologie de Montpellier, CSS-Université Montpellier, Montpellier, France

ENZA MARIA VALENTE

Section of Neuroscience, Department of Medicine and Surgery, University of Salerno, Salerno, Italy and Neurogenetics Unit, CSS-Mendel Institute, Rome, Italy

16.1 INTRODUCTION

Parkinson's disease (PD) represents the second-most common neurodegenerative disorder after Alzheimer disease, with a prevalence of 1% to 2% in people older than age 60 and almost 5 million individuals affected worldwide (Dorsey et al., 2007; Nussbaum and Ellis, 2003). In addition to age, the incidence of PD, estimated to be around 8-18/100,000 person-years, also depends on gender, with an increased risk in men (de Lau and Breteler, 2006).

Clinically, PD is characterized by the four cardinal symptoms of resting tremor, bradykinesia, rigidity, and postural instability. These are imputable to a progressive degeneration of dopaminergic neurons within the substantia nigra pars compacta (SNc), and the consequent dopamine depletion in the striatum (Dauer and Przedborski, 2003). Moreover, typical proteinaceous inclusions known as Lewy bodies (LB), containing α-synuclein, ubiquitin, and other neurofilaments, are usually observed in the cytoplasm of surviving neurons (Forno, 1996; Spillantini et al., 1997). The motor manifestations are often accompanied by nonmotor events, including cognitive and

The Functions, Disease-Related Dysfunctions, and Therapeutic Targeting of Neuronal Mitochondria, First Edition. Edited by Valentin K. Gribkoff, Elizabeth A. Jonas and J. Marie Hardwick.

neurobehavioral impairment, olfactory defects, dysautonomia, and sleep disturbance, indicating that the neurodegenerative process also extends to other brain areas (Jankovic, 2008).

Although slowly progressive, PD has an irreversible outcome, and available treatments are currently unable to arrest or delay the process of neuronal loss in the early stages of the disease; both pharmacological (L-Dopa and dopamine-agonists) and surgical (Deep Brain Stimulation) approaches remain merely symptomatic (palliative) and are often associated with treatment-related side effects (Schapira, 2007a). In this light, understanding the molecular basis of PD would represent a fundamental achievement to promote therapeutic advances and greatly impact global health care.

To date, it has been shown that mitochondrial dysfunction, oxidative stress and the intracellular accumulation of unwanted components such as aggregated proteins and damaged organelles, because of the impairment of intracellular clearance systems, represent the main pathogenetic steps leading to PD. These findings are concurrently highlighting the central role of mitochondria in PD pathogenesis, and the fact that distinct pathways may eventually converge at mitochondria to promote neuronal function and survival (Valente et al., 2012).

16.2 ROLE OF MITOCHONDRIA IN SPORADIC PD

16.2.1 Complex I Deficiency and mtDNA Defects

The development of typical PD symptoms in people contaminated with the opioid drug 1-methyl-4-phenyl-1,2,3,6-tetrahydropyridine (MPTP) represented the first association between PD pathogenesis and mitochondria; in fact, in its active oxidized form (MPP+), this synthetic compound is internalized by the dopamine transporter and is able to selectively impair mitochondrial complex I activity (NADH:ubiquinone oxidoreductase), leading to altered oxidative phosphorylation, ROS production, and dopaminergic neuronal death. Indeed, the experimental injection of MPTP in squirrel monkeys also resulted in parkinsonism responsive to L-dopa, the dopamine precursor that is currently used to alleviate PD symptoms (Langston et al., 1984). Similar to MPTP, the agriculture chemicals rotenone and paraquat, as well as the natural compound annonacin—known to result in parkinsonism by destroying the SNc dopaminergic neurons—also impair the mitochondrial respiratory chain by inhibiting complex I activity (Blesa et al., 2012).

Further confirming its involvement in PD pathogenesis, a significant complex I deficiency was observed in the SNc of about 30% patients with sporadic PD (Parker, Parks, and Swerdlow, 2008; Schapira, 2007b). Several hypotheses were formulated to explain this finding at the molecular level. To date, the most plausible explanation relates to the identification of a marked accumulation of somatic mitochondrial DNA (mtDNA) mutations in the SNc with age; in fact, among the 13 proteins encoded by the mtDNA, six represent complex I subunits, which means that genetic alterations of the mtDNA would eventually result in a marked impairment of the activity of this complex (Hauser and Hastings, 2013). Supporting this notion, when mitochondria derived from patients with PD are transplanted in cells containing no mtDNA, the resulting cybrids displayed a significant decrease in complex I activity, increased ROS production, and a higher susceptibility to MPTP-induced apoptosis (Swerdlow et al., 1996). Further, SNc neurons from postmortem brains displayed an age-dependent increase of mtDNA deletions that was significantly higher in patients with PD compared to age-matched controls (Bender et al., 2006). Mutations in mtDNA were also observed in transgenic mice expressing a dysfunctional version of the mitochondrial DNA polymerase γ (POLG), resulting in respiratory chain impairment and premature aging (Kujoth et al., 2005; Trifunovic et al., 2004). Interestingly, some families with pathogenic POLG mutations were reported to manifest several neurological alterations, including loss of SNc dopaminergic neurons and the development of parkinsonism (Davidzon et al., 2006; Synofzik et al., 2010).

Mice with deletions in the mitochondrial transcription factor TFAM, another fundamental gene required for the maintenance of mtDNA, also displayed dopaminergic neuronal death and motor dysfunctions, further confirming that failure in mtDNA integrity may be associated with PD pathogenesis (Ekstrand et al., 2007). In a different mouse model, the expression of a mitochondrial specific endonuclease, able to generate double-strand breaks in mtDNA, induced a progressive degeneration of SNc dopaminergic neurons and a behavioral phenotype associated with striatal dysfunction (Pickrell et al., 2011).

An alternative explanation for the complex I deficiency observed in PD came from a recent genomewide meta-analysis in samples from patients with PD and healthy controls that identified 10 novel gene sets involved in the regulation of mitochondrial electron transport and glucose metabolism. These sets of genes, all under the control of the fundamental transcription factor PGC-1α, were all down-regulated in patients with PD, suggesting that the complex I defects may result from extensive alterations in neuronal bioenergetics. In fact, transduction of rat dopaminergic neurons with adenovirus carrying human PGC-1α was able to counteract the neuronal loss induced by either overexpression of mutant A53T α-synuclein or rotenone administration (Zheng et al., 2010).

16.2.2 Oxidative Stress and ROS Production

In eukaryotic cells, mitochondrial respiration represents the main source of reactive oxygen species (ROS), produced in the form of superoxide anion ($O_2^{\cdot-}$) by the electron transport chain complexes. Under normal conditions, several mitochondrial enzymes, such as manganese superoxide dismutase (MnSOD) and other peroxidases, are able to convert $O_2^{\cdot-}$ to hydrogen peroxide (H_2O_2), in turn reducing H_2O_2 to water and oxygen. If such fundamental antioxidant functions are insufficient to balance ROS production, the resulting oxidative stress may compromise the main cellular components, including proteins, DNA and lipids, eventually leading to cell death (Hauser and Hastings, 2013). Accordingly, high levels of oxidized macromolecules have been found in the SNc of patients with PD (Floor and Wetzel, 1998), and increased oxidative stress was also observed following complex I impairment as a result of MPP+ or rotenone administration (Esteves et al., 2009; Hasegawa et al., 1990; Sherer et al., 2003;

Votyakova and Reynolds, 2005). Moreover, a significant decrease in glutathione (GSH) levels, a key antioxidant molecule able to counteract the toxic effect of free radicals, was selectively observed in the SNc of patients with PD (Perry, Godin, and Hansen, 1982; Sian et al., 1994). Because GSH depletion seems to occur early in the pathogenetic process, the loss of antioxidant capacity could fuel ROS production and induce mtDNA damage and respiratory chain defects, thus impairing mitochondrial function and accelerating the progression of neurodegeneration (Dexter et al., 1994).

ROS in the form of hydroxyl radicals (HO) are selectively generated in the SNc of patients with PD following the interaction between H_2O_2 and iron. Iron is strongly enriched in the dopaminergic neurons of PD brains probably because of the increased expression of the iron transporter Transferrin (Hirsch et al., 1991; Mastroberardino et al., 2009; Oakley et al., 2007; Sofic et al., 1988). In support of this notion, the administration of iron chelators is able to reduce oxidative stress, ameliorating mitochondrial function and promoting neuroprotection (Weinreb et al., 2010).

The metabolism of dopamine also contributes to the higher level of oxidative stress found in SNc dopaminergic neurons, both by producing H_2O_2 and through the formation of oxidized dopamine quinone (DAQ); the expression of DAQ is highly increased in PD brains. The in vitro treatment of functional mitochondria with DAQ was found to produce a strong enhancement of mitochondrial depolarization and opening of the mitochondrial permeability transition pore (Berman and Hastings, 1999). Moreover, DAQ is also able to irreversibly modify the Cysteine residues of GSH and other key cellular proteins, thus affecting the antioxidant capacity and ultimately neuronal viability (Rabinovic and Hastings, 1998; Spencer et al., 1998). Proteomic studies on neuronal cell lines and isolated brain mitochondria subjected to dopamine exposure revealed many DAQ-modified mitochondrial proteins, including complex I subunits, phospho-creatine kinase and isocitrate dehydrogenase (Van Laar et al., 2009); of note, the PD-associated proteins α-synuclein, Parkin and DJ-1 are also known to be chemically altered by DAQ (Conway et al., 2001; LaVoie et al., 2005; Van Laar et al., 2009).

16.3 MITOCHONDRIAL DYSFUNCTION IN MONOGENIC PD

At present, PD is universally recognized as a heterogeneous and complex disorder, with predisposing genetic and environmental causes. The development of parkinsonism in individuals affected by the 1918 flu pandemic who suffered from lethargic encephalitis, and in subjects exposed to the synthetic drug MPTP, long supported the environmental hypothesis; moreover, several stress conditions, such as traumatic head injury or exposure to toxins, pesticides and heavy metals, have been also linked to PD pathogenesis in many epidemiologic studies (Di Monte, 2003).

The genetic contribution to PD etiology only emerged in the last two decades, with the identification of several loci and genes causal for autosomal dominant (ADPD) or recessive (ARPD) forms of parkinsonism. Even if these monogenic forms currently account only for a small percentage of patients (about 10%), proteins encoded by PD-associated genes have been found to play a crucial role in key intracellular pathways involved in neurodegeneration, suggesting that both idiopathic (sporadic) and familial forms could share the same pathogenetic mechanisms. In general, proteins encoded by AD PD genes (e.g., SNCA and LRKK2) tend to be toxic when mutated or overexpressed, whereas impairment of ARPD gene products (e.g., Parkin, PINK1, DJ-1) may lead to neurodegeneration as a result of a loss of function. Importantly, increasing evidence now indicates that different PD-associated proteins are implicated in the same pathways, many of which are directly involved in the maintenance of mitochondrial integrity and function (Valente et al., 2012). Regulation of mitochondrial quality control assumes particular importance in dopaminergic neurons, because of their high level of energy demand and the need to face dopamine-induced ROS production; hence, mutations in the PARK genes could severely impair mitochondrial homeostasis and ultimately result in the massive neuronal loss observed in patients with PD.

16.3.1 Autosomal Dominant PD

To date, only mutations in the SNCA, LRRK2, and VPS35 genes have been clearly associated with AD PD; several other genes, including EIF4G1, HTRA2, UCHL1, and GIGYF2, have been also suggested to cause AD PD, but their role in PD pathogenesis still remains uncertain (Puschmann, 2013).

16.3.1a SNCA (α-synuclein)

The *SNCA* gene encodes a small 140 amino acid protein, termed *α-synuclein*, characterized by a strong propensity to form insoluble aggregates, that represents the most abundant fibrillar component of the LB (Spillantini et al., 1997). The identification of the *SNCA* p.A53T mutation in a large Italian-Greek family with AD PD (the Contursi kindred) represented the first milestone in PD genetics (Polymeropoulos et al., 1997). Only a few additional missense mutations were subsequently identified despite extensive screenings of large patient cohorts, making SNCA point mutations a rare cause of PD and parkinsonism. Interestingly, SNCA gene duplications or triplications were more frequently identified as causative of PD with a clear correlation between the number of SNCA copies and the phenotypic severity, suggesting a dose-dependent effect of α-synuclein overexpression in the human brain (Fuchs et al., 2008; Singleton et al., 2003). Further supporting this notion, polymorphisms in SNCA regulatory elements that were found to robustly increase α-synuclein levels, including NACP-Rep1 in the promoter region and variants within the 3'-UTR, were found to be associated with a higher risk of developing PD (Winkler et al., 2007).

Both point mutations and whole gene amplifications of *SNCA* would increase the tendency of α-synuclein to oligomerize and then form fibrils, resulting in neurotoxicity and ultimately cell death (Karpinar et al., 2009). Interestingly, levels of α-synuclein appear to increase with age, in line with the increased incidence of PD in aged people (Collier, Kanaan, and Kordower, 2011).

In healthy brain, α-synuclein is strongly expressed at the presynaptic level, in close proximity to synaptic vesicles, where it appears to play a role in neurotransmitter release and regulation of synaptic transmission by interacting with the SNARE complex (Kahle, 2008). Indeed, PD-associated mutations result in a severe impairment of synaptic function, likely as a result of altered morphology, distribution and recycling of synaptic vesicles (Chung et al., 2009; Nemani et al., 2010; Scott et al., 2010). Based on the most plausible model, mutant α-synuclein oligomers would form porelike structures on the membrane of synaptic vesicles, leading to neurotransmitter leakage in the cytosol; in the case of dopamine leakage from vesicles, this could affect not only synaptic transmission, but might also promote the formation of DAQ, resulting in ROS production, mitochondrial dysfunction and cell death (Mosharov et al., 2009). Moreover, it was recently shown that mutant α-synuclein could also interfere with cytoskeleton dynamics, indirectly influencing mitochondrial transport along axonal processes (Chung et al., 2009; Sousa et al., 2009).

In addition to these indirect roles on mitochondrial function and dynamics, several groups reported that a percentage of mutant α-synuclein resides within mitochondria (either at the inner or the outer mitochondrial membrane), where it may cause an impairment of complex I similar to that observed in sporadic PD after exposure to mitochondrial toxins (Devi et al., 2008; Liu et al., 2009a; Loeb et al., 2010). Mitochondrial alterations resulting from abnormal α-synuclein accumulation enhance ROS production and promote oxidative stress, thus contributing to cytochrome c release and neuronal death (Botella, Bayersdorfer, and Schneuwly, 2008; Hsu et al., 2000; Xu et al., 2002). Moreover, a recent work described a novel localization of α-synuclein at the mitochondria-associated ER membranes (MAM), where it could regulate mitochondrial morphology; PD-related α-synuclein point mutations were found to impair its localization at MAM, leading to reduced tethering between ER and mitochondria and increased mitochondrial fragmentation (Guardia-Laguarta et al., 2014).

Studies on animal models also support a role for α-synuclein in contributing to mitochondrial damage. In fact, SNCA transgenic mice displayed relevant changes in mitochondrial morphology, with an increased level of fragmentation that is usually associated with mitochondrial dysfunction and neurodegeneration (Kamp et al., 2010; Martin et al., 2006; Nakamura et al., 2011); neuronal death induced by mutant α-synuclein could also result from massive mitochondrial degradation because of excessive activation of mitophagy (Choubey et al., 2011). On the other hand, the overexpression of both the transcription factor PGC-1α and the mitochondrial chaperone TRAP1 were found to be able to ameliorate α-synuclein-induced toxicity and neuronal loss (Butler et al., 2012; Zheng et al., 2010).

16.3.1b Leucine-Rich Repeats 2 (LRRK2)

To date, mutations in the *LRRK2* gene represent the most frequent cause of familial PD, with a frequency of the most common mutation (p.G2019S) reaching up to 40% in selected populations (Healy et al., 2008; Paisan-Ruiz et al., 2004; Zimprich et al., 2004). Common variants at the LRRK2 locus were also associated with an increased risk to develop sporadic PD (Satake et al., 2009; Simon-Sanchez et al., 2009). Despite this, the physiological role of LRRK2 is still unclear and additional studies are necessary to better understand its involvement in PD pathogenesis.

In mammalian brain, the LRRK2 protein is highly expressed in the striatum and cerebral cortex (Galter et al., 2006; Higashi et al., 2007). At the cellular level, it was found to be associated with different vesicular structures and intracellular membranes (Biskup et al., 2006; West et al., 2007). Biochemically, LRRK2 is a multidomain protein of 2527 amino acids, containing both a GTPase and a serine/threonine kinase activity (Mata et al., 2006); several protein interaction domains are also present, indicating that LRRK2 could act as a scaffold for the assembly of different macromolecular complexes. In light of its complexity, LRRK2 is involved in many cellular pathways aimed at regulating cytoskeleton organization, synaptic vesicle trafficking and neuronal process outgrowth. LRRK2 could be also implicated in the regulation of UPS and autophagy pathways, as suggested by the accumulation of protein inclusions in the brains of patients with *LRRK2* mutations (Giasson et al., 2006; Ross et al., 2006).

The LRRK2 protein was also found to localize at the outer mitochondrial membrane, where it was implicated in the regulation of mitochondrial morphology, dynamics, and function (Tsika and Moore, 2012); in different studies, LRRK2 mutations were variably associated with increased mitochondrial elongation or fragmentation, the latter probably the result of a Drp-1 dependent mechanism (Mortiboys et al., 2010; Niu et al., 2012; Wang et al., 2012). In this case, the ability of LRRK2 to promote fission correlated with a decreased mitochondrial membrane potential and ATP production, increased ROS and neuronal death, a phenotype that could be reversed by inhibiting Drp-1 or by promoting mitochondrial fusion (Wang et al., 2012). Transgenic mice harboring the LRRK2 *G2019S* mutation display an atypical accumulation of aggregated mitochondria, consistent with impaired mitophagy (Ramonet et al., 2011), whereas other studies revealed that LRRK2 may enhance mitophagy in response to defects in calcium homeostasis (Cherra et al., 2013). Moreover, overexpression of the G2019S mutant protein in α-synuclein A53T transgenic mice was found to exacerbate the mitochondrial defects observed in this model (Lin et al., 2009). Further confirming its role in mitochondrial homeostasis, LRRK2 was also shown to interact with PD-associated proteins Parkin, DJ-1 and PINK1, well known to regulate the mitochondrial stress response at multiple levels (Venderova et al., 2009).

16.3.1c Vacuolar Protein Sorting 35 (VPS35)

Mutations in the *VPS35* gene were identified in 2011 by whole exome sequencing as causative of AD PD, although their overall frequency is quite low (0.1 to 1%). The VPS35 mRNA is highly expressed in many tissues, including the brain, and encodes a 92kDa protein involved in the recycling of membrane complexes and intracellular protein sorting (Chen et al., 2013; Vilarino-Guell et al., 2011; Zimprich et al., 2011).

To date, evidence suggests that the role of VPS35 in PD may involve two distinct pathways: regulation of Wnt signaling and iron transport. The Wnt/β-catenin pathway is known to play an essential function in the maturation of dopaminergic neurons, thus representing a potential key mechanism in PD pathogenesis; indeed, severe impairment of this signal transduction was observed in different PD models subjected to the mitochondrial toxin MPTP and 6-OHDA. *VPS35* mutations, leading to defective Wnt secretion, would reduce the activation of the Wnt/β-catenin pathway, increasing dopaminergic neuron vulnerability and neuronal loss. On the other hand, *VPS35* mutations could affect iron metabolism by disrupting the correct sorting of the iron transporter DMT1. This would result in increased iron levels within dopaminergic neurons, and ultimately cell death (Deng, Gao, and Jankovic, 2013). Despite these findings, the biological impact of *VPS35* mutations on PD pathogenesis remains largely unknown, and further studies are necessary to explore in depth the effect of VPS35 on the different intracellular transport pathways. For instance, it has been shown that VPS35 is recruited to mitochondria-derived vesicles that are responsible for cargo transport from the mitochondria to the peroxisomes, a key step in the reduction of peroxide species that is required to prevent ROS production (Braschi et al., 2010).

16.3.2 Autosomal Recessive PD

Autosomal recessive forms of PD are primarily caused by homozygous or compound heterozygous mutations in three major genes: Parkin, PINK1 and DJ-1 (Bonifati et al., 2003; Kitada et al., 1998; Valente et al., 2004). The corresponding proteins have distinct enzymatic activities and were shown to play multiple neuroprotective functions, partly independent from each other, partly interacting within the same pathways. Mutations in these genes determine a loss-of-function phenotype, reducing the neuroprotective effect of the encoded proteins and finally resulting in increased neuronal death. Of note, all these proteins are constitutively localized in mitochondria or relocate to the organelle in response to several parkinsonian stressors, thus indicating their crucial role in maintaining mitochondrial homeostasis, function and quality control (Trempe and Fon, 2013). As a consequence, PD-associated pathogenetic mutations usually lead to severe alterations in neuronal bioenergetics, because of decreased mitochondrial membrane potential and oxidative phosphorylation, impaired free radical scavenger activity and oxidative damage (Corti, Lesage and Brice, 2011; Exner et al., 2012).

16.3.2a PARK2 (Parkin)

Mutations in the PARK2 gene represent the most common cause of early-onset AR PD, representing about half of recessive familial forms (Kitada et al., 1998). More than 170 different mutations, including exonic deletion or multiplication, small deletions and insertions, missense and truncating mutations, have been reported so far, often associated with a significant loss of SNc dopaminergic neurons (Mori et al., 1998). Interestingly, the frequency of PARK2 mutations inversely correlates with the age of onset, reaching about 80% in patients with PD younger than 20 years (Puschmann, 2013).

The PARK2 gene encodes for Parkin, a 465 amino acid protein with intrinsic E3 ubiquitin ligase activity, that mediates the K48-linked conjugation of one or multiple ubiquitin molecules to selective substrates, targeting them for proteasomal degradation (Shimura et al., 2000). The ubiquitin-proteasome system (UPS) represents a key quality control mechanism within cells, preventing the unwanted accumulation of misfolded protein and damaged organelles; indeed, its impairment has been implicated in the pathogenesis of many neurodegenerative disorders, including PD (Lehman, 2009). Several mutations in the PARK2 gene were found to impair the E3 ligase activity of Parkin, resulting in inefficient substrate clearance and subsequent protein aggregation (Shimura et al., 2000). Of note, the E3 ubiquitin ligase activity of Parkin is not limited to target proteins for proteasomal degradation, but can also catalyze both mono-ubiquitination and K27- or K63-linked poly-ubiquitination of distinct substrates, resulting in the regulation of several proteasome-independent cellular processes such as transcriptional regulation, protein trafficking and signal transduction (Doss-Pepe, Chen, and Madura, 2005; Hampe et al., 2006; Matsuda et al., 2006; Mukhopadhyay and Riezman, 2007). Several findings also point to a role for Parkin in protecting cells from different mitochondrial stressors; in fact, Parkin-deficient cells display a significant decrease in mitochondrial membrane potential, increased ROS production and stress-induced apoptosis (Moore, 2006; Pilsl and Winklhofer, 2012). Moreover, a strong reduction of the antioxidant capacity and a concomitant increase in lipid peroxidation was observed in the brains of Parkin-KO mice (Palacino et al., 2004). The ability of Parkin to attenuate oxidative damage may result from its physical and functional interaction with mtDNA, sustaining efficient repair and activating the key mitochondrial transcription factor TFAM (Kuroda et al., 2006; Narendra et al., 2008).

Relevant insights into the neuroprotective activity of Parkin came from extensive studies in *Drosophila* that revealed its fundamental role in regulating the dynamic properties of the mitochondrial network. Parkin-deficient flies displayed motor defects, muscle degeneration, loss of dopaminergic neurons and decreased lifespan, a phenotype clearly associated with profound aberrations in mitochondrial morphology, including enlarged/elongated or swollen organelles, fragmented cristae and loss of the outer mitochondrial membrane (Cha et al., 2005; Greene et al., 2003; Whitworth et al., 2005). This phenotype was also observed in PINK1-KO flies and, interestingly, it could be rescued in both models by Parkin overexpression, thus indicating the existence of a common pathway with Parkin genetically downstream of PINK1 (Clark et al., 2006; Park et al., 2006; Yang et al., 2006). The same abnormalities in mitochondrial morphology could also be corrected by increasing the dosage of the pro-fission factor dDrp1 or dFis1 or by down-regulating the profusion factors dOpa1 or dMfn1, suggesting that the PINK1-Parkin pathway might be involved in activating mitochondrial fragmentation (Deng et al., 2008; Park, Lee, and Chung, 2009; Poole et al., 2008; Yang et al., 2008).

The proper balance between mitochondrial fission and fusion represents a fundamental quality control mechanism, required for the efficient distribution of mtDNA and respiratory complexes throughout the mitochondrial network and for the selective removal of damaged organelles through the autophagy pathway. The latter process, termed *mitophagy*, is often preceded by an increased mitochondrial fragmentation, leading to "daughter" mitochondria with unbalanced membrane potential: the healthy organelle will rejoin the network, whereas the depolarized one would be excluded from refusion, become stationary and finally undergo degradation (Ashrafi and Schwarz, 2013; Twig and Shirihai, 2011).

According to this scenario, several groups have recently demonstrated that mitochondrial fission triggered by the PINK1-Parkin axis could effectively facilitate the selective removal of dysfunctional mitochondria through mitophagy (Geisler et al., 2010; Narendra et al., 2010; Vives-Bauza et al., 2010); following mitochondrial depolarization, PINK1 recruits cytosolic Parkin on the outer mitochondrial membrane of damaged mitochondria, where it mediates the proteasomal degradation of several substrates, leading to inhibition of fusion and mitochondrial arrest (through K48-linked ubiquitination of Mfn and Miro, respectively). This first step allows for the "quarantine" of depolarized mitochondria, and once completed, Parkin would mediate the K27- or K63-linked ubiquitination of distinct targets (i.e., VDAC and p62), initiating a signaling cascade necessary to direct the autophagy machinery to damaged "quarantined" organelles and promote their elimination (Ashrafi and Schwarz, 2013). In a separate model, Parkin was found to directly recruit Ambra1, an activator of the key pro-autophagic protein Beclin1, resulting in the formation of autophagosomes around dysfunctional mitochondria (Van Humbeeck et al., 2011). In response to mitochondrial depolarization, Parkin is also able to promote mitochondrial biogenesis, promoting the replacement of damaged organelles with newly synthesized functional mitochondria; to this aim, Parkin also induces the degradation of PARIS, a transcriptional repressor of the mitochondrial biogenesis factor PGC-1α, leading to increased expression of target genes (Shin et al., 2011).

16.3.2b PTEN-Induced Kinase1 (PINK1)

Homozygous mutations in the PINK1 gene were first identified in three consanguineous families from Italy and Spain affected by early-onset PD (Valente et al., 2004). Missense mutations, small insertions/deletions, and multiexon deletions were also reported, making PINK1 the second most common cause of AR PD, with a frequency ranging from 1 to 10% worldwide.

PINK1 is highly expressed in the SNc and displays immunoreactivity in LB-like neuronal inclusions. Moreover, KO mice display a phenotype characterized by reduced nigro-striatal dopamine release, impaired synaptic plasticity and diffuse mitochondrial dysfunction, albeit in the absence of clear neurodegeneration (Kitada 2007). Further, PINK1 depletion in the SH-SY5Y dopaminergic cell line was found to trigger α-synuclein aggregation by altering proteasome activity (Liu et al., 2009b), whereas the overexpression of PINK1 in *Drosophila* is able to suppress α-synuclein-induced toxicity (Todd and Staveley, 2008).

The PINK1 gene encodes a ubiquitously expressed 581 amino acid protein, characterized by some homologies with members of the Ca^{2+}/calmodulin-dependent serine/threonine kinase family. Interestingly, PINK1 is the unique protein associated with familial PD having a specific mitochondria targeting sequence, triggering a wealth of studies aimed at demonstrating its direct role in the regulation of mitochondrial function (Valente et al., 2012).

In healthy cells, PINK1 is constitutively imported in the mitochondria and efficiently degraded through a two-step process, involving some mitochondrial proteases (i.e., MPP and PARL) and, ultimately, the N-end rule proteasomal pathway

(Yamano and Youle, 2013). In this manner, PINK1 levels are maintained at low levels in basal conditions. Conversely, specific stress conditions, which are able to compromise mitochondrial homeostasis, such as loss of membrane potential or accumulation of misfolded proteins in the mitochondrial matrix, are able to block PINK1 mitochondrial import, resulting in its stabilization at the outer mitochondrial membrane. This selective accumulation flags damaged organelles as potentially dangerous, promptly activating Parkin recruitment and the mitophagy cascade. Importantly, PINK1 stabilization could also make the kinase available for other substrates, leading to a wider neuroprotective response that proceeds in parallel with the elimination of dysfunctional mitochondria. In fact, upon treatment with the mitochondrial uncoupler CCCP, PINK1 was found to phosphorylate the pro-survival protein Bcl-xL and impair its pro-apoptotic cleavage, thus protecting cells from apoptosis (Arena et al., 2013). Moreover, other known or putative targets of PINK1 reside within mitochondria and may contribute to counteracting stress-induced mitochondrial damage. For instance, it has been previously demonstrated that in H_2O_2-treated cells PINK1 is able to phosphorylate the mitochondrial chaperone TRAP1, leading to reduced cytochrome c release and increased cell survival (Pridgeon et al., 2007). Similarly, the PINK1-dependent phosphorylation of the mitochondrial protease HtrA2 enhances survival of cells subjected to oxidative stress (Plun-Favreau et al., 2007). Through a proteomic analysis in PINK1-overexpressing cells, Rakovic and colleagues (2011) recently identified four putative PINK1 interactors localized in mitochondria, including the major mitochondrial chaperone mortalin/GRP75, known to play several pro-survival functions in response to different apoptotic stimuli. Interestingly, mutations in the GRP75 gene are known to increase the risk of developing PD and reduced GRP75 expression was observed in brains from patients with PD. The other interactors, namely HSP60, LRPPRC and TUFM, could also be linked to PD pathology in light of their involvement in the regulation of mitochondrial respiratory chain activity (Rakovic et al., 2011). In line with these findings, PINK1 was recently found to interact with key subunits of the electron transport chain and to sustain mitochondrial complex activity and ATP production (Gautier, Kitada, and Shen, 2008; Grunewald et al., 2009; Liu et al., 2009b; Morais et al., 2014). Moreover, PINK1 is also involved in the regulation of calcium buffering, thus preventing cell death induced by mitochondrial calcium overload (Gandhi et al., 2009; Marongiu et al., 2009). The maintenance of calcium homeostasis plays a crucial role in SNc dopaminergic neurons, as they preferentially use calcium channels to sustain their activity. The disruption of both respiratory chain activity and calcium metabolism by PINK1 mutations is expected to significantly increase ROS production and oxidative stress, and is often accompanied by a reduced expression of glutathione and other antioxidant enzymes (Abramov et al., 2011; Hoepken et al., 2007; Wood-Kaczmar et al., 2008).

16.3.2c DJ-1

A vast genomic deletion and a missense p.L166P mutation in the DJ-1 gene were first reported by Bonifati et al. (2003) in a Dutch and Italian family, respectively. Subsequent studies identified a few more mutations and deletions; however DJ-1 pathogenic variants are overall extremely rare, accounting for less than 1% of AR PD (Puschmann, 2013).

The DJ-1 gene encodes a 189 amino acid protein, which is a member of a large family of molecular chaperones highly conserved across species (Bandyopadhyay and Cookson, 2004; Lucas and Marin, 2007). Although the DJ-1 protein was previously implicated in several biological processes, its exact biochemical function is still largely unknown. Currently, the most widely supported hypothesis suggests that DJ-1 plays a "signaling" role during cellular responses to oxidative stress, likely by modulating the expression of genes involved in the antioxidant reaction (Cookson, 2010). In cells subjected to oxidative stress, the sulfhydryl group of a conserved Cysteine residue (C106 in human DJ-1) reacts with ROS and undergoes oxidization to sulfinic acid, leading DJ-1 to accumulate at mitochondria where it exerts its neuroprotective activity (Canet-Aviles et al., 2004; Kinumi et al., 2004). Supporting this notion, the C106 substitution to the nonoxidizable Alanine (DJ-1 C106A) was shown to impair the mitochondrial localization of DJ-1, resulting in decreased protection against oxidative stress; similarly, overexpressed C106A DJ-1 failed to protect cells from rotenone toxicity (Blackinton et al., 2009). Rat brains systemically exposed to rotenone, as well as postmortem brains from patients with sporadic PD, also displayed a significant accumulation of oxidized DJ-1 (Bandyopadhyay and Cookson, 2004; Betarbet et al., 2006; Mitsumoto and Nakagawa, 2001). Animal models further confirmed a key role for DJ-1 in regulating the cellular response against oxidative stress. In fact, SNc neurons from DJ-1 KO mice showed severe ROS-induced mitochondrial damage (Guzman et al., 2010) and higher sensitivity to MPTP toxicity (Kim et al., 2005). DJ-1 down-regulation in *Drosophila* and in *Caenorhabditis elegans* exacerbated cell death by toxins able to induce oxidative stress (Hao, Giasson, and Bonini, 2010; van der Brug et al., 2008; Ved et al., 2005). In different cellular models, loss of DJ-1 was also shown to alter several aspects of mitochondrial dynamics and function. Both mouse embryonic fibroblasts from DJ-1 KO mice and SH-SY5Y cells subjected to DJ-1 silencing had increased mitochondrial fragmentation, respiratory chain defects, increased ROS production and impaired mitophagy (Krebiehl et al., 2010; Thomas et al., 2011). Interestingly, DJ-1 over-expression was able to restore the mitochondrial defects observed in PINK1 but not in Parkin-null flies, indicating that DJ-1 would act downstream of PINK1, but independently from Parkin (Hao et al., 2010). Even if all these findings clearly suggest a role for DJ-1 in maintaining mitochondrial health following oxidative damage, no evident changes in the nigrostriatal pathway have been found to date in the absence of functional DJ-1 (Hauser and Hastings, 2013).

16.4 CONCLUSIONS

Convincing arguments, coming from environmental and genetic studies, indicate mitochondrial dysfunction as a common hallmark of both sporadic and familial PD. The failure of mitochondrial homeostasis plays a crucial role in the pathogenetic process, leading to excessive ROS production, mtDNA mutations, respiratory chain defects, and neuronal death. To avoid this collapse, tightly regulated mitochondrial quality control pathways continuously ensure the maintenance of mitochondrial health, excluding the damaged organelles from the network, providing for their degradation and, at the same time, sustaining the replacement with newly synthesized mitochondria. Any breakdown in this dynamic process, resulting from high levels of oxidative stress, exposure to toxins or genetic causes, fatally leads to the accumulation of dysfunctional mitochondria; these dysfunctional mitochondria are not only unable to fulfill neuronal energy demands, but are also a source of extremely dangerous components (i.e. ROS, cytochrome c, etc.), that eventually lead to irreversible damage and neuronal death. In this light, therapeutic strategies aimed to support mitochondrial function in the early stages of the pathogenic process would be helpful to reduce the rate of neuronal death in PD and possibly in other neurodegenerative disorders.

REFERENCES

Abramov, A. Y., et al., 2011. Bioenergetic consequences of PINK1 mutations in Parkinson disease. *PLoS One*, 6, p.e25622.

Arena, G., et al., 2013. PINK1 protects against cell death induced by mitochondrial depolarization, by phosphorylating Bcl-xL and impairing its pro-apoptotic cleavage. *Cell Death Differ*, 20, pp.920–930.

Ashrafi, G. and Schwarz, T. L., 2013. The pathways of mitophagy for quality control and clearance of mitochondria. *Cell Death Differ*, 20, pp.31–42.

Bandyopadhyay, S. and Cookson, M. R., 2004. Evolutionary and functional relationships within the DJ1 superfamily. *BMC Evol Biol*, 4, p.6.

Bender, A., et al., 2006. High levels of mitochondrial DNA deletions in substantia nigra neurons in aging and Parkinson disease. *Nat Genet*, 38, pp.515–517.

Berman, S. B. and Hastings, T. G., 1999. Dopamine oxidation alters mitochondrial respiration and induces permeability transition in brain mitochondria: Implications for Parkinson's disease. *J Neurochem*, 73, pp.1127–1137.

Betarbet, R., et al., 2006. Intersecting pathways to neurodegeneration in Parkinson's disease: Effects of the pesticide rotenone on DJ-1, alpha-synuclein, and the ubiquitin-proteasome system. *Neurobiol Dis*, 22, pp.404–420.

Biskup, S., et al., 2006. Localization of LRRK2 to membranous and vesicular structures in mammalian brain. *Ann Neurol*, 60, pp.557–569.

Blackinton, J., et al., 2009. Formation of a stabilized cysteine sulfinic acid is critical for the mitochondrial function of the parkinsonism protein DJ-1. *J Biol Chem*, 284, pp.6476–6485.

Blesa, J., et al., 2012. Classic and new animal models of Parkinson's disease. *J Biomed Biotechnol*, 2012, pp.845618.

Bonifati, V., et al., 2003. Mutations in the DJ-1 gene associated with autosomal recessive early-onset parkinsonism. *Science*, 299, pp.256–259.

Botella, J. A., Bayersdorfer, F. and Schneuwly, S., 2008. Superoxide dismutase overexpression protects dopaminergic neurons in a Drosophila model of Parkinson's disease. *Neurobiol Dis*, 30, pp.65–73.

Braschi, E., et al., 2010. Vps35 mediates vesicle transport between the mitochondria and peroxisomes. *Curr Biol*, 20, pp.1310–1315.

Butler, E. K., et al., 2012. The mitochondrial chaperone protein TRAP1 mitigates alpha-Synuclein toxicity. *PLoS Genet*, 8, p.e1002488.

Canet-Aviles, R. M., et al., 2004. The Parkinson's disease protein DJ-1 is neuroprotective due to cysteine-sulfinic acid-driven mitochondrial localization. *Proc Natl Acad Sci U S A*, 101, pp.9103–9108.

Cha, G. H., et al., 2005. Parkin negatively regulates JNK pathway in the dopaminergic neurons of Drosophila. *Proc Natl Acad Sci U S A*, 102, pp.10345–10350.

Chen, Y., et al., 2013. VPS35 Asp620Asn and EIF4G1 Arg1205His mutations are rare in Parkinson disease from Southwest China. *Neurobiol Aging*, 34, pp.1709.e7–1709.e8.

Cherra, S. J., 3rd, et al., 2013. Mutant LRRK2 elicits calcium imbalance and depletion of dendritic mitochondria in neurons. *Am J Pathol*, 182, pp.474–484.

Choubey, V., et al., 2011. Mutant A53T alpha-synuclein induces neuronal death by increasing mitochondrial autophagy. *J Biol Chem*, 286, pp.10814–10824.

Chung, C. Y., et al., 2009. Dynamic changes in presynaptic and axonal transport proteins combined with striatal neuroinflammation precede dopaminergic neuronal loss in a rat model of AAV alpha-synucleinopathy. *J Neurosci*, 29, pp.3365–3373.

Clark, I. E., et al., 2006. Drosophila pink1 is required for mitochondrial function and interacts genetically with parkin. *Nature*, 441, pp.1162–1166.

Collier, T. J., Kanaan, N. M. and Kordower, J. H., 2011. Ageing as a primary risk factor for Parkinson's disease: evidence from studies of non-human primates. *Nat Rev Neurosci*, 12, pp.359–366.

Conway, K. A., et al., 2001. Kinetic stabilization of the alpha-synuclein protofibril by a dopamine-alpha-synuclein adduct. *Science*, 294, pp.1346–1349.

Cookson, M. R., 2010. DJ-1, PINK1, and their effects on mitochondrial pathways. *Mov Disord*, 25 Suppl 1, pp.S44–S48.

Corti, O., Lesage, S. and Brice, A., 2011. What genetics tells us about the causes and mechanisms of Parkinson's disease. *Physiol Rev*, 91, pp.1161–1218.

Dauer, W. and Przedborski, S., 2003. Parkinson's disease: Mechanisms and models. *Neuron*, 39, pp.889–909.

Davidzon, G., et al., 2006. Early-onset familial parkinsonism due to POLG mutations. *Ann Neurol*, 59, pp.859–862.

de Lau, L. M. and Breteler, M. M., 2006. Epidemiology of Parkinson's disease. *Lancet Neurol*, 5, pp.525–535.

Deng, H., et al., 2008. The Parkinson's disease genes pink1 and parkin promote mitochondrial fission and/or inhibit fusion in Drosophila. *Proc Natl Acad Sci U S A*, 105, pp.14503–14508.

Deng, H., Gao, K. and Jankovic, J., 2013. The VPS35 gene and Parkinson's disease. *Mov Disord*, 28, pp.569–575.

Devi, L., et al., 2008. Mitochondrial import and accumulation of alpha-synuclein impair complex I in human dopaminergic neuronal cultures and Parkinson disease brain. *J Biol Chem*, 283, pp.9089–9100.

Dexter, D. T., et al., 1994. Indices of oxidative stress and mitochondrial function in individuals with incidental Lewy body disease. *Ann Neurol*, 35, pp.38–44.

Di Monte, D. A., 2003. The environment and Parkinson's disease: is the nigrostriatal system preferentially targeted by neurotoxins? *Lancet Neurol*, 2, pp.531–538.

Dorsey, E. R., et al., 2007. Projected number of people with Parkinson disease in the most populous nations, 2005 through 2030. *Neurology*, 68, 384–386.

Doss-Pepe, E. W., Chen, L. and Madura, K., 2005. Alpha-synuclein and parkin contribute to the assembly of ubiquitin lysine 63-linked multiubiquitin chains. *J Biol Chem*, 280, pp.16619–16624.

Ekstrand, M. I., et al., 2007. Progressive parkinsonism in mice with respiratory-chain-deficient dopamine neurons. *Proc Natl Acad Sci U S A*, 104, pp.1325–1330.

Esteves, A. R., et al., 2009. Oxidative stress involvement in alpha-synuclein oligomerization in Parkinson's disease cybrids. *Antioxid Redox Signal*, 11, pp.439–448.

Exner, N., et al., 2012. Mitochondrial dysfunction in Parkinson's disease: Molecular mechanisms and pathophysiological consequences. *EMBO J*, 31, pp.3038–3062.

Floor, E. and Wetzel, M.G., 1998. Increased protein oxidation in human substantia nigra pars compacta in comparison with basal ganglia and prefrontal cortex measured with an improved dinitrophenylhydrazine assay. *J Neurochem*, 70, pp.268–275.

Forno, L. S., 1996. Neuropathology of Parkinson's disease. *J Neuropathol Exp Neurol*, 55, pp.259–272.

Fuchs, J., et al., 2008. Genetic variability in the SNCA gene influences alpha-synuclein levels in the blood and brain. *FASEB J*, 22, pp.1327–1334.

Galter, D., et al., 2006. LRRK2 expression linked to dopamine-innervated areas. *Ann Neurol*, 59, pp.714–719.

Gandhi, S., et al., 2009. PINK1-associated Parkinson's disease is caused by neuronal vulnerability to calcium-induced cell death. *Mol Cell*, 33, pp.627–638.

Gautier, C. A., Kitada, T. and Shen, J., 2008. Loss of PINK1 causes mitochondrial functional defects and increased sensitivity to oxidative stress. *Proc Natl Acad Sci U S A*, 105, pp.11364–11369.

Geisler, S., et al., 2010. PINK1/Parkin-mediated mitophagy is dependent on VDAC1 and p62/SQSTM1. *Nat Cell Biol*, 12, pp.119–131.

Giasson, B. I., et al., 2006. Biochemical and pathological characterization of Lrrk2. *Ann Neurol*, 59, pp.315–322.

Greene, J. C., et al., 2003. Mitochondrial pathology and apoptotic muscle degeneration in Drosophila parkin mutants. *Proc Natl Acad Sci U S A*, 100, pp.4078–4083.

Grunewald, A., et al., 2009. Differential effects of PINK1 nonsense and missense mutations on mitochondrial function and morphology. *Exp Neurol*, 219, pp.266–273.

Guardia-Laguarta, C., et al., 2014. alpha-Synuclein is localized to mitochondria-associated ER membranes. *J Neurosci*, 34, pp.249–259.

Guzman, J. N., et al., 2010. Oxidant stress evoked by pacemaking in dopaminergic neurons is attenuated by DJ-1. *Nature*, 468, pp.696–700.

Hampe, C., et al., 2006. Biochemical analysis of Parkinson's disease-causing variants of Parkin, an E3 ubiquitin-protein ligase with monoubiquitylation capacity. *Hum Mol Genet*, 15, pp.2059–2075.

Hao, L. Y., Giasson, B. I. and Bonini, N. M., 2010. DJ-1 is critical for mitochondrial function and rescues PINK1 loss of function. *Proc Natl Acad Sci U S A*, 107, pp.9747–9752.

Hasegawa, E., et al., 1990. 1-Methyl-4-phenylpyridinium (MPP+) induces NADH-dependent superoxide formation and enhances NADH-dependent lipid peroxidation in bovine heart submitochondrial particles. *Biochem Biophys Res Commun*, 170, pp.1049–1055.

Hauser, D. N. and Hastings, T. G., 2013. Mitochondrial dysfunction and oxidative stress in Parkinson's disease and monogenic parkinsonism. *Neurobiol Dis*, 51, pp.35–42.

Healy, D. G., et al., 2008. Phenotype, genotype, and worldwide genetic penetrance of LRRK2-associated Parkinson's disease: A case-control study. *Lancet Neurol*, 7, pp.583–590.

Higashi, S., et al., 2007. Localization of Parkinson's disease-associated LRRK2 in normal and pathological human brain. *Brain Res*, 1155, pp.208–219.

Hirsch, E. C., et al., 1991. Iron and aluminum increase in the substantia nigra of patients with Parkinson's disease: An X-ray microanalysis. *J Neurochem*, 56, pp.446–451.

Hoepken, H. H., et al., 2007. Mitochondrial dysfunction, peroxidation damage and changes in glutathione metabolism in PARK6. *Neurobiol Dis*, 25, pp.401–411.

Hsu, L. J., et al., 2000. alpha-synuclein promotes mitochondrial deficit and oxidative stress. *Am J Pathol*, 157, pp.401–410.

Jankovic, J., 2008. Parkinson's disease: Clinical features and diagnosis. *J Neurol Neurosurg Psychiatry*, 79, pp.368–376.

Kahle, P. J., 2008. alpha-Synucleinopathy models and human neuropathology: Similarities and differences. *Acta Neuropathol*, 115, pp.87–95.

Kamp, F., et al., 2010. Inhibition of mitochondrial fusion by alpha-synuclein is rescued by PINK1, Parkin and DJ-1. *EMBO J*, 29, pp.3571–3589.

Karpinar, D. P., et al., 2009. Pre-fibrillar alpha-synuclein variants with impaired beta-structure increase neurotoxicity in Parkinson's disease models. *EMBO J*, 28, pp.3256–3268.

Kim, R. H., et al., 2005. Hypersensitivity of DJ-1-deficient mice to 1-methyl-4-phenyl-1,2,3, 6-tetrahydropyrindine (MPTP) and oxidative stress. *Proc Natl Acad Sci U S A*, 102, pp.5215–5220.

Kinumi, T., et al., 2004. Cysteine-106 of DJ-1 is the most sensitive cysteine residue to hydrogen peroxide-mediated oxidation in vivo in human umbilical vein endothelial cells. *Biochem Biophys Res Commun*, 317, pp.722–728.

Kitada, T., et al., 1998. Mutations in the parkin gene cause autosomal recessive juvenile parkinsonism. *Nature*, 392, pp.605–608.

Kitada, T., et al., 2007. Impaired dopamine release and synaptic plasticity in the striatum of PINK1-deficient mice. *Proc Natl Acad Sci U S A*, 104, pp.11441–11446.

Krebiehl, G., et al., 2010. Reduced basal autophagy and impaired mitochondrial dynamics due to loss of Parkinson's disease-associated protein DJ-1. *PLoS One*, 5, p.e9367.

Kujoth, G. C., et al., 2005. Mitochondrial DNA mutations, oxidative stress, and apoptosis in mammalian aging. *Science*, 309, pp.481–484.

Kuroda, Y., et al., 2006. Parkin enhances mitochondrial biogenesis in proliferating cells. *Hum Mol Genet*, 15, pp.883–895.

Langston, J. W., et al., 1984. 1-Methyl-4-phenylpyridinium ion (MPP+): identification of a metabolite of MPTP, a toxin selective to the substantia nigra. *Neurosci Lett*, 48, pp.87–92.

LaVoie, M. J., et al., 2005. Dopamine covalently modifies and functionally inactivates parkin. *Nat Med*, 11, pp.1214–1221.

Lehman, N. L., 2009. The ubiquitin proteasome system in neuropathology. *Acta Neuropathol*, 118, 329–347.

Lin, X., et al., 2009. Leucine-rich repeat kinase 2 regulates the progression of neuropathology induced by Parkinson's-disease-related mutant alpha-synuclein. *Neuron*, 64, pp.807–827.

Liu, G., et al., 2009a. alpha-Synuclein is differentially expressed in mitochondria from different rat brain regions and dose-dependently down-regulates complex I activity. *Neurosci Lett*, 454, pp.187–192.

Liu, W., et al., 2009b. PINK1 defect causes mitochondrial dysfunction, proteasomal deficit and alpha-synuclein aggregation in cell culture models of Parkinson's disease. *PLoS One*, 4, p.e4597.

Loeb, V., et al., 2010. The transgenic overexpression of alpha-synuclein and not its related pathology associates with complex I inhibition. *J Biol Chem*, 285, pp.7334–7343.

Lucas, J. I. and Marin, I., 2007. A new evolutionary paradigm for the Parkinson disease gene DJ-1. *Mol Biol Evol*, 24, pp.551–561.

Marongiu, R., et al., 2009. Mutant Pink1 induces mitochondrial dysfunction in a neuronal cell model of Parkinson's disease by disturbing calcium flux. *J Neurochem*, 108, pp.1561–1574.

Martin, L. J., et al., 2006. Parkinson's disease alpha-synuclein transgenic mice develop neuronal mitochondrial degeneration and cell death. *J Neurosci*, 26, pp.41–50.

Mastroberardino, P. G., et al., 2009. A novel transferrin/TfR2-mediated mitochondrial iron transport system is disrupted in Parkinson's disease. *Neurobiol Dis*, 34, pp.417–431.

Mata, I. F., et al., 2006. LRRK2 in Parkinson's disease: Protein domains and functional insights. *Trends Neurosci*, 29, pp.286–293.

Matsuda, N., et al., 2006. Diverse effects of pathogenic mutations of Parkin that catalyze multiple monoubiquitylation in vitro. *J Biol Chem*, 281, pp.3204–3209.

Mitsumoto, A. and Nakagawa, Y., 2001. DJ-1 is an indicator for endogenous reactive oxygen species elicited by endotoxin. *Free Radic Res*, 35, pp.885–893.

Moore, D. J., 2006. Parkin: A multifaceted ubiquitin ligase. *Biochem Soc Trans*, 34, pp.749–753.

Morais, V. A., et al., 2014. PINK1 loss-of-function mutations affect mitochondrial complex I activity via NdufA10 ubiquinone uncoupling. *Science*, 344, pp.203–207.

Mori, H., et al., 1998. Pathologic and biochemical studies of juvenile parkinsonism linked to chromosome 6q. *Neurology*, 51, pp.890–892.

Mortiboys, H., et al., 2010. Mitochondrial impairment in patients with Parkinson disease with the G2019S mutation in LRRK2. *Neurology*, 75, pp.2017–2020.

Mosharov, E. V., et al., 2009. Interplay between cytosolic dopamine, calcium, and alpha-synuclein causes selective death of substantia nigra neurons. *Neuron*, 62, pp.218–229.

Mukhopadhyay, D. and Riezman, H., 2007. Proteasome-independent functions of ubiquitin in endocytosis and signaling. *Science*, 315, pp.201–205.

Nakamura, K., et al., 2011. Direct membrane association drives mitochondrial fission by the Parkinson disease-associated protein alpha-synuclein. *J Biol Chem*, 286, pp.20710–20726.

Narendra, D., et al., 2008. Parkin is recruited selectively to impaired mitochondria and promotes their autophagy. *J Cell Biol*, 183, pp.795–803.

Narendra, D. P., et al., 2010. PINK1 is selectively stabilized on impaired mitochondria to activate Parkin. *PLoS Biol*, 8, p.e1000298.

Nemani, V. M., et al., 2010. Increased expression of alpha-synuclein reduces neurotransmitter release by inhibiting synaptic vesicle reclustering after endocytosis. *Neuron*, 65, pp.66–79.

Niu, J., et al., 2012. Leucine-rich repeat kinase 2 disturbs mitochondrial dynamics via Dynamin-like protein. *J Neurochem*, 122, pp.650–658.

Nussbaum, R. L. and Ellis, C. E., 2003. Alzheimer's disease and Parkinson's disease. *N Engl J Med*, 348, pp.1356–1364.

Oakley, A. E., et al., 2007. Individual dopaminergic neurons show raised iron levels in Parkinson disease. *Neurology*, 68, pp.1820–1825.

Paisan-Ruiz, C., et al., 2004. Cloning of the gene containing mutations that cause PARK8-linked Parkinson's disease. *Neuron*, 44, pp.595–600.

Palacino, J. J., et al., 2004. Mitochondrial dysfunction and oxidative damage in parkin-deficient mice. *J Biol Chem*, 279, pp.18614–18622.

Park, J., et al., 2006. Mitochondrial dysfunction in Drosophila PINK1 mutants is complemented by parkin. *Nature*, 441, pp.1157–1161.

Park, J., Lee, G. and Chung, J., 2009. The PINK1-Parkin pathway is involved in the regulation of mitochondrial remodeling process. *Biochem Biophys Res Commun*, 378, pp.518–523.

Parker, W. D., Jr., Parks, J. K. and Swerdlow, R. H., 2008. Complex I deficiency in Parkinson's disease frontal cortex. *Brain Res*, 1189, pp.215–218.

Perry, T. L., Godin, D. V. and Hansen, S., 1982. Parkinson's disease: a disorder due to nigral glutathione deficiency? *Neurosci Lett*, 33, pp.305–310.

Pickrell, A. M., et al., 2011. Striatal dysfunctions associated with mitochondrial DNA damage in dopaminergic neurons in a mouse model of Parkinson's disease. *J Neurosci*, 31, pp.17649–17658.

Pilsl, A. and Winklhofer, K. F., 2012. Parkin, PINK1 and mitochondrial integrity: emerging concepts of mitochondrial dysfunction in Parkinson's disease. *Acta Neuropathol*, 123, pp.173–188.

Plun-Favreau, H., et al., 2007. The mitochondrial protease HtrA2 is regulated by Parkinson's disease-associated kinase PINK1. *Nat Cell Biol*, 9, pp.1243–1252.

Polymeropoulos, M. H., et al., 1997. Mutation in the alpha-synuclein gene identified in families with Parkinson's disease. *Science*, 276, pp.2045–2047.

Poole, A. C., et al., 2008. The PINK1/Parkin pathway regulates mitochondrial morphology. *Proc Natl Acad Sci U S A*, 105, pp.1638–1643.

Pridgeon, J. W., et al., 2007. PINK1 protects against oxidative stress by phosphorylating mitochondrial chaperone TRAP1. *PLoS Biol*, 5, p.e172.

Puschmann, A., 2013. Monogenic Parkinson's disease and parkinsonism: clinical phenotypes and frequencies of known mutations. *Parkinsonism Relat Disord*, 19, pp.407–415.

Rabinovic, A. D. and Hastings, T. G., 1998. Role of endogenous glutathione in the oxidation of dopamine. *J Neurochem*, 71, pp.2071–2078.

Rakovic, A., et al., 2011. PINK1-interacting proteins: Proteomic analysis of overexpressed PINK1. *Parkinsons Dis*, 2011, p.153979.

Ramonet, D., et al., 2011. Dopaminergic neuronal loss, reduced neurite complexity and autophagic abnormalities in transgenic mice expressing G2019S mutant LRRK2. *PLoS One*, 6, p.e18568.

Ross, O. A., et al., 2006. Lrrk2 and Lewy body disease. *Ann Neurol*, 59, pp.388–393.

Satake, W., et al., 2009. Genome-wide association study identifies common variants at four loci as genetic risk factors for Parkinson's disease. *Nat Genet*, 41, pp.1303–1307.

Schapira, A. H., 2007a. Treatment options in the modern management of Parkinson disease. *Arch Neurol*, 64, pp.1083–1088.

Schapira, A. H., 2007b. Mitochondrial dysfunction in Parkinson's disease. *Cell Death Differ*, 14, pp.1261–1266.

Scott, D. A., et al., 2010. A pathologic cascade leading to synaptic dysfunction in alpha-synuclein-induced neurodegeneration. *J Neurosci*, 30, pp.8083–8095.

Sherer, T. B., et al., 2003. Mechanism of toxicity in rotenone models of Parkinson's disease. *J Neurosci*, 23, pp.10756–10764.

Shimura, H., et al., 2000. Familial Parkinson disease gene product, parkin, is a ubiquitin-protein ligase. *Nat Genet*, 25, pp.302–305.

Shin, J. H., et al., 2011. PARIS (ZNF746) repression of PGC-1alpha contributes to neurodegeneration in Parkinson's disease. *Cell*, 144, pp.689–702.

Sian, J., et al., 1994. Alterations in glutathione levels in Parkinson's disease and other neurodegenerative disorders affecting basal ganglia. *Ann Neurol*, 36, pp.348–355.

Simon-Sanchez, J., et al., 2009. Genome-wide association study reveals genetic risk underlying Parkinson's disease. *Nat Genet*, 41, pp.1308–1312.

Singleton, A. B., et al., 2003. alpha-Synuclein locus triplication causes Parkinson's disease. *Science*, 302, p.841.

Sofic, E., et al., 1988. Increased iron (III) and total iron content in post mortem substantia nigra of parkinsonian brain. *J Neural Transm*, 74, p.199–205.

Sousa, V. L., et al., 2009. {alpha}-synuclein and its A30P mutant affect actin cytoskeletal structure and dynamics. *Mol Biol Cell*, 20, p.3725–3739.

Spencer, J. P., et al., 1998. Conjugates of catecholamines with cysteine and GSH in Parkinson's disease: possible mechanisms of formation involving reactive oxygen species. *J Neurochem*, 71, p.2112–2122.

Spillantini, M. G., et al., 1997. Alpha-synuclein in Lewy bodies. *Nature*, 388, pp.839–840.

Swerdlow, R. H., et al., 1996. Origin and functional consequences of the complex I defect in Parkinson's disease. *Ann Neurol*, 40, pp.663–671.

Synofzik, M., et al., 2010. Sustained dopaminergic response of parkinsonism and depression in POLG-associated parkinsonism. *Mov Disord*, 25, pp.243–245.

Thomas, K. J., et al., 2011. DJ-1 acts in parallel to the PINK1/parkin pathway to control mitochondrial function and autophagy. *Hum Mol Genet*, 20, pp.40–50.

Todd, A. M. and Staveley, B. E., 2008. Pink1 suppresses alpha-synuclein-induced phenotypes in a Drosophila model of Parkinson's disease. *Genome*, 51, pp.1040–1046.

Trempe, J. F. and Fon, E.A., 2013. Structure and function of Parkin, PINK1, and DJ-1, the three musketeers of neuroprotection. *Front Neurol*, 4, p.38.

Trifunovic, A., et al., 2004. Premature ageing in mice expressing defective mitochondrial DNA polymerase. *Nature*, 429, pp.417–423.

Tsika, E. and Moore, D. J., 2012. Mechanisms of LRRK2-mediated neurodegeneration. *Curr Neurol Neurosci Rep*, 12, 251–260.

Twig, G. and Shirihai, O. S., 2011. The interplay between mitochondrial dynamics and mitophagy. *Antioxid Redox Signal*, 14, pp.1939–1951.

Valente, E. M., et al., 2004. Hereditary early-onset Parkinson's disease caused by mutations in PINK1. *Science*, 304, pp.1158–1160.

Valente, E. M., et al., 2012. Molecular pathways in sporadic PD. *Parkinsonism Relat Disord*, 18 Suppl 1, pp.S71–S73.

van der Brug, M. P., et al., 2008. RNA binding activity of the recessive parkinsonism protein DJ-1 supports involvement in multiple cellular pathways. *Proc Natl Acad Sci U S A*, 105, pp.10244–10249.

Van Humbeeck, C., et al., 2011. Parkin interacts with Ambra1 to induce mitophagy. *J Neurosci*, 31, pp.10249–10261.

Van Laar, V. S., et al., 2009. Proteomic identification of dopamine-conjugated proteins from isolated rat brain mitochondria and SH-SY5Y cells. *Neurobiol Dis*, 34, pp.487–500.

Ved, R., et al., 2005. Similar patterns of mitochondrial vulnerability and rescue induced by genetic modification of alpha-synuclein, parkin, and DJ-1 in Caenorhabditis elegans. *J Biol Chem*, 280, pp.42655–42668.

Venderova, K., et al., 2009. Leucine-rich repeat kinase 2 interacts with Parkin, DJ-1 and PINK-1 in a Drosophila melanogaster model of Parkinson's disease. *Hum Mol Genet*, 18, pp.4390–4404.

Vilarino-Guell, C., et al., 2011. VPS35 mutations in Parkinson disease. *Am J Hum Genet*, 89, pp.162–167.

Vives-Bauza, C., et al., 2010. PINK1-dependent recruitment of Parkin to mitochondria in mitophagy. *Proc Natl Acad Sci U S A*, 107, pp.378–383.

Votyakova, T. V. and Reynolds, I. J., 2005. Ca2+-induced permeabilization promotes free radical release from rat brain mitochondria with partially inhibited complex I. *J Neurochem*, 93, pp.526–537.

Wang, X., et al., 2012. LRRK2 regulates mitochondrial dynamics and function through direct interaction with DLP1. *Hum Mol Genet*, 21, pp.1931–1944.

Weinreb, O., et al., 2010. Neuroprotective multifunctional iron chelators: from redox-sensitive process to novel therapeutic opportunities. *Antioxid Redox Signal*, 13, pp.919–949.

West, A. B., et al., 2007. Parkinson's disease-associated mutations in LRRK2 link enhanced GTP-binding and kinase activities to neuronal toxicity. *Hum Mol Genet*, 16, pp.223–232.

Whitworth, A. J., et al., 2005. Increased glutathione S-transferase activity rescues dopaminergic neuron loss in a Drosophila model of Parkinson's disease. *Proc Natl Acad Sci U S A*, 102, pp.8024–8029.

Winkler, S., et al., 2007. alpha-Synuclein and Parkinson disease susceptibility. *Neurology*, 69, pp.1745–1750.

Wood-Kaczmar, A., et al., 2008. PINK1 is necessary for long term survival and mitochondrial function in human dopaminergic neurons. *PloS one*, 3, p.e2455.

Xu, J., et al., 2002. Dopamine-dependent neurotoxicity of alpha-synuclein: A mechanism for selective neurodegeneration in Parkinson disease. *Nat Med*, 8, pp.600–606.

Yamano, K. and Youle, R. J., 2013. PINK1 is degraded through the N-end rule pathway. *Autophagy*, 9, pp.1758–1769.

Yang, Y., et al., 2006. Mitochondrial pathology and muscle and dopaminergic neuron degeneration caused by inactivation of Drosophila Pink1 is rescued by Parkin. *Proc Natl Acad Sci U S A*, 103, pp.10793–10798.

Yang, Y., et al., 2008. Pink1 regulates mitochondrial dynamics through interaction with the fission/fusion machinery. *Proc Natl Acad Sci U S A*, 105, pp.7070–7075.

Zheng, B., et al., 2010. PGC-1alpha, a potential therapeutic target for early intervention in Parkinson's disease. *Sci Transl Med*, 2, p.52ra73.

Zimprich, A., et al., 2004. Mutations in LRRK2 cause autosomal-dominant parkinsonism with pleomorphic pathology. *Neuron*, 44, pp.601–607.

Zimprich, A., et al., 2011. A mutation in VPS35, encoding a subunit of the retromer complex, causes late-onset Parkinson disease. *Am J Hum Genet*, 89, pp.168–175.

17

THERAPEUTIC TARGETING OF NEURONAL MITOCHONDRIA IN BRAIN INJURY

HEATHER M. YONUTAS, EDWARD D. HALL AND PATRICK G. SULLIVAN

Department of Anatomy and Neurobiology, Spinal Cord and Brain Injury Research Center, University of Kentucky Chandler College of Medicine, Lexington, Kentucky

17.1 INTRODUCTION

Mitochondria are an integral component of the function of almost all cells, and particularly neurons, because of high neuronal energy demands. The study of these microscopic organelles has revealed fundamental insights into the workings of a cell that cover a broad spectrum from genetics to biophysics to cell biology and have demonstrated a relationship to the onset of various chronic neurodegenerative diseases. Mitochondria have also been linked to the pathology seen following acute brain trauma caused by strokes and brain injury and have profound implications as a key organelle in the initiation of various cell death pathways. For these reasons, mitochondria occupy a central position in the search for new and effective therapies in these disorders.

Neuronal homeostasis and function have been found to be dependent on mitochondrial bioenergetics. As the proclaimed "power plant" of the cell, mitochondria provide most of the energy needed for the cell to function properly (Chance and Smith, 1955; Chance and Williams, 1955a, 1955b, 1955c, 1955d; Chance et al., 1955, 1973; Chance, Sies, and Boveris, 1979; Chancey and Zatz, 1955). As mitochondria become dysfunctional, cellular dysfunction soon follows. Therefore, intense recent interest

The Functions, Disease-Related Dysfunctions, and Therapeutic Targeting of Neuronal Mitochondria, First Edition. Edited by Valentin K. Gribkoff, Elizabeth A. Jonas and J. Marie Hardwick.

has focused on the discovery of pharmaceutical interventions targeting the mitochondrial dysfunction that is associated with various diseased states. Many therapeutic interventions that target mitochondrial function have shown positive outcome measures in various neuronal disease models, although the translation of these findings to human disease has yet to be fully realized. One disease model that has been linked to mitochondrial dysfunction and has shown improved outcome with mitochondrial-specific interventions is traumatic brain injury (TBI).

In this chapter we will discuss mitochondrial bioenergetics and their importance to proper brain function. With a focus on TBI, the importance of mitochondria in cell death pathways initiated by an injury to the central nervous system (CNS) will be reviewed, and we will highlight findings that indicate that interventions targeting mitochondria are able to improve histological and functional outcomes following injury in animal models.

17.2 MITOCHONDRIA BIOENERGETICS

The study of mitochondria dates back to the mid-1800s, when researchers observed these previously unknown but characteristically structured subcellular organelles and described them with words such as *chondros*, *Korn*, and *grain*. As staining techniques improved, there was a shift in the terminology. In 1898, the Greek term *mitos*, which refers to their threadlike internal appearance, combined with chondros, which is from the Greek for "granule," led to the eventual coining of the new word "mitochondrion" (review by Chowdry 1918, quoted in Lehninger, 1965 and Scheffler, 2008). Remarkably, in the 1890s, Altman termed these intracellular granules *bioplasts* and proposed they were autonomous, forming bacterial-like colonies within their host cells (Scheffler, 2008). Based on our current knowledge and the evolutionary theory of a prokaryotic and eukaryotic symbiosis formed billions of years ago, this was indeed an astute observation (Wallace, 1992, 2007, 2010).

To understand the importance of mitochondrial dysfunction in TBI, we need to address normal mitochondrial function as it relates to the CNS. In the cells of higher-order animals, 95% of the common energy currency, ATP, is produced via oxidative phosphorylation within mitochondria. Mitochondria are intracellular organelles with a dual (inner and outer) membrane system, each responsible for specific functions. The outer mitochondrial membrane (OMM) contains many transporter proteins tasked with the import and export of ions and proteins necessary to maintain mitochondrial function (Nicholls and Ferguson, 2002). The inner mitochondrial membrane (IMM) has many folds, termed *cristae*, which increase the surface area available for mitochondrial respiration and are the site of electron transport and oxidative phosphorylation (OXPHOS). The space enclosed by the IMM is termed the *matrix* and contains enzymes involved in cellular metabolism and calcium regulation.

Within the cristae there are five protein complexes that comprise the electron transport chain (ETC). In terms of thermodynamics and electrochemical properties, each step in the ETC is bound to free-energy changes as each component cycles through oxidation and reduction reactions. The components of the ETC include complex I, II, III, IV, as well as the mobile carriers, coenzyme Q10 and cytochrome C. Complex I, NADH-ubiquinone oxidoreductase, which is embedded within the IM, converts NADH to NAD+ by accepting an electron into the Fe-S center of the protein. As a by-product of this electron donation, a proton is translocated from the matrix into the intermembrane space (Wake et al., 2007), which is located between the inner and outer membranes. Complex II, succinate dehydrogenase, in addition to its function as an ETC protein, is also a key component of the Krebs cycle, which converts the glycolytic product pyruvate into substrates for the ETC. Complex II uses the conversion of succinate to accept electrons from FADH2 into the ETC with no translocation of protons from the matrix to the IMS. Complexes I and II transfer their electrons to the ubiquinone, coenzyme Q10, located within the IMM. These electrons are then passed to Complex III, ubiquinone-cytochrome-C oxidoreductase, via the Q-cycle, resulting in a proton being translocated into the IMM space. Cytochrome C then accepts this electron and transports it to Complex IV, cytochrome-C oxidase; again translocating a proton into the IMM space. It is at Complex IV that oxygen plays its vital role as the final electron acceptor for the ETC; O_2 molecules are reduced by these electrons, forming H_2O. The protons that have been translocated into the IMM space generate a chemiosmotic gradient/membrane potential, $\Delta\Psi$, which is used by Complex V (ATP synthase), to facilitate the phosphorylation of adenosine diphosphate (ADP) into adenosine triphosphate (ATP) for use as an energy source for cellular processes.

Although OXPHOS is a critically important process, it does create toxic by-products. Because of this, mitochondria are known as one of the cell's major supplier of reactive oxygen species (ROS) and reactive nitrogen species (RNS). During OXPHOS, complex IV will release an electron to molecular oxygen, O_2, allowing for the eventual formation of H_2O. However, electron "slippage" can also occur. This occurs as a result of inefficiencies within Complexes I, II, or III. Once an electron slips, or is released, from the ETC, it can reduce O_2 or react with proteins, nucleic acids, and fatty acids within the mitochondria and mitochondrial membrane, leading to damage. This process is often referred to as metabolic-mediated oxidative damage.

Following the release of an electron from the ETC, mitochondrial manganese superoxide dismutase (MnSOD) also known as SOD2, a mitochondrial-located metalloenyzme that is a crucial for the dismutation of this highly reactive $O_2^{\bullet-}$ into H_2O_2, can neutralize the reactive $O_2^{\bullet-}$ formed. However H_2O_2 is also a strong oxidizing agent. Therefore, glutathione peroxidase, which is also located within the mitochondria, can convert H_2O_2 into neutral H_2O (Jastroch et al., 2010; Moreira et al., 2010).

Although the exact mechanism is not completely understood, cells within the brain seem to be more vulnerable to oxidative stress when compared to cells in most other organs. This has been attributed to the fact that the brain is an organ that consumes approximately 20% of the body's total oxygen, greatly disproportionate to its percentage of body weight, and has naturally lower levels of endogenous antioxidant activity relative to other tissues (Floyd and Hensley, 2002; Halliwell and Gutteridge, 1985; Shulman et al., 2004). Additionally, neurons have high energy demands compared to most other cell types. Among other reasons, the Na^+/K^+ ATPase pumps located within neuronal membranes function to maintain the membrane potential necessary for action potentials to be generated for intercellular communication. It is estimated that one third of the body's entire ATP production is used by these pumps (Buono and Kolkhorst, 2001). Because neurons primarily use oxidative phosphorylation to generate their ATP pools, they proportionately produce more metabolic-mediated oxidative stress, which can be detrimental to the cell over time.

In addition to producing 95% of the cell's energy currency, ATP, and being a major source of oxidative stress, mitochondria also play an integral role in the buffering of cytosolic calcium (Ca^{2+}; see also contribution by Usachev, Chapter 4 of this volume). This is important because Ca^{2+} is a potent intracellular signaling molecule with both beneficial and detrimental roles in neurons. In the brain, during normal cellular interactions, neurons become activated following the binding of various neurotransmitters to their respective metabotropic or inotropic receptors. One particularly ubiquitous and important neurotransmitter that has been implicated in the Ca^{2+} signaling pathway and has an important role in the secondary injury cascade of TBI is the amino acid glutamate. Glutamate binds to many receptors; one of most important is the ionotropic α-amino-3-hydroxy-5-methyl-4-isoxazolepropionic acid (AMPA) receptor. Once glutamate is bound, these ionotropic receptors move to an open conformation leading to the influx of Na^+ causing a local depolarization. This local depolarization leads to the displacement of Mg^{2+} blockage in another glutamate receptor class, the N-methyl-D-aspartate (NMDA) receptor. Once Mg^{2+} is removed, the NMDA receptor has a higher probability of being in an "open" confirmation; this ionotropic receptor allows Ca2+ to enter the neuron. Glutamate-induced depolarization produced by AMPA receptors therefore results indirectly in Ca^{2+} entry via NMDA receptors and by the family of voltage-dependent Ca^{2+} channels.

As long as the NMDA receptors and any voltage-dependent Ca^{2+} channels activated by glutamate release remain in an open confirmation, Ca^{2+} will continue to enter the neuron. At lower concentrations this Ca^{2+} has been shown to regulate the response to neurotransmitter release and has been shown to regulate the Krebs cycle through activation of pyruvate dehydrogenase (PDH), NAD+-dependent isocitrate dehydrogenase (ICDH) and the α-ketoglutarate dehydrogenase complex (KGDHC). This leads to increased NADH production, which can, in turn, lead to increased ATP production by the ETC. Ca^{2+}-mediated activation has also been shown to increase Complex V activity, resulting in increased ATP synthesis as well (Sullivan et al., 2005; Yi and Hazell, 2006). However, if Ca^{2+} levels remain high for too long, detrimental effects will ensue. For instance, sustained high Ca^{2+} concentrations in the cytosol can affect synaptic stability and function through the activation of various calcineurins and calpains and cause cytoskeletal components to break down, resulting in detrimental effects on neuronal viability (Supnet and Bezprozvanny, 2010)

As transient increases in cytosolic Ca^{2+} levels occur, mitochondria are able to sequester Ca^{2+} in an electrogenic fashion, under normal conditions effectively removing it from the cytosol. As mentioned previously, the ETC generates a negative membrane potential ($\Delta\Psi$) within the mitochondria. Ca^{2+} uniporters, located in the inner mitochondrial membrane, use this $\Delta\Psi$ generated by the ETC to conduct Ca^{2+} ions down their electrochemical gradient into the matrix. As the Ca^{2+} uniporter continues to translocate Ca^{2+} into the matrix, the $\Delta\Psi$ becomes diminished leading to further activation of the ETC and an attempt to restore the depleted potential. This process results in an intimate relationship between mitochondrial bioenergetics, Ca^{2+} homeostasis and ATP production (see Chapter 4 of this volume). In a state of extreme neuronal stress, such as following an injury to the CNS, this relationship can become aberrant, leading to the generation of increased oxidative stress and the activation of the mitochondrial permeability transition pore, mPTP.

As mentioned previously, in a resting state mitochondria maintain a membrane potential of ~ −180mV. As mitochondrial matrix Ca^{2+} concentrations, oxidative stress and damage to ETC proteins increases, $\Delta\Psi$ can become depleted. Once reduced to ~ −100 to −120mV, the ETC will begin to slow down, stop and potentially start running backward. Once this occurs, ATP production will decrease, and if the ETC begins to run backward, ATP will be consumed by ATP hydrolysis. This leads to increased generation of ROS and RNS (Sullivan, Sebastian, and Hall, 2011). The increase in ROS and RNS can then lead to increased oxidative damage within the mitochondria. Mitochondrial dysfunction is further exacerbated as the damage to lipid, protein and nucleic acids continues. Eventually, the redox potential within the mitochondrion changes and crucial mitochondrial proteins like the adenine nucleotide translocator (ANT) undergo oxidative modifications. This occurs specifically as a thiol oxidation of ANT, resulting in a conformational change that promotes cyclophilin D (CypD) binding. Once ANT, an inner membrane mitochondrial protein, is bound to cypD, conformational changes occur that induce ANT, cypD, and many other proteins within the complex to bind with the voltage-dependent anion channel (VDAC) (Belzacq et al., 2002; Halestrap, 2006; Varanyuwatana and Halestrap, 2012). VDAC is an outer membrane mitochondrial protein that, once bound to the inner mitochondrial membrane protein complex including Cyclophilin D, ANT, and many other proteins, forms the mitochondrial permeability transition pore, mPTP (for additional discussion of the structure and function of the mPTP, see Chapters 1 to 3 of this volume).

The formation of the mPTP exposes the inner mitochondrial matrix to the cytosol, allowing the mitochondria to release apoptotic initiating factors like cytochrome C, SMAC/diablo and AIF which all promote cell death. Additionally, mitochondria can also initiate another route of cell death called necrosis. This occurs as a result of aberrant changes in the concentration of ions like chloride across the mitochondrial membrane. These ionic concentration differences allow water to freely diffuse across the mitochondrial membrane, leading to swelling of the inner membrane resulting from increases in osmotic pressure. ATP production decreases because of dysfunction within the ETC as the $\Delta\Psi$ decreases and oxidative stress increases. This causes a complete destruction of the mitochondria as a result of the swelling of the inner mitochondrial membrane. Eventually these mitochondria will burst leading to the initiation of cell death.

17.3 TRAUMATIC BRAIN INJURY

Based on estimates from the Centers for Disease Control and Prevention, approximately 1.7 million brain injuries occur annually in the United States, resulting in a cost of approximately $76 billion in direct and indirect health-care costs.

These numbers do not include the number of US residents who sustain a mild TBI and are not seen by physicians. With such a large affected population and the high costs of caring for these patients, the need for pro-regenerative or neuroprotective therapeutic agents for the treatment of this disease is great. However to date there is still no drug approved by the Food and Drug Administration (FDA) for its treatment. Although significant effort has gone into the discovery of drugs to treat TBI, many problems and failures have been encountered along the way because of the complicated biochemical pathways that are initiated following the primary injury.

17.3.1 Models of TBI

TBI often occurs in two separate phases. The first phase is the primary injury. This is the initial occurrence of the injury where an object releases kinetic energy via mechanical force directed onto the cranium. This primary injury leads to the disruption of blood vessels, shearing of axons and some immediate cell death (LaPlaca et al., 2007). This primary insult within the brain causes cellular membranes around the injury to become permeable to various proteins, ions, and other molecules initiating the secondary injury phase (Tate, Garcia, and LaPlaca, 2007). To date, there are a number of known risk factors that increase the chances of sustaining a TBI following some degree of cranial trauma. These include age, drug usage, nutrition, and psychosocial status. The best practice to avoid the primary injury is prevention, and although prevention techniques have improved over the years, mechanical insults to the brain still occur at an alarming rate.

Following the primary injury, within the first seconds, to minutes, to days and now seemingly expanding to months and years later is the second phase of a TBI called the *secondary injury*. The secondary injury cascade is defined by a complicated set of events including alterations of endogenous neurochemical mechanisms, ischemic insult, massive edema, increases in oxidative stress, and mitochondrial dysfunction which all lead to progressive damage (Yi and Hazell, 2006). It is the progressive damage that researchers and physicians have targeted when investigating potential therapeutic interventions following injury. It is thought that with the advent of effective therapeutic intervention, improved functional outcomes could be obtained as a result of decreased tissue dysfunction and neurodegeneration.

In the clinical setting, TBIs present in many shapes and forms. The heterogeneous nature of clinical TBI causation has sparked the invention of many different experimental injury models (Saatman et al., 2008). Rotational injuries, penetrating injuries, anoxic or hypoxic injury, contusion injury, as well as combination injuries, have all been documented in a clinical setting. Therefore, researchers have multiple experimental models of TBI that are used as a means of inducing specific injuries to reproduce what is seen clinically. Although no model has been shown to be superior, to have predictive validity, given that no successful therapy has yet been developed, several models will be discussed. However, we will concentrate on contusion injury, modeled by controlled cortical impact, because of the significant degree of injury and apparent mitochondrial dysfunction that is induced, and the large number of studies performed using these models.

The experimental injury models can be either static, which would be a compression injury where a force is applied and never removed, or dynamic. Dynamic injury models are the most commonly studied forms of injury. These dynamic injuries can either be the result of an indirect force being applied to the brain, for example induced by an explosion, which is modeled by a blast wave injury model, or by a direct injury model. Direct injury models can be further classified by whether the experimental injuries are penetrating or non-penetrating. Two well-studied models of dynamic, direct, penetrating injuries are the fluid percussion model and the controlled cortical impact model (Cernak, 2005).

The first model discussed is the blast wave injury model. This model has become an area of intense interest due to the marked functional changes produced by blasts or explosions in soldiers, often only detected following their return to the United States and other allied countries from war theaters over the last 14 years. The pathology of this form of injury is still under investigation because discrete anatomical changes have yet to be observed in many of these cases. Despite the absence of an obvious severe injury, functional changes, including behavioral changes, are observed. Following the primary insult, shock waves from the explosion can pass through the brain tissue leading to minute neuroanatomical changes. Additionally, compression of blood vessels can also occur leading to changes in blood flow to the brain, possibly inducing a short-lasting hypoxic state. Lastly, as the blast wave hits the cranium, a complex injury can occur in which the brain is pushed in the direction of the movement of the shock wave, compressing itself along the opposite side of the cranial cavity, then the brain will move back in the opposite direction within the cranial cavity, impacting the skull on the other side. Therefore such an insult, if powerful enough, can cause a coup and contracoup injury impacting both sides of the brain. Although no marked mitochondrial dysfunction has been documented with this model, further investigation is continuing.

The next model of brain injury discussed will be the fluid percussion injury (FPI) model. In this model, water confined in a tube is impinged on by a weight with a specific kinetic energy. This generates a pressure pulse in the water. The water is then rapidly injected into the cranial cavity, through a Luer cap placed in the skull, providing a mechanical force to the brain that is diffuse in nature (Dixon et al., 1987). Following injury, animals experience apnea, and the severity of the injury is measured by the "righting time," the time required for the animal to return to its normal stance and posture. Depending on the location of this injury, either midline or lateral, and the severity, this injury does not produce any overt histopathology or contusion at the site of injury, however with lateral fluid percussions, apparent cavitation and cell death is markedly observed within the ventricular region out to 1 year post-injury. As a model, FPI produces an injury that is thought to mimic a rotational injury of the brain, which can be the result of a clinically traumatic event such as a car accident or sports related injury.

The last model discussed produces a much more focal contusion compared to the diffuse injury seen with FPI. This is the *controlled cortical impact* (CCI) model (Povlishock et al., 1994). In this model damage to the blood–brain barrier (BBB) and cytoskeletal damage in the cortex and hippocampal formation has also been seen

(Brustovetsky and Dubinsky, 2000). CCI uses a pneumatic piston with a specific velocity, depth and dwell time to induce an injury. The severity of the injury can be adjusted based on these parameters. This injury provides a nonpenetrating deformation of the cortex leading to a hematoma that has a similar histopathology to a clinical contusion injury. Immediately following the injury, a bruise occurs that begins to produce progressive neurodegeneration that can extend up to years (Kumar et al., 2013).

17.3.2 Secondary Injury Cascade of TBI

Following the primary injury a large secondary injury cascade becomes activated. This secondary injury cascade arises as a result of alterations in excitatory amino acids (EAAs) such as glutamate (Rose et al., 2002; Yamamoto et al., 1999), increases in oxidative stress (Hall et al., 2004; Marklund et al., 2001; Tavazzi et al., 2005), the disruption of Ca^{2+} homeostasis (Sullivan, Thompson, and Scheff, 1999; Xiong et al., 1997a), and mitochondrial dysfunction (Davis et al., 2008; Hall et al., 2005; Lifshitz et al., 2004; Opii et al., 2007; Pandya et al., 2007; Pandya, Pauly, and Sullivan, 2009; Pandya, Nukala, and Sullivan, 2013; Readnower et al., 2011; Sauerbeck et al., 2011; Singh et al., 2006a; Sullivan et al., 1998, 2002, 2005). All of these can individually or in combination contribute to increased tissue damage at the site of the injury. It is believed that by targeting these secondary injury cascades, improved histological and functional outcomes could be produced.

Although we do not yet have a thorough understanding of the secondary injury cascade, the implications of abnormally high EAA levels around the site of the injury are generally accepted. The most common of these EAAs, as discussed previously, is glutamate. High glutamate levels are thought to be a result of the breakdown of the BBB, excessive synaptic release from neurons, nearby hyperexcited astrocytes and excitatory amino acid forming micropores within the cell membrane (Bullock et al., 1998). Once in the synaptic cleft, these EAAs are able to activate multiple receptors including AMPA and NMDA receptors. Activation of these receptors leads to Ca^{2+} influx into the cell. As mentioned previously, at low to moderate levels, this intracellular Ca^{2+}, $[Ca^{2+}]_i$, is used as a potent signaling molecule, important in many functions such as increased metabolism, neurotransmitter release and alterations of the cytoskeleton. However, following an injury, EAA concentrations around the injury site increase to pathogenic levels, leading to drastic increases in Ca^{2+} influx which has the potential of leading to excitotoxic cell death. The most important organelles involved in the rapid buffering of this Ca^{2+}, to avoid neuronal cell damage or death, are the mitochondria.

As mentioned previously, mitochondria have low-affinity, high capacity Ca^{2+} pumps in their inner membrane that allow for rapid sequestration of $[Ca^{2+}]_i$. These pumps use electrochemical gradients generated by the ETC translocation of protons across the inner membrane to create and maintain a membrane potential, $\Delta\Psi$. Although the primary function of $\Delta\Psi$ is to drive ATP synthesis, it also allows for electrogenic movement of $[Ca^{2+}]_i$ into the mitochondria. As the mitochondria buffer Ca^{2+}, a reduction in $\Delta\Psi$ occurs. The heightened intracellular Ca^{2+} levels place excessive demand on the ability of mitochondria to sequester Ca^{2+}, eventually resulting in mitochondrial dysfunction that has been documented for up to at least 48 hours post-injury (Sullivan et al., 2011). It has been shown that cortical mitochondrial Ca^{2+} levels in sham-operated rats range around ~25nmol Ca^{2+}/mg. However, after TBI, mitochondrial Ca^{2+} levels increase nearly 60% to levels near ~40nmol Ca^{2+}/mg (Pandya et al., 2009). This increase leads to a drastic change in $\Delta\Psi$, eventually causing a disruption of mitochondrial homeostasis. Because $[Ca^{2+}]_i$ can reach pathogenic levels following injury, extreme $\Delta\Psi$ reduction can occur. Consequently ETC-mediated ATP synthesis is hindered, leading to increased ROS production. ROS can then oxidize mitochondrial proteins leading to further mitochondrial dysfunction and the eventual initiation of cell death pathways.

As pointed out, the cascade of posttraumatic mitochondrial ROS generation and oxygen radical reactions begins in response to rapid elevations in intracellular Ca^{2+} immediately following the primary mechanical injury to the brain. Reduction of O_2 by a single electron will produce a superoxide radical, $O_2^{\bullet-}$, which is considered to be a modestly reactive primordial radical that can potentially react with other molecules to give rise to much more reactive, and thus more potentially damaging radical species. The reason that $O_2^{\bullet-}$ is only modestly reactive is that it can act as either an oxidant by stealing an electron from another oxidizable molecule or it can act as a reductant, meaning it can donate its unpaired electron to another radical species, i.e. an electron-donating antioxidant.

Although $O_2^{\bullet-}$ itself is less reactive than •OH radical, its reaction with nitric oxide (•NO) radical forms the highly reactive oxidizing agent, peroxynitrite (PN), ONOO-. This reaction, $O_2^{\bullet-}$ + •NO → ONOO-, occurs with a very high, diffusion-limited rate constant. Subsequently, at physiological pH, ONOO- can either undergo protonation to form peroxynitrous acid, ONOOH, or it can react with CO_2 to form nitrosoperoxocarbonate, $ONOOCO_2$. The ONOOH can break down to form highly reactive nitrogen dioxide, ONOOH → $•NO_2$ + •OH. Alternatively, the $ONOOCO_2$ can decompose into $•NO_2$ and carbonate radical $ONOOCO_2$ → $•NO_2$ + $•CO_3$.

Increased production of reactive free radicals, which is a defining characteristic of oxidative stress in the injured brain, has been shown to cause oxidative damage to cellular, including mitochondrial, lipids and proteins leading to functional compromise, and possibly cell death in both the microvascular and brain parenchymal cells. The major form of radical-induced oxidative damage involves oxidative attack on cell membrane polyunsaturated fatty acids, triggering the process of lipid peroxidation (LP), which has three distinct chemical phases: initiation, propagation and termination. The LP is initiated when a highly reactive, (i.e., electron-seeking), oxygen radical (e.g., •OH, $•NO_2$ or $•CO_3$), reacts with membrane polyunsaturated fatty acids such as arachidonic acid, linoleic acid, eicosapentaenoic acid, or docosahexaenoic acid, resulting in disruptions in membrane integrity. Specifically, initiation of LP occurs when a highly electrophilic radical steals the hydrogen electron from an allylic carbon of the peroxidizable polyunsaturated fatty acid. The allylic carbon is susceptible to free radical attack because it is surrounded by two relatively electronegative double bonds, which tend to pull one carbon electron away from the hydrogen electron it is paired with. Consequently, a reactive free radical can pull the hydrogen electron off of the carbon because the commitment of the carbon electron to staying paired with it has been weakened by the surrounding double bonds. This reaction results in the original radical being quenched while the polyunsaturated fatty acid, L, becomes a lipid radical, L• as a result of its having lost an electron.

In the subsequent propagation step, the unstable L• reacts with O_2 to form a lipid peroxyl radical, LOO•. The LOO• in turn abstracts a hydrogen atom from an adjacent polyunsaturated fatty acid yielding a lipid hydroperoxide, LOOH, and a second L•, which sets off a series of propagation "chain" reactions. These propagation reactions are terminated in the third step when the peroxidizable substrate becomes depleted or a lipid radical reacts with another radical or radical scavenger to yield relatively stable nonradical end products. One of those end products that is often used to measure LP is the 3 carbon-containing malondialdehyde (MDA), which is mainly a stable nontoxic compound that when measured represents an LP indicator. In contrast, two highly toxic aldehyde products of LP are 4-hydroxynonenal (4-HNE) or 2-propenal (acrolein) both of which have been well characterized in experimental brain or spinal cord injury models (Bains and Hall, 2012; Hall, Vaishnav and Mustafa, 2010; Hamann and Shi, 2009). These latter two aldehyde LP end products can covalently bind to basic amino acids such as lysine, histidine, arginine, or cysteine in cellular proteins, altering their structural and functional properties.

In addition to LP, free radicals can cause various forms of oxidative protein damage. A major mechanism involves carbonylation by reaction of various free radicals with susceptible amino acids. Secondly, $•NO_2$ can nitrate the 3 position of tyrosine residues in proteins forming 3-nitrotyrosine (3-NT) which is a specific footprint of PN-induced cellular damage. Similarly, lipid peroxyl radicals (LOO•) can promote tyrosine nitration by producing initial oxidation, the loss of an electron, which would enhance the ability of $•NO_2$ to nitrate the phenyl ring (Mustafa et al., 2010). Thus, in summary, LP can cause cellular damage by destroying the integrity of the phospholipid bilayer as well as membrane-localized proteins by modification carbonylation, nitration or binding of aldehyde LP end products such as 4-HNE or acrolein.

Mitochondrial oxidative stress and resulting dysfunction plays an especially critical role in the posttraumatic cell death cascade in injured brain (Lifshitz et al., 2003; Singh et al., 2006b; Sullivan et al., 1999; Xiong et al., 1997a) or spinal cord (Azbill et al., 1997; Sullivan et al., 2007). The initiator of this mitochondrial dysfunction involves a rapidly evolving postinjury intracellular accumulation of Ca^{2+} ions, which causes the mitochondria to attempt to sequester the excess Ca^{2+}, which in turn causes mitochondrial respiratory dysfunction, resulting in reduced OXPHOS (Lifshitz et al., 2003; Singh et al., 2006b; Sullivan et al., 1998; Xiong et al., 1997a) and Ca^{++} buffering capacity (Lifshitz et al., 2003; Sullivan et al., 1999) and mitochondrial failure resulting from mitochondrial permeability transition (MPT) (Sullivan et al., 2002, 2005). In addition to the potential activation of mitochondrial caspase-dependent and caspase-independent cell death cascades, this mitochondrial failure causes a disruption of synaptic function, and indeed synaptic mitochondria are more susceptible than nonsynaptic mitochondria (Sullivan et al., 1998).

The disruption of respiratory function is preceded by, or at least coincident with, an increase in mitochondrial free radical production or free radical-triggered LP-mediated oxidative damage (Azbill et al., 1997; Singh et al., 2006b; Sullivan et al., 2007).

Evidence has accumulated which shows that a particularly important oxidant that is being formed by injured brain or spinal cord mitochondria is PN (Bao and Liu, 2002, 2003; Deng et al., 2007; Sullivan et al., 2007; Xiong, Rabchevsky, and Hall, 2007; Xiong and Hall, 2009). The relationship of PN generation to mitochondrial dysfunction in the injured spinal cord has been recently documented. Studies have shown that the timing of post-spinal cord injury mitochondrial dysfunction resulting in respiratory and Ca^{2+} buffering impairment is preceded by an increase in PN-induced LP, 3-NT, protein nitration, and 4-HNE and protein carbonyl content in mitochondrial proteins. Nitric oxide and a mitochondrial nitric oxide synthase (NOS) isoform, mtNOS, have been shown to be present in mitochondria (Lopez-Figueroa et al., 2000; Zanella et al., 2002). Exposure of mitochondria to excessive Ca^{2+}, which is known to cause dysfunction, leads to PN generation, which in turn triggers mitochondrial Ca^{2+} release. In other words this limits their Ca^{2+} uptake or buffering capacity (Bringold, Ghafourifar, and Richter, 2000). Application of each of the three PN forms in vitro ($ONOO^-$, ONOOH, and $ONOOCO_2$) has been found to deplete mitochondrial antioxidant stores and to cause protein nitration, which is the hallmark of PN oxidative damage (Valdez et al., 2000). The loss of mitochondrial function and the increase in oxidative damage markers, including 3-NT, is antagonized by early in vivo post-TBI treatment with the PN radical scavenger tempol (Deng et al., 2007). Furthermore, exposure of noninjured mitochondria to PN quickly leads to respiratory dysfunction that is antagonized by tempol pretreatment, along with a reduction in nitration of mitochondrial proteins (Xiong, Singh, and Hall, 2009).

Additionally, the neurotoxic LP end products 4-HNE or acrolein are also able to produce a concentration-related attenuation of the respiratory function of normal brain mitochondria (Vaishnav et al., 2010). In those experiments, acrolein was observed to be 10 times more potent than 4-HNE. However, scavenging of these carbonyl-containing compounds with phenelzine has been shown to attenuate mitochondrial respiratory depression along with a reduction in the levels of aldehyde-modified mitochondrial proteins (Singh et al., 2013).

As mitochondria remain in a dysfunctional state because of a dysregulation of Ca^{2+} and increased ROS concentrations, the ETC will reach a point where it can no longer maintain the necessary $\Delta\Psi$ needed to allow Complex V to function properly. As the mitochondria remain in a dysfunctional state with a decreased $\Delta\Psi$ hovering around −100 to −120mV, which is a drastic decrease from the normal membrane potential of ~ −180mV, the electron transport systems will begin to slow down or even halt. Reactive superoxide molecules will continue to be released within the mitochondria (Sullivan et al., 2011). As this occurs, crucial mitochondrial proteins like ANT undergo oxidation modifications, specifically a thiol oxidation, resulting in a conformational change that promotes cyclophilin D (cypD) binding with ANT. Once ANT, an inner membrane mitochondrial protein, is bound to cypD, a conformational change occurs to promote ANT, cypD, and many other proteins within the complex to bind with the voltage-dependent anion channel (VDAC) (Belzacq et al., 2002; Halestrap, 2006; Varanyuwatana and Halestrap, 2012). VDAC is an outer membrane mitochondrial protein that when bound to an inner mitochondrial membrane protein complex including cypD, ANT, and many other proteins, forms a complex with at least some of the characteristics of the mitochondrial permeability transition pore (mPTP). Formation of the mPTP exposes the inner mitochondrial matrix to the cytosol, allowing for the release of apoptotic initiating factors like Cytochrome C, SMAC/diablo and AIF, which all promote cell death. Neuroprotective agents that target mitochondrial dynamics and mPTP formation may improve neuronal function in many disease states and therefore the mPTP is a potential target for the development of treatments for TBI.

17.4 PHARMACEUTICAL INTERVENTIONS

Brain tissue undergoing the secondary injury cascade following TBI, specifically neuroinflammation, free radical and other reactive species accumulation, and mitochondrial dysfunction, is compromised and at risk for cell death through various apoptotic and necrotic cell death pathways. Pharmacological intervention should, with the right agent(s), minimize this injury cascade and increase tissue sparing and decrease functional deficits, providing a neural substrate for rehabilitation.

17.4.1 Targeting Mitochondrial Dysfunction

Some of the most promising neuroprotective agents following TBI are drugs that target mitochondrial dysfunction. However, as in experimental and clinical treatment of stroke, time of delivery is clearly a critical factor. Mitochondrial-targeting drugs seem to work best when delivered within the first hour or two of injury; in experimental TBI, however, some can be effective when given up to 8 hours post-injury. This is thought to be a result of decreasing the formation of the mPTP within the injured tissue, leading to decreased neuronal death and increased cortical sparing. Because of this, any therapeutic intervention that targets mitochondrial dysfunction following injury to the CNS needs to be given as quickly after injury as possible (Sullivan et al., 2011).

One way to avoid the initiation of the mPTP is through decreasing the Ca^{2+} load in the mitochondria. It has been shown that the use of mitochondrial uncouplers such as carbonilcyanide p-triflouromethoxyphenylhydrazone (FCCP) can produce benefit following experimental injury. These uncouplers are weak lipophilic acids that are able to deplete the membrane potential and thereby decrease the electromotive force that allows Ca^{2+} to enter the mitochondria. Work from our group has found that treating animals with FCCP post-injury will lead to improved histological and functional outcomes following injury (Pandya et al., 2009).

As mentioned previously, TBI-induced oxidative stress leads to protein dysfunction as the result of oxidative modifications. Oxidative stress with PDH has been documented following injury, resulting in decreased substrates for the ETC. Additionally, glucose-mediated depression has been seen in both animals and humans following TBI that is dependent on the injury severity (Bergsneider et al., 1997; Yoshino et al., 1991). Therefore, with metabolic enzymatic dysfunction and decreased metabolic substrates, a state of acute metabolic crisis has been shown to occur

(Wright et al., 2013). Alternative biofuels like ketones and beta-hydroxybutyric acid (BHB), which are substrates that easily feed into the ETC, have become a promising area of research (Sullivan et al., 2004). Additionally, the use of ketone bodies in various animal models have been shown to decrease free radical production through increases in glutathione, as well as increasing the expression of various endogenous mitochondrial uncoupling proteins (Davis et al., 2008; Sullivan et al., 2004; Ziegler et al., 2003). Additionally, increased ATP production has also been seen following postinjury BHB treatment. With the resulting markedly improved mitochondrial bioenergetics in these experimental conditions, cortical sparing and functional recovery has been seen (Appelberg, Hovda and Prins, 2009; Prins, Fujima and Hovda, 2005).

17.4.2 Targeting Oxidative Stress

Consistent with the preceding discussion of the role of mitochondrial membrane LP in posttraumatic secondary neuronal injury, several pharmacological compounds with antioxidant properties have been shown to be neuroprotective in neurotrauma models. The pyrrolopyrimidine compound U-101033 is a highly potent and selective LP inhibitor that possesses excellent BBB permeability and neuronal penetration, and has been shown to be neuroprotective in ischemic stroke and TBI models (Hall et al., 1997). In the CCI TBI model in rats, U-101033, administered in two intravenous bolus doses at 5 minutes and 2 hours post-injury, produced a significant protection of mitochondrial respiration and calcium homeostasis 6 hours and 14 days after TBI (Xiong et al., 1997b). In a subsequent study Xiong and colleagues (1998) found that mitochondrial protection was enhanced if the antioxidant U-101033 was combined with the N-type calcium channel blocker SNX-111, supporting the notion that posttraumatic mitochondrial respiratory dysfunction is a result of a combination of ROS-induced LP and disruption of Ca^{2+} buffering capacity. A second highly lipophilic and selective LP inhibitor, the 2-methylaminochroman U-83836 (Hall et al., 1991), has similarly been documented to reduce mitochondrial respiratory failure and to maintain Ca^{2+} buffering capacity in the CCI TBI paradigm together with a reduction in the 4-HNE and 3-NT content of mitochondrial proteins (Mustafa et al., 2010). This effect is associated with an attenuation of Ca^{2+}-mediated proteolytic degradation of brain axonal neurofilaments, and the temporal window (i.e., maximal postinjury treatment delay) for this neuroprotective action extends out to 12 hours after TBI (Mustafa et al., 2011).

Other antioxidant drugs that are currently being explored in acute neurotrauma models include a collection of compounds from various pharmacological classes. Each of these compounds contains a hydrazine moiety (-NH-NH_2) that enables them to scavenge the neurotoxic LP-derived aldehydes 4-HNE and acrolein by covalently reacting with their carbonyl functional group. One of our groups has studied the effects of the hydrazine-containing compound phenelzine, which has long been used as an antidepressant and more recently documented to react with 4-HNE and acrolein (Galvani et al., 2008). Recently published results have reported the ability of the hydrazine-containing phenelzine to protect isolated rat brain mitochondria from the respiratory depressant effects of 4-HNE, together with a concentration-related attenuation of 4-HNE modified mitochondrial proteins. Subsequent in vivo studies in the rat CCI TBI model have found that a single 10 mg/kg subcutaneous dose of phenelzine can also reduce early (3-hour post-TBI) mitochondrial respiratory failure as well as reducing cortical lesion volume at 14 days post-injury. Further experiments are ongoing to determine dose-response, therapeutic window, and optimum treatment regimen.

17.4.3 Interventions with Multiple Targets

Lastly, it is thought that due to the complicated nature of the secondary injury cascade, the best pharmaceutical interventions will likely target multiple steps in this cascade. Cyclosporin A (CSA) is a drug that has been investigated for use in TBI and has shown great promise. CSA had multiple indications, including immune suppression, inhibition of mPTP conductance and indirect antioxidant properties. As an indirect antioxidant, CSA is not able to directly scavenger reactive radicals or aldehydes. Rather, CSA inhibition of MPT in TBI models secondarily reduces mitochondrial free radical formation and decreases the further accumulation of 4-HNE and 3-NT together with decreased lesion volume. Its temporal window, the longest time following the injury and before administration of the first dose that still demonstrates efficacy, may be as long as 8 hours post-TBI (Mbye et al., 2008; Okonkwo et al., 1999; Okonkwo and Povlishock, 1999; Sullivan et al., 1999, 2011; Sullivan, Thompson, and Scheff, 2000). This effect is totally unrelated to immunosuppression because the nonimmunosuppressive cyclosporin A analog NIM811 is able to duplicate the posttraumatic mitochondrial and neuroprotective effects of CSA (Mbye et al., 2008, 2009). A Phase 2 dose escalation and safety study in 40 patients with severe TBI has suggested that CSA treatment initiated within the first 8 hours and continued for 3 days may improve their functional recovery (Hatton et al., 2008). However, additional clinical trials are needed to determine the reliability of this effect.

Another drug that has shown promise as a neuroprotective agent following TBI is pioglitazone. This drug has been approved by the FDA for the treatment of type 2 diabetes, and targets various peroxisome proliferating activating receptor (PPAR) isoforms. PPAR is a nuclear receptor family that has been shown to modulate the expression of inflammatory cytokines and to modulate mitochondrial bioenergetics. This drug has shown promise in various animal models of TBI, leading to decreased neuroinflammation and decreased mitochondrial dysfunction. It has been shown that pioglitazone is able to increase cortical sparing and improve functional recovery following injury.

17.5 CONCLUSION

Mitochondrial dysfunction is an important therapeutic target for TBI and should also be important for the treatment of other acute conditions such as stroke. By improving mitochondrial bioenergetics, which will decrease cell death, improved cortical and functional outcomes should result. Although much more work is still needed in this area, the results in animal models and preliminary human trials have been promising.

REFERENCES

Appelberg, K. S., Hovda, D. A. and Prins, M. L., 2009. The effects of a ketogenic diet on behavioral outcome after controlled cortical impact injury in the juvenile and adult rat. *J Neurotrauma*, 26, pp.497–506.

Azbill, R. D., et al., 1997. Impaired mitochondrial function, oxidative stress and altered antioxidant enzyme activities following traumatic spinal cord injury. *Brain Res*, 765, pp.283–290.

Bains, M. and Hall, E. D., 2012. Antioxidant therapies in traumatic brain and spinal cord injury. *Biochim Biophys Acta*, 1822, pp.675–684.

Bao, F. and Liu, D., 2002. Peroxynitrite generated in the rat spinal cord induces neuron death and neurological deficits. *Neuroscience*, 115, pp.839–849.

Bao, F. and Liu, D., 2003. Peroxynitrite generated in the rat spinal cord induces apoptotic cell death and activates caspase-3. *Neuroscience*, 116, pp.59–70.

Belzacq, A. S., et al., 2002. The adenine nucleotide translocator in apoptosis. *Biochimie*, 84, pp.167–176.

Bergsneider, M., et al., 1997. Cerebral hyperglycolysis following severe traumatic brain injury in humans: a positron emission tomography study. *J Neurosurg*, 86, pp.241–251.

Bringold, U., Ghafourifar, P. and Richter, C., 2000. Peroxynitrite formed by mitochondrial NO synthase promotes mitochondrial Ca2+ release. *Free Radic Biol Med*, 29, pp.343–348.

Brustovetsky, N. and Dubinsky, J. M., 2000. Dual responses of CNS mitochondria to elevated calcium. *J Neurosci*, 20, pp.103–113.

Bullock, R., et al., 1998. Factors affecting excitatory amino acid release following severe human head injury. *J Neurosurg*, 89, pp.507–518.

Buono, M. J. and Kolkhorst, F. W., 2001. Estimating ATP resynthesis during a marathon run: a method to introduce metabolism. *Advances in Physiology Education*, 25, pp.70–71.

Cernak, I., 2005. Animal models of head trauma. *NeuroRx*, 2, pp.410–422.

Chance, B. and Smith, L., 1955. Respiratory pigments of Rhodospirillum rubrum. *Nature*, 175, pp.803–806.

Chance, B. and Williams, G. R., 1955a. Respiratory enzymes in oxidative phosphorylation. IV. The respiratory chain. *J Biol Chem*, 217, pp.429–438.

Chance, B. and Williams, G. R., 1955b. Respiratory enzymes in oxidative phosphorylation. I. Kinetics of oxygen utilization. *J Biol Chem*, 217, pp.383–393.

Chance, B. and Williams, G. R., 1955c. Respiratory enzymes in oxidative phosphorylation. II. Difference spectra. *J Biol Chem*, 217, pp.395–407.

Chance, B. and Williams, G. R., 1955d. Respiratory enzymes in oxidative phosphorylation. III. The steady state. *J Biol Chem*, 217, pp.409–427.

Chance, B., et al., 1955. Respiratory enzymes in oxidative phosphorylation. V. A mechanism for oxidative phosphorylation. *J Biol Chem*, 217, pp.439–451.

Chance, B., et al., 1973. Basic principles of tissue oxygen determination from mitochondrial signals. *Adv Exp Med Biol*, 37A, pp.277–292.

Chance, B., Sies, H. and Boveris, A., 1979. Hydroperoxide metabolism in mammalian organs. *Physiol Rev*, 59, pp.527–605.

Chancey, R. L. and Zatz, L. M., 1955. Sporadic acute anicteric hepatitis associated with upper respiratory infection. *J Am Med Assoc*, 158, pp.1013–1016.

Davis, L. M., et al., 2008. Fasting is neuroprotective following traumatic brain injury. *J Neurosci Res*, 86, pp.1812–1822.

Deng, Y., et al., 2007. Temporal relationship of peroxynitrite-induced oxidative damage, calpain-mediated cytoskeletal degradation and neurodegeneration after traumatic brain injury. *Exp Neurol*, 205, pp.154–165.

Dixon, C. E., et al., 1987. A fluid percussion model of experimental brain injury in the rat. *J Neurosurg*, 67, pp.110–119.

Floyd, R. A. and Hensley, K., 2002. Oxidative stress in brain aging. Implications for therapeutics of neurodegenerative diseases. *Neurobiol Aging*, 23, pp.795–807.

Galvani, S., et al., 2008. Carbonyl scavenger and antiatherogenic effects of hydrazine derivatives. *Free Radic Biol Med*, 45, pp.1457–1467.

Halestrap, A. P., 2006. Calcium, mitochondria and reperfusion injury: A pore way to die. *Biochem Soc Trans*, 34, pp.232–237.

Hall, E. D., et al., 1991. U-78517F: a potent inhibitor of lipid peroxidation with activity in experimental brain injury and ischemia. *J Pharmacol Exp Ther*, 258, pp.688–694.

Hall, E. D., et al., 1997. Pyrrolopyrimidines: novel brain-penetrating antioxidants with neuroprotective activity in brain injury and ischemia models. *J Pharmacol Exp Ther*, 281, pp.895–904.

Hall, E. D., et al., 2004. Peroxynitrite-mediated protein nitration and lipid peroxidation in a mouse model of traumatic brain injury. *J Neurotrauma*, 21, pp.9–20.

Hall, E. D., et al., 2005. Spatial and temporal characteristics of neurodegeneration after controlled cortical impact in mice: more than a focal brain injury. *J Neurotrauma*, 22, pp.252–265.

Hall, E. D., Vaishnav, R. A. and Mustafa, A. G., 2010. Antioxidant therapies for traumatic brain injury. *Neurotherapeutics*, 7, pp.51–61.

Halliwell, B. and Gutteridge, J. M. C., 1985. Oxygen radicals in the nervous system. *Trends Neurosci*, 8, pp.22–26.

Hamann, K. and Shi, R., 2009. Acrolein scavenging: a potential novel mechanism of attenuating oxidative stress following spinal cord injury. *J Neurochem*, 111, pp.1348–1356.

Hatton, J., et al., 2008. Dosing and safety of cyclosporine in patients with severe brain injury. *J Neurosurg*, 109, pp.699–707.

Jastroch, M., et al., 2010. Mitochondrial proton and electron leaks. *Essays Biochem*, 47, pp.53–67.

Kumar, A., et al., 2013. Traumatic brain injury in aged animals increases lesion size and chronically alters microglial/macrophage classical and alternative activation states. *Neurobiol Aging*, 34, pp.1397–1411.

LaPlaca, M. C., et al., 2007. CNS injury biomechanics and experimental models. *Prog Brain Res*, 161, pp.13–26.

Lehninger, A. L., 1965. *The mitochondrion; molecular basis of structure and function.* New York: W. A. Benjamin.

Lifshitz, J., et al., 2003. Structural and functional damage sustained by mitochondria after traumatic brain injury in the rat: Evidence for differentially sensitive populations in the cortex and hippocampus. *J Cereb Blood Flow Metab*, 23, pp.219–231.

Lifshitz, J., et al., 2004. Mitochondrial damage and dysfunction in traumatic brain injury. *Mitochondrion*, 4, pp.705–713.

Lopez-Figueroa, M. O., et al., 2000. Direct evidence of nitric oxide presence within mitochondria. *Biochem Biophys Res Commun*, 272, pp.129–133.

Marklund, N., et al., 2001. Free radical scavenger posttreatment improves functional and morphological outcome after fluid percussion injury in the rat. *J Neurotrauma*, 18, pp.821–832.

Mbye, L. H., et al., 2008. Attenuation of acute mitochondrial dysfunction after traumatic brain injury in mice by NIM811, a non-immunosuppressive cyclosporin A analog. *Exp Neurol*, 209, pp.243–253.

Mbye, L. H., et al., 2009. Comparative neuroprotective effects of cyclosporin A and NIM811, a nonimmunosuppressive cyclosporin A analog, following traumatic brain injury. *J Cereb Blood Flow Metab*, 29, pp.87–97.

Moreira, P. I., et al., 2010. Mitochondria: a therapeutic target in neurodegeneration. *Biochim Biophys Acta*, 1802, pp.212–220.

Mustafa, A. G., et al., 2010. Mitochondrial protection after traumatic brain injury by scavenging lipid peroxyl radicals. *J Neurochem*, 114, pp.271–280.

Mustafa, A. G., et al., 2011. Pharmacological inhibition of lipid peroxidation attenuates calpain-mediated cytoskeletal degradation after traumatic brain injury. *J Neurochem*, 117, pp.579–588.

Nicholls, D. G. and Ferguson, S. J., 2002. Bioenergetics. London: Academic Press.

Okonkwo, D. O., et al., 1999. Cyclosporin A limits calcium-induced axonal damage following traumatic brain injury. *Neuroreport*, 10, pp.353–358.

Okonkwo, D. O. and Povlishock, J.T., 1999. An intrathecal bolus of cyclosporin A before injury preserves mitochondrial integrity and attenuates axonal disruption in traumatic brain injury. *J Cereb Blood Flow Metab*, 19, pp.443–451.

Opii, W. O., et al., 2007. Proteomic identification of oxidized mitochondrial proteins following experimental traumatic brain injury. *J Neurotrauma*, 24, pp.772–789.

Pandya, J. D., et al., 2007. Post-injury administration of mitochondrial uncouplers increases tissue sparing and improves behavioral outcome following traumatic brain injury in Rodents. *J Neurotrauma*, 24, pp.798–811.

Pandya, J. D., Pauly, J. R. and Sullivan, P. G., 2009. The optimal dosage and window of opportunity to maintain mitochondrial homeostasis following traumatic brain injury using the uncoupler FCCP. *Exp Neurol*, 218, pp.381–389.

Pandya, J. D., Nukala, V. N. and Sullivan, P. G., 2013. Concentration dependent effect of calcium on brain mitochondrial bioenergetics and oxidative stress parameters. *Front Neuroenergetics*, 5, p.10.

Povlishock, J. T., et al., 1994. Workshop on animal models of traumatic brain injury. *J Neurotrauma*, 11, pp.723–732.

Prins, M. L., Fujima, L. S. and Hovda, D. A., 2005. Age-dependent reduction of cortical contusion volume by ketones after traumatic brain injury. *J Neurosci Res*, 82, pp.413–420.

Readnower, R. D., et al., 2011. Post-injury administration of the mitochondrial permeability transition pore inhibitor, NIM811, is neuroprotective and improves cognition after traumatic brain injury in rats. *J Neurotrauma*, 28, pp.1845–1853.

Rose, M. E., et al., 2002. Regulation of interstitial excitatory amino acid concentrations after cortical contusion injury. *Brain Res*, 943, pp.15–22.

Saatman, K. E., et al., 2008. Classification of traumatic brain injury for targeted therapies. *J Neurotrauma*, 25, pp.719–738.

Sauerbeck, A., et al., 2011. Pioglitazone attenuates mitochondrial dysfunction, cognitive impairment, cortical tissue loss, and inflammation following traumatic brain injury. *Exp Neurol*, 227, pp.128–135.

Scheffler, I. E., 2008. *Mitochondria*. New York: John Wiley & Sons, Inc.

Shulman, R. G., et al., 2004. Energetic basis of brain activity: Implications for neuroimaging. *Trends Neurosci*, 27, pp.489–495.

Singh, I. N., et al., 2006a. Time course of post-traumatic mitochondrial oxidative damage and dysfunction in a mouse model of focal traumatic brain injury: implications for neuroprotective therapy. *J Cereb Blood Flow Metab*, 26, pp.1407–1418.

Singh, I. N., et al., 2006b. Time course of post-traumatic mitochondrial oxidative damage and dysfunction in a mouse model of focal traumatic brain injury: implications for neuroprotective therapy. *J Cereb Blood Flow Metab*, 26, pp.1407–1418.

Singh, I. N., et al., 2013. Phenelzine mitochondrial functional preservation and neuroprotection after traumatic brain injury related to scavenging of the lipid peroxidation-derived aldehyde 4-hydroxy-2-nonenal. *J Cereb Blood Flow Metab*, 33, pp.593–599.

Sullivan, P. G., et al., 1998. Traumatic brain injury alters synaptic homeostasis: implications for impaired mitochondrial and transport function. *J Neurotrauma*, 15, pp.789–798.

Sullivan, P. G., Thompson, M. B. and Scheff, S. W., 1999. Cyclosporin A attenuates acute mitochondrial dysfunction following traumatic brain injury. *Exp Neurol*, 160, pp.226–234.

Sullivan, P. G., Thompson, M. and Scheff, S. W., 2000. Continuous infusion of cyclosporin A postinjury significantly ameliorates cortical damage following traumatic brain injury. *Exp Neurol*, 161, pp.631–637.

Sullivan, P. G., et al., 2002. Cytochrome c release and caspase activation after traumatic brain injury. *Brain Res*, 949, pp.88–96.

Sullivan, P. G., et al., 2004. The ketogenic diet increases mitochondrial uncoupling protein levels and activity. *Ann Neurol*, 55, pp.576–80.

Sullivan, P. G., et al., 2005. Mitochondrial permeability transition in CNS trauma: Cause or effect of neuronal cell death? *J Neurosci Res*, 79, pp.231–239.

Sullivan, P. G., et al., 2007. Temporal characterization of mitochondrial bioenergetics after spinal cord injury. *J Neurotrauma*, 24, pp.991–999.

Sullivan, P. G., Sebastian, A. H. and Hall, E.D., 2011. Therapeutic window analysis of the neuroprotective effects of cyclosporine A after traumatic brain injury. *J Neurotrauma*, 28, pp.311–318.

Supnet, C. and Bezprozvanny, I., 2010. Neuronal calcium signaling, mitochondrial dysfunction, and Alzheimer's disease. *J Alzheimers Dis*, 20 Suppl 2, pp.S487–S498.

Tate, C. C., Garcia, A. J. and LaPlaca, M. C., 2007. Plasma fibronectin is neuroprotective following traumatic brain injury. *Exp Neurol*, 207, pp/13–22.

Tavazzi, B., et al., 2005. Cerebral oxidative stress and depression of energy metabolism correlate with severity of diffuse brain injury in rats. *Neurosurgery*, 56, pp.582–589; 582-589.

Vaishnav, R. A., et al., 2010. Lipid peroxidation-derived reactive aldehydes directly and differentially impair spinal cord and brain mitochondrial function. *J Neurotrauma*, 27, pp.1311–1320.

Valdez, L. B., et al., 2000. Reactions of peroxynitrite in the mitochondrial matrix. *Free Radic Biol Med*, 29, pp.349–356.

Varanyuwatana, P. and Halestrap, A. P., 2012. The roles of phosphate and the phosphate carrier in the mitochondrial permeability transition pore. *Mitochondrion*, 12, pp.120–125.

Wake, D. J., et al., 2007. Effects of peroxisome proliferator-activated receptor-alpha and -gamma agonists on 11beta-hydroxysteroid dehydrogenase type 1 in subcutaneous adipose tissue in men. *J Clin Endocrinol Metab*, 92, pp.1848–1856.

Wallace, D. C., 1992. Mitochondrial genetics: A paradigm for aging and degenerative diseases? *Science*, 256, pp.628–632.

Wallace, D. C., 2007. Why do we still have a maternally inherited mitochondrial DNA? Insights from evolutionary medicine. *Annu Rev Biochem*, 76, pp.781–821.

Wallace, D. C., 2010. Mitochondrial DNA mutations in disease and aging. *Environ Mol Mutagen*, 51, pp.440–450.

Wright, M. J., et al., 2013. Early metabolic crisis-related brain atrophy and cognition in traumatic brain injury. *Brain Imaging Behav*, 7, pp.307–315.

Xiong, Y., et al., 1997a. Mitochondrial dysfunction and calcium perturbation induced by traumatic brain injury. *J Neurotrauma*, 14, pp.23–34.

Xiong, Y., et al., 1997b. Amelioration of mitochondrial function by a novel antioxidant U-101033E following traumatic brain injury in rats. *J Neurotrauma*, 14, pp.907–917.

Xiong, Y., et al., 1998. Mitochondrial dysfunction after experimental traumatic brain injury: Combined efficacy of SNX-111 and U-101033E. *J Neurotrauma*, 15, pp.531–544.

Xiong, Y., Rabchevsky, A. G. and Hall, E. D., 2007. Role of peroxynitrite in secondary oxidative damage after spinal cord injury. *J Neurochem*, 100, pp.639–649.

Xiong, Y. and Hall, E. D., 2009. Pharmacological evidence for a role of peroxynitrite in the pathophysiology of spinal cord injury. *Exp Neurol*, 216, pp.105–114.

Xiong, Y., Singh, I. N. and Hall, E. D., 2009. Tempol protection of spinal cord mitochondria from peroxynitrite-induced oxidative damage. *Free Radic Res*, 43, pp.604–612.

Yamamoto, T., et al., 1999. CSF and ECF glutamate concentrations in head injured patients. *Acta Neurochir Suppl*, 75, pp.17–19.

Yi, J. H. and Hazell, A. S., 2006. Excitotoxic mechanisms and the role of astrocytic glutamate transporters in traumatic brain injury. *Neurochem Int*, 48, pp.394–403.

Yoshino, A., et al., 1991. Dynamic changes in local cerebral glucose utilization following cerebral conclusion in rats: Evidence of a hyper- and subsequent hypometabolic state. *Brain Res*, 561, pp.106–119.

Zanella, B., et al., 2002. Mitochondrial nitric oxide localization in H9c2 cells revealed by confocal microscopy. *Biochem Biophys Res Commun*, 290, pp.1010–1014.

Ziegler, D. R., et al., 2003. Ketogenic diet increases glutathione peroxidase activity in rat hippocampus. *Neurochem Res*, 28, pp.1793–1797.

18

THE USE OF FIBROBLASTS FROM PATIENTS WITH INHERITED MITOCHONDRIAL DISORDERS FOR PATHOMECHANISTIC STUDIES AND EVALUATION OF THERAPIES

DEVORAH SOIFERMAN AND ANN SAADA

Monique and Jacques Roboh Department of Genetic Research and the Department of Genetic and Metabolic Diseases, Hadassah-Hebrew University Medical Center Jerusalem, Israel

18.1 INTRODUCTION

Mitochondria are essential organelles present in nucleated, eukaryotic cells. They perform a vital role in many cellular pathways including apoptosis, calcium, and iron homeostasis, steroid and heme biosynthesis, urea formation, fatty acid and amino acid oxidation, and innate immunity; however their main function is to supply the majority of cellular energy in the form of adenosine triphosphate (ATP) via oxidative phosphorylation (OXPHOS) performed by the mitochondrial respiratory chain (MRC). The MRC is embedded in the mitochondrial inner membrane and is comprised of approximately 90 proteins organized into five multi-subunit enzymatic protein complexes (I-V) and two electron carriers: coenzyme Q10 (CoQ10) and cytochrome *c*. Complexes I to IV transfer electrons derived from the oxidation of nutrients to molecular oxygen via the reducing equivalents NADH and $FADH_2$. Complexes I, II, and IV concomitantly form a proton gradient across the mitochondrial inner

The Functions, Disease-Related Dysfunctions, and Therapeutic Targeting of Neuronal Mitochondria, First Edition. Edited by Valentin K. Gribkoff, Elizabeth A. Jonas and J. Marie Hardwick.

membrane, which is used by complex V (ATP synthase) to form ATP. Mitochondrial MRC function depends on two genomes, both the nuclear genome and the mitochondrial genome (mtDNA). The small 16.5kb mtDNA encodes 13 MRC subunits of complexes I, III, IV, and V, as well as 22 tRNAs and 2 rRNAs needed for the separate, mitochondrial translation system. The nuclear DNA (nDNA) encodes all complex II subunits and the remaining MRC subunits. In addition, several hundred proteins that serve as assembly factors, are involved in cofactor synthesis, in mitochondrial DNA replication, transcription, repair and nucleotide synthesis, in mitochondrial translation, and in regulation of mitochondrial dynamics are all encoded by nuclear genes (DiMauro and Schon, 2003; Goldstein et al., 2012). Inherited mitochondrial disorders (MiDis) may thus arise either from maternally inherited mutations in the mtDNA or mutations in the nDNA, inherited in a Mendelian fashion. Regardless of the molecular basis, the mutations may cause isolated defects in a single MRC complex or combined defects involving several complexes (typically with intact complex II) (Di Donato, 2009). MiDis directly affecting MRC causing OXPHOS dysfunction are common, with an estimated prevalence of 1:500 children and 1:10000 adults (Schaefer et al., 2004; Skladal, Halliday, and Thorburn, 2003). This group of disorders presents with a wide range of symptoms, varying from relatively mild myopathy or diabetes in adulthood to debilitating and fatal encephalomyopathies with or without cardiac and liver dysfunction presenting at birth (Di Mauro and Schon, 2003). Symptoms may manifest in a single tissue, or as multisystemic disorders, although tissues with high energy demands, such as brain, muscle, and liver are preferentially affected. The central nervous system is frequently involved because it comprises some of the body's most metabolically demanding cells that are highly dependent on ATP produced by OXPHOS. Moreover, ATP has direct neuro- and co-transmitter effects via purinoreceptors in both the central and peripheral nervous systems (Benarroch, 2010).

Over the years it has become clear that mitochondria play a significant role in neuronal function, and neurodegenerative diseases may result as a consequence of mitochondrial OXPHOS dysfunction (Breuer et al., 2013). Furthermore, ATP deficit may not be the only trigger. Additional pathological features of mitochondrial dysfunction of primary MiDis, including oxidative stress and disrupted signal transduction, are also observed in mitochondria of the more common neurodegenerative diseases such as Huntington's, Alzheimer's and Parkinson's disease and amyotrophic lateral sclerosis (ALS). Thus, knowledge gained from studying primary MiDis may also be pertinent in understanding the pathophysiology of many neurodegenerative diseases (Schon and Manfredi, 2003). The following sections will briefly discuss mitochondrial encephalopathies with emphasis on the use of primary patient-derived fibroblasts for pathomechanistic studies and in the evaluation of treatment options. An in-depth discussion of the clinical presentation and molecular characterization of mitochondrial disorders and the involvement of OXPHOS in neurodegenerative diseases is beyond the scope of this chapter, and the reader is kindly referred to several excellent reviews on these subjects (Boczonadi and Horvath, 2014; Breuer et al., 2013; Chinnery and Hudson, 2013; Copeland, 2012; DiMauro et al., 2013; Schon and Manfredi, 2003; Suomalainen, 2013).

18.1.1 Identification of Mitochondrial Disorders

Identification of a MiDi often involves three parallel investigations: clinical, biochemical, and genetic. Clinical suspicion of mitochondrial disease is based on common presenting symptoms that are typically progressive or recurrent, although partial recovery may occur after decline. Symptoms may be aggravated by a metabolic stressor such as infection, fever, or fasting. Encephalomyopathy, or encephalopathy in combination with hepatopathy, is one of the marked neurological symptoms of mitochondrial disease. Other common neurological features include myopathy, neuropathy, and leuko-encephalopathy. Additional "red-flags" include stroke, basal ganglia lesions, epilepsy, cognitive decline, ataxia, ocular signs, and sensorineural hearing loss. Exercise intolerance and diabetes may also be indicative of mitochondrial disease (Breuer et al., 2013; Di Donato, 2009; DiMauro and Schon, 2003).

18.1.1a Biochemical Investigation

Biochemical screening tests may include a variety of blood and urine biomarkers including: metabolic acidosis; elevated plasma lactate, pyruvate, alanine, and beta-hydroxybutyrate; decreased plasma carnitine and CoQ10; and increased ketones and Krebs cycle intermediates in urine. However these markers are not always present in MiDis and they are generally non-specific because they can be detected also in other metabolic derangements. Lately FGF21 has proven to be a useful biomarker in blood, especially for MiDis involving muscle (Suomalainen, 2013).

To verify MRC dysfunction a tissue biopsy, preferably obtained from a diseased tissue such as muscle or liver, is warranted. Tissue homogenates and isolated mitochondria may be analyzed for complex I-V enzymatic activities, oxygen consumption, and CoQ10 content, while histological examination may reveal the presence of ragged-red fibers (mitochondrial proliferation), complex II (succinate dehydrogenase) and IV (cytochrome *c* oxidase). Enzymatic MRC defects and defects in pyruvate metabolism (pyruvate dehydrogenase) may also be detected in skin fibroblasts (DiMauro and Schon, 2003; DiMauro, 2013). However, fibroblasts can only be used for diagnostic purposes if these dividing cells express the phenotype and normal results can therefore not exclude mitochondrial dysfunction in other tissues. If the phenotype is clearly present in the fibroblasts, they may also be used for a variety of genetic and biochemical manipulations and screening for potential treatments (this will be further discussed).

18.1.1b Molecular Investigation

Confirmation of the molecular etiology of MiDis is challenging. Currently, only half of biochemically confirmed mitochondrial disorders reach molecular diagnosis, because of the two distinct but interacting genomes that contribute proteins necessary for normal mitochondrial function. More than 250 disease-causing mutations have been identified in the mtDNA as well as a number of mtDNA deletions. Notably the manifestation of mtDNA mutation depends on the percentage of mutated mtDNA (mutant load/heteroplasmy) in the various tissues (Chinnery and Hudson, 2013).

An increasing number of disease-causing mutations in the (>70) nDNA encoded MRC subunits and their chaperones have been reported. Notably all of the proteins involved in mtDNA replication, transcription, and protein translation are nDNA encoded and mutations in these genes can induce multiple mtDNA deletions, mtDNA depletion syndromes and disorders with impaired mtDNA translation (Boczonadi and Horvath, 2014; Copeland, 2012).

Moreover, mitochondrial encephalopathies may be caused by mutations in genes that encode for proteins involved in CoQ10 synthesis, mitochondrial biogenesis, fusion and fission, maintenance of the mitochondrial inner membrane lipid milieu, and solute transport across the inner membrane (DiMauro, 2013). Approximately 1500 nDNA-encoded proteins have been detected in the mitochondria; many of them have an as-yet unidentified role in normal mitochondrial function (Pagliarini et al., 2008). To date, more than 100 mutations in nuclear-encoded genes have been identified as pathogenic for MiDis. Advances in sequencing technology coupled with the steep decrease in cost mean that whole-exome sequencing has become more commonplace in the clinical setting, so this number is only set to rise (Carroll et al., 2013).

18.1.2 Pathomechanism of Mitochondrial Disorders

The obvious consequence of respiratory defect is reduced cellular ATP. However, in addition to decreased energy supply, other derangements are clearly present, such as reduced mitochondrial membrane potential and increased reactive oxygen species (ROS) formation. Under normal physiological conditions, 1% to 2% of molecular oxygen consumed by mammalian cells is metabolized to ROS via electron leakage from the MRC. Therefore, mitochondria are a major source of endogenous ROS in human and animal cells (Jezek and Hlavatá, 2005). A dysfunctional MRC increases ROS production (mainly via complexes I and III) and many of the pathogenic mtDNA mutations and nuclear-encoded mitochondrial disorders are linked to ROS overproduction (Golubitzky et al., 2011; Soiferman et al., 2014; Taylor and Turnbull, 2005). Excess ROS may cause direct oxidative damage to nucleic acids, proteins and lipids, which may in turn lead to DNA mutations, accumulation of unfolded or misfolded proteins and membrane structural instability, further exacerbating the OXPHOS dysfunction (vicious cycle) (Fig. 18.1) ROS can activate a variety of signaling pathways, including opening of the mitochondrial transition pore (mPTP) and initiation of the apoptotic cascasdes (Desagher and Martinou, 2000; Lee and Wei, 2007; Martindale and Holbrook, 2002; Orrenius, Gogvadze, and Zhivotovsky, 2007). Several studies support the correlation between apoptosis and the pathogenesis of mtDNA-associated mitochondrial diseases, similar to that observed in the brains of patients with neurodegenerative diseases (Auré et al., 2006; Su et al., 2002; Umaki et al., 2002). Reduced mitochondrial membrane potential also disrupts calcium homeostasis. Calcium is an important signaling ion involved in the regulation of many physiological and pathological cellular pathways; over the past two decades mitochondria have emerged as sensors and regulators of calcium signaling (Gunter et al., 1994; Rizzuto et al., 2012). Calcium homeostasis is especially crucial in

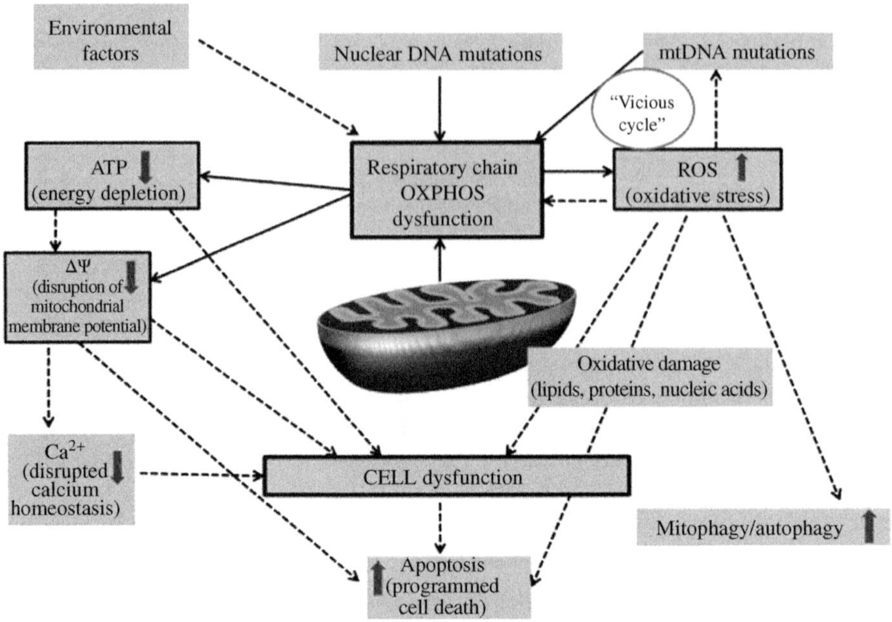

FIGURE 18.1 Representation of the pathomechanistic role of mitochondrial respiratory chain defects and subsequent mitochondrial and cellular dysfunction in the development of mitochondrial disorders. Dashed lines represent secondary events. mtDNA, mitochondrial DNA; OXPHOS, oxidative phosphorylation.

electrically active cells such as muscle and neurons that experience large calcium influxes during synaptically mediated depolarizations (Fig.18.1).

18.1.3 Treatment of Mitochondrial Disorders

Despite advances in identification, characterization and elucidation of MiDi pathomechanisms and the accumulation of vast amounts of data, studies focused on treatment are lagging behind. The extreme variability of OXPHOS disorders on a clinical, biochemical, and genetic level makes the investigation of therapeutic strategies difficult and intricate. Currently only primary CoQ10 defects and mitochondrial neurogastointestinal encephalomyopathy (MNGIE) are fully treatable by CoQ supplementation and bone marrow transplantation respectively (Hirano, Garone, and Quinzii, 2012), whereas a variety of vitamins, cofactors, amino acids, antioxidants, inducers of mitochondrial biogenesis, and chemical chaperones have been proposed to treat additional mitochondrial disorders (Fig. 18.2).

To date, fewer than 20 clinical trials for potential treatments of mitochondrial disorders have been reported or are in progress; molecules evaluated include dichloroacetate (DCA), arginine and the CoQ analogs idebenone and EPI-743, with uncertain outcome (Kerr, 2013). Obviously, any pharmacological strategy has to correct multiple levels of mitochondrial dysfunction including energy production, mitochondrial

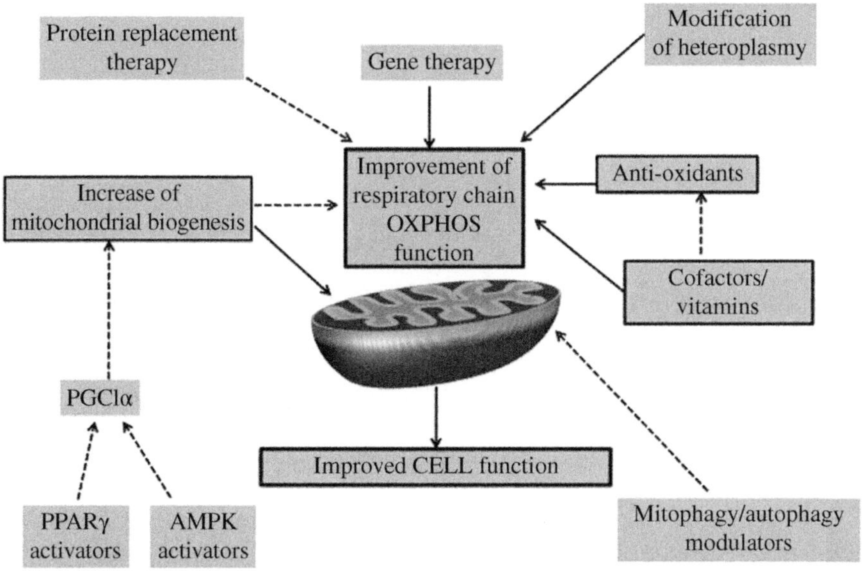

FIGURE 18.2 Summary of some therapeutic strategies for improving mitochondrial respiratory chain function in mitochondrial disorders. Dashed lines represent secondary events. OXPHOS, oxidative phosphorylation.

biogenesis, oxidative stress, mitochondrial morphology, calcium homeostasis, and apoptosis. The separate evaluation of these parameters is complex in patients. Moreover, even if the trial is well designed, the validation of the different treatments is hampered by difficulty in recruiting sufficient numbers of patients, genetic heterogeneity of the patients, the lack of appropriate control subjects, and the fact that patients often concurrently take a cocktail of supplements in the hope of enhancing mitochondrial function (Stacpoole, 2011). Consequently fewer than 5% of reported trials of mitochondrial therapeutics involved more than five patients. Positive outcomes were more frequently reported by flawed studies, for example trials that were nonrandomized, nonblinded, or based on biomarkers of unproven clinical significance. This potential publication bias may have deleterious consequences for caregivers and patients. There is a need for improved trial design and the development of new, accurate mitochondrial disease biomarkers (Pfeffer et al., 2013).

18.1.4 Models of Mitochondrial Disorders

Although the mitochondrion originated from bacteria, over time it diverged and developed unique properties. The structural simplicity of the bacterial electron transport machinery in comparison with eukaryotic counterparts, their short generation time and relative ease of gene manipulation provide a useful system in which to study protein structure and function, provided the mitochondrial counterpart is similar. However, many mitochondrial proteins are not present in bacteria and the prokaryotic systems lack complexity, thus manifestations of human diseases are virtually impossible to mimic and the current models are mostly of eukaryotic origin.

18.1.4a Microorganisms and Animals

The yeast *Saccharomyces cerevisiae* was the early model of choice for mitochondrial studies because the organism's ability to survive as either haploid or diploid and to vacillate between two metabolic states means that mutations that disrupt mitochondrial function in other species may be maintained and studied in *S. cerevisiae*. An additional advantage is the ease of genetic manipulation; yeast "complementation assays" are often used to confirm the roles of genes and mutations in human disorders. The simplicity of this model is also a drawback; *S. cerevisiae* lack, for example, the supernumerary subunits and assembly factors of MRC complex I (Ogilvie, Kennaway, and Shoubridge, 2005). The genome of the nematode *Caenorhabditis elegans* is amenable to manipulation and several databases and a public strain repository facilitate study of this low-cost animal system. Additionally, a high degree of evolutionary conservation of the five electron transport chain complexes has been predicted, however their exact composition has not yet been finalized. *Drosophila melanogaster*'s conservation of basic cellular processes with mammalian systems, record of modeling human diseases, short generation time, and compatibility with many cutting edge genetic tools make it an attractive model system for studying human mitochondrial diseases (Rea et al., 2010).

Rodent models provide another way of assessing potential mitochondrial therapies. Transgenic mice can be ordered commercially. However, the process is still time-consuming and expensive; therefore, it is unfeasible to produce an animal model for each and every individual defect (Dogan and Trifunovic, 2011). Moreover, treatment response does not always correlate with that observed in humans (Djouadi and Bastin, 2011).

18.1.4b Tissue Culture

Individual patient's cells in tissue culture are a convenient, accessible model and enable the study of private defects. Lymphocytes are readily available but they are difficult to manipulate experimentally. Additionally, the phenotype is often not expressed, as exemplified by the finding that a severe complex II mutation resulted in only a minor decrease in lymphoblastoid succinate dehydrogenase activity (Levitas et al., 2010).

Myoblasts are most likely to express the phenotype observed in muscle, but it is generally not feasible to derive myoblasts from each diagnostic muscle biopsy because most of the muscle tissue is used up for enzymatic, pathological, and molecular workup and the culturing of myoblasts is expensive and demanding. Fibroblasts are an alternative to myoblasts and are easily obtained during a muscle biopsy or after. Skin fibroblasts are in general the most accessible and renewable source of cells for both diagnosis and research (Saada, 2014).

The major drawback with fibroblasts in culture is that they sometimes fail to maintain the diseased phenotype. Nonetheless, primary fibroblasts derived from many patients do express mitochondrial dysfunction, although the defect is sometimes unmasked only under stressful growth conditions, for example in culture medium devoid of glucose or serum (Iuso et al., 2006; Robinson, 1996; Taanman, Muddle, and Muntau, 2003).

Studies based on primary patient fibroblasts with exact molecular characterization have contributed immensely to the deciphering of new biochemical pathways and interactions, including assembly models of complex I and complex IV of the MRC (Dieteren, Koopman, and Nijtmans, 2009; Lazarou et al., 2009), aspects of mitochondrial translation (Carroll et al., 2013; Emdadul Haque et al., 2008; Smeitink et al., 2006), RNA import (Vedrenne et al., 2012), mtDNA content (Ashley et al., 2008; Pontarin et al., 2012; Saada, 2008), and mitochondrial membrane structure and organization (Mayr et al., 2012; Schlame, 2013).

Furthermore, for the study of mtDNA mutations, fibroblasts are easily manipulated to generate transmitochondrial cybrids. These hybrid cells are obtained by fusing enucleated patient cells to a host cell line devoid of endogenous mtDNA (ρ^0) and result in varying proportions of the patient's mutated mtDNA on a "normal, standard" nuclear background. This system is useful for studying the effects of heteroplasmy, varying proportions of mutant and wild-type mtDNA in each cell, and the threshold effect of the mutation, although it does not model the interplay between the patient's nuclear and mtDNA (Schon, DiMauro, and Hirano, 2012).

Induced pluripotent stem cells (iPSCs) generated from a variety of somatic cells including patient fibroblasts are multipurpose basic research and clinical tools. These cells are used both to model disorders and to investigate therapeutic avenues including drug screening and cell therapy. However, several issues with reprogramming and genomic instability must be overcome before they can be applied in a clinical setting (Wang and Doering, 2012).

18.2 PATHOMECHANISTIC STUDIES OF MITOCHONDRIAL DISORDERS IN PATIENTS' FIBROBLASTS

Research performed in fibroblasts and cybrids derived from patients affected by mitochondrial disorders with neurological involvement has been instrumental in the understanding of disease pathomechanisms on the cellular level. The following sections will discuss the use of patient fibroblasts in elucidating pathomechanism and screening for beneficial molecules in fibroblasts of patients with mitochondrial disorders. To date, several main interconnected pathways of insult have been described as a result of defective oxidative phosphorylation in patient cells, which we will expand upon briefly (see Fig. 18.1).

18.2.1 Reduced Cellular ATP

Patient fibroblasts harboring a nuclear DNA-related mutation generally display reduced ATP content and decreased ATP synthesis capacity (Golubitzky et al., 2011; Shepherd et al., 2006; Soiferman et al., 2014). Under increased energy demands this is even more evident, as was observed in myoclonic epilepsy with ragged red fibers (MERRF) and mitochondrial encephalomyopathy, lactic acidosis, and stroke-like episodes (MELAS) fibroblasts, which showed significantly reduced ability to maintain ATP/ADP ratio under these conditions (James et al., 1999). Further research disclosed that in patient cells and cybrids with mtDNA mutations, ATP synthesis rate corresponded to mutant (heteroplasmic) load (Porteous et al., 1998).

18.2.2 Increased Oxidative Stress

Increased ROS levels were detected in fibroblasts derived from patients with a variety of OXPHOS disorders, including isolated MRC complex I deficiency, mtDNA replication, and translation disorders and MELAS (Golubitzky et al., 2011; Soiferman et al., 2014; Voets et al., 2012). A study in fibroblasts from patients with chronic progressive external ophthalmoplegia (CPEO) showed that mitochondrial dysfunction-induced oxidative stress could increase microtubule protein acetylation and alter mitochondrial distribution. Evidence was also found that ROS produced by defective mitochondria can inhibit the ubiquitin-proteasome pathway and protease activity. This disruption of the protein degradation system causes accumulation of aberrant proteins within the mitochondria, thus rendering the affected cells more susceptible to apoptosis. Additionally, oxidative stress was found to modulate phosphorylation of mitochondrial proteins (Wu, Y.T., et al., 2010). Another study reported an imbalance in the mRNA, protein, and enzyme activity levels of antioxidant enzymes in MERRF patient fibroblasts. ROS were also shown to damage proteins including prohibitin, Lon protease and aconitase. Also noted was a dramatic increase in gene expression and activity of matrix metalloproteinase 1 (MMP1), which may play a role in the cytoskeleton remodeling that contributes to the muscle weakness and atrophy commonly seen in these patients (Wu, S.B., et al., 2010).

18.2.3 Reduction of Mitochondrial Membrane Potential

The availability of specific mitochondrial membrane-potential sensitive fluorescent dyes such as JC-1 and tetramethylrhodamine esters has made it possible to estimate mitochondrial membrane potential in situ in tissue cultures from MiDi patients. Decreased mitochondrial membrane potential in MELAS and MERRF fibroblasts compared to controls and in fibroblast from patients with Barth's syndrome. Fibroblasts from patients harboring mutations in various nuclear-encoded mitochondrial proteins also displayed significant decreases in mitochondrial membrane potential (James et al., 1999; Karkucinska-Wieckowska et al., 2013; Soiferman et al., 2014). Even though these studies obviously show a decrease in membrane potential, the nature of the alteration is unclear. It might be a primary result of MRC dysfunction causing increased permeability of the mitochondrial inner membrane or a secondary, protective mechanism attempting to reduce ROS formation.

18.2.4 Disruption of Calcium Homeostasis

Decreased mitochondrial membrane potential in MELAS cells led to elevated resting intracellular calcium and impaired the cells' ability to normally sequester calcium influxes induced by depolarization (Rothman, 1999). However, cybrids derived from MERRF and NARP (neuropathy, ataxia and retinitis pigmentosa) patients showed no abnormality in cytosolic Ca^{2+} (Brini et al., 1999). Mouse embryonic stem cells derived from cybrids harboring pathogenic mtDNA mutations and differentiated into neurons displayed normal response to a single glutamate application, but repeated stimulation revealed markedly distorted Ca^{2+} transients in response to later stimuli in a train (Trevelyan et al., 2010). The distortion of the Ca^{2+} signal may be expected to affect brain function and to explain the appearance of neuronal pathology, with or without accompanying neuronal death. Over-accumulation of Ca^{2+} may increase net mitochondrial ROS emission or limit the ability of mitochondria to scavenge ROS (Adam-Vizi and Starkov, 2010) and can induce mitochondrial permeability transition pore opening and apoptotic cell death (Baumgartner et al., 2009).

18.2.5 Coenzyme Q10 Deficiency

Although defects in CoQ synthesis leading to primary CoQ10 deficiency do not involve mutations in MRC subunits, these disorders are clearly MiDis: fibroblasts derived from patients with primary CoQ deficiency displayed secondary decreased activities and expression of respiratory chain enzymes, decreased mitochondrial membrane potential, increased ROS production and activation of mPTP (Rodríguez-Hernández et al., 2009). Reciprocally, primary mitochondrial diseases may exhibit secondary decreases in CoQ10 levels, as was demonstrated in MELAS fibroblasts (Cotán et al., 2011). Moderately decreased CoQ10 levels in fibroblasts from six unrelated riboflavin-responsive multiple acyl-CoA dehydrogenation (RR-MADD) patients were also observed, associated with increased levels of mitochondrial reactive oxygen species (Cornelius et al., 2013).

18.2.6 Mitochondrial Dynamics and Mitophagy

Mutations in genes encoding proteins involved in the regulation of mitochondrial morphology, such as OPA1 and MFN1, have been identified in mitochondrial disorders and neurodegenerative diseases alike. Disturbed mitochondrial morphology and fragmentation is frequently associated with both primary and secondary OXPHOS dysfunction. Dysfunctional and distorted mitochondria are also associated with increased autophagy and mitophagy (Scorrano, 2013). Increased mitophagy was detected in fibroblasts from patients affected by both primary and secondary CoQ10 deficiency; preventing mitophagy promoted apoptosis, thus selectively removing damaged mitochondria by mitophagy may have a protective role in some diseases (Cotán et al., 2011). Conversely, CoQ10 supplementation improved MRC activity and normalized raised mitophagy levels in MELAS fibroblasts. This suggests that although mitophagy may play a protective role in this disorder, excessive mitophagy, mitochondrial loss, and autophagosome accumulation may impair cellular respiration and induce apoptosis (Cotán et al., 2011; Rodríguez-Hernández et al., 2009).

18.3 EVALUATION OF THERAPEUTIC OPTIONS USING PATIENT DERIVED FIBROBLASTS

Primary patient fibroblasts are a feasible model for screening potential therapies, especially on an individual basis. Using a relatively small number of cells, a variety of compounds can be assessed concurrently for their effects on multiple mitochondrial parameters, including MRC enzymatic activity, cell growth, ROS production, ATP production, oxygen consumption, microscopic evaluation of mitochondrial content, and morphology (Golubitzky et al., 2011; Saada, 2014; Soiferman et al., 2014).

The following sections exemplify the individualized response of fibroblasts derived from patients with mitochondrial disorders to various treatments including vitamins, antioxidants and assorted pharmacological treatments (Andreux, Houtkooper and Auwerx, 2013) as well as different approaches to genetic manipulation (see Fig. 18.2).

18.3.1 Pharmacological Approaches

Pharmacological approaches include the evaluation of small molecules functioning as enzyme cofactors, electron carriers, antioxidants, inducers of mitochondrial biogenesis, and chemical chaperones. Some of these compounds have mixed roles, functioning both as cofactors and antioxidants, and some may also exert their function by altering gene expression. The following paragraphs will only mention a few selected compounds that were evaluated on MiDi patients' cells.

18.3.1a Vitamins and Antioxidants

The rationale for screening vitamins is that they frequently function as MRC substrates and co-enzymes or as antioxidants and electron donors. Riboflavin, a precursor of flavin mononucleotide (FMN) present in MRC complex I, was beneficial to fibroblasts derived from patients with complex I defects because of mutations in NDUFS2, ACAD9, and AIF and in MELAS (Garrido-Maraver et al., 2012; Gerards et al., 2011; Saada, 2011). Nicotinamide (NAM), an amide form of vitamin B3, is readily converted to conenzyme β-nicotinamide-adenine dinucleotide (NAD^+) through the salvage pathway. NAM supplementation led to a decrease in mitochondrial content and an increase in mitochondrial membrane potential by activating autophagy and inducing mitochondrial fragmentation. NAD^+ treatment also prevented rontenone-induced mitochondrial depolarization, apoptosis and necrosis (Hong et al., 2014; Jang, Kang and Hwang, 2012). Menadione (vitamin K) and ascorbate (vitamin C), although not directly linked to the MRC, are potential electron donors to complexes II and IV, respectively, and were reported to have beneficial effects on fibroblasts with complex I and CoQ defects (Saada, 2011). In EFT patient fibroblasts ascorbate significantly reduced ROS production and increased growth, ATP content and complex IV activity, in line with its dual function as antioxidant and CIV electron donor (Soiferman et al., 2014). Ascorbate also normalized superoxide production and oxidative stress-induced cell death in CoQ2-deficient fibroblasts (López et al., 2010). Trolox, a water-soluble vitamin E derivative, dramatically reduced ROS levels and mitigated complex I deficiency, with the most beneficial effects observed in patient fibroblasts in which primarily the expression of complex I was reduced and not the activity (Koopman et al., 2008). Additionally, mitochondrial membrane potential and calcium homeostasis were restored (Distelmaier et al., 2009). Moreover, as vitamins are regarded to be a safe food additive, they seem a feasible treatment option for some mitochondrial disorders. Selenite, an inorganic form of selenium, which plays a pivotal role in the cellular antioxidant defense system, reduced ROS and increased expression of antioxidant enzymes in NARP cybrids and increased ATP and respiratory chain subunit levels (Wojewoda, Duszyński, and Szczepanowska, 2011).

N-acetyl cysteine (NAC) is a precursor of intracellular cysteine and glutathione that exerts its antioxidant effects by stimulating cytosolic enzymes participating in glutathione metabolism and through direct interaction with radical species. NAC supplementation reduced ROS production in fibroblasts derived from EFTs, MRPS22, and TRMU patients (Soiferman et al., 2014) and decreased expression levels of harmful metallomatrix proteins in MERRF fibroblasts (De Flora et al., 2001).

Mitochondrial targeted antioxidants contain antioxidant moieties such as ubiquinone (MitoQ) and tocopherol (MitoE) covalently bound to a lipophilic triphenylphosphonium (TPP) cation. Because of the large mitochondrial membrane potential, the antioxidant-conjugated cations are accumulated within the mitochondria. However, as decreased mitochondrial membrane potential is a hallmark of many mitochondrial disorders, this method may not be effective in all cases of mitochondrial dysfunction (Smith et al., 2008).

Coenzyme Q10 is essential for cellular energy production, transferring electrons from complex I, complex II, the electron transfer flavoprotein ubiquinone reductase (fatty acid oxidation) and 3-glycerolphosphate shuttle (glycolysis) to complex III. CoQ10 and its analogs are used therapeutically by virtue of their capabilities as electron carriers, antioxidants or both. Primary CoQ10 deficiency is caused by mutations in genes encoding CoQ10 biosynthetic proteins and is one of the few mitochondrial disorders that actually can be effectively cured by CoQ10 supplementation (Hirano et al., 2012). In contrast, in fibroblasts from patients with primary CoQ synthesis deficiencies, the short chain analogues CoQ2 and idebenone proved ineffective in improving cellular energetics (López et al., 2010).

Recently, secondary CoQ10 deficiency has been reported in cells derived from patients with other, nonrelated OXPHOS mutations. CoQ10 supplementation was found to partially restore defective parameters including respiratory chain enzyme activities, CoQ10 levels, mitochondrial membrane depolarization, and autophagosome elimination in both MELAS and MERRF fibroblasts and cybrids (De la Mata et al., 2012; Rodríguez-Hernández et al., 2009). CoQ10, but not riboflavin, supplementation normalized depleted CoQ10 in fibroblasts derived from riboflavin-responsive multiple acyl-CoA dehydrogenation deficiency (RR-MADD) patients (Cornelius et al., 2013). The efficiency of the short-chain CoQ analog idebenone in LHON patient cells is still under debate because the compound increased complex I activity but was found to have contradictory effects on mitochondrial respiration (Angebault et al., 2011).

18.3.1b Induction of Mitochondrial Biogenesis

Theoretically, reduced OXPHOS could be compensated for by increasing mitochondrial content in the cell. Mitochondrial biogenesis may be induced by pharmacological activation of the peroxisome proliferator activated receptor gamma (PPARγ) coactivator 1α (PGC1α) axis. PGC1α is induced by PPARγ and adenosine monophosphate dependent protein kinase (AMPK). AMPK may in itself be activated directly by sirtuin 1 (SIRT1) or indirectly by inhibition of cyclic AMP phosphodiesterases. Recently there has been increasing evidence that estrogen receptors can regulate mitochondrial biogenesis and MRC expression (Chen et al., 2009).

The PPAR agonist bezafibrate was shown to increase mRNA and protein levels and respiratory chain enzyme activities in complex I, III, and IV -deficient cells (Bastin et al., 2008). Bezafibrate also improved parameters of mitochondrial function in fibroblasts from complex I-deficient patients and some translation disorders, as did aminoimidazole carboxamide riboside (AICAR), an AMPK agonist (Golubitzky et al., 2011; Soiferman et al., 2014). The PPARγ agonist pioglitazone restored mitochondrial content and expression of mitochondrial biogenesis regulators, neutralized oxidative damage to DNA and proteins and reversed the bioenergetic defect in a neurodegenerative disease (Morató et al., 2013). This drug is already FDA approved for treatment of type II diabetes and concerns about associated health risks have recently been allayed.

Controversy surrounds the effects of resveratrol, a polyphenolic phytoalexin. The compound has been shown to act mainly in a SIRT1- and AMPK-independent manner, instead affecting the estrogen receptor (ER) and estrogen-related receptor alpha (ERR α) signaling pathways. Some studies reported that resveratrol supplementation increased the amount of mutated proteins, stimulated residual enzyme activities, increased oxygen consumption rates and decreased lactate production in moderate OXPHOS-deficient fibroblasts. However, other groups found that resveratrol exerted negligible or even detrimental effects on the treated patient fibroblasts (De Paepe et al., 2014; Golubitzky et al., 2011; Lopes Costa et al., 2013). Thus, the benefits of inducing mitochondrial biogenesis may vary, depending on the OXPHOS defect and the axis that is stimulated.

18.3.1c Improving mtDNA Maintenance and Translation

Mitochondrial DNA depletion syndromes caused by defective mtDNA maintenance are frequently the result of mutations in genes associated with mitochondrial deoxyribonucleotide (dNTP) metabolism. Examples include the decreased mtDNA content and imbalanced mitochondrial dGTP and dCTP pools in quiescent patient fibroblasts bearing DGUOK mutations (Saada, 2008; Taanman et al., 2003) and the limited dCTP availability that accounts for mtDNA depletion in MNGIE (González-Vioque et al., 2011). Supplementation of the lacking dNTPs or respective deoxyribonucleotides restored dNTP balance and mtDNA content; thus, therapy approaches focusing on replenishing the deficient substrates should be explored.

Pharmacological approaches to nuclear encoded disorders affecting the mitochondrial translation system are absent and in contrast these disorders may be adversely affected by translation-targeted antibiotics (Jones et al., 2009). An exception is the reversible infantile liver failure resulting from TRMU mutations. The gene encodes an enzyme that catalyzes mitochondrial tRNA thiouridylation, whose activity may be compromised by the limited availability of cysteine in the neonatal period. Interestingly, L-cysteine, a supplement readily available as a food additive, improved the OXPHOS-defective phenotype in TRMU fibroblasts (Boczonadi et al., 2013).

18.3.2 Genetic Manipulation

The molecular basis of inherited mitochondrial disorders (MiDis) is varied and the mode of inheritance depends on whether they are caused by mutations in the mtDNA or mutations in the nuclear genome, thus genetic approaches to therapy vary depending on the origin of the mutation.

18.3.2a Mitochondrial DNA Mutations

Mitochondrial DNA disorders are generally heteroplasmic, with mutated and wild type mitochondrial genomes present simultaneously in the same cell; the pathogenic threshold is generally >70% mutant mtDNA. Therefore in principle, there is no need to totally eliminate the mutation but a partial expression of wild type genes or a shift of the heteroplasmy toward the wild-type mtDNA above the threshold would suffice. Gene therapy–based approaches to ameliorate mtDNA-associated pathologies can be divided into allotopic or antigenomic strategies.

Allotopic expression is the expression of mtDNA-encoded genes in the nuclear-cytosolic compartment and their subsequent mitochondrial import. This method involves the recoding of mitochondrial genes to adapt them to the universal code and the addition of a DNA sequence encoding a mitochondrial localization signal. Nuclear expression and mitochondrial targeting of recombinant transgenic tRNAs was able to rescue the MELAS phenotype in transmitochondrial cybrid cells (Karicheva et al., 2011). However, the mtDNA-encoded MRC proteins are highly hydrophobic and this characteristic limits the efficiency of their import into the mitochondria. One approach to overcoming this challenge is based on localizing the mRNA to the mitochondrial surface. This technique rescued respiratory chain defects in fibroblasts harboring mtDNA mutations in subunits of complex I or complex V (Bonnet et al., 2007). Application of this method by electroporation prevented impairment of visual function in a rat model of LHON and may pave the way toward a treatment for patients suffering from optic neuropathies with a mitochondrial etiology (Ellouze et al., 2008). An additional tactic, the appendage of a 20-ribonucleotide stem-loop sequence from the H1 RNA component of the human RNaseP enzyme, has been found to be effective for both coding mRNA fusion transcripts and noncoding RNAs such as tRNA (Wang et al., 2012).

Antigenomic approaches aim to reduce the level of heteroplasmy by a variety of methods. A small decrease in mutational load may be enough to rescue the biochemical phenotype, with the potential of improving the clinical outcome if the mutational load passes under the pathogenic threshold. One strategy is the mitochondrial targeting of restriction endonucleases or transcription activator-like effector nucleases that specifically cleave mutant mtDNA molecules (Bacman et al., 2013;

Gammage et al., 2014). Another approach is synthetic peptide nucleic acid (PNA) molecules that inhibit mutant mtDNA replication; these were found to be efficient in vitro, but incapable of interacting with mtDNA in vivo (Muratovska et al., 2001; Taylor et al., 1997). However, oligoribonucleotides complementary to the mutant mtDNA region were imported into the mitochondria and successfully reduced the proportion of mtDNA harboring a large deletion associated with Kearns Sayre Syndrome and improved mitochondrial translation in a cybrid model (Comte et al., 2013). Metabolically and exercise-induced approaches are two other methods of heteroplasmy shifting (Desquiret-Dumas et al., 2012; Smith and Lightowlers, 2011).

18.3.2b Protein-Replacement Therapy

Enzyme replacement therapy has gained attention and been successful in lysosomal disorders although this approach is yet to be employed for MidDis. Allogeneic haematopoietic stem cell transplantation, a form of "enzyme replacement therapy" to restore thymidine phosphorylase activity in blood to reduce excess thymidine, is a promising treatment for MNGIE that has already reached the clinical arena (Hirano et al., 2012). Clearly this is not suitable for most primary MiDis because the hematopoietic system is not the target tissue. Still, some additional approaches have been investigated in primary studies, mainly using fibroblasts of patients with mitochondrial disorders.

Expression of the ascidium alternative oxidase (AOX) corrected the various consequences of the complex IV deficiency in fibroblasts derived from a COX15-deficient patient (Dassa et al., 2009). Application of recombinant lipoamide dehydrogenase (pyruvate dehydrogenase subunit E3), SCO2 (copper and Complex IV chaperone), and NDUFAF4 (assembly factors of complex IV and I, respectively) fused with human immunodeficiency virus trans-activator of transcription (HIV-TAT) enabled delivery of the proteins to the mitochondrial matrix and resulted in restored OXPHOS function (Foltopoulou et al., 2010; Marcus et al., 2013; Rapoport et al., 2008). Moreover, these results have been supported by an in-vivo model, providing proof of concept for the feasibility of TAT-mediated protein replacement therapy in mitochondrial diseases (Rapoport et al., 2011).

18.4 CONCLUSION

The consideration that mitochondria participate in such a wide range of neurological disorders, from fatal neonatal diseases to neurodegenerative disorders of old age, invokes a sense of amazement. Mitochondria perform varied functions that are crucial for cell survival, including ROS generation, calcium homeostasis and apoptosis, in addition to the standard role as ATP providers. Pathogenesis of any mitochondrial disease is likely to involve all these functions to some extent at some time point. Fibroblasts derived from patients with mitochondrial disorders are suitable models for investigating both disease pathomechanism and potential therapies. These findings may also enhance our understanding of and lead to appropriate treatments for the classical neurodegenerative disorders.

ACKNOWLEDGMENTS

This work was supported by the Manackerman Charitable Trust Fund UK.

REFERENCES

Adam-Vizi, V. and Starkov, A. A., 2010. Calcium and mitochondrial reactive oxygen species generation: how to read the facts. *J Alzheimers Dis.* 20 Suppl 2:S413–S426.

Andreux, P. A., Houtkooper, R. H. and Auwerx, J., 2013. Pharmacological approaches to restore mitochondrial function. *Nat Rev Drug Discov*, 12, pp.465–483.

Angebault, C., et al., 2011. Idebenone increases mitochondrial complex I activity in fibroblasts from LHON patients while producing contradictory effects on respiration. *BMC Res Notes*, 4, p.557.

Ashley, N., et al., 2008. Depletion of mitochondrial DNA in fibroblast cultures from patients with POLG1 mutations is a consequence of catalytic mutations. *Hum Mol Genet*, 17, pp.2496–2506.

Auré, K., et al., 2006. Apoptosis in mitochondrial myopathies is linked to mitochondrial proliferation. *Brain*, 129, pp.1249–1259.

Bacman, S. R., et al., 2013. Specific elimination of mutant mitochondrial genomes in patient-derived cells by mitoTALENs. *Nat Med*, 19, pp.1111–1113.

Bastin, J., et al., 2008. Activation of peroxisome proliferator-activated receptor pathway stimulates the mitochondrial respiratory chain and can correct deficiencies in patients' cells lacking its components. *J Clin Endocrinol Metab*, 93, pp.1433–1441.

Baumgartner, H. K., et al., 2009. Calcium elevation in mitochondria is the main Ca^{2+} requirement for mitochondrial permeability transition pore (mPTP) opening. *J Biol Chem*, 284, pp.20796–20803.

Benarroch, E. E., 2010. Adenosine triphosphate: a multifaceted chemical signal in the nervous system. *Neurology*, 74, pp.601–607.

Boczonadi, V., et al., 2013. Altered 2-thiouridylation impairs mitochondrial translation in reversible infantile respiratory chain deficiency. *Hum Mol Genet*, 22, pp.4602–4615.

Boczonadi, V. and Horvath, R., 2014. Mitochondria: Impaired mitochondrial translation in human disease. *Int J Biochem Cell Biol*, 48, pp.77–84.

Bonnet, C., et al., 2007. Allotopic mRNA localization to the mitochondrial surface rescues respiratory chain defects in fibroblasts harboring mitochondrial DNA mutations affecting complex I or V subunits. *Rejuvenation Res*, 10, pp.127–144.

Breuer, M. E., et al., 2013. The role of mitochondrial OXPHOS dysfunction in the development of neurologic diseases. *Neurobiol Dis*, 51, pp.27–34.

Brini, M., et al., 1999. A calcium signaling defect in the pathogenesis of a mitochondrial DNA inherited oxidative phosphorylation deficiency. *Nat Med*, 5, pp.951–954.

Carroll, C. J., et al., 2013. Whole-exome sequencing identifies a mutation in the mitochondrial ribosome protein MRPL44 to underlie mitochondrial infantile cardiomyopathy. *J Med Genet*, 50, pp.151–159.

Chen, J. Q., et al., 2009. Regulation of mitochondrial respiratory chain biogenesis by estrogens/estrogen receptors and physiological, pathological and pharmacological implications. *Biochim et Biophys Acta*, 1793, pp.1540–1570.

Chinnery, P. F. and Hudson, G., 2013. Mitochondrial genetics. *Br Med Bull*, 106, pp.135–159.

Comte, C., et al., 2013. Mitochondrial targeting of recombinant RNAs modulates the level of a heteroplasmic mutation in human mitochondrial DNA associated with Kearns Sayre Syndrome. *Nucleic Acids Res*, 41, pp.418–433.

Copeland, W. C., 2012. Defects in mitochondrial DNA replication and human disease. *Crit Rev Biochem Molr Biol*, 47, pp.64–74.

Cornelius, N., et al., 2013. Secondary coenzyme Q10 deficiency and oxidative stress in cultured fibroblasts from patients with riboflavin responsive multiple Acyl-CoA dehydrogenation deficiency. *Hum Mol Genet*, 22, pp.3819–3827.

Cotán, D., et al., 2011. Secondary coenzyme Q10 deficiency triggers mitochondria degradation by mitophagy in MELAS fibroblasts. *FASEB J*, 25, pp.2669–2687.

Dassa, E. P., et al., 2009. Expression of the alternative oxidase complements cytochrome c oxidase deficiency in human cells. *EMBO Mol Med*, 1, pp.30–36.

De Flora, S., et al., 2001. Mechanisms of N-acetylcysteine in the prevention of DNA damage and cancer, with special reference to smoking-related end-points. *Carcinogenesis*, 22, pp.999–1013.

De la Mata, M., et al., 2012. Recovery of MERRF fibroblasts and cybrids pathophysiology by coenzyme Q10. *Neurotherapeutics*, 9, pp.446–463.

De Paepe, B., et al., 2014. Effect of resveratrol on cultured skin fibroblasts from patients with oxidative phosphorylation defects. *Phytother Res*, 28, pp.312–316.

Desagher, S. and Martinou, J. C., 2000. Mitochondria as the central control point of apoptosis. *Trends Cell Biol*, 10, pp.369–377.

Desquiret-Dumas, V., et al., 2012. Metabolically induced heteroplasmy shifting and l-arginine treatment reduce the energetic defect in a neuronal-like model of MELAS. *Biochim Biophys Acta*, 1822, pp.1019–1029.

Di Donato, S., 2009. Multisystem manifestations of mitochondrial disorders. *J Neurol*, 256, pp.693–710.

Dieteren, C. E., Koopman, W. J. and Nijtmans, L.G., 2009. Tracing human mitochondrial complex I assembly by use of GFP-tagged subunits. *Methods Enzymol*, 456, pp.133–151.

DiMauro, S. and Schon, E. A., 2003. Mitochondrial respiratory-chain diseases. *N Engl J Med*, 348, pp.2656–2668.

DiMauro, S., et al., 2013. The clinical maze of mitochondrial neurology. *Nat Rev Neurol*, 9, pp.429–444.

Distelmaier, F., et al., 2009. The antioxidant Trolox restores mitochondrial membrane potential and Ca^{2+} -stimulated ATP production in human complex I deficiency. *J Mol Med (Berl)*, 87, pp.515–522.

Djouadi, F. and Bastin, J., 2011. Species differences in the effects of bezafibrate as a potential treatment of mitochondrial disorders. *Cell Metab*, 14, pp.715–716.

Dogan, S. A. and Trifunovic A., 2011. Modelling mitochondrial dysfunction in mice. *Physiol Res*, 60 (Suppl. 1), pp.S61–S70.

Ellouze, S., et al., 2008. Optimized allotopic expression of the human mitochondrial ND4 prevents blindness in a rat model of mitochondrial dysfunction. *Am J Hum Genet*, 83, pp.373–387.

Emdadul Haque M., et al., 2008. The effect of mutated mitochondrial ribosomal proteins S16 and S22 on the assembly of the small and large ribosomal subunits in human mitochondria. *Mitochondrion*, 8, pp.254–261.

Foltopoulou, P. F., et al., 2010. Intracellular delivery of full length recombinant human mitochondrial L-Sco2 protein into the mitochondria of permanent cell lines and SCO2 deficient patient's primary cells. *Biochim Biophys Acta*, 1802, pp.497–508.

Gammage, P. A., et al., 2014. Mitochondrially targeted ZFNs for selective degradation of pathogenic mitochondrial genomes bearing large-scale deletions or point mutations. *EMBO Mol Med*, 6, pp.458–466.

Garrido-Maraver, J., et al., 2012. Screening of effective pharmacological treatments for MELAS syndrome using yeasts, fibroblasts and cybrid models of the disease. *Br J Pharmacol*, 167, pp.1311–1328.

Gerards, M., et al., 2011. Riboflavin-responsive oxidative phosphorylation complex I deficiency caused by defective ACAD9: new function for an old gene. *Brain*, 134, pp.210–219.

Goldstein, A., Bhatia, P. and Vento, J. M., 2012. Update on nuclear mitochondrial genes and neurologic disorders. *Semin Pediatr Neurol*, 19, pp.181–193.

Golubitzky, A., et al., 2011. Screening for active small molecules in mitochondrial complex I deficient patient's fibroblasts, reveals AICAR as the most beneficial compound. *PLoS One*, 6, p.e26883.

González-Vioque, E., et al., 2011. Limited dCTP availability accounts for mitochondrial DNA depletion in mitochondrial neurogastrointestinal encephalomyopathy (MNGIE). *PLoS Genet*, 7, p.e1002035.

Gunter, T. E., et al., 1994. Mitochondrial calcium transport: physiological and pathological relevance. *Am J Physiol*, 267, pp.C313–C339.

Hirano, M., Garone, C. and Quinzii, C. M., 2012. CoQ(10) deficiencies and MNGIE: Two treatable mitochondrial disorders. *Biochim Biophys Acta*, 1820, pp.625–631.

Hong, Y., et al., 2014. NAD(+) treatment prevents rotenone-induced apoptosis and necrosis of differentiated PC12 cells. *Neurosci Lett*, 560, pp.46–50.

Iuso, A., et al., 2006. Dysfunctions of cellular oxidative metabolism in patients with mutations in the NDUFS1 and NDUFS4 genes of complex I. *J Biol Chem*, 281, pp.10374–10380.

James, A. M., et al., 1999. Decreased ATP synthesis is phenotypically expressed during increased energy demand in fibroblasts containing mitochondrial tRNA mutations. *Eur J Biochem*, 259, pp.462–469.

Jang, S., Kang, H. T. and Hwang, E. S., 2012. Nicotinamide-induced mitophagy: event mediated by high NAD+/NADH ratio and SIRT1 protein activation. *J Biol Chem*, 287, pp.19304–19314.

Jezek, P. and Hlavatá, L., 2005. Mitochondria in homeostasis of reactive oxygen species in cell, tissues, and organism. *Int J Biochem Cell Biol*, 37, pp.2478–2503.

Jones, C. N., et al., 2009. Antibiotic effects on mitochondrial translation and in patients with mitochondrial translational defects. *Mitochondrion*, 9, pp.429–437.

Karicheva, O. Z., et al. 2011. Correction of the consequences of mitochondrial 3243A>G mutation in the MT-TL1 gene causing the MELAS syndrome by tRNA import into mitochondria. *Nucleic Acids Res*, 39, pp.8173–8186.

Karkucinska-Wieckowska, A., et al., 2013. Left ventricular noncompaction (LVNC) and low mitochondrial membrane potential are specific for Barth syndrome. *J Inherit Metab Dis*, 36, pp.929–937.

Kerr, D. S., 2013. Review of clinical trials for mitochondrial disorders: 1997–2012. *Neurotherapeutics*, 10, pp.307–319.

Koopman, W. J., et al., 2008. Mitigation of NADH: Ubiquinone oxidoreductase deficiency by chronic Trolox treatment. *Biochim Biophys Acta*, 1777, pp.853–859.

Lazarou, M., et al., 2009. Assembly of nuclear DNA-encoded subunits into mitochondrial complex IV, and their preferential integration into supercomplex forms in patient mitochondria. *FEBS J*, 276, pp.6701–6713.

Lee, H. C. and Wei, Y. H., 2007. Oxidative stress, mitochondrial DNA mutation, and apoptosis in aging. *Exp Biol Med (Maywood)*, 232, pp.592–606.

Levitas, A., et al., 2010. Familial neonatal isolated cardiomyopathy caused by a mutation in the flavoprotein subunit of succinate dehydrogenase. *Eur J Hum Genet*, 18, pp.1160–1165.

Lopes, Costa A., et al., 2014. Beneficial effects of resveratrol on respiratory chain defects in patients' fibroblasts involve estrogen receptor and estrogen-related receptor alpha signaling. *Hum Mol Genet*, 23, pp.2106–2119.

López, L. C., et al., 2010. Treatment of CoQ(10) deficient fibroblasts with ubiquinone, CoQ analogs, and vitamin C: time- and compound-dependent effects. *PLoS One*, 5, p.e11897.

Marcus, D., et al., 2013. Replacement of the C6ORF66 assembly factor (NDUFAF4) restores complex I activity in patient cells. *Mol Med*, 19, pp.124–134.

Martindale, J. L. and Holbrook, N. J., 2002. Cellular response to oxidative stress: Signaling for suicide and survival. *J Cell Physiol*, 192, pp.1–15.

Mayr, J. A., et al., 2012. Lack of the mitochondrial protein acylglycerol kinase causes Sengers syndrome. *Am J Hum Genet*, 90, pp.314–320.

Morató, L., et al., 2013. Pioglitazone halts axonal degeneration in a mouse model of X-linked adrenoleukodystrophy. *Brain*, 136, pp.2432–2443.

Muratovska, A., et al., 2001. Targeting peptide nucleic acid (PNA) oligomers to mitochondria within cells by conjugation to lipophilic cations: Implications for mitochondrial DNA replication, expression and disease. *Nucleic Acids Res*, 29, pp.1852–1863.

Ogilvie, I., Kennaway, N. G. and Shoubridge, E.A., 2005. A molecular chaperone for mitochondrial complex I assembly is mutated in a progressive encephalopathy. *J Clin Invest*, 115, pp.2784–2792.

Orrenius, S., Gogvadze, V. and Zhivotovsky, B., 2007. Mitochondrial oxidative stress: implications for cell death. *Annu Rev Pharmacol Toxicol*, 47, pp.143–183.

Pagliarini, D. J., et al., 2008. A mitochondrial protein compendium elucidates complex I disease biology. *Cell*, 134, pp.112–123.

Pfeffer, G., et al., 2013. New treatments for mitochondrial disease-no time to drop our standards. *Nat Rev Neurol*, 9, pp.474–481.

Pontarin, G., et al., 2012. Mammalian ribonucleotide reductase subunit p53R2 is required for mitochondrial DNA replication and DNA repair in quiescent cells. *Proc Natl Acad Sci U S A*, 109, pp.13302–13307.

Porteous, W. K., et al. 1998. Bioenergetic consequences of accumulating the common 4977-bp mitochondrial DNA deletion. *Eur J Biochem*, 257, pp.192–201.

Rapoport, M., et al., 2008. TAT-mediated delivery of LAD restores pyruvate dehydrogenase complex activity in the mitochondria of patients with LAD deficiency. *Mol Ther*, 16, pp.691–697.

Rapoport, M., et al., 2011. Successful TAT-mediated enzyme replacement therapy in a mouse model of mitochondrial E3 deficiency. *J Mol Med (Berl)*, 89, pp.161–170.

Rea, S. L., et al., 2010. Bacteria, yeast, worms, and flies: exploiting simple model organisms to investigate human mitochondrial diseases. *Dev Disabil Res Rev*, 16, pp.200–218.

Rizzuto, R., et al., 2012. Mitochondria as sensors and regulators of calcium signaling. *Nat Rev Mol Cell Biol*, 13, pp.566–578.

Robinson, B. H., 1996. Use of fibroblast and lymphoblast cultures for detection of respiratory chain defects. *Methods Enzymol*, 264, pp.454–464.

Rodríguez-Hernández, A., et al., 2009. Coenzyme Q deficiency triggers mitochondria degradation by mitophagy. *Autophagy*, 5, pp.19–32.

Rothman, S. M., 1999. Mutations of the mitochondrial genome: Clinical overview and possible pathophysiology of cell damage. *Biochem Soc Symp*, 66, pp.111–122.

Saada, A., 2008. Mitochondrial deoxyribonucleotide pools in deoxyguanosine kinase deficiency. *Mol Genet Metab*, 95, pp.169–173.

Saada, A., 2011. The use of individual patient's fibroblasts in the search for personalized treatment of nuclear encoded OXPHOS diseases. *Mol Genet Metab*, 104, pp.39–47.

Saada, A., 2014. Mitochondria: Mitochondrial OXPHOS (dys) function ex vivo. The use of primary fibroblasts. *Int J Biochem Cell Biol*, 48, pp.60–65.

Schaefer, A.M., et al., 2004. The epidemiology of mitochondrial disorders—past, present and future. *Biochim Biophys Acta*, 1659, pp.115–120.

Schlame, M., 2013. Cardiolipin remodeling and the function of tafazzin. *Biochim Biophys Acta*, 1831, pp.582–588.

Schon, E. A. and Manfredi, G., 2003. Neuronal degeneration and mitochondrial dysfunction. *J Clin Invest*, 111, pp.303–312.

Schon, E. A., DiMauro, S. and Hirano, M., 2012. Human mitochondrial DNA: Roles of inherited and somatic mutations. *Nat Rev Genet*, 13, pp.878–890.

Scorrano, L., 2013. Keeping mitochondria in shape: a matter of life and death. *Eur J Clin Invest*, 43, pp.886–893.

Shepherd, R. K., et al., 2006. Measurement of ATP production in mitochondrial disorders. *J Inherit Metab Dis*, 29, pp.86–91.

Skladal, D., Halliday, J. and Thorburn, D. R., 2003. Minimum birth prevalence of mitochondrial respiratory chain disorders in children. *Brain*, 126, pp.1905–1912.

Smeitink, J. A., et al., 2006. Distinct clinical phenotypes associated with a mutation in the mitochondrial translation elongation factor EFTs. *Am J Hum Genet*, 79, pp.869–877.

Smith, R. A., et al., 2008. Mitochondria-targeted antioxidants in the treatment of disease. *Ann N Y Acad Sci*, 1147, pp.105–111.

Smith, P. M. and Lightowlers, R. N., 2011. Altering the balance between healthy and mutated mitochondrial DNA. *J Inherit Metab Dis*, 34, pp.309–313.

Soiferman, D., et al., 2014. The effect of small molecules on nuclear-encoded translation diseases. *Biochimie*, 100, pp.184–191.

Stacpoole, P. W., 2011. Why are there no proven therapies for genetic mitochondrial diseases? *Mitochondrion*, 11, pp.679–685.

Su, J. H., et al., 2002. Caspase-cleaved amyloid precursor protein and activated caspase-3 are co-localized in the granules of granulovacuolar degeneration in Alzheimer's disease and Down's syndrome brain. *Acta Neuropathol*, 104, pp.1–6.

Suomalainen, A., 2013. Fibroblast growth factor 21: a novel biomarker for human muscle-manifesting mitochondrial disorders. *Expert Opin Med Diagn*, 7, pp.313–317.

Taanman, J. W., Muddle, J. R. and Muntau, A. C., 2003. Mitochondrial DNA depletion can be prevented by dGMP and dAMP supplementation in a resting culture of deoxyguanosine kinase-deficient fibroblasts. *Hum Mol Genet*, 12, pp.1839–1845.

Taylor, R. W., et al., 1997. Selective inhibition of mutant human mitochondrial DNA replication in vitro by peptide nucleic acids. *Nat Genet*, 15, pp.212–215.

Taylor, R. W. and Turnbull, D. M., 2005. Mitochondrial DNA mutations in human disease. *Nat Rev Genet*, 6, pp.389–402.

Trevelyan, A. J., et al., 2010. Mitochondrial DNA mutations affect calcium handling in differentiated neurons. *Brain*, 133, pp.787–796.

Umaki, Y., et al., 2002. Apoptosis-related changes in skeletal muscles of patients with mitochondrial diseases. *Acta Neuropathol*, 103, pp.163–170.

Vedrenne, V., et al., 2012. Mutation in PNPT1, which encodes a polyribonucleotide nucleotidyltransferase, impairs RNA import into mitochondria and causes respiratory-chain deficiency. *Am J Hum Genet*, 91, pp.912–918.

Voets, A. M., et al., 2012. Patient-derived fibroblasts indicate oxidative stress status and may justify antioxidant therapy in OXPHOS disorders. *Biochim Biophys Acta*, 1817, pp.1971–1978.

Wang, G., et al., 2012. Correcting human mitochondrial mutations with targeted RNA import. *Proc Natl Acad Sci U S A*, 109, pp.4840–4845.

Wang, H. and Doering, L. C., 2012. Induced pluripotent stem cells to model and treat neurogenetic disorders. *Neural Plast*, 2012, p.346053.

Wojewoda, M., Duszyński, J. and Szczepanowska, J., 2011. NARP mutation and mtDNA depletion trigger mitochondrial biogenesis which can be modulated by selenite supplementation. *Int J Biochem Cell Biol*, 43, pp.1178–1186.

Wu, Y. T., et al., 2010a. Mitochondrial respiratory dysfunction-elicited oxidative stress and posttranslational protein modification in mitochondrial diseases. *Ann N Y Acad Sci*, 1201, pp.147–156.

Wu, S. B., et al., 2010b. Mitochondrial DNA mutation-elicited oxidative stress, oxidative damage, and altered gene expression in cultured cells of patients with MERRF syndrome. *Mol Neurobiol*, 41, pp.256–266.

INDEX

The Functions, Disease-Related Dysfunctions, and Therapeutic Targeting of Neuronal Mitochondria, First Edition. Edited by Valentin K. Gribkoff, Elizabeth A. Jonas and J. Marie Hardwick.

Printed and bound by CPI Group (UK) Ltd, Croydon, CR0 4YY

17/09/2026

14964284-0003